College Algebra:

Historical Notes

Tze-San Lee

Applied Math Press, LLC

College Algebra: Historical Notes

Cover designer: Ana, books-design.com

Editor: Ken, FirstEditing.com,

Copyeditor: Applied Math Press

For corrections or other information, contact publisher by e-mail or regular mail:

Applied Math Press, LLC, 914 Rebecca Street, Lilburn, GA 30047

e-mail: appliedmathpress@gmail.com.

Library of Congress Control Number: 2015934565

ISBN-13: 978-0692393116

Applied Math Press, Atlanta, GA

www.appliedmathpress.com

Acknowledgments

The first draft of this book was written during the period of 1999-2001 while I was still affiliated with Western Illinois University (WIU). After getting a job offer from the Centers for Disease Controls and Prevention (CDC), I took an early retirement from the WIU and moved to Atlanta to work as a mathematical statistician for the CDC in 2002. The book was then left untouched for the past 12 years. After retiring from the CDC in May 2014, I returned to pick up where I left off to finish the final polish on this book.

I would like to thank all of my students. Without the challenging questions they raised in the classroom, I would not be able to write this book. I'm also grateful for the kind assistance I received from my wife (Ting-Ting) and my son (John), both of whom made a contribution to the typing of the manuscript. Incidentally, all the graphical figures in this book are plotted by using the "Mathematica" software.

Preface

Algebra may be divided into "classical algebra" (theory of equations and analytic geometry) and "modern algebra" (abstract and linear algebra). Classical algebra was developed over a period of 4000 years, while modern algebra has only appeared since the 19th century. Although Babylonian mathematicians began to solve linear and/or quadratic equations around 1900-1600 B.C., equations were merely expressed rhetorically, that is, only in words and no symbols were used. The complete use of symbols as finally seen in the work of R. Descartes in 1637 A.D.

This book is comprised of two parts. Starting with the notion of variable, the main focus in Part I is solving the equations for the roots, while building up the notion of "function" from the relationship between the "dependent" and the "independent" variables is the emphasis of Part II. Since the notion of function is more abstract than that of the equation, I present them separately. Hopefully, a big contrast between "dependent versus independent variables" and "equation versus function" will be made clear.

Background materials are introduced in the first four chapters in Part I. The degree of difficulty of solving equations increases when both the number of equations and the number of variables are increased, so we begin with solving a single equation in one variable (Chapters 5 and 6), and then move to solving a single inequality in one variable (Chapters 7 and 8). Chapter 9 handles the problem with the combination of inequality and absolute value, while Chapter 10 deals with a single inequality in two variables. Chapter 11 deals with proportions and ratios, while Chapter 12 studies the topic of a quadratic equation in two variables. Built upon the notion of function, Chapters 13-15 are simply an extension of Chapters 2-4. Chapter 16 covers very important types of functions, exponential and logarithmic functions. Chapter 17 introduces the concept of matrices that are the building block for the subject of linear algebra. Chapter 18 deals with more classical topics on sequences and series, and Chapter 19 covers the topic of probability that constitutes a solid base for studying the subject of statistics.

This book has some features that are distinctly different from other textbooks on college algebra. It's based on my years of teaching experience to design a book suitable for all (average/above average) students to learn the subject. The book is written concisely from a panoramic viewpoint. Only materials that are essential to the understanding of the concept and the computational techniques in algebra are presented. Exercises for mechanical drilling are kept at a minimum.

In fact, I adopt a lecture-note style in the book. The entire book is centered at the mathematical definitions. Each definition (or concept) is added with the author's experienced comments and illustrated by the appropriate examples. Historical footnotes provide the context to show that all ideas in mathematics did not come easily. Instead, they were worked out by many great mathematicians over many years and living in different countries. References are provided to help those students who wish to widen their scope by pursuing further readings.

The author's aim is to convince the students that mastering the subject of college algebra is indeed the first step towards building a successful engineering/science career. This book can be used as a formal textbook, a reference supplement to the adopted textbook, or for the purpose of being self-taught.

Tze-San Lee
Atlanta, Georgia
March 31, 2015

Table of Contents

Table of Contents

Table of Figures

Table of Figures

Table of Figures

List of Notation

$\aleph_0$	countably infinite, 3
c	uncountably infinite, 3
N	the set of positive integers, 3
I	the set of all integers, 3
$\|a\|$	the absolute value of a number a, 4
$d(a,b)$	the distance between two numbers a and b, 4
$\bar{z}$	the complex conjugate of a complex number z, 9
$\Rightarrow$	imply, 38
$P(x, y)$	a point P in the Cartesian plane of R^2, 78
$d(P, Q)$	distance between two points P and Q, 79
$\in$	belongs to ot member of, 169
R	the set of real numbers, 169
$p_n(x)$	a polynomial function of x with degree n, 193
$r(x) = \dfrac{p_n(x)}{q_m(x)}$	a rational function of x, 253
$x \to a$	x approaches a, 265
$x \to a^+$	x approaches a from the right side of a, 265
$x \to a^-$	x approaches a from the left side of a, 265
a^x	an exponential function of x with the base a, 283
$\sinh(x)$	the hyperbolic sine function of x, 292-293
$\cosh(x)$	the hyperbolic cosine function of x, 292-293
$\log_a x$	a logarithmic function of x with the base a, 294
R^n	n-dimensional Euclidean space, 325
ERO	elementary row operation, 325
$A_{m\times n}$	a matrix of size $m\times n$, 333
$\vec{x}$	a column vector in R^n, 333
A^{T}	the transpose of the matrix A, 336
$[A\|\vec{b}]$	the augmented matrix of A and $\vec{b}$, 328
$\det(A)$	determinant of the matrix A, 345-346
A^{-1}	the inverse of a matrix A, 364
a_n	a sequence, 379
n!	n factorial, 382

List of Notation

$\sum_{n=0}^{N} a_n$	a series, 387
$\binom{n}{k}$	take k objects out of n objects without replacement, 427
n(A)	the number of elements in a set A, 466
$P(E)$	the probability of an event E, 466
$\cup$	union, 472
$\cap$	intersection, 472
$\overline{E}$	the complement of an event E, 472
$P(E\|F)$	the conditional probability of E given F, 485

Part I

Equations, Expressions, and Inequalities

Chapter 1
Variables, Exponents, and Complex Numbers

In this chapter we're going to review some elementary terms that are useful in this course. Each term is accompanied with not only historical footnotes so that you know where the notion came from, but also the connection with other topics to be discussed later.

1.1 Set and Variables

Definition 1.1.1
A set is a collection of objects.

Remarks
1. Apparently, the notion of a set appears trivial. Yet, this term was only recognized quite recently in 1895 by G. Cantor (1845-1918, Germany). Built upon the notion of set, Cantor asserts that there are an infinite number of transfinite numbers if infinity is regarded as a transfinite number, a modern theory on the (cardinality) order of infinity (References 1, pp. 273-275; 2 & 6).
2. The set is used everywhere; for example, it is used to express the solution set to an equation (Chapters 5-7, 9, and 12). Also the domain and the range of a function in Chapter 13 of Part II.

Definition 1.1.2
A variable is a letter x that represents any number in a set of numbers.

Remarks
1. It is no surprise that many students have a hard time making sense of symbol used in algebra, because it took mathematicians millennia of working verbally with algebra until the concepts were familiar enough to meaningfully describe with using symbols.
2. Egyptian mathematicians used the word "aha" to represent the unknown quantity. As a result, their equations were expressed

rhetorically. F. Viète (1540-1603, French) introduced at the end of 16th century the idea of representing known and unknown numbers by letters who used vowels to indicate the unknowns and consonants for the known quantities, while R. Descartes (1596-1650, French) invented the convention of representing unknown in equations by *x, y, and z* and known by *a, b, and c* (Reference 3, p.19).

3. Only the independent variable is considered in Part I, while the dependent variable that is related to the notion of function is studied in Chapter 13 in Part II.

Example 1.1.1

Let N denote the set of all natural numbers, i.e., $N = \{1, 2, 3, 4, \ldots\}$. If we let the variable x be any natural number, then N can also be expressed by using the set builder notation as

$$N = \{x \mid x \text{ is a natural number}\}.$$

Remark

According to Cantor's concept, the cardinality of N, namely the total number of elements in N, is the first transfinite number, denoted by $\aleph_0$, called the countably infinite.

Example 1.1.2

Let I denote the set of all integers, i.e., $I = \{\ldots, -3, -2, -1, 0, 1, 2, 3\ldots\}$. However, if we let the variable x be any integer, then I can also be expressed by using the set builder notation as

$$I = \{x \mid x \text{ is an integer}\}.$$

Remark

It has been proved that the cardinality of I is the same as that of N. Even the set of all rational numbers has the same cardinality as that of N. Only the cardinality of the set of all real numbers is greater than $\aleph_0$ and denoted by **c**, the uncountably infinite (Reference 6).

Definition 1.1.3

A constant is a specific number.

Remark

Evidently, a constant is not a variable. In general, the letters a, b, and c are used to denote a constant, whereas the letters x, y, and z are used to denote a variable.

1.2 Absolute Value

Definition 1.2.1
Any number is comprised of two parts, its magnitude and its sign.

Example 1.2.1
The number – 2 has a magnitude of 2 and its sign is negative.

Definition 1.2.2
The absolute value of variable x, denoted by $|x|$, is its magnitude by disregarding its sign, namely

$|x| = x$, if x is positive or 0;
$|x| = -x$, if x is negative.

Remarks
1. If $|x| = 5$, then $x = \pm 5$ because $|5| = 5 = |-5|$. Hence, in general, if $|x| = c$, $x = \pm c$, where c is a positive constant. This property will be used in solving linear equations in Chapter 5 and inequalities in Chapter 9.
2. J. R. Argand (1768-1822, French) introduced the term "module," meaning “unit of measure” in French, in 1806 specifically for the *complex* absolute value and it was borrowed into English in 1866 as the Latin equivalent "modulus." The term "absolute value" has been used in this sense from at least 1806 in French and 1857 in English. The notation $|x|$ was introduced by K. Weierstrass (1815-1897, Germany) in 1841. Other names for absolute value include "the numerical value" and "the magnitude" (Reference 4).

Definition 1.2.3
Let a and b be two real numbers. Then the distance between a and b is defined as

$$d(a, b) = |b - a|.$$

Example 1.2.2
The distance between –3 and 5 is $d(-3, 5) = |5 - (-3)| = |8|$.

(Arithmetic) Operational Rules for Absolute Value
Let x and y be two real numbers.

1. Addition Rule

$$|x+y| \le |x| + |y|. \tag{1.2.1}$$

2. Multiplication Rule

$$|x \cdot y| = |x| \cdot |y|. \tag{1.2.2}$$

3. Division Rule

$$\left|\frac{x}{y}\right| = \frac{|x|}{|y|},\ y \ne 0. \tag{1.2.3}$$

Remarks

1. Note that addition rule is an inequality. When both x and y are positive or negative numbers, equality holds. In all other cases, the inequality of "strictly less than" holds.
2. When x and y are the vectors $\vec{x}$ and $\vec{y}$ in Chapter 17 and the absolute value replaced by the length (or norm) of the vector (see remark 2 of Definition17.2.2), the addition rule is known as the triangle inequality arising from the fact that the sum of the length of two sides in a triangle is greater than the length of the third side.

Exercises

Evaluate each of the following distances between two given numbers or other operations:

1. $d(-4, 7) = ?$ **2.** $d(-3, -5) = ?$ **3.** $d(6, -8) = ?$

4. $|2 - \sqrt{5}| = ?$ **5.** $|(-3) \cdot 5| = ?$ **6.** $8 - |-15| = ?$

7. $\left|-\frac{1}{2}\right| - \left|\frac{1}{3}\right| = ?$ **8.** $\left|\frac{1}{4} - \frac{1}{3}\right| = ?$ **9.** $\left|(-\frac{4}{3}) \cdot (-\frac{7}{5})\right| = ?$

10. $\left|\dfrac{\frac{3}{4} - \frac{4}{5}}{\frac{6}{7} - \frac{7}{8}}\right| = ?$

1.3 Exponents, Roots, and Radicals

Definition 1.3.1
For any positive integer n and a real number a, define $a^n = a\times a\times a\times\ldots\times a$, namely, a times itself n times, is called the nth power of a, in that a is known as the base and n as the exponent.

Remark
If the base a is replaced by a variable x, x^n is called the power function in Chapter 13. They are very useful in constituting a polynomial function of degree n in Chapter 14. Both Chapters 13 and 14 are covered in Part II.

Example 1.3.1
$2^1 = 2,\ \ 2^2 = 2\times 2 = 4,\ \ 2^3 = 2\times 2\times 2 = 8,\ \ 3^4 = 3\times 3\times 3\times 3 = 81.$

Definition 1.3.2
For any nonzero real number a, $a^0 = 1$.

Example 1.3.2
$5^0 = 1$, $(2/3)^0 = 1$. However, 0^0 is undefined.

Definition 1.3.3
For a nonzero real number a, i.e., $a \neq 0$ and any positive integer n,
$a^{-n} = \dfrac{1}{a^n}$.

Example 1.3.3

$2^{-2} = 1/2^2 = 1/4$, $2^{-3} = 1/2^3 = 1/8$, $3^{-4} = 1/3^4 = 1/81$.

Definition 1.3.4
For any positive integer n and a real number a, the nth root of a is a number b defined as $b = \sqrt[n]{a} = a^{\frac{1}{n}}$. If n is even, a is restricted to be a positive real number.

Remarks

1. For the rational exponents of $a > 0$, define

$$a^{\frac{m}{n}} = (\sqrt[n]{a})^m = \sqrt[n]{a^m},$$

where m and n are positive integers.

2. In mathematics, an algebraic number is a number that is a root of a nonzero polynomial equation in one variable with rational coefficients. Any number that can be obtained through using a finite number of arithmetic operations and taking the nth root, where n is a positive integer is an algebraic number. However, there are algebraic numbers that are not obtained in this manner. All these numbers are roots to polynomial equations of degree ≥ 5. This is the so-called the Galois' theory that will be mentioned again in Section 14.5 of Part II (Reference 5).

(Arithmetic) Operational Rules for Rational Exponents

Let r and s be rational numbers.

1. Product Rule

$$a^r \cdot a^s = a^{r+s}.$$

2. Quotient Rule

$$\frac{a^r}{a^s} = a^{r-s}.$$

3. Power of the Power Rule

$$(a^r)^s = a^{r \cdot s}.$$

4. Power of the Product Rule

$$(a \cdot b)^r = a^r \cdot b^r.$$

5. Power of the Quotient Rule

$$(\frac{a}{b})^r = \frac{a^r}{b^r}, \ b \neq 0.$$

Examples 1.3.4

By using the (arithmetic) operational rules, calculate the following:

a) $2^3 \cdot 2^5 = 2^{3+5} = 2^8 = 256$.

b) $2^3/2^5 = 2^{3-5} = 2^{-2} = 1/2^2 = 1/4$.

c) $(3^2)^4 = 3^{2\times 4} = 3^8 = 6561$.

d) $(x \cdot y^2)^3 = x^3(y^2)^3 = x^3y^6$.

e) $(3x^3y^4/5x^2y)^2 = (3xy^3/5)^2 = 3^2x^2y^6/5^2 = 9x^2y^6/25$.

Example 1.3.5

Simplify each of the following:

a) $\sqrt{x} \cdot \sqrt[3]{x} = ?$

b) $\dfrac{\sqrt[3]{x^2}}{\sqrt[4]{x^3}} = ?$

c) $(\sqrt[3]{x^2})^{\frac{3}{4}} = ?$

d) $\sqrt[3]{x^3 y^6} = ?$

e) $\sqrt{\dfrac{x^4}{y^6}} = ?$ $(y \neq 0.)$

Solution

a) $\sqrt{x} \cdot \sqrt[3]{x} = x^{\frac{1}{2}} \cdot x^{\frac{1}{3}} = x^{\frac{1}{2}+\frac{1}{3}} = x^{\frac{5}{6}} = \sqrt[6]{x^5}$.

b) $\dfrac{\sqrt[3]{x^2}}{\sqrt[4]{x^3}} = \dfrac{x^{\frac{2}{3}}}{x^{\frac{3}{4}}} = x^{\frac{2}{3}-\frac{3}{4}} = x^{-\frac{1}{12}} = \dfrac{1}{\sqrt[12]{x}}$.

c) $(\sqrt[3]{x^2})^{\frac{3}{4}} = (x^{\frac{2}{3}})^{\frac{3}{4}} = x^{\frac{2}{3}\times\frac{3}{4}} = x^{\frac{1}{2}} = \sqrt{x}$.

d) $\sqrt[3]{x^3 y^6} = (x^3 y^6)^{\frac{1}{3}} = (x^3)^{\frac{1}{3}} \times (y^6)^{\frac{1}{3}} = x^{\frac{1}{3}\times 3} \times y^{6\cdot\frac{1}{3}} = x^1 y^2 = xy^2$.

e) $\sqrt{\dfrac{x^4}{y^6}} = (\dfrac{x^4}{y^6})^{\frac{1}{2}} = \dfrac{x^{4\cdot\frac{1}{2}}}{y^{6\cdot\frac{1}{2}}} = \dfrac{x^2}{y^3}$.

Exercises

Simplify each of the following by using the operational rules for exponents:

1. $27^{\frac{4}{3}} = ?$

2. $(-128)^{\frac{2}{7}} = ?$

3. $\sqrt[3]{2} \cdot \sqrt[4]{2} = ?$

4. $4^{-1} + 4^{-2} = ?$

5. $3\sqrt{18} - 12\sqrt{8} = ?$

6. $2\sqrt{48} + \sqrt{75} = ?$

7. $\sqrt[4]{x^3} \cdot \sqrt[4]{x} \cdot \sqrt{x} = ?$

8. $(x^2 y^3)^{-5} = ?$

9. $4x \cdot \sqrt[3]{27x^9 y^{12}} = ?$ **10.** $\left(\dfrac{x^{-\frac{1}{2}} \cdot y^{\frac{1}{2}}}{x^{-\frac{1}{3}} \cdot y^{\frac{1}{3}}}\right)^{-2} = ?$

1.4 Complex numbers

Definition 1.4.1
The letter i, called the imaginary unit, represents the square root of – 1, i.e., $i = \sqrt{-1}$.

Remarks
1. Nowadays an instructor usually uses the following rationale as a motivation for an introduction of the imaginary unit "i" by claiming that there exists no real number x whose square is – 1, that is, $x^2 = -1$ has no real solution.
2. Yet, the discovery of the complex number has a long history. The first clear recognition of imaginaries was made by Mahāvīra (Indian) in the ninth century with a remark that a negative number has no square root. Next, G. Cardano (1501-1576, Italian) regarded imaginaries as fictitious, but used them formally without raising any question as to the legitimacy of the formalism when he presented the Tartaglia-Cardano's formula (Theorem 14.5.4) for solving cubic equations in closed form. Finally, C. Wessel (1745-1818, Norwegian) took the final step to produce a consistent, useful interpretation of complex numbers in 1797.
3. In addition, C. F. Gauss (1777-1855, Germany) took as the subject of his doctoral dissertation (1799) a proof of the fundamental theorem of algebra: "An algebraic equation has a root of the form $a + bi$, a, b real" (Reference 1, pp. 167-180). This fundamental theorem of algebra will be covered in Section 14.6 of Part II.

Definition 1.4.2
A complex number is a number of the form $a + bi$, where a and b are real numbers and i is called the imaginary unit.

Remarks

1. When $b = 0$, a complex number reduces to a real number. This implies that the set of all real numbers is a subset of complex numbers. Geometrically, the set of all real numbers can be represented as a (one dimensional) line, while the set of complex numbers as a (two-dimensional) plane.
2. Many mathematicians contributed to the full development of complex numbers. G. Cardano (1501-1576, Italian) conceived of complex numbers in 1545. The (arithmetic) operational rules for complex numbers were developed by R. Bombelli (1526-1572, Italian). A more abstract formalism were further developed by W. R. Hamilton (1805-1865, Irish) who extended this abstraction to the theory of quaternions (Reference 7).

Definition 1.4.3

Let $z = a + bi$ be a complex number. Then a is called the real part and b the imaginary part of z.

Definition 1.4.4

The absolute value of a complex number $z = a + bi$ is $|z| = \sqrt{a^2 + b^2}$.

Example 1.4.1

The power of i has the following property: for $n = 1, 2, \ldots$

$$i^1 = i,\ i^2 = -1,\ i^3 = -i,\ i^4 = -1,\ i^{4n+1} = i,\ i^{4n+2} = -1,$$
$$i^{4n+3} = -i,\ i^{4n+4} = 1.$$

Definition 1.4.5

Let $z = a + bi$ be a complex number. Then $\bar{z} = a - bi$ is called the complex conjugate of z.

Example 1.4.2

(a) The complex conjugate of $2 - 3i$ is $2 + 3i$.
(b) The complex conjugate of $-3 + 4i$ is $-3 - 4i$.

Example 1.4.3

(a) The absolute value of $z = 2 - 3i$ is $|z| = \sqrt{2^2 + (-3)^2} = \sqrt{13}$.

(b) The absolute value of $z = -3 + 4i$ is $|z| = \sqrt{(-3)^2 + 4^2} = \sqrt{25} = 5$.

(Arithmetic) Operational Rules for Complex Numbers

Let $z = a + bi$ and $w = c + di$ be two complex numbers. Then we have the following rules:

1. Addition Rule

$z + w = (a + c) + (b + d)i.$

$z - w = (a - c) + (b - d)i.$

2. Scalar Multiplication Rule

$k \cdot z = ka + kbi$, where k is a real number.

3. Multiplication Rule

$z \cdot w = ac - bd + (ad + bc)i$

4. Division Rule

$$\frac{z}{w} = \frac{ac + bd}{c^2 + d^2} + \frac{bc - ad}{c^2 + d^2} i$$

Remarks

1. From the multiplication rule, we have that $|z|^2 = z \cdot \bar{z}$.
2. The division rule for complex number is equivalent to rationalizing the denominator by multiplying both the numerator and the denominator with $\overline{w}$, the complex conjugate of w.

Example 1.4.4

Simplify each of the following by using the (arithmetic) operational rules:

(a) $(4 - 2i) - (5 + 4i) = ?$

(b) $(5 - 3i)(1 + 2i) = ?$

(c) $\dfrac{6 + 2i}{3 - 5i} = ?$

Solution

(a) $(4 - 2i) - (5 + 4i) = 4 - 5 + (-2 - 4)i = -1 - 6i$.

(b) $(5 - 3i)(1 + 2i) = 5 \cdot 1 - (-3) \cdot 2 + (5 \cdot 2 + (-3) \cdot 1)i = -1 + 7i$.

(c) $\dfrac{6 + 2i}{3 - 5i} = \dfrac{(6 + 2i)(3 + 5i)}{(3 - 5i)(3 + 5i)} = \dfrac{6 \cdot 5 - 2 \cdot 5 + (6 \cdot 5 + 2 \cdot 3)i}{3^2 + 5^2} = \dfrac{8 + 36i}{34}$

$= \dfrac{8}{34} + \dfrac{36}{34} i = \dfrac{4}{17} + \dfrac{18}{17} i.$

Exercises

1. Simplify each of the following by using the operational rules:
(a) $5i \cdot 2i = ?$, (b) $(3i)^5 = ?$, (c) $\sqrt{-9} \cdot \sqrt{-16} = ?$,
(d) $\sqrt{-a} \cdot \sqrt{-b} = ?$, $a > 0$, $b > 0$.

2. Write the complex conjugate of the following:
(a) $-5 + 7i$, (b) $-4i$, (c) $3 - 8i$, (d) 9.

3. Find the absolute value of the following:
(a) $-3 + 5i$, (b) $4i$, (c) $4 - 7i$, (d) -6.

4. Simplify each of the following by using the operational rules:
(a) $(3 + 6i) - (-2 - 4i) = ?$ (b) $(5 - 3i) - 8 = ?$
(c) $(-2 + 3i)(2 + 2i) = ?$ (d) $(4 + 5i)(-4 + 5i) = ?$
(e) $3 \cdot (5 - 3i) = ?$ (f) $\dfrac{5 + 2i}{4 - 3i} = ?$
(g) $\dfrac{2 + 5i}{4i} = ?$.

1.5 References

1. Bell, E. T. (1945). The Development of Mathematics, Dover Publications, Inc., New York.
2. Muir, J. (1961). Of Men & Numbers: The Story of Great Mathematicians, Dover Publications, Inc., New York.
3. Sorell, T. (2000). Descartes: A Very Short Introduction, Oxford University Press, New York.
4. http://en.wikipedia.org/wiki/Absolute_value.
5. http://en.wikipedia.org/wiki/Algebraic_number.
6. http://en.wikipedia.org/wiki/Cantor's_first_uncountability_proof.
7. http://en.wikipedia.org/wiki/Complex_number.

Chapter 2
Polynomial Expressions

Solving a polynomial equation of degree n is of central importance in algebra. Solving a polynomial equation of degree one and two is covered in Chapters 5 and 6 in Part I. To solve a polynomial equation of degree higher than two is a challenging task that will be covered in Chapter 14 in Part II. However, in this chapter we're going to learn first some basic arithmetic operational rules with polynomial expressions.

2.1 Monomial Expression and Degree

Definition 2.1.1
A monomial (or term) expression is a constant, a variable, or a product of constants and/or variables.

Definition 2.1.2
A polynomial expression is a monomial expression or a sum of monomial expressions.

Remark
The term "polynomial expression" can be simplified to "polynomial," provided that you're able to discern the difference between "polynomial equation" (Chapter 5) and "polynomial function" (Chapter 14 in Part II).

Example 2.1.1
$3x^4 + 2x^3 + 5x^2 + 6x + 8$ is a polynomial expression in a variable x.

Example 2.1.2
$x^2 + 2xy + 3y^2$ is a polynomial expression in two variables x and y.

Example 2.1.3

$x^2y + y^2z + z^2x + 2xy + 2yz + 2xz + 5$ is a polynomial expression in three variables x, y, and z.

Definition 2.1.3
The degree of a monomial expression is the sum of exponents of all its variables.

Definition 2.1.4
The degree of a nonzero polynomial expression is the highest degree among its monomial expressions.

Definition 2.1.5
A polynomial expression of the form $a_n x^n + a_{n-1} x^{n-1} + ... + a_1 x + a_0$ is called a polynomial (expression) of degree n in x, where the leading coefficient a_n is not zero.

Remarks
1. The elegant and practical notation we use here was only developed beginning in the 15th century.
2. So long as you grasp the contrast between the terminology "polynomial expression" in Part I and "polynomial function" in Part II, it is fine to call the "polynomial expression" as the "polynomial" in short.

Example 2.1.4
The degree of the polynomial expressions in Examples 2.1.1-3 are 4, 2, and 3, respectively.

Exercises

Find the degree of each of the following:

1. $x^2y^5z^3 + 2xy^2z^4$.

2. $2x^5 - x^{16} + x - x^4$.

3. $x^2y^2 + xy^3 + y^4$.

4. $x^4 - 3xy^5 + y^7$.

5. $3x^2y^7 + 2xy^9 - 4x^6y^6 + 10$.

2.2 (Arithmetic) Operational Rules for Polynomial Expressions

Definition 2.2.1
Two or more terms are said to be "similar" if (i) they contain precisely the same variables and (ii) each variable has the same exponent in each term.

Definition 2.2.2
The constant factor of a simplified monomial expression is called the numerical coefficient (or coefficient) of the monomial expression.

The following are (arithmetic) operational rules for polynomial expressions:

1. **Addition Rule**
   To add or subtract two polynomial expressions is to add or sutract the coefficients of their similar terms. (2.2.1)
2. **Scalar (or Constant) Multiplication Rule**
   To multiply a polynomial expression by a constant is to multiply the coefficients of every term in polynomial expression by this constant. (2.2.2)
3. **Multiplication Rule**
   To multiply two polynomial expressions is to multiply each term of the first polynomial expression by each term of the second polynomial expression and then simplify it by using the operations rules for exponents. (2.2.3)
4. **Division Rule**
   To divide one polynomial expression, called the dividend, by another polynomial expression of lower degree, called the divisor is to use long division and continue until the remainder is at least one degree lower than the divisor. (2.2.4)

Example 2.2.1
Simplify by using Eq. (2.2.1), the addition rule:

$$(x^2 - 2xy + 2y^2 + 3z) - (3x^2 - xy - y^2 - z) = ?.$$

Solution

$$
\begin{aligned}
&(x^2-2xy+2y^2+3z)-(3x^2-xy-y^2-z)\\
&=x^2-2xy+3y^2+3z-3x^2+xy+y^2+z\\
&=(x^2-3x^2)+(-2xy+xy)+(3y^2+y^2)+(3z+z)\\
&=-2x^2-xy+4y^2+4z\,.
\end{aligned}
$$

Example 2.2.2

Simplify by using Eq. (2.2.2), the scalar multiplication rule:

$$3\cdot(2x^2y^3+5x^3y^2)=?$$

Solution

$$
\begin{aligned}
&3\cdot(2x^2y^3+5x^3y^2)\\
&=(3\cdot 2)\cdot x^2y^3+(3\cdot 5)\cdot x^3y^2\\
&=6x^2y^3+15x^3y^2\,.
\end{aligned}
$$

Example 2.2.3

Simplify by using Eq. (2.2.3), the multiplication rule:

$$(x^2+xy+2y^2)\cdot(3x+4y)=?$$

Solution

$$
\begin{aligned}
&(x^2+xy+2y^2)\cdot(3x+4y)\\
&=x^2\cdot 3x+xy\cdot 3x+2y^2\cdot 3x+x^2\cdot 4y+xy\cdot 4y+2y^2\cdot 4y\\
&=3x^3+3x^2y+6xy^2+4x^2y+4xy^2+8y^3\\
&=3x^3+(3x^2y+4x^2y)+(6xy^2+4xy^2)+8y^3\\
&=3x^3+7x^2y+10xy^2+8y^3\,.
\end{aligned}
$$

Example 2.2.4

Simplify by using Eq. (2.2.4), the long division: $\dfrac{x^5-3x^2+2x-5}{x^2+3}=?$

Solution

By using the long division,

$$
\begin{array}{r@{}l}
 & \;x-6\\
x^2+3\,\big) & \overline{x^3-3x^2+2x-5}\\
 & \underline{x^3+3x^2}\\
 & -6x^2+2x-5
\end{array}
$$

$$\frac{-6x^2-18}{2x+13}$$

Hence,

$$\frac{x^3-3x^2+2x-5}{x^2+3}=x-6+\frac{2x+13}{x^2+3}.$$

Exercises

Simplify by using the operational rules:

1. $(x^3+3x^2+2)+(x^2-4x+1)=?$

2. $(3x^2+5x-2)-5(x^2-2x+8)=?$

3. $(3x^2y+4+xy^2)-(2x^2y-3xy^2)=?$

4. $3xy(x^2-2xy+4y^2)=?$

5. $(2x-5)(4x+3)=?$

6. $(x+y)^2=?$

7. $(x-y)^2=?$

8. $(x+4)^3=?$

9. $(x-5)^2=?$

10. $(x+y)(x^2-xy+y^2)=?$

11. $(x-y)(x^2+xy+y^2)=?$

12. $\dfrac{2x^3-3x^2+4x-5}{x^2}=?$, where $x\neq 0$.

13. $\dfrac{2x^2+7x+10}{x+3}=?$, where $x\neq -3$.

14. $\dfrac{2x^3+11x^2-15x+10}{2x-1}=?$, where $x\neq\dfrac{1}{2}$.

15. $\dfrac{4x^3+x^2-x-3}{x^2+x+1}=?$

Chapter 3
Factoring Polynomial Expressions

To know how to factor polynomial expressions is vital to the success of solving a polynomial equation of degree higher than one. In this chapter, we're going to learn some basic factoring techniques.

3.1 Finding Common Factors

The simplest of all factoring techniques is to find the greatest common factor among all its terms in a polynomial expression.

Definition 3.1.1
The greatest common factor (GCF) among the terms of a polynomial expression is the product of the GCF of the coefficients among all the terms and the lowest power of every variable that occurs in every term of a polynomial expression.

Example 3.1.1
Factor the polynomial expression: $4x^5 + 8x^4 - 12x^2$.
Solution
Since

$$4x^5 = 2^2 \cdot x^5,$$
$$8x^4 = 2^3 \cdot x^4,$$
$$12x^2 = 2^2 \cdot 3 \cdot x^2,$$

the GCF of $4x^5$, $8x^4$, and $12x^2$ is $2^2 \cdot x^2 = 4x^2$. Therefore,

$$4x^5 + 8x^4 - 12x^2 = 4x^2\left(\frac{4x^5}{4x^2} + \frac{8x^4}{4x^2} - \frac{12x^2}{4x^2}\right)$$
$$= 4x^2(x^3 + 2x^2 - 3).$$

Example 3.1.2

Factor the polynomial: $4x^2y^2 - 12xy^3$.

Solution

Since

$$4x^2y^2 = 2^2 \cdot x^2 \cdot y^2,$$

$$12xy^3 = 2^2 \cdot 3 \cdot x \cdot y^3,$$

the GCF of $4x^2y^2$ and $12xy^3$ is $2^2 \cdot x \cdot y^2 = 4xy^2$. Therefore,

$$4x^2y^2 - 12xy^3 = 4xy^2\left(\frac{4x^2y^2}{4xy^2} - \frac{12xy^3}{4xy^2}\right) = 4xy^2(x - 3y).$$

Exercises

Factor each of the following polynomials:

1. $3x + 6 = ?$

2. $3ax + 6a = ?$

3. $3x^2 + 15x = ?$

4. $4x^3 + 12x^2 + 18x = ?$

5. $4y^5 - 10y^3 + 6y = ?$

6. $4(x+3) + x(x+3) = ?$

7. $3(x-1) - 9(x-1)^2 = ?$

8. $x^2(x-2) - x(x-2) = ?$

9. $2x^2y + 8xy^2 = ?$

10. $8xy^5 - 10x^4y^3 + 6x^6y^2 = ?$

3.2 Special Factoring Formula

Four commonly used factoring formulas are given as follows:

1. The Difference between Two Squares

$$a^2 - b^2 = (a+b)(a-b). \qquad (3.2.1)$$

2. The Perfect Square

$$a^2 + 2ab + b^2 = (a+b)^2; \qquad (3.2.2a)$$

$$a^2 - 2ab + b^2 = (a-b)^2. \qquad (3.2.2b)$$

3. The Difference between Two Cubes

$$a^3 - b^3 = (a-b)(a^2 + ab + b^2). \qquad (3.2.3a)$$

4. The Sum between Two Cubes

$$a^3 + b^3 = (a+b)(a^2 - ab + b^2). \qquad (3.2.3b)$$

Remark

All the above formulas can be proven by expanding the right side of the equation that equals the left side of the equation.

Example 3.2.1

Factor $50x^2 - 72 = ?$

Solution

$$50x^2 - 72 = 2(25x^2 - 36) = 2[(5x)^2 - 6^2] = 2(5x+6)(5x-6)$$

(Using Eq. (3.2.1)).

Example 3.2.2

Factor $36y^2 + 24y + 4 = ?$

Solution

$$36y^2 + 24y + 4 = 4(9y^2 + 6y + 1) = 4[(3y)^2 + 2\cdot 3y \cdot 1 + 1^2]$$

$$= 4(3y+1)^2$$ (Using Eq. (3.2.2a)).

Example 3.2.3

Factor $x^3 - 8 = ?$

Solution

$$x^3 - 8 = x^3 - 2^3 = (x-2)(x^2 + x\cdot 2 + 2^2) = (x-2)(x^2 + 2x + 4)$$

(Using Eq. (3.2.3a)).

Example 3.2.4

Factor $x^3 + 8 = ?$

Solution

$$x^3 + 8 = x^3 - 2^3 = (x+2)(x^2 - x\cdot 2 + 2^2) = (x+2)(x^2 - 2x + 4)$$

(Using Eq. (3.2.3b)).

Example 3.2.5

Factor $16x^4 - y^4 = ?$

Solution

$$16x^4 - y^4 = (4x^2)^2 - (y^2)^2 = (4x^2 + y^2)(4x^2 - y^2$$

$$= (4x^2 + y^2)[(2x)^2 - y^2] = (4x^2 + y^2)(2x + y)(2x - y).$$

Exercises

Factor each of the following polynomials completely:

1. $x^2 - 16 = ?$ **2.** $25 - 4x^2 = ?$ **3.** $3x^2 - 75y^2 = ?$

4. $4x^2 - (y+3)^2 = ?.$ **5.** $4y^2 + 4y + 1 = ?$

6. $4x^2 - 12xy + 9y^2 = ?$ **7.** $x^3 - 12x^2 + 36x = ?$

8. $24x^3 - 3 = ?$ **9.** $54 - 2y^3 = ?$

10. $8z^3 + 27 = ?$ **11.** $128w^3 + 250 = ?$

12. $x^4 - 16 = ?$ **13.** $3x^4 + 6x^2y^2 + 3y^4 = ?$

14. $3x^4 - 6x^2y^2 + 3y^4 = ?$ **15.** $6x^3y^2 - 12x^2y^2 + 6xy^2 = ?$

3.3 Methods for Factoring a Quadratic Polynomial Expression

Definition 3.3.1
A polynomial expression in x of degree two is called a quadratic polynomial expression in x.

There are two general methods for factoring a quadratic polynomial in x as shown below.

1. Method of Trial and Error

Step 1. Check to see if the leading coefficient of the quadratic polynomial is positive. If yes, do nothing. If no, factor the negative sign outside the entire polynomial.

Step 2. Factor the degree-two term into the product of two linear factors. Do a similar thing on the constant (or degree-zero) term. Then, list in pair, a combination of the factors of the degree-two term and the constant term.

Step 3. Attach the positive or negative sign to the factors of the constant term according the following rule: If the constant term is a positive number, attach both of the factored number with the same sign, namely, either both positive or both negative.

Otherwise, both of the factored numbers of the constant term has to be of opposite signs, namely, if one assigns a positive sign to one factored number, then the other factored number has to be assigned with a negative sign.

Step 4. Find the sum of all the cross-products between the factored terms of the degree-two term and the constant term for all possible combinations. If the sum matches the term of degree one in the given polynomial expression, then the factoring process is completed. Otherwise, the method of trial and error is unable to factor the given polynomial.

2. Method of Completing the Squares

Step 1. Check to see if the leading coefficient (or the coefficient of the degree-two term) is one. If not, factor this leading coefficient outside the polynomial by dividing all the remaining terms in the polynomial with this leading coefficient.

Step 2. Take a square of half of the coefficient of the current degree-one term, add this term, and subtract this term from the constant term. Then the first three terms will become a perfect square of a linear polynomial expression.

Step 3. If the simplified constant term is a negative number, then use the special factoring formula $a^2 - b^2 = (a+b)(a-b)$ to factor it into the product of two linear polynomial expressions. Otherwise, the quadratic polynomial expression cannot be factored into the product of two linear polynomial expressions with real-coefficient, although it can be factored into the product of two linear polynomial expressions with complex coefficients.

Remarks

1. The method of trial and error is usually limited to the quadratic polynomial with integer coefficients. Even with integer coefficients, the method of trial and error may not work sometimes.
2. Yet, the method of completing the squares always works by either giving you the factored solution or indicating no solution.

Example 3.3.1

Factor by the method of trial and error: $x^2 + 5x + 6$.

Solution

Step 1. Factor the degree-two term and the constant term as follows:

$$x^2 = x \cdot x, \quad 6 = 1 \cdot 6 = 2 \cdot 3.$$

Step 2. List all possible combinations between factors of the degree-two term and the factors of the constant term as shown below:

$$\begin{matrix} x & 1 \\ x & 6 \\ \multicolumn{2}{c}{(1)} \end{matrix} \qquad \begin{matrix} x & 2 \\ x & 3 \\ \multicolumn{2}{c}{(2)} \end{matrix}$$

Step 3. Since the constant term is a positive number, 6, the pairs in Step 2 can be expanded with the attached sign as follows:

$$\begin{matrix} x+1 & x-1 & x+2 & x-2 \\ x+6 & x-6 & x+3 & x-3 \\ (1a) & (1b) & (2a) & (2b) \end{matrix}$$

Due to that the coefficient of the degree-one term is a positive number 5, so we disregard the pairs (1b) and (2b).

Step 4. The sum of the cross product between the pairs of (1a) and (2a) is $3x + 2x = 5x$ that matches the degree-one term of the given polynomial expression. Hence,

$$x^2 + 5x + 6 = (x+2)(x+3).$$

Remark

The method of completing the squares also works for the given polynomial expression.

Example 3.3.2

Factor by the method of trial and error: $5x^2 - 13x - 6$.

Solution

Step 1. Factor the degree-two term and the constant term as follows:

$$5x^2 = x \cdot 5x,$$
$$6 = 1 \cdot 6 = 2 \cdot 3.$$

Step 2. List all possible combinations between the degree-two term and the constant term as shown below:

$$\begin{matrix} x & 1 \\ 5x & 6 \\ \multicolumn{2}{c}{(1)} \end{matrix} \qquad \begin{matrix} x & 6 \\ 5x & 1 \\ \multicolumn{2}{c}{(2)} \end{matrix} \qquad \begin{matrix} x & 2 \\ 5x & 3 \\ \multicolumn{2}{c}{(3)} \end{matrix} \qquad \begin{matrix} x & 3 \\ 5x & 2 \\ \multicolumn{2}{c}{(4)} \end{matrix}$$

Step 3. Since the sign of the constant term – 6 is negative, the assigned signs to the factors of the constant term must be opposite. This expands the four pairs in Step 2 into eight pairs:

$x \quad -1$	$x \quad +1$	$x \quad -6$	$x \quad +6$
$5x \quad +6$	$5x \quad -6$	$5x \quad +1$	$5x \quad -1$
(1a)	(1b)	(2a)	(2b)
$x \quad +2$	$x \quad -2$	$x \quad +3$	$x \quad -3$
$5x \quad -3$	$5x \quad +3$	$5x \quad -2$	$5x \quad +2$
(3a)	(3b)	(4a)	(4b)

Step 4. Find the sum of all the cross-products between two linear factors for all eight pairs. Only (4b) gives the result of $2x - 15x = -13x$ that matches the degree-one term of the given polynomial expression. Hence,

$$5x^2 - 13x - 6 = (x-3)(5x+2).$$

Remark

If the sign of the degree-two term of the given polynomial is negative, then you should factor the negative sign outside the entire polynomial expression and work with the transformed polynomial expression. Let us suppose that the given polynomial expression be $-5x^2 + 13x + 6$. Then we need to work with $-(5x^2 - 13x - 6)$ instead and focus on the expression within the parenthesis.

Example 3.3.3

Factor by using the method of completing the squares: $x^2 + 3x + 1$.

Solution

Since the leading coefficient of $x^2 + 3x + 1$ is 1, we can proceed immediately to factor the given polynomial expression by the method of completing the squares as follows:

$$x^2 + 3x + 1 = x^2 + 3x + (\frac{3}{2})^2 - (\frac{3}{2})^2 + 1$$

$$= (x + \frac{3}{2})^2 - \frac{9}{4} + 1 = (x + \frac{3}{2})^2 - \frac{5}{4}$$

$$= (x + \frac{3}{2})^2 - (\frac{\sqrt{5}}{2})^2$$

$$= (x + \frac{3}{2} + \frac{\sqrt{5}}{2})(x + \frac{3}{2} - \frac{\sqrt{5}}{2})$$

$$=(x+\frac{3+\sqrt{5}}{2})(x+\frac{3-\sqrt{5}}{2}).$$

Remark
The method of trial and error does not work for this example.

Example 3.3.4
Factor by using the method of completing the squares:

$$-2x^2+8x+4.$$

Solution
Since the leading coefficient of $-2x^2+8x+4$ is -2, we need to factor -2 outside the entire polynomial and then proceed with the method of completing the square as follows:

$$-2x^2+8x+4$$
$$=-2(x^2-4x-2)$$
$$=-2[x^2-4x+(\frac{-4}{2})^2-(\frac{-4}{2})^2-2]$$
$$=-2[(x-2)^2-6]=-2[(x-2)^2-(\sqrt{6})^2]$$
$$=-2(x-2+\sqrt{6})(x-2-\sqrt{6}).$$

Remark
The method of trial and error does not work for this example either.

Example 3.3.5
Factor completely: $64y^4-244y^2+225=?$

Solution
By letting $x=y^2$, the given polynomial reduces to a quadratic polynomial in x of the form $64y^4-244y^2+225$. By applying the method of trial and error, we have

$$64y^4-244y^2+225=64x^2-244x+225$$
$$=(4x-9)(16x-25)=(4y^2-9)(16y^2-25)$$

(substituting $x=y^2$ back)

$$=[(2y)^2-3^2][(4y)^2-5^2]$$
$$=(2y+3)(2y-3)(4y+5)(4y-5)\text{ (Use Eq. (3.2.1))}.$$

Exercises

In each of the following problems, factor the given quadratic function completely:

1. $3x^2 + 4x + 1 = ?$

2. $3x^2 + 11x + 6 = ?$

3. $6x^2 + 11x + 4 = ?$

4. $8x^2 + 6x - 5 = ?$

5. $10 - 11x - 6x^2 = ?$

6. $21 - 22x - 8x^2 = ?$

7. $12x^3 + x^2 - 6x = ?$

8. $15x^3 + 17x^2 - 4x = ?$

9. $x^2 - 2x - 4 = ?$

10. $x^2 + 4x + 1 = ?$

11. $2x^2 - 4x - 1 = ?$

12. $3x^2 + 6x - 2 = ?$

13. $2x^2 - 3xy - 2y^2 = ?$

14. $6x^2 - 5xy - 6y^2 = ?$

15. $4x^4 - 8x^2 + 3 = ?$

Chapter 4
Rational Expressions

Just like a rational number is the ratio of two integers, a rational expression is the division between two polynomials. A rational expression becomes a rational function that will be covered in Chapter 15 in Part II, once its domain is specified.

4.1 Rational Expression

Definition 4.1.1

A mathematical expression of the form $\frac{P}{Q}$ is called a rational expression, where P and $Q(\neq 0)$ are polynomial expressions. Furthermore, P is called the numerator and Q the denominator of the rational expression $\frac{P}{Q}$.

Remarks

1. The definition of a rational expression is similar to that of a rational number, $\frac{m}{n}$, except the integers m and $n(\neq 0)$ are replaced by the polynomial expressions P and Q.
2. A polynomial expression is a special case of a rational expression when $Q = 1$.
3. There is no concept of degree for a rational expression.
4. For special type of rational function, please see remark 4 with Definition 15.1.1 in Part II.

Example 4.1.1

(a) $\frac{x^2-7}{x^3+5x}$ is a rational expression in one variable x.

(b) $\frac{7x-3y}{x^2+y^2}$ is a rational expression in two variables x and y.

(c) $\frac{3xyz}{x^2+y^2+z^2}$ is a rational expression in three variables x, y, and z.

Definition 4.1.2

Two rational expressions $\frac{P}{Q}$ and $\frac{R}{S}$ are said to be equivalent if their cross-products are equal, namely, $P \cdot S = Q \cdot R$.

Definition 4.1.3

To simplify a rational expression is to reduce it to an equivalent expression so that its numerator and denominator has no common factors other than 1 or -1.

Example 4.1.2

Simplify each of the following rational expressions:

(a) $\frac{9x^3y^2z^4}{-15xy^4z^7} = \frac{9}{-15} \cdot \frac{x^3}{x} \cdot \frac{y^2}{y^4} \cdot \frac{z^4}{z^7}$ $(x \neq 0, y \neq 0, z \neq 0)$

$= -\frac{3}{5} \cdot x^{3-1} \cdot y^{2-4} \cdot z^{4-7} = -\frac{3}{5} \cdot x^2 \cdot y^{-2} \cdot z^{-3} = -\frac{3}{5} \cdot x^2 \cdot \frac{1}{y^2} \cdot \frac{1}{z^3}$

$= -\frac{3x^2}{5y^2z^3}$.

(b) $\frac{x^5(y+z)^2}{x^3(y+z)^3} = \frac{x^5}{x^3} \cdot \frac{(y+z)^2}{(y+z)^3}$ $(x \neq 0, y \neq -z)$

$= x^{5-3} \cdot (y+z)^{2-3} = x^2 \cdot (y+z)^{-1} = x^2 \cdot \frac{1}{y+z} = \frac{x^2}{y+z}$.

(c) $\frac{x^2-5x+6}{x^2+x-6} = \frac{(x-2)(x-3)}{(x-2)(x+3)}$ $(x \neq -3, x \neq 2)$

$= \frac{x-3}{x+3}$.

Exercises

Simplify each of the following rational expressions:

1. $\dfrac{6a^3b^2}{8ac^3} = ?\ (a \neq 0, c \neq 0)$

2. $\dfrac{9x^4y^3}{21x^2y^4} = ?\ (x \neq 0, y \neq 0)$

3. $\dfrac{6x^3y^3 + 9x^2y^2}{3xy^2} = ?\ (x \neq 0, y \neq 0)$

4. $\dfrac{4x + 4y}{9x^2 - 9y^2} = ?\ (x \neq \pm y)$

5. $\dfrac{9x^3y}{5x^2y^3 + 12xy^4} = ?\ (x \neq 0, y \neq 0, 5x + 4y \neq 0)$

6. $\dfrac{x^2 - 9}{x^2 - 5x + 6} = ?\ (x \neq 2, x \neq 3)$

7. $\dfrac{x^2 - 4}{x^3 - 8} = ?\ (x \neq 2)$

8. $\dfrac{x^3 + a^3}{a^2 - x^2} = ?\ (x \neq \pm a)$

9. $\dfrac{x^3 - x^2 - 6x}{x^4 - 16} = ?\ (x \neq \pm 2)$

10. $\dfrac{x^3 - 4x}{x^6 - 64} = ?\ (x \neq \pm 2)$

4.2 Least Common Denominator

Definition 4.2.1

A polynomial expression P is the least common multiple (LCM) of a set of polynomials if the following two requirements are met:

1. each polynomial expression in the set divides P, and
2. any polynomial expression divisible by all the polynomial expressions in the set is also divisible by P.

Remark

To find the LCM, factor the polynomial expressions completely and take all distinct factors, each to the greatest power to that it appears in any of the given polynomial expressions.

Example 4.2.1
Find the LCM of x^2-2x-8, x^2-4x, and x^2+2x-3.
Solution
Since

$$x^2-2x-8=(x+2)(x-4),$$
$$x^2-4x=x(x-4),$$
$$x^2+2x-3=(x-1)(x+3),$$

we have

$$\text{LCM }=x(x-4)(x-1)(x+2)(x+3).$$

Example 4.2.2
Find the LCM of $3(x^2z-y^2z)$, $9(xz^2+x^2z)$, and $12(x^2y-yz^2)$.
Solution
Since

$$3(x^2z-y^2z)=3z(x+y)(x-y)$$
$$9(xz^2+x^2z)=3^2\cdot xz(x+z),$$
$$12(x^2y-yz^2)=2^2\cdot 3\cdot y(x+z)(x-z),$$

we have

$$\text{LCM }=2^2\cdot 3^2\cdot xyz(x+y)(x-y)(x+z)(x-z).$$

Definition 4.2.2
The least common denominator (LCD) of several rational expressions is the LCM of all different denominators.

Remark
Finding the LCD is very important if we need to add two rational expressions with different denominators. We need to change two given rational expressions into the equivalent expressions with the same LCD so that their numerators can be added together through the use of addition rule for polynomial expressions.

Example 4.2.3

(a) Find the LCD of

$$\frac{x+2}{x^2+3x-4} \text{ and } \frac{x+5}{x^2-3x+2}.$$

(b) Convert the two rational expressions in part (a) into the equivalent rational expression with this LCD as the denominator.

Solution

(a) Since the denominators of two given rational expression are

$$x^2+3x-4=(x-1)(x+4) \text{ and } x^2-3x+2=(x-1)(x-2),$$

$$\text{LCD } =(x-1)(x-2)(x+4).$$

(b) $$\frac{x+2}{x^2+3x-4}=\frac{(x+2)(x-2)}{(x-1)(x+4)(x-2)}=\frac{x^2-4}{(x-1)(x-2)(x+4)}.$$

Also,

$$\frac{x+5}{x^2-3x+2}=\frac{(x+5)(x+4)}{(x-1)(x-2)(x+4)}=\frac{x^2+9x+20}{(x-1)(x-2)(x+4)}.$$

Exercises

In each of the following set of rational expressions, find (a) the LCD, and (b) the equivalent rational expressions with this LCD as the denominator:

1. $\{\frac{8x+5}{15x}, \frac{2x+3}{9x^2}\}$.

2. $\{\frac{3x}{x-2}, \frac{5}{x+3}\}$.

3. $\{\frac{3x+5}{x^2-4}, \frac{2-5x}{x^2-x-6}\}$.

4. $\{\frac{-1}{x^2-ax}, \frac{x+a}{x^2(x-a)}\}$.

5. $\{\frac{x-2}{3x^2}, \frac{4}{5y}, \frac{y-x}{2xy^2}\}$.

4.3 (Arithmetic) Operational Rules for Rational Expressions

Let $\frac{P}{Q}$ and $\frac{R}{S}$ be two rational expressions, where P, $Q \neq 0$, R, and $S \neq 0$ are polynomial expressions. Then,

1. Addition/Subtraction Rule

$$\frac{P}{Q}+\frac{R}{S}=\frac{PS+QR}{QS};\ \frac{P}{Q}-\frac{R}{S}=\frac{PS-QR}{QS}. \qquad (4.3.1)$$

2. Scalar Multiplication Rule

$$k\cdot\frac{P}{Q}=\frac{kP}{Q}, \text{ where } k \text{ is a constant.} \qquad (4.3.2)$$

3. Multiplication Rule

$$\frac{P}{Q}\cdot\frac{R}{S}=\frac{PR}{QS}. \qquad (4.3.3)$$

4. Division Rule

$$\frac{P}{Q}\div\frac{R}{S}=\frac{PS}{QR}. \qquad (4.3.4)$$

Remarks

1. For the addition rule, if Q and S have common factors, then the LCD of Q and S, rather than $Q{\cdot}S$, will be used.
2. As a convention, the result of the operation is always simplified.

Example 4.3.1

Simplify by using the addition rule:

$$\frac{2x-5}{x^2+2x-3}-\frac{3-4x}{x^2+3x-4}=?$$

Solution

$$\frac{2x-5}{x^2+2x-3}-\frac{3-4x}{x^2+3x-4}$$

$$=\frac{2x-5}{(x-1)(x+3)}-\frac{3-4x}{(x-1)(x+4)}$$

$$=\frac{(2x-5)(x+4)}{(x-1)(x+3)(x+4)}-\frac{(3-4x)(x+3)}{(x-1)(x+3)(x+4)}$$

$$=\frac{2x^2+3x-20}{(x-1)(x+3)(x+4)}-\frac{-4x^2-9x+9}{(x-1)(x+3)(x+4)}$$

$$=\frac{2x^2+3x-20-(-4x^2-9x+9)}{(x-1)(x+3)(x+4)}$$

$$=\frac{2x^2+3x-20+4x^2+9x-9}{(x-1)(x+3)(x+4)}$$

$$=\frac{6x^2+12x-29}{(x-1)(x+3)(x+4)} \quad (x\neq 1,-3,\text{ or }-4)$$

$$=\frac{6x^2+12x-29}{x^3+6x^2+5x-12}.$$

Example 4.3.2

Simplify by using Eqs. (4.3.1) and (4.3.2), the addition and scalar multiplication rules:

$$3\cdot\frac{5x-2}{x+1}-5\cdot\frac{3x-4}{x+3}=?$$

Solution

$$3\cdot\frac{5x-2}{x+1}-5\cdot\frac{3x-4}{x+3}$$

$$=\frac{3\cdot(5x-2)}{x+1}-\frac{5\cdot(3x-4)}{x+3}$$

$$=\frac{15x-6}{x+1}-\frac{15x-20}{x+3}$$

$$=\frac{(15x-6)(x+3)}{(x+1)(x+3)}-\frac{(x+1)(15x-20)}{(x+1)(x+3)}$$

$$=\frac{15x^2+39x-18}{(x+1)(x+3)}-\frac{15x^2-5x-20}{(x+1)(x+3)}$$

$$=\frac{15x^2+39x-18-(15x^2-5x-20)}{(x+1)(x+3)}$$

$$=\frac{15x^2+39x-18-15x^2+5x+20}{(x+1)(x+3)}$$

$$=\frac{44x+2}{(x+1)(x+3)}=\frac{44x+2}{x^2+4x+3}.(x\neq -1,\text{ or }-3)$$

Example 4.3.3

Simplify by using Eqs. (4.3.1) and (4.3.3), the addition and multiplication rules:

$$(x-2+\frac{3}{x+2})\cdot(1+\frac{1}{x+1})=? \quad (x\neq -1,\text{ or }-2)$$

Solution

$$(x-2+\frac{3}{x+2})\cdot(1+\frac{1}{x+1})$$
$$=[\frac{(x-2)(x+2)}{x+2}+\frac{3}{x+2}]\cdot[\frac{1\cdot(x+1)}{x+1}+\frac{1}{x+1}]$$
$$=(\frac{x^2-4+3}{x+2})\cdot(\frac{x+1+1}{x+1})$$
$$=\frac{x^2-1}{x+2}\cdot\frac{x+2}{x+1}=\frac{(x+1)(x-1)}{x+2}\cdot\frac{x+2}{x+1}=x-1.$$

Example 4.3.4
Simplify by using Eqs. (4.3.1) and (4.3.4), the addition and division rules:

$$(x-\frac{3x-4}{x-1})\div(x-\frac{2}{x-1})=?$$

Solution

$$(x-\frac{3x-4}{x-1})\div(x-\frac{2}{x-1})$$
$$=[\frac{x\cdot(x-1)-(3x-4)}{x-1}]\div[\frac{x\cdot(x-1)-2}{x-1}]$$
$$=(\frac{x^2-x-3x+4}{x-1})\div(\frac{x^2-x-2}{x-1})$$
$$=\frac{x^2-4x+4}{x-1}\cdot\frac{x-1}{x^2-x-2}\quad(x\neq 1)$$
$$=\frac{x^2-4x+4}{x^2-x-2}=\frac{(x-2)^2}{(x+1)(x-2)}=\frac{x-2}{x+1}\quad(x\neq -1,\text{ or }2).$$

Exercises

Simplify each of the following by using the operational rules for rational expressions:

1. $\frac{x^2}{x+y}-\frac{y^2}{x+y}=?\quad(x\neq -y)$

2. $\frac{x}{x-3}-\frac{5}{x+3}=?\quad(x\neq 3,\text{ or }-3)$

3. $2\cdot\frac{x}{x+1}+3\cdot\frac{4}{x-1}=?$ $(x\neq 1, \text{or} -1)$

4. $\frac{2}{x^2-x-2}+\frac{3}{x^2+3x+2}=?$ $(x\neq -2, -1, \text{or } 2)$

5. $5\cdot\frac{x+1}{x^2-x-2}-2\cdot\frac{x-1}{x^2+3x+2}=?$ $(x\neq -2, -1, \text{or } 2)$

6. $(2+\frac{3}{x})\cdot\frac{x^2}{4x^2-9}=?$ $(x\neq -\frac{3}{2}, \frac{3}{2}, \text{or } 0)$

7. $(x-\frac{2}{x-1})\cdot(x+\frac{1}{x-2})=?$ $(x\neq 1, \text{or } 2)$

8. $(\frac{2}{x+1}-\frac{3}{x-1})\div(\frac{x}{x^2-1})=?$ $(x\neq -1, \text{or } 1)$

9. $(2x+\frac{3x}{x+1})\div(4x-\frac{5}{x+2})=?$ $(x\neq -2, \text{or} -1)$

10. $(\frac{x^2+y^2}{2xy}-1)\div(\frac{3x^2-3y^2}{4x^3y^2})=?$ $(x\neq 0, y\neq 0, x\neq \pm y)$

Chapter 5
A Linear Equation in One Variable

In this chapter we're concerned with finding the root of a linear equation in a variable x. Then, we show how to apply it to solve some word problems.

5.1 (Arithmetic) Operational Rules

Definition 5.1.1

An equation of the form $a_n x^n + a_{n-1} x^{n-1} + \dots + a_1 x + a_0 = 0$ is called a polynomial equation of degree n in x, where $a_n \neq 0$, and all a_i, $i =$ 1, 2, …,n, are real constants.

Remarks

1. Finding a solution to a polynomial equation in x is vital to the advanced mathematics that can be employed in solving real world problems.
2. Although most Egyptian mathematics dealt with arithmetic, mainly with applications to the measurement of geometric figures, there were forerunners of several topics on algebra, except that they used the word "aha" rather than the letter x for the unknown. "Aha" means "heap" or "quantity" (Reference 1, pp. 30-33).

Definition 5.1.2

A solution to an equation, which is called a root of the given equation, is a number that satisfies the given equation.

Definition 5.1.3

A polynomial equation of degree one in x is called a linear equation in x.

Remark
A linear equation has a unique root.

Before learning how to find the root to the given equation, we need to familiarize ourselves with the following (arithmetic) operational rules for equation. Then, the idea of finding a solution to an equation is to reduce the given equation to another simpler equivalent equation. Repeat this process as many times as needed, until it leads to the desired solution.

(Arithmetic) Operational Rules for Equations
Let a and b denote the left side and the right side of an equation and c be any real number or any expression involving a variable.

1. Addition Rule
Given $a = b$. Then $a + c = b + c$. (5.1.1)

2. Subtraction Rule
Given $a = b$. Then $a - c = b - c$. (5.1.2)

3. Multiplication Rule
Given $a = b$. Then $a \cdot c = b \cdot c$. (5.1.3)

4. Division Rule
Given $a = b$. Then $\frac{a}{c} = \frac{b}{c}$, where $c \neq 0$. (5.1.4)

5. Power Rule
If $a = b$, then $a^n = b^n$, where n is a positive integer. (5.1.5)

Remarks
1. Basically, the above rules say that you are free to add, subtract, multiply, divide, raising power, and taking the nth root, as long as you apply the same operation on both sides of the given equation. Then, you will get the equivalent equation in the sense that the solution to the new equation is the same as the given equation. The only restriction is that you are not permitted to divide both sides of the equation by a zero number.
2. The above rules hold for any polynomial equation, not limited to linear equations considered in this chapter.

5.2 Solving a Linear and Related Equation

Solving a linear equation in a single variable x is very straightforward as shown in the following example.

Example 5.2.1 (A genuine linear equation)
Solve $x - 5 = 4x + 10$ for the unknown x.
Solution
From the given equation,

$$x - 5 + 5 = 4x + 10 + 5, \qquad \text{(Eq. (5.1.1))}$$
$$\Rightarrow \quad x = 4x + 15, \qquad \text{(Simplifying)}$$
$$\Rightarrow \quad x - 4x = 4x + 15 - 4x, \qquad \text{(Eq. (5.1.1))}$$
$$\Rightarrow \quad -3x = 15, \qquad \text{(Simplifying)}$$
$$\Rightarrow \quad \frac{-3x}{-3} = \frac{15}{-3}, \qquad \text{(Eq. (5.1.4))}$$
$$\Rightarrow \quad x = -5. \qquad \text{(Simplifying)}$$

Hence, the solution set of the root to the given equation is $\{-5\}$.

Remarks
1. Once you're familiar with the (arithmetic) operational rules, you can manipulate these rules in a short-cut way for the sake of collecting the likely terms on the same side by moving the terms from one-side of the equation to the other side by changing the sign of the moved term with the opposite sign. For this example, we can do it as follows:

   $$5 - 10 = 4x - x, \Rightarrow -5 = 3x, \Rightarrow x = -5/3.$$
2. In the next example we combine the solution of a linear equation with the absolute value of a variable.

Example 5.2.2 (A "disguised linear" equation)
Solve $\quad |\,3x + 1\,| = 4 \quad$ for the unknown x.
Solution
By removing the absolute value symbol on $3x + 1$, we have

$$3x + 1 = \pm\, 4.$$

When $3x + 1 = 4$, we have

$$3x + 1 = 4, \Rightarrow 3x + 1 - 1 = 4 - 1, \Rightarrow 3x = 3,$$

$$\Rightarrow \frac{3x}{3} = \frac{3}{3}, \Rightarrow x = 1.$$

When $3x + 1 = -4$ we have

$$3x + 1 = -4, \Rightarrow 3x = -4 - 1 = -5, , \Rightarrow x = \frac{-5}{3} = -\frac{5}{3}.$$

Hence, the solution set of the roots is $\{-\frac{5}{3}, 1\}$.

Example 5.2.3 (A disguised linear equation involving the absolute value)

Solve $\quad | 3x - 5 | = 4x + 1 \quad$ for the unknown x.

Solution

Since the absolute value on the left side is always nonnegative, we need to impose this constraint on the right side, namely, $4x + 1 \geq 0$, or $x \geq -\frac{1}{4}$. Now, removing the absolute value on the left side, we have

$$3x - 5 = \pm (4x + 1).$$

When $3x - 5 = 4x + 1$, we have

$$3x - 5 = 4x + 1, \Rightarrow 3x - 5 - 3x - 1 = 4x + 1 - 3x - 1,$$
$$\Rightarrow -6 = x, \text{ or } x = -6.$$

But, $x = -6$ does not satisfy the constraint $x \geq -\frac{1}{4}$. Hence, $x = -6$ is discarded.

When $3x - 5 = -(4x + 1)$, we have

$$3x - 5 + 4x + 5 = -4x - 1 + 4x + 5, \Rightarrow 7x = 4, \Rightarrow x = \frac{4}{7}.$$

Since $x = \frac{4}{7}$ satisfies the constraint $x \geq -\frac{1}{4}$. Hence, the solution set of the root is $\{\frac{4}{7}\}$.

Example 5.2.4 (A disguised linear equation involving rational expressions)

Solve $\quad \frac{3}{x+2} = \frac{2}{x+3} \quad$ for the unknown x.

Solution

By multiplying both sides of the equation by their LCD = $(x+2)(x+3)$, we obtain

$$(x+2)(x+3)\cdot\frac{3}{x+2}=(x+2)(x+3)\cdot\frac{2}{x+3},\ (x\neq -2,-3)$$

$$3(x+3)=2(x+2),\Rightarrow 3x+9=2x+4,\Rightarrow$$

$$3x-2x=4-9,\Rightarrow x=-5.$$

Since $-5\neq -2$ or -3, the solution set is $\{-5\}$.

Example 5.2.5 (A disguised linear equation involving a square root symbol)

Solve $\sqrt{2x-3}=5$ for the unknown x.

Solution

First, the expression under the square root symbol is required to be nonnegative. Therefore, the constraints on x are that $2x-3\geq 0$, or $x\geq\frac{3}{2}$. Next, by taking Eq. (5.1.5) on both sides of the equation, we obtain

$$(\sqrt{2x-3})^2=5^2,\Rightarrow [(2x-3)^{\frac{1}{2}}]^2=25,\Rightarrow$$

$$(2x-3)^{\frac{1}{2}\cdot 2}=25,\Rightarrow (2x-3)^1=25,\Rightarrow$$

$$2x-3=25,\Rightarrow 2x-3+3=25+3,\Rightarrow$$

$$2x=28,\Rightarrow \frac{2x}{2}=\frac{28}{2},\Rightarrow x=14.$$

Since $14\geq\frac{3}{2}$, the solution set is $\{14\}$.

Exercises

Solve each of the following (genuine/disguised) linear equations:

1. $3x-2=13$.

2. $2.6-0.4y=1.2$.

3. $7z-3=5z+9$.

4. $\frac{3x}{2}-\frac{x}{4}-8=0$.

5. $\frac{3x+5}{2}=\frac{4x-7}{3}$.

6. $2(5x+4)-3(2x-5)=28$.

7. $2(x-1)-(5-2x)=3(2x+3)$.

8. $\frac{y+2}{3}+\frac{5(3-y)}{6}=2$.

9. $0.08(z-5)-0.15(4-3z)=0.03(z-4)$.

10. $|2x-7|=5$. **11.** $|2x+3|=x-1$.

12. $\frac{2x+3}{x-5}=4 \ \ (x \neq 5)$. **13.** $\frac{2x+1}{2x-3}=1 \ \ (x \neq \frac{3}{2})$.

14. $\sqrt{2x-5}=3 \ (x \geq \frac{5}{2})$.

15. $\sqrt{3x-10}=\sqrt{x-2} \ (x \geq \frac{10}{3})$.

5.3 Word Problems

Example 5.3.1

Three times a number decreased by 37 is 50. Find the number.

Solution

Since we do not know this number, we use the variable x to represent this unknown number.Then, by translating the sentence of "3-times a number decreased by 37 is 50" into an equation, we have

$$3 \cdot x - 37 = 50.$$

By solving the above linear equation,

$$3x - 37 + 37 = 50 + 37, \Rightarrow 3x = 87, \Rightarrow$$

$$\frac{3x}{3} = \frac{87}{3}, \Rightarrow x = 29.$$

Hence, the number is 29.

Example 5.3.2

Steve's age is three times that of his son, Philip. The sum of their ages is 48. How old are they?

Solution

Since Philip's age is unknown, let his unknown age be x. Then, Steve's age is $3x$. Since the sum of their ages is 48, we have

$$3x + x = 48, \Rightarrow 4x = 48, \Rightarrow x = 12.$$

Also, $3x = 36$.

Hence, Steve's age is 36 and Philip's age is 12.

Example 5.3.3

Teresa bought \$18.70 worth of \$0.33 and \$0.55 stamps. If she bought 50 stamps in all, howmany stamps of each kind did she buy?

Solution

There are two kinds of stamps, \$0.33 and \$0.55, bought by Teresa. Although we do not know the number of each kind, we are told the total number of stamps she bought is 50. So, if we let x be the unknown number of \$0.33 stamps bought by Teresa, then the unknown number of \$0.55 stamps is $50 - x$. Then, using x of \$0.33 stamps and $50 - x$ of \$0.55 stamps to set up an equation from the given information we are told in the first sentence, we obtain

$$x \cdot 0.33 + (50 - x) \cdot 0.55 = 18.70.$$

By solving the above linear equation in x, we have

$$0.33x + 50 \cdot 0.55 - 0.55x = 18.70, \Rightarrow$$
$$-0.22x + 27.50 = 18.70, \Rightarrow$$
$$-0.22x = 18.70 - 27.50 = -8.80,$$
$$\frac{-0.22x}{-0.22} = \frac{-8.8}{-0.22}, \Rightarrow x = 40.$$

Also, $50 - x = 50 - 40 = 10$.

Hence, Teresa bought 40 of \$0.33 stamps and 10 of \$0.55 stamps.

Example 5.3.4

Tom invests some money to buy a 6-month certificate of deposit (CD) from a bank at the 5 percent annual interest, and his wife, Susan, invests some money to buy a 6-month CD from an e-bank at the 6 percent annual interest. If their total investment is \$10,000 and their total return in interest after 6 months is \$560, how much money does each invest?

Solution

Let x be the unknown amount of money invested by Tom. Then $\$10,000 - x$ is the amount of money invested by Susan. Since their total return in interest is \$560, we set up an equation for this information as follows:

$$x \cdot 0.05 + (10{,}000 - x) \cdot 0.06 = 560.$$

By solving the above linear equation,

$$0.05x + 600 - 0.06x = 560, \Rightarrow$$
$$-0.01x = 560 - 600 = -40, \Rightarrow$$
$$\frac{-0.01x}{-0.01} = \frac{-40}{-0.01}, \Rightarrow x = 4{,}000,$$

Also, $10{,}000 - x = 10{,}000 - 4{,}000 = 6{,}000$.
Hence, the amount of money invested, respectively, by Tom and Susan are \$4,000 and \$6,000.

Example 5.3.5
The perimeter of a rectangle is 26 feet. Twice the length is 2 feet less than five times the width.Find the dimensions of the rectangle.
Solution
Let x be the unknown length of the rectangle. Then, since the perimeter of the rectangle is 26 feet, $\frac{1}{2}(26-2x)=13-x$ is the width of the rectangle. From the given information in the second sentence, we set up an equation as follows:

$$2\cdot x = 5\cdot(13-x)-2, \Rightarrow\ 2x = 65-5x-2, \Rightarrow$$
$$2x = 63-5x, \Rightarrow\ 2x+5x = 63-5x+5x, \Rightarrow$$
$$7x = 63, \Rightarrow \frac{7x}{7} = \frac{63}{7}, \Rightarrow\ x = 9.$$
$$\text{Also, } \frac{1}{2}(26-2x) = \frac{1}{2}(26-2\cdot 9) = \frac{1}{2}\cdot 8 = 4.$$

Hence, the length and the width of the rectangle are 9 feet and 4 feet, respectively.

Exercises

1. The sum of twice a number and 11 is 45. Find the number.

2. Twice a number decreased by 15 is 35. Find the number.

3. The sum of three consecutive integers is 39. Find the smallest integer.

4. One number is three times of a second number. The sum of two numbers is 36. Find the two numbers.

5. One-third of a number is 5 more than one-fourth of the number. Find the number.

6. The absolute value of a number decreased by 4 is 7. Find the numbers.

7. The absolute value of a number increased by 2 equals to twice the number decreased by 6. Find the number.

8. The square root of a number decreased by 5 is 8. Find the number.

9. Mary's age is twice that of Beth's. The sum of their ages is 36. How old are they?

10. In 10 years Bill will be three-half as old as he is now. How old is he?

11. Tom has $6.50 in nickels and dimes. If there are 90 coins in all, how many of each coin does Tom have?

12. Wendy and Jill both earn $5.45 per hour by working part time at a supermarket in their hometown. One weekend Jill worked 6 hours longer than Wendy. Together they earned $141.7. How long did each work?

13. Beth received $20,000 from a trust fund and invested in two ways. Some is invested at 8% and the rest at 10%. How much is invested at each rate if she receives $1900 simple interest per year?

14. A house was sold for $130,000. How much does the owner make if the realtor's commission is 7% of the selling price?

15. Together, a house and the lot cost $175,000. The house costs seven times as much as the lot decreased by $15,000. Find the cost of the house.

16. The price of a 19-inch television set has been discounted 30%. The sale price is $140. Find the original list price of the television set.

17. A rectangle is four times as long as it is wide, and the perimeter is 120 inches. Find the dimensions of the rectangle.

18. The length of a rectangle building is 100 feet longer than its width. If the perimeter of the building is 600 feet. Find the dimensions of the building.

19. How many gallons of water must be added to 5 gallons of 6% solution of salty water to make a 2% salt solution?

20. A trip to visit the Disney World sponsored by the student's union at a certain university costs $250 per student. If there had been 5 more students, it would cost each student $25 less. How many students took the trip?

5.4 Reference

1. Bunt, L. N. H., Jones, P. S., and Bedient, J. D. (1988). The Historical Roots of Elementary Mathematics, Dover Publications Inc., New York.

Chapter 6
A Quadratic Equation in One Variable

In this chapter we're going to learn how to solve a quadratic equation in a single variable. Then, we show how to apply it to the word problems.

Historically, Babylonian mathematicians, as early as 3000 B.C., could solve problems relating the areas and the sides of rectangles. Later on, geometric methods were used by to solve quadratic equations by Egyptian, Greek, Chinese, and Indian mathematicians. The Egyptian papyrus (dating back to the Middle Kingdom, 2050–1650 B.C.) contains solutions to a two-term quadratic equations. Chinese mathematicians from 200 B.C. used geometric method of dissection to solve quadratic equations with positive roots. Rules of quadratic equations were given in the book "Jiu-Zhang Suan-Shu (九章算术) (The Nine Chapters on the Mathematical Art)" (Reference 1). These early geometric methods do not appear to have had a general formula. Euclid (365-257 B.C., Greek) produced a more abstract geometric method around 300 B.C. In 628 Indian mathematician Brahmagupta (598-665) gave the first explicit (although still not completely general) solutions of the quadratic equations $ax^2 + bx = c$. In 1545, G. Cardano (1501-1576, Italian) compiled the works related to the quadratic/cubic equations in his book "Ars Magna." The quadratic formula covering all cases was first obtained by S. Stevin (1548-1620, Flemish) in 1594. The formula in the form (Eqs. (6.3.1-3)) we know today was published in 1637 by R. Descartes (1596-1650, French) in his book "La Géométrie" (Reference 2).

To solve a polynomial equation of degree greater than two, we have to learn two new formulas, which are covered in Chapter 14 in Part II: Tartaglia-Cardano's formula for cubic equations (Theorem 14.5.4) and Ferrari's formula for quartic equations (Theorem 14.5.5).

6.1 Solve Quadratic Equations by Factoring

Definition 6.1.1

A polynomial equation of degree two in x is called a quadratic equation in a variable x. Ordinarily, a general form of a quadratic equation in x is denoted by $ax^2 + bx + c = 0$, where $a \neq 0$, b and c are real constants.

Remarks

1. For a definition of a polynomial equation of degree n, see Section 5.1. A quadratic equation in x has at most two distinct real roots.
2. Two methods are generally employed in solving a quadratic equation. One method is the factoring technique. You try to factor the given equation into the product of two linear factors. Then, using the zero-factor rule to find the roots. The other method is to use a quadratic formula that is always applicable to solve any quadratic equation in one variable.

Additional Rules for Quadratic Equations:

1. Zero-Factor Rule

Given that $A \cdot B = 0$. Then $A = 0$ or $B = 0$. (6.1.1)

2. Square-Root Rule

Given that $A^2 = B^2$. Then $A = \pm\sqrt{B^2} = \pm|B|$. If B is a positive number, then $A = \pm B$. (6.1.2)

Remarks

1. The square-root rule is obtained from Eq. (3.2.1), the difference-between-two-squares rule given in Section 3.2.
2. Solving a quadratic equation by factoring is built upon the zero-factor rule.
3. All (arithmetic) operational rules given in Section 5.1 are applicable in this chapter.

Example 6.1.1

Solve for the unknown x by the method of factoring:

$$2x^2 - 5x - 3 = 0.$$

Solution

By using the method of trial and errors given in Section 3.3, we obtain

$$2x^2 - 5x - 3 = 0, \Rightarrow (2x + 1)(x - 3) = 0.$$

By the zero factor rule,

$$2x + 1 = 0, \text{ or } x - 3 = 0, \Rightarrow x = -\frac{1}{2}, \text{ or } x = 3.$$

Hence, the solution set of the roots is $\{-\frac{1}{2}, 3\}$.

Example 6.1.2

Solve for the unknown x by the square-root rule: $(x - 2)^2 = 5$.

Solution

By using Eq. (6.1.2),

$$(x - 2)^2 = 5, \Rightarrow x - 2 = \pm\sqrt{5}, \Rightarrow x = 2 \pm \sqrt{5}.$$

Hence, the solution set of the roots is $\{2 - \sqrt{5}, 2 + \sqrt{5}\}$.

Exercises

Solve each of the following quadratic equations:

1. $x^2 + 5x + 6 = 0$. **2.** $5x^2 - 13x - 6 = 0$. **3.** $6x^2 + 11x + 4 = 0$.
4. $8x^2 + 6x - 5 = 0$. **5.** $2x^2 - 12x + 18 = 0$. **6.** $2x^2 = 18$.
7. $5x^2 = 16$. **8.** $(x - 2)^2 = 9$. **9.** $(\frac{x-1}{2})^2 = 25$.
10. $(\frac{3x-5}{2})^2 = 7$.

6.2 Completing the Square

Example 6.2.1

Solve for the unknown x by the method of completing the square:

$$x^2 - 2x - 4 = 0.$$

Solution

By following the method of completing the square given in Section 3.3,

$$x^2 - 2x - 4 = 0, \Rightarrow x^2 - 2x + (\frac{-2}{2})^2 - (\frac{-2}{2})^2 - 4 = 0, \Rightarrow$$
$$x^2 - 2x + (-1)^2 - (-1)^2 - 4 = 0, \Rightarrow (x - 1)^2 - 1 - 4 = 0, \Rightarrow$$
$$(x - 1)^2 - 5 = 0, \Rightarrow (x - 1)^2 = 5, \Rightarrow$$
$$(x - 1) = \pm\sqrt{5}, \Rightarrow x = 1 \pm\sqrt{5}.$$

Hence, the solution set of the roots is $\{1 \pm\sqrt{5}\}$.

Example 6.2.2
Solve for the unknown x by the method of completing the square:
$$x^2 + 2x + 4 = 0.$$

Solution
By following the method of completing the square given in Section 3.3,

$$x^2 + 2x + 4 = 0, \Rightarrow x^2 + 2x + (\frac{2}{2})^2 - (\frac{2}{2})^2 + 4 = 0, \Rightarrow$$
$$x^2 + 2x + 1^2 - 1^2 + 4 = 0, \Rightarrow (x + 1)^2 + 3 = 0, \Rightarrow (x + 1)^2 = -3.$$

But $(x + 1)^2$ is always nonnegative for any real x; Hence, it is impossible to equal to -3. Therefore, there are no real roots as solutions to the given quadratic equation.

Remark
Although the given equation in this example has no real roots, it does have complex roots as shown in Section 6.4.

Exercises

Solve each of the following equations for the real roots by the method of completing the squares:

1. $x^2 + 5x + 6 = 0$.
2. $2x^2 - 3x - 2 = 0$.
3. $5x^2 - 13x - 6 = 0$.
4. $6x^2 + 11x + 4 = 0$.
5. $8x^2 + 6x - 5 = 0$.
6. $x^2 + x - 1 = 0$.
7. $x^2 + 4x + 2 = 0$.
8. $x^2 - 5x + 2 = 0$.
9. $2x^2 - 3x - 1 = 0$.
10. $3x^2 + 5x - 1 = 0$.

6.3 The Quadratic Formula

Given: $ax^2 + bx + c = 0$, where $a \neq 0$. Let $D = b^2 - 4ac$, where D is called the quadratic discriminant. Then,

Case (i): If $D > 0$, the given equation has two distinct real roots given by

$$x = \frac{-b \pm \sqrt{D}}{2a}. \qquad (6.3.1)$$

Case (ii): If $D = 0$, the given equation has a double root given by

$$x = -\frac{b}{2a}. \qquad (6.3.2)$$

Case (iii): If $D < 0$, the given equation has no real solutions. Instead, it has two distinct complex roots given by

$$x = \frac{-b \pm i\sqrt{-D}}{2a}, \qquad (6.3.3)$$

where $i = \sqrt{-1}$ is the unit of complex number.

Remarks

1. The quadratic formula is obtained by using the method of completing the square given in Section 3.3.
2. A double root is a root of multiplicity 2 (Definition 14.3.5).

Example 6.3.1

Solve for the unknown x by using the quadratic formula:

$$2x^2 - 5x - 3 = 0.$$

Solution

Step 1. Let $a = 2$, $b = -5$, $c = -3$. Compute the quadratic discriminant D as follows:

$$D = b^2 - 4ac = (-5)^2 - 4 \cdot 2 \cdot (-3) = 25 + 24 = 49.$$

Step 2. Since $D = 49 > 0$, there are two distinct real roots given by using Eq. (6.3.1)

$$x = \frac{-b \pm \sqrt{D}}{2a} = \frac{-(-5) \pm \sqrt{49}}{2 \cdot 2} = \frac{5 \pm 7}{4} = -\frac{1}{2}, \text{or } 3.$$

Hence, the solution set of the roots is $\{3, -\frac{1}{2}\}$.

Remarks

1. The solution set here is exactly the same as that of Example 6.1.1.

2. This is because the quadratic formula is always applicable for solving any quadratic equation.

Example 6.3.2

Solve for the unknown x by using the quadratic formula:

$$2x^2 - 12x + 18 = 0.$$

Solution

Step 0. We notice that the given quadratic equation has a common factor 2 among all of its three terms. It should be simplified first as follows:

$$2 \cdot (x^2 - 6x + 9) = 0.$$

Since $2 \neq 0$, we have by the zero-factor rule

$$x^2 - 6x + 9 = 0.$$

Step 1. Now, let $a = 1$, $b = -6$, and $c = 9$. Compute the quadratic discriminant D as follows:

$$D = b^2 - 4ac = (-6)^2 - 4 \cdot 1 \cdot 9 = 36 - 36 = 0.$$

Step 2. Since $D = 0$, there is a double root given by using Eq. (6.3.2)

$$x = \frac{-b}{2a} = \frac{-(-6)}{2 \cdot 1} = \frac{6}{2} = 3.$$

Hence, the solution set of the roots is $\{3\}$, where $x = 3$ is a double root.

Example 6.3.3

Solve for the unknown x by using the quadratic formula:

$$3x^2 + 5x - 1 = 0.$$

Solution

Step 1. Let $a = 3$, $b = 5$, $c = -1$. Compute the quadratic discriminant D as follows:

$$D = b^2 - 4ac = 5^2 - 4 \cdot 3 \cdot (-1) = 25 + 12 = 37.$$

Step 2. Since $D = 37 > 0$, there are two distinct real roots given by using Eq. (6.3.1)

$$x = \frac{-b \pm \sqrt{D}}{2a} = \frac{-5 \pm \sqrt{37}}{2 \cdot 3} = \frac{-5 \pm \sqrt{37}}{6} = -\frac{5}{6} \pm \frac{\sqrt{37}}{6}.$$

Hence, the solution set of the roots is $\{ -\frac{5}{6} \pm \frac{\sqrt{37}}{6} \}$.

Remark

The solution set for this example is exactly the same as problem 10 in Exercises for Section 6.2.

Exercises

Solve each of the following quadratic equations for the unknown x by using the quadratic formula given by Eq. (6.3.1):

1. Repeat all the problems given in Exercises for Sections 6.1 and 6.2.

2. $3x^2 + 2 = 7x$. **3.** $2x^2 = 4x - 1$. **4.** $(x - 4)(x - 5) = 25$.

5. $3(x + 1)(x - 2) + (2x - 1)(x + 3) = 1$.

6.4 Complex Roots

As we know from the quadratic formula, the quadratic equation has two distinct complex roots if the quadratic discriminant D is negative. Two examples are given in this section. One is straightforward, while the other needs additional simplification first before using the quadratic formula.

Example 6.4.1

Solve for the known x by using the quadratic formula:

$$x^2 + 2x + 3 = 0.$$

Solution

Step 1. Let $a = 1$, $b = 2$, $c = 3$. Compute the quadratic discriminant D as follows:

$$D = b^2 - 4ac = 2^2 - 4 \cdot 1 \cdot 3 = 4 - 12 = -8.$$

Step 2. Since $D = -8 < 0$, there are two distinct complex roots given by using Eq. (6.3.3)

$$x = \frac{-b \pm i\sqrt{-D}}{2a} = \frac{-2 \pm i\sqrt{-(-8)}}{2 \cdot 1} = \frac{-2 \pm 2\sqrt{2}i}{2} = -1 \pm \sqrt{2}i.$$

Hence, the solution set of the roots is $\{-1 \pm \sqrt{2}i\}$.

Example 6.4.2

Solve for the known x by using the quadratic formula:

$$(3x + 1)(x - 2) - (x - 4)(x + 6) = 4.$$

Solution

Step 0. The given equation needs to be simplified as follows:

$$(3x + 1)(x - 2) - (x - 4)(x + 6) = 4, \Rightarrow$$
$$3x^2 + x - 6x - 2 - (x^2 - 4x + 6x - 24) = 4, \Rightarrow$$
$$3x^2 - 5x - 2 - (x^2 + 2x - 24) = 4, \Rightarrow$$
$$2x^2 - 7x + 22 = 4, \Rightarrow$$
$$2x^2 - 7x + 22 - 4 = 4 - 4, \Rightarrow$$
$$2x^2 - 7x + 18 = 0.$$

Step 1. Let $a = 2$, $b = -7$, $c = 18$. Compute the quadratic discriminant D as follows:

$$D = b^2 - 4ac = (-7)^2 - 4{\cdot}2{\cdot}18 = 49 - 144 = -95.$$

Step 2. Since $D = -95 < 0$, there are two distinct complex roots given by using Eq. (6.3.3)

$$x = \frac{-b \pm i\sqrt{-D}}{2a} = \frac{-(-7) \pm i\sqrt{-(-95)}}{2 \cdot 2} = \frac{7 \pm i\sqrt{95}}{4} = \frac{7}{4} \pm \frac{\sqrt{95}}{4} i \quad .$$

Hence, the solution set of the roots is $\{\frac{7}{4} \pm \frac{\sqrt{95}}{4} i\}$.

Exercises

Solve each of the following quadratic equations by using the quadratic formula:

1. $2x^2 + x + 3 = 0.$ **2.** $3x^2 = x - 1.$ **3.** $4x^2 - 3x = -5.$

4. $(\frac{5x-1}{2}) = -10.$ **5.** $2(x + 1)(x + 2) + 3(x - 2)(x - 3) = 1.$

6.5 Solve the Quadratic and Related Equations

In this sections several example are given to illustrate how to solve some equation that are not genuine quadratic equations, but can be reduced or simplified to quadratic equations.

Example 6.5.1

Solve a cubic polynomial equation for the unknown x:

$2x^3 - 5x^2 + 3x = 0.$

Solution

First, we notice that the given equation has a common factor of x in all of its three terms. Hence, we proceed to solve it as follows:

$$x \cdot (2x^2 - 5x + 3) = 0, \Rightarrow x = 0, \text{ or } 2x^2 - 5x + 3 = 0.$$

Then, solving the second equation by using the method of factoring,

$$2x^2 - 5x + 3 = 0, \Rightarrow (2x - 3)(x - 1) = 0, \Rightarrow$$

$$2x - 3 = 0, \text{ or } x - 1 = 0, \Rightarrow x = \frac{3}{2}, \text{ or } x = 1.$$

Hence, the solution set of the roots is $\{0, 1, \frac{3}{2}\}$.

Example 6.5.2

Solve a quartic polynomial equation for the unknown x:

$$x^4 - 8x^2 + 12 = 0.$$

Solution

By letting $y = x^2$, the given equation reduces to

$$y^2 - 8y + 12 = 0.$$

By using the method of factoring, we have

$$(y - 2)(y - 6) = 0, \Rightarrow y = 2, \text{ or } y = 6.$$

By substituting $y = x^2$ back, we obtain

$$x^2 = 2, \text{ or } x^2 = 6, \Rightarrow x = \pm\sqrt{2}, \text{ or } x = \pm\sqrt{6}.$$

Hence, the solution set of the roots is $\{\pm\sqrt{2}, \pm\sqrt{6}\}$.

Example 6.5.3

Solve an equation related to rational expressions for the unknown x:

$$\frac{3}{x-2} + \frac{4}{x+1} = 2 \quad (x \neq -1, 2).$$

Solution

First, we use the addition rule for rational expressions to simplify the left side of the given equation. Then, eliminate the denominator by using the multiplication rule for equations. After that, it becomes a quadratic equation. The step-by-step solution is given as follows:

$$\frac{3(x+1)}{(x-2)(x+1)} + \frac{4(x-2)}{(x-2)(x+1)} = 2, (x \neq -1, \text{ or } 2) \Rightarrow$$

$$\frac{3x+3+4x-8}{(x-2)(x+1)}=2, \Rightarrow \quad \frac{7x-5}{(x-2)(x+1)}=2, \Rightarrow$$

$$(x-2)(x+1)\cdot\frac{7x-5}{(x-2)(x+1)}=(x-2)(x+1)\cdot 2, \Rightarrow$$

$$7x-5=2x^2-2x-4, \Rightarrow \ 2x^2-9x+1=0.$$

By using the quadratic formula to solve the above quadratic equation, we obtain

$$x=\frac{-(-9)\pm\sqrt{(-9)^2-4\cdot 2\cdot 1}}{2\cdot 2}=\frac{9\pm\sqrt{73}}{4}.$$

Since $\frac{9\pm\sqrt{73}}{4}\neq -1$ or 2, the solution set of the roots is $\{\frac{9\pm\sqrt{73}}{4}\}$.

Example 6.5.4

Solve an equation related with the square-root expression for the unknown x: $\quad 2\sqrt{x-1}=1+\sqrt{x+2}$.

Solution

Since the expressions inside the square-root symbol is required to be nonnegative, the following constraints has to be imposed on the roots of the given equation:

$$x-1\geq 0 \text{ and } x+2\geq 0 \Rightarrow x\geq 1 \text{ and } x\geq -2.$$

Since $x\geq 1$ always implies $x\geq -2$, the two constraints reduces to a single constraint of $x\geq 1$. By using Eq. (6.1.2) on both sides of the given equation, we obtain

$$(2\sqrt{x-1})^2=(1+\sqrt{x+2})^2, \Rightarrow$$

$$2^2\cdot(\sqrt{x-1})^2=1^2+2\cdot 1\cdot\sqrt{x=2}+(\sqrt{x+2})^2, \Rightarrow$$

$$4(x-1)=1+2\sqrt{x+2}+x+2=x+3+2\sqrt{x+2}, \Rightarrow$$

$$4x-4-x-3=2\sqrt{x+2}, \Rightarrow$$

$$3x-7=2\sqrt{x+2}, \Rightarrow$$

$$(3x-7)^2=(2\sqrt{x+2})^2, \Rightarrow$$

$$(3x)^2+2\cdot 3x\cdot(-7)+(-7)^2=2^2\cdot(\sqrt{x+2})^2, \Rightarrow$$

$$9x^2-42x+49=4(x+2)=4x+8, \Rightarrow$$

$$9x^2-42x+49-4x-8=0, \Rightarrow$$

$$9x^2-46x+41=0.$$

By using the quadratic formula, Eq. (6.3.1), we have

$$x=\frac{-46\pm\sqrt{(-46)^2-4\cdot 9\cdot 41}}{2\cdot 9}=\frac{46\pm\sqrt{640}}{18}$$
$$=\frac{46\pm 8\sqrt{10}}{18}=\frac{2(23\pm 4\sqrt{10})}{18}=\frac{23\pm 4\sqrt{10}}{9}.$$

Since $\frac{23\pm 4\sqrt{10}}{9}\geq 1$, the solution set of the roots is $\{\frac{23\pm 4\sqrt{10}}{9}\}$.

Example 6.5.5

Find the real roots to an equation in x that is involved with the absolute value:

$$|x^2-3x+1|=x+2.$$

Solution

Since the left side of the given equation has the absolute value symbol, this requires the right side of the given equation to be nonnegative, i.e., $x+2\geq 0$, or $x\geq -2$. Therefore, the roots to the given equation is required to satisfy the constraint $x\geq -2$.

Now, by removing the absolute value symbol on the left side of the given equation, we need to attach the plus or minus sign to the right side of the equation, i.e.,

$$x^2-3x+1=\pm(x+2).$$

Case 1. When $x^2-3x+1=x+2$, we have

$$x^2-3x+1-x-2=x+2-x-2=0,$$
$$x^2-4x-1=0.$$

By using the quadratic formula, Eq. (6.3.1), we obtain

$$x=\frac{-(-4)\pm\sqrt{(-4)^2-4\cdot 1\cdot(-1)}}{2\cdot 1}=\frac{4\pm\sqrt{20}}{2}=\frac{4\pm 2\sqrt{5}}{2}=2\pm\sqrt{5}.$$

Since $x=2\pm\sqrt{5}>-2$, the solution set is $\{2\pm\sqrt{5}\}$.

Case 2. When $x^2-3x+1=-(x+2)$, we have

$$x^2-3x+1=-x-2,$$
$$x^2-3x+1+x+2=-x-2+x+2=0,$$
$$x^2-2x+3=0.$$

Since the quadratic discriminant $D=(-2)^2-4\cdot 1\cdot 3=-8<0$, there are no real roots to the above equation. By combining the result of Case 1 and Case 2, the solution set for the given equation is $\{2\pm\sqrt{5}\}$.

Exercises

Solve each of the following equations for the unknown x:

1. $6x^3 - 15x^2 - 3x = 0$.

2. $x^4 - 16 = 0$.

3. $(2x - 1)^4 - 3(2x - 1)^2 + 2 = 0$.

4. $2x + 3 = \frac{1}{x}$, $(x \neq 0)$.

5. $\frac{2}{x-3} + \frac{1}{x-2} = 3$, $(x \neq 2 \text{ or } 3)$.

6. $\frac{2x}{x-1} + \frac{3}{x+2} = 1$, $(x \neq -2 \text{ or } 1)$.

7. $\frac{2x}{x^2 - 3x + 2} + \frac{3}{x^2 - 5x + 6} = \frac{4}{x^2 - 4x + 3}$, $(x \neq 1, 2, \text{ or } 3)$.

8. $\sqrt{x-2} = 2 + \sqrt{3x-1}$, $(x \geq 2)$.

9. $\sqrt{x-1} + 2\sqrt{x-2} - 3 = 0$, $(x \geq 2)$.

10. $|2x^2 - 4x + 3| = 5$.

6.6 Word Problems

Example 6.6.1

Find three positive consecutive integers such that the product of the first two is 62 greater than the third.

Solution

Let x be the first integer. Then the other two integers are $x + 1$ and $x + 2$. Since the product of the first two integers is 62 greater than the third, we have

$$x \cdot (x + 1) - (x + 2) = 62, \Rightarrow x^2 + x - x - 2 = 62, \Rightarrow$$
$$x^2 - 2 = 62, \Rightarrow x^2 = 62 + 2, \Rightarrow x^2 = 64, \Rightarrow x = \pm 8.$$

Since $x = -8$ is not a positive integer, $x = -8$ is discarded. Hence, the three consecutive integers are $\{8, 9, 10\}$.

Example 6.6.2

The length of a rectangle is 9 inches more than the width. If the area is 90 square inches, find the dimensions of the rectangle.

Solution

Let the width of the rectangle be x inches, where $x > 0$. Then it follows that its length is $x + 9$ inches. Since its area is 90 in^2, we have

$$x \cdot (x + 9) = 90, \Rightarrow x^2 + 9x - 90 = 0, \Rightarrow (x - 6)(x + 15) = 0, \Rightarrow$$
$$x - 6 = 0, \text{ or } x + 15 = 0, \Rightarrow x = 6, \text{ or } x = -15.$$

Since the width of a rectangle cannot be a negative number, $x = -15$ is discarded. Hence, the width and the length of the rectangle are 6 inches and 15 inches, respectively.

Example 6.6.3

The product of Judy's hourly wage and Mary's hourly wage is \$168. The sum of their hourly wages is \$26. If Judy's hourly wage is higher than Mary's wage, what are their hourly wages?

Solution

Let x be Judy's hourly wage. Since the sum of their hourly wages is \$26, this means that Mary's wage is $26 - x$. Now, since the product of their wages is \$168, we obtain

$$x \cdot (26 - x) = 168.$$

By solving the above quadratic equation, we have

$$26x - x^2 = 168, \Rightarrow x^2 - 26x + 168 = 0, \Rightarrow$$
$$(x - 12)(x - 14) = 0, \Rightarrow x - 12 = 0 \text{ or } x - 14 = 0. \Rightarrow$$
$$x = 12 \text{ or } x = 14.$$

Since Judy's wage is higher than Mary's wage, we conclude that Judy earns \$14 per hour, while Mary earns \$12 per hour.

Example 6.6.4

Two cars, each traveling at a constant speed, leave St. Louis headed for Chicago, approximately 300 miles away. One car travels 15 miles per hour faster and arrives 1 hour earlier than the other car. What is the speed of the faster car?

Solution

Let x be the speed of the slower car, where $x > 0$. Then the speed of the faster car is $x + 15$. Since the faster car arrives 1 hour earlier, we set up an equation for this information as follows:

$$\frac{300}{x} - \frac{300}{x+15} = 1.$$

By following Example 6.5.3, we proceed to solve the above equation as follows:

$$\frac{300(x+15)}{x(x+15)}-\frac{300x}{x(x+15)}=1, \Rightarrow \frac{300(x+15)-300x}{x(x+15)}=1,$$

$\Rightarrow$

$$\frac{300x+4{,}500-300x}{x(x+15)}=1, \Rightarrow \frac{4{,}500}{x(x+15)}=1, \Rightarrow$$

$$x(x+15)\cdot\frac{4{,}500}{x(x+15)}=x(x+15)\cdot 1, \Rightarrow 4500=x^2+15x,$$

$\Rightarrow$

$x^2 + 15x - 4500 = 0, \Rightarrow (x - 60)(x + 75) = 0. \Rightarrow$

$x - 60 = 0$ or $x + 75 = 0, \Rightarrow x = 60$ or $x = -75$.

Since x, the speed of the slower car, cannot be a negative number, the root $x = -75$ is discarded. Hence, $x = 60$ is the speed of the slower car. Consequently, the speed of the faster car is 75 miles per hour.

Exercises

1. The sum of two numbers is 16 and their product is 63. Find these numbers.

2. Six times a certain positive number is 27 less than its square. Find this number.

3. When three times a positive number decreased by 2 is multiplied by the number increased by 3, the result is 18. Find this number.

4. Find a positive number that is 5 more than its reciprocal.

5. The perimeter of a rectangle is 26 feet. The area of the rectangle is 40 square feet. What are the dimensions of the rectangle?

6. The area of a rectangular wood-paneled wall in the basement of house is 136 square feet. If the length were increased by 3 feet and the height were increased by 2 feet, the area would be 200 square feet. What are the dimensions of the wall?

7. The product of a father's age and his son's age is 848. The sum of their ages is 69. What is the father's age? What is his son's age?

8. A man rowed a boat at a constant speed. He finds that it takes him 1 hour longer to make a 4-mile trip upstream than it does downstream. If the current flows at 2 miles per hour, how fast can the man row in still water?

9. A plane travels at a constant speed between two cities 2200 miles apart. When the head wind is 40 miles an hour, it reaches its destination 30 minutes later. What is the speed of the plane?

10. A man did a job for $240. It took him 8 hours longer than he expected and thus he earned $5.00 an hour less than he anticipated. How long did he expect it would take to do the job?

6.7 References

1. Ho, P. Y. (1985). Li, Qi and Shu: An Introduction to Science and Civilization in China, Dover Publications, Inc., New York.
2. http://en.wikipedia.org/wiki/Quadratic_equation.

Chapter 7
A Linear Inequality in One Variable

In this chapter we deal with solving a linear inequality in a single variable. Then, we show how to solve it mixed with the absolute value. Throughout history until modern times, Mathematics cannot be separated from astronomy. Aristarchus (280-260 B.C., Greek) introduced the use of inequalities to keep track of the range of errors for his estimates of the distance from the earth to the moon and sun. In fact, the use of inequalities to bound errors has become common place for users of computers today (Reference 1, pp. 92-125).

7.1 Intervals

Definition 7.1.1

(a) The set of all real numbers x such that $a < x < b$ is called an open interval, denoted by (a, b), where a and b, called the left and right end-point of the interval, respectively, are real numbers. Equivalently, (a, b) is expressed as

$$(a, b) = \{x \mid a < x < b, x\text{: a real number}\}.$$

(b) The set of all real numbers x such that $a \leq x \leq b$ is called a closed interval, denoted by $[a, b]$, where a and b are real numbers. Equivalently, $[a, b]$ is expressed as

$$[a, b] = \{x \mid a \leq x \leq b, x\text{: a real number}\}.$$

(c) The set of all real numbers x such that $a \leq x < b$ is called a half-closed interval, denoted by $[a, b)$, where a and b are real numbers. Equivalently, $[a, b)$ is expressed as

$$[a, b) = \{x \mid a \leq x < b, x\text{: a real number}\}.$$

(d) The set of all real numbers x such that $a < x \leq b$ is called a half-open interval, denoted by $(a, b]$, where a and b are real numbers. Equivalently, $(a, b]$ is expressed as

$$(a, b] = \{x \mid a < x \leq b, x\text{: a real number}\}.$$

Remarks

1. For half-closed and half-open intervals, the judging criterion used in its definition is relied on the left end-point of the interval.
2. In the 3rd century B. C. Archimedes (287-212 B.C., Greek) calculated upper and lower bounds for the irrational number π with the result $227/71 < \pi < 22/7$. Nowadays intervals are used to measure rounding errors associated with floating-point numbers (Reference 2).

Definition 7.1.2

An interval is said to be bounded if both its left and right end-points are real numbers. Otherwise, it is said to be unbounded.

Definition 7.1.3

An unbounded interval with just one of the end-points being a real number is called a half-line. If the left end-point of a half-line is $-\infty$, then it is called a left half-line. If the right end-point of a half-line is $+\infty$, it is called a right half-line. Furthermore, if the end-point of a half-line, being a real number, is included in the half-line, it is called a closed half-line. Otherwise the half-line is called an open half-line.

Example 7.1.1

Express each of the following inequalities in x by using the intervals, where x is a variable.

(a) $-\frac{1}{2} < x \leq 10$.

(b) $-3 < x$.

(c) $-10 \leq x < 20$.

(d) $21 \geq x$.

(e) $1 < x < 2$.

(f) $12 < x \leq 18$.

Solution

(a) $(-\frac{1}{2}, 10]$.

(b) $(-3, +\infty)$.

(c) $[-10, 20)$.

(d) $(-\infty, 21]$.

(e) (1, 2).
(f) (12, 18].

Example 7.1.2

(a) [– 2, 1] $= \{x \mid -2 \le x \le 1$, x: a real number$\}$ is a closed interval.
(b) (– 3, 2) $= \{x \mid -3 < x < 2$, x: a real number$\}$ is an open interval.
(c) [– 5, 3) $= \{x \mid -5 \le x < 3$, x: a real number$\}$ is a half-closed interval.
(d) (1, 4] $= \{x \mid 1 < x \le 4$, x: a real number$\}$ is a half-opened interval.
(e) (– ∞, 3] $= \{x \mid -\infty < x \le 3$, x: a real number$\}$ is a closed left-half-line.
(f) [– 1, + ∞) $= \{x \mid -1 \le x < +\infty$, x: a real number$\}$ is a closed right-half-line.
(g) (– ∞, – 2) $= \{x \mid -\infty < x < 2$, x: a real number$\}$ is an open left-half-line.
(h) (– 4, + ∞) $= \{x \mid -4 < x < +\infty$, x: a real number$\}$ is an open right-half-line.

Exercises

1. Express each of the following inequalities in x by using the intervals, where x is a variable: (a) $-1 \le x \le 3$, (b) $5 < x < 10$, (c) $x \le -17$, (d) $x > -\frac{4}{3}$, (e) $9 \ge x$, (f) $1 \le x < 8$, (g) $12 \ge x > 3$, (h) $x \ge -6$.

2. Classify each of the following bounded/unbounded intervals as one of the eight categories: (i) closed interval, (ii) open interval, (iii) half-closed interval, (iv) half-open interval, (v) closed right-half-line, (vi) open right-half-line, (vii) closed left-half-line, (viii) open left-half-line:
(a) (0, 2], (b) [– 5, 6], (c) (– ∞, 40], (d) (– 10, 20), (e) (– ∞, 10), (f) $(\frac{1}{3}, \frac{1}{2})$, (g) $(\frac{1}{4}, +\infty)$, (h) $[-2, \frac{1}{6})$, (i) $[-\frac{1}{5}, +\infty)$, (j) (– ∞, 0).

7.2 Operational Rules for Inequalities

Let a, b, and c be real numbers.

1. Addition Rule

If $a < b$, then $a + c < b + c$ or $a - c < b - c$. (7.2.1)

2. Multiplication Rule

If $a < b$, then $a \cdot c < b \cdot c$, if $c > 0$ or $a \cdot c > b \cdot c$, if $c < 0$. (7.2.2)

3. Division Rule

If $a < b$, then $\frac{a}{c} < \frac{b}{c}$, if $c > 0$ or $\frac{a}{c} > \frac{b}{c}$, if $c < 0$. (7.2.3)

4. Transitive Rule

If $a < b$ and $b < c$, then $a < c$. (7.2.4)

Remarks

1. For both the multiplication and the division rules, the direction of inequality must be reversed if you either multiply or divide by a negative nonzero real number c.
2. All the rules still hold if the strict inequality "<" (or ">") is replaced by the non-strict inequality "≤" (or "≥").

7.3 Solve a Linear Inequality

Definition 7.3.1

A polynomial inequality of degree n in x is a polynomial equation of degree n in x with the equality symbol "=" replaced by any of the inequality symbols: >, <, ≤, or ≥.

Definition 7.3.2

A solution to a polynomial inequality of degree n in x is a real number that satisfies the given inequality. The solution set to the given polynomial inequality is the set of all solutions.

Definition 7.3.3

A polynomial inequality of degree one in x is called a linear inequaity in a variable x.

Remark

In contrast to a linear equation in x that has just one solution, a linear inequality in x has infinitely many solutions. To express the solution set of these infinitely many solutions, either use the interval or graphing it on the real line.

Example 7. 3.1
Solve a single linear inequality for the unknown x:

$$3(x-2)+2(x+1)\geq 7x+3.$$

Solution
By applying the operational rules for inequalities, we proceed to solve the given inequality as follows:

$3x - 6 + 2x + 2 \geq 7x + 3,$ (simplification)
$5x - 4 \geq 7x + 3,$ (simplification)
$5x - 4 - 5x - 3 \geq 7x + 3 - 5x - 3,$ (Eq. (7.2.1))
$-7 \geq 2x$ (simplification)
$-\frac{7}{2} \geq x$, or $x \leq -\frac{7}{2}.$ (Eq. (7.2.3))

Hence, the solution set is a closed left-half-line $(-\infty,-\frac{7}{2}]$.

Example 7.3.2
Solve a double linear inequality for the unknown x:

$$-3 \leq 3 - 5x < 4.$$

Solution
By applying the operational rules, we obtain

$-3 - 3 \leq 3 - 5x - 3 < 4 - 3,$ (Eq. (7.2.1))
$-6 \leq -5x < 1,$ (simplification)
$\frac{-6}{-5} \geq \frac{-5x}{-5} > \frac{1}{-5},$ (Eq. (7.2.3))
$\frac{6}{5} \geq x > -\frac{1}{5}.$ (simplification)

Hence, the solution set is a half-open interval $(-\frac{1}{5},\frac{6}{5}]$.

Exercises

Solve each of the following inequalities for the unknown x:

1. $5x - 2 \le 7$. **2.** $\frac{1}{2}x - \frac{1}{3} > \frac{5}{12}$. **3.** $4x - 1 < 3 - 2x$.

4. $3(x - 5) \ge 6 + 2(7 - 2x)$. **5.** $2(2x + 1) - 3(5x + 4) > 1$.

6. $\frac{1}{3} < 2x + \frac{1}{2} < 2$. **7.** $-2 < 2x - 3 \le 4$.

8. $7 \ge 5 - 2x \ge 3$. **9.** $2 \le \frac{3-2x}{6} < 3$.

10. $-2 < 2(x - 1) + 3(2x + 3) \le 10$.

7.4 Absolute Value and Inequalities

Operational Rules for Inequalities with the Absolute Value
Let X be any mathematical expression and a be a positive number.

1. The Less-Than Rule

Given that $|X| < a$. Then $-a < X < a$. (7.4.1)

2. The Greater-Than Rule

Given that $|X| > a$. Then $X > a$, or $X < -a$. (7.4.2)

Remarks

1. Both rules still hold when the strict inequality "<" is replaced by the non-strict inequality "≤ " or ">" by "≥".
2. Many probabilistic or statistical laws are expressed in the form of the inequality with the absolute value. For example, the famous Chebyshev's Law is expressed as

   $\Pr(|X - \mu_X| \le k \cdot \sigma_X) \ge 1 - \frac{1}{k^2}$, $k > 1$, where μ_X and σ_X are the mean and the standard deviation of a random variable X (Reference 3).

Example 7.4.1

Solve the following inequality for the unknown x:

$$|3x - 4| \le 5.$$

Solution

By using Eq. (7.4.1), we have

$$-5 \leq 3x - 4 \leq 5, \Rightarrow -5 + 4 \leq 3x - 4 + 4 \leq 5 + 4 \Rightarrow$$
$$-1 \leq 3x \leq 9, \Rightarrow \frac{-1}{3} \leq \frac{3x}{3} \leq \frac{9}{3}, \Rightarrow -\frac{1}{3} \leq x \leq 3.$$

Hence, the solution set is a closed interval $[-\frac{1}{3}, 3]$.

Example 7.4.2

Solve the following inequality for the unknown x:

$$\left|\frac{3x+1}{4}\right| > 2.$$

Solution

By using Eq. (7.4.2), we have

$$\frac{3x+1}{4} > 2, \text{ or } \frac{3x+1}{4} < -2.$$

Case 1. When $\frac{3x+1}{4} > 2$, we have

$$4 \cdot \frac{3x+1}{4} > 4 \cdot 2, \Rightarrow 3x + 1 > 8, \Rightarrow$$
$$3x + 1 - 1 > 8 - 1, \Rightarrow 3x > 7, \Rightarrow \frac{3x}{3} > \frac{7}{3}, \Rightarrow$$
$$x > \frac{7}{3}.$$

Case 2. When $\frac{3x+1}{4} < -2$, we have

$$4 \cdot \frac{3x+1}{4} < 4 \cdot (-2), \Rightarrow 3x + 1 < -8, \Rightarrow$$
$$3x + 1 - 1 < -8 - 1, \Rightarrow 3x < -9, \Rightarrow \frac{3x}{3} < \frac{-9}{3}, \Rightarrow$$
$$x < -3.$$

By combining the results of Cases 1 and 2, the solution set is $\{(\frac{7}{3}, +\infty) \cup (-\infty, -3)\}$.

Exercises

Solve each of the following inequalities with the absolute value for the unknown x:

1. $|x-2|<3$.

2. $|2x-1|\le 4$.

3. $\left|\frac{x}{2}+3\right|<5$.

4. $\left|\frac{2x+1}{3}\right|\le 2$.

5. $|3x|+\frac{1}{4}<\frac{1}{3}$.

6. $|x+1|>2$.

7. $|2x+3|\ge 4$.

8. $\left|\frac{3x}{4}-\frac{1}{2}\right|\ge\frac{1}{5}$.

9. $\left|\frac{x}{3}\right|-\frac{1}{2}>\frac{1}{4}$.

10. $\left|\frac{2x+7}{5}\right|>3$.

7.5 References

1. Resnikoff, H. L. and Wells, R. O. (1984). Mathematics in Civilization, Dover Publications, Inc., New York.
2. http://en.wikipedia.org/wiki/Interval_arithmetic.
3. http://en.wikipedia.org/wiki/Chebyshev's_inequality.

Chapter 8
A Quadratic Inequality in One Variable

In this chapter we're going to learn how to solve a quadratic inequality in a single variable. Then, we show how to solve it mixed with the absolute value.

8.1 Solve a Quadratic Inequality

Definition 8.1
A polynomial inequality of degree two in x is called a quadratic inequality in a variable x.

Remarks
1. For a definition of a polynomial inequality of degree n, see Section 7.1.
2. Unlike a linear inequality, a quadratic inequality in x may not have any real solution at all if the associated quadratic equation has merely complex roots. See Example 8.1.2.
3. All operational rules for inequalities given in Section 7.2 are also applicable to solving a quadratic inequality.

Additional rules for quadratic inequalities
Let a and b be two real number with $a < b$.

1. **The Negative Factor-Product Rule**
   Given that $(x-a)(x-b)<0$. Then $a<x<b$. (8.1.1)
2. **The Positive Factor-Product Rule**
   Given that $(x-a)(x-b)>0$. Then $x<a$ or $x>b$. (8.1.2)

Remarks

1. The validity of the above rules can be asserted by choosing a specific number from the solution set to check if it satisfies the given inequality.
2. The rules remain to be true when the strict inequality is replaced by the corresponding non-strict inequality, i.e., "<" is replaced by "≤", or " >" by "≥".
3. If a quadratic inequality can not be factored into the product of two linear factors with real roots, then the given quadratic inequality has no real solutions.
4. Note that a and b are the roots of $(x-a)(x-b)=0$. Therefore, to solve a quadratic inequality, we need to find the roots of its equation counterpart. Then use either one of the rules.

Example 8.1.1

Solve the following inequality for the unknown x:

$$x^2 - x \leq 6.$$

Solution

First, we need to put the given inequality into the standard form, i.e., the right side of the inequality has to be zero. Then, using the factoring technique given in Chapter 6, we have

$$x^2 - x - 6 \leq 6 - 6, \qquad \text{(Eq. (7.2.1))}$$

$$x^2 - x - 6 \leq 0, \qquad \text{(simplification)}$$

$$(x+2)(x-3) \leq 0, \qquad \text{(factoring)}$$

$$[x-(-2)][x-3] \leq 0.$$

Since $-2 < 3$, we obtain by using Eq. (8.1.1) $-2 \leq x \leq 3$.
Hence, the solution set is a closed interval $[-2, 3]$.

Example 8.1.2

Solve the following inequality for the unknown x:

$$x^2 + 4x + 4 < 0.$$

Solution

By using the factoring technique, we have

$$(x+2)^2 < 0.$$

But, for any real number x, $(x+2)^2$ is always nonnegative that contradicts to the above inequality. Hence, the solution set is empty, i.e., it has no solution.

Remark

The graph of the given function $f(x) = x^2 + 4x + 4$ is a parabola with the vertex at the point $(-2, 0)$. To graph a quadratic function, see Chapter 14 in Part II.

Example 8.1.3

Solve the following inequality for the unknown x:

$$2x^2 - 4x + 2 > 0.$$

Solution

By using the factoring technique, we have

$$2(x^2 - 2x + 1) > 0, \qquad \text{(simplification)}$$

$$2(x-1)^2 > 0, \qquad \text{(factoring)}$$

$$\frac{2(x-1)^2}{2} > \frac{0}{2}, \qquad \text{(Eq. (7.2.3))}$$

$$(x-1)^2 > 0.$$

Notice that for any $x \neq 1$, $(x-1)^2$ is positive that satisfies the above inequality. For $x = 1$, $(x-1)^2 = 0$ that does not satisfies the above inequality. Hence, the solution set is $\{x|\ x \neq 1, x\text{: any real number}\} = (-\infty, 1) \cup (1, +\infty)$.

Example 8.1.4

Solve the following inequality for the unknown x:

$$(2x-1)(x-3) \geq 7.$$

Solution

First we need to simplify the given inequality into the standard form and then proceed to solve it as follows:

$$(2x-1)(x-3) \geq 7,$$

$$2x^2 - x - 6x + 3 \geq 7$$

$$2x^2 - 7x + 3 \geq 7,$$

$$2x^2 + 7x + 3 - 7 \geq 7 - 7,$$

$$2x^2 - 7x - 4 \geq 0$$

$$(2x+1)(x-4) \geq 0,$$

$$2(x + \frac{1}{2})(x-4) \geq 0,$$

$$(x + \frac{1}{2})(x-4) \geq 0,$$

$$[x - (-\frac{1}{2})][x-4] \geq 0.$$

Since $-\frac{1}{2} < 4$, we obtain by using Eq. (8.1.2) $x \le -\frac{1}{2}$, or $x \ge 4$.

Hence, the solution set is $\{ (-\infty, -\frac{1}{2}] \cup [4, +\infty) \}$.

Exercises

Solve each of the following quadratic inequalities for the unknown x:

1. $x^2 + 5x + 6 \ge 0$. **2.** $x^2 \le 8$. **3.** $x^2 - 6x + 9 \le 0$.

4. $2x^2 > 18$. **5.** $4 + 3x - x^2 \le 0$. **6.** $2x^2 + 1 > 5x$.

7. $2 > x^2 + 4x$. **8.** $2x^2 + x < -3$.

9. $(x-1)(x-2) < 3(x+2)(x+3)$.

10. $2(x+1)(x+2) + 3(x-2)(x-3) \ge 1$.

8.2 Solve the Quadratic and Related Inequalities

Example 8.2.1

Solve the following polynomial inequality of degree four for the unknown x:

$$x^4 - 8x^2 + 12 \ge 0.$$

Solution

By letting $y = x^2$, the given inequality reduces to

$y^2 - 8y + 12 \ge 0$,

$(y-2)(y-6) \ge 0$, (factoring)

$y \le 2$ or $y \ge 6$, (Eq. (8.2.1))

$x^2 \le 2$, or $x^2 \ge 6$, (since $y = x^2$)

$x^2 - 2 \le 0$, or $x^2 - 6 \ge 0$,

$(x+\sqrt{2})(x-\sqrt{2}) \le 0$, or $(x+\sqrt{6})(x-\sqrt{6}) \ge 0$,

$-\sqrt{2} \le x \le \sqrt{2}$; or $x \le -\sqrt{6}$, $x \ge \sqrt{6}$.

Hence, the solution set is $\{ (-\infty, -\sqrt{6}] \cup [-\sqrt{2}, \sqrt{2}] \cup [\sqrt{6}, +\infty) \}$.

Example 8.2.2

Solve the following inequality related to the absolute value for the unknown x:

$$|x^2 - 3x + 1| < 1.$$

Solution

By using Eq. (7.4.1) given in Section 7.4, we obtain

$$-1 < x^2 - 3x + 1 < 1,$$

By breaking the above inequality into two separate ones, we have

$$-1 < x^2 - 3x + 1, \text{ and } x^2 - 3x + 1 < 1, \Rightarrow$$
$$-1 + 1 < x^2 - 3x + 1 + 1, \text{ and } x^2 - 3x + 1 - 1 < 1 - 1, \Rightarrow$$
$$0 < x^2 - 3x + 2, \text{ and } x^2 - 3x < 0, \Rightarrow$$
$$0 < (x-1)(x-2), \text{ and } x(x-3) < 0, \Rightarrow$$
$$x < 1, \text{ or } x > 2, \text{ and } 0 < x < 3.$$

For those values of x satisfying the above two solution sets simultaneously, we have

$$[(-\infty,1) \cup (2,+\infty)] \cap (0,3)$$
$$= [(-\infty,1) \cap (0,3)] \cup [(2,+\infty) \cap (0,3)]$$
$$= (0,1) \cup (2,3).$$

Hence, the solution set is $\{ (0,1) \cup (2,3) \}$.

Example 8.2.3

Solve the following inequality involving the square root expression for the unknown x:

$$\sqrt{x^2 - 2x - 3} \geq 1.$$

Solution

By the definition of the square root, the radicand $x^2 - 2x - 3$ is required to be nonnegative that is equivalent to solving

$$x^2 - 2x - 3 \geq 0, \Rightarrow (x+1)(x-3) \geq 0, \Rightarrow x \leq -1, \text{ or } x \geq 3.$$

Therefore, we need to impose the constraints on x that are either $x \leq -1$ or $x \geq 3$. Now, we are ready to solve the given inequality. Since both sides of the given inequality are nonnegative, we obtain by taking the square power on both sides

$$(\sqrt{x^2 - 2x - 3})^2 \geq 1^2, \Rightarrow x^2 - 2x - 3 \geq 1, \Rightarrow$$
$$x^2 - 2x - 4 \geq 0, \Rightarrow [x - (1+\sqrt{5})][x - (1-\sqrt{5})] \geq 0,$$

(Using the method of completing the square to factor) $\Rightarrow$

$$x \leq 1 - \sqrt{5}, \text{ or } x \geq 1 + \sqrt{5}.$$

Since the above solutions satisfy both of the constraints $x \le -1$ or $x \ge 3$, the solution set is $\{ (-\infty, 1-\sqrt{5}] \cup [1+\sqrt{5}, +\infty) \}$.

Example 8.2.4

Solve the following inequality involving the rational expression for the unknown x:

$$\frac{1}{2x-1} \ge 4.$$

Solution

First, we need to convert the given inequality into a polynomial inequality. To get rid of $2x - 1$ in the denominator of the left side of the given inequality, we have to multiply both sides of the inequality by $(2x-1)^2$ that is positive for any real number $x \ne \frac{1}{2}$ so that the direction of the inequality remains unchanged. Now, let us proceed to solve it as follows:

$$(2x-1)^2 \cdot \frac{1}{2x-1} \ge (2x-1)^2 \cdot 4, \ x \ne \frac{1}{2}, \Rightarrow$$
$$(2x-1)^{2-1} \ge 4(4x^2 - 4x + 1), \Rightarrow$$
$$2x - 1 \ge 16x^2 - 16x + 4, \Rightarrow$$
$$0 \ge 16x^2 - 18x + 5.$$

To solve the above quadratic inequality, let us solve the roots of $16x^2 - 18x + 5 = 0$. By using the quadratic formula, we obtain

$$x = \frac{-(-18) \pm \sqrt{(-18)^2 - 4 \cdot 16 \cdot 5}}{2 \cdot 16} = \frac{18 \pm \sqrt{4}}{32} = \frac{5}{8}, \text{ or } \frac{1}{2}.$$

Since $\frac{1}{2} < \frac{5}{8}$, the solution set obtained by using Eq. (8.2.1) is $\{ [\frac{1}{2}, \frac{5}{8}] \}$.

Exercises

Solve each of the following genuine/disguised quadratic inequalities for the unknown x:

1. $x^4 - 3x^2 > 10$.

2. $2(x-1)^4 - 5 \le 9(x-1)^2$.

3. $2x + 3\sqrt{x} \le -1 \ (x \ge 0)$.

4. $2\sqrt[3]{x^2} + 3 \ge -5\sqrt[3]{x}$.

5. $|3x^2 + x + 2| \le 4$.

6. $|2x^2 + 4x + 1| > 2$.

7. $\sqrt{x^2 - x - 6} \ge 2$.

8. $\sqrt{5x^2 - 3x - 2} < 1$.

9. $\dfrac{2x}{x^2 + 1} \le 3$.

10. $2 - \dfrac{1}{\sqrt{x}} > 4\sqrt{x} \quad (x \ne 0)$.

Chapter 9
Linear Equations in Two Variables

In this chapter we're going to learn how solve a system of linear equations in two variables. Then we show how to apply it to solve the word problems.

9.1 The Cartesian Plane

Definition 9.1.1
The Cartesian plane is formed by two real number lines intersecting at right angles, as shown in Figure 9.1.1. The horizontal line is called the x-axis, while the vertical line is called the y-axis. The point of intersection of the two axes is called the origin O. The axes separate the plane into four regions called quadrants. These quadrants are identified by numbers. The upper right-hand quadrant is called the first quadrant, or Quadrant I (Figure 9.1.1). In a counterclockwise direction, the other quadrants are numbered II, III, and IV, respectively.

Remarks

1. To each point in the plane there corresponds an ordered pair of numbers (x, y). Conversely, corresponding to each ordered pair of number (x, y) is a point in the plane. To avoid a confusion with the open interval symbol defined in Sec. 7.1, a capital letter P will be attached as a prefix for the ordered pair, e.g., $P(x, y)$.
2. By the way, Quadrant I, II, III and IV can be, respectively, characterized as follows:

    Quadrant I: $\{(x, y)| \, x \geq 0, y \geq 0\}$,
    Quadrant II: $\{(x, y)| \, x \leq 0, y \geq 0\}$,
    Quadrant III: $\{(x, y)| \, x \leq 0, y \leq 0\}$,
    Quadrant IV: $\{(x, y)| \, x \geq 0, y \leq 0\}$.

3. R. Descartes (1596-1650, French) introduced the coordinate system to bring the analytic tools of algebra into the visual immediacy of geometry that created a branch of mathematics, called the analytic geometry. He published his ideas in 1657 in a treatise called La Géométry. Yet, Menaechmus (375-325 B.C., Greek), who is reportedly a tutor of Alexander the Great, solved problems and proved theorems by using a method that had a strong resemblance to the use of coordinates and it has sometimes been maintained that he had introduced analytic geometry (Reference 2).

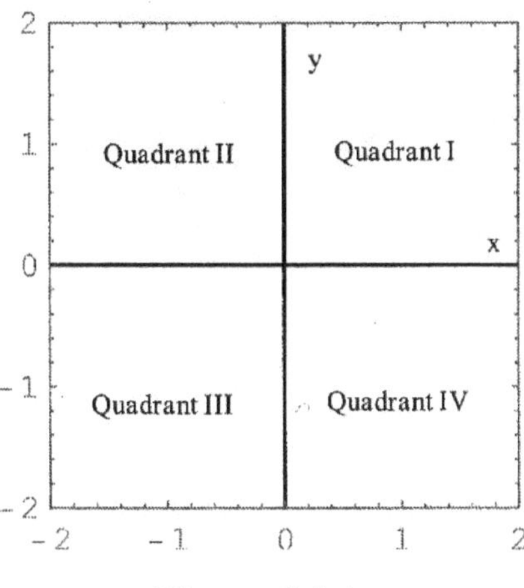

Figure 9.1.1

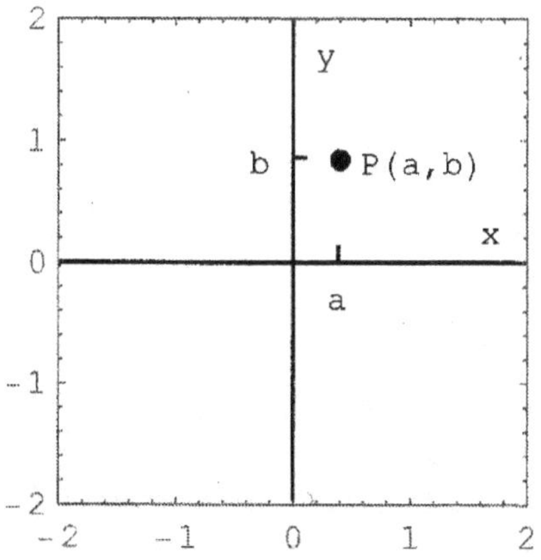

Figure 9.1.2

Definition 9.1.2
The first number a of a point $P(a, b)$ is called the x-coordinate (or the abscissa) of a point P, while the second number b is called the y-coordinate (or the ordinate) of the point P.

Remarks
1. The x-coordinate of a point $P(a, b)$ tells how far to the left or right the point P is from the y-axis, and its y-coordinate tells how far up or down a point P is from the x-axis, as shown in Figure 9.1.2.
2. The points sitting on the x-axis all have 0 as their y-coordinate and are of the form $P(a, 0)$, where a is a real number. The points sitting on the y-axis are of the form $Q(0, b)$, where b is a real number. The origin is denoted by $O(0, 0)$.

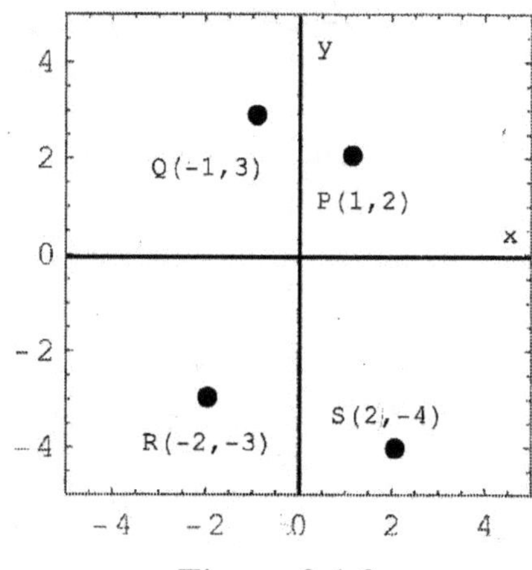

Figure 9.1.3

Example 9.1.1
Plot the points $P(1, 2)$, $Q(-1, 3)$, $R(-2, -3)$, and $S(2, -4)$ in the Cartesian plane.
Solution
To plot the point $P(1, 2)$, we move rightward horizontally from the origin $O(0, 0)$ along the x-axis to a point $(1, 0)$ and then move upward vertically from a point $(1, 0)$ to the desired location of the point $P(1, 2)$ as shown in Figure 9.1.3. The other three points Q, R, and S can be plotted in a similar way (Figure 9.1.3).

Remark
Notice that the four points *P*, *Q*, *R*, and *S* lie in Quadrant I, II, III, and IV, respectively.

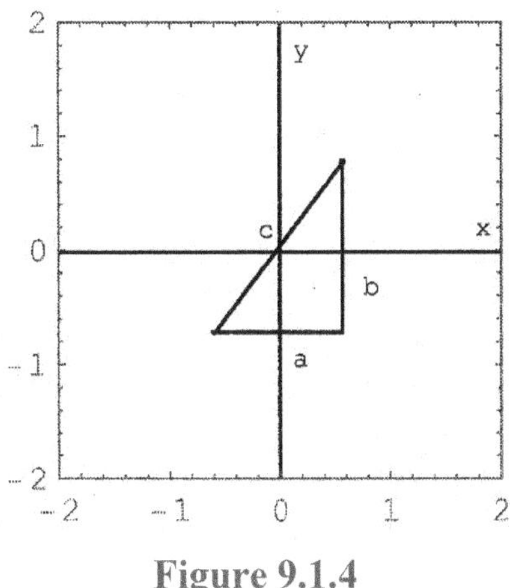

Figure 9.1.4

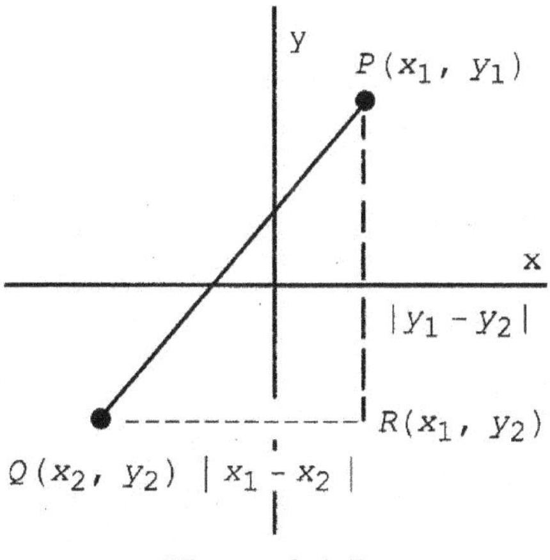

Figure 9.1.5

Theorem 9.1.1 (The Distance Formula)
The distance $d(P,Q)$ between any two points $P(x_1, y_1)$ and $Q(x_2, y_2)$ in the Cartesian plane is given by

$$d(P,Q)=\sqrt{(x_2-x_1)^2+(y_2-y_1)^2}=\sqrt{(x_1-x_2)^2+(y_1-y_2)^2}\,. \qquad (9.1.1)$$

Proof

Assume that $x_1 \neq x_2$, and $y_1 \neq y_2$, namely, $P(x_1, y_1)$ and $Q(x_2, y_2)$ do not lie on the same horizontal or vertical line. With two points P and Q, a right triangle ΔPQR can be formed, as shown in Figure 9.1.5. The length of the side $\overline{PR}$ is $d(y_1, y_2)$, the length of the side $\overline{QR}$ is $d(x_1, x_2)$, and the length of the hypotenuse $\overline{PQ}$ is $d(P, Q)$. By the Pythagorean Theorem, we have

$$[d(P,Q)]^2=(x_2-x_1)^2+(y_2-y_1)^2.$$

By taking the square root on both sides of the above equation, we obtain

$$d(P,Q)=\pm\sqrt{(x_2-x_1)^2+(y_2-y_1)^2}\,.$$

Since the distance $d(P, Q)$ has to be a nonnegative number, the minus sign must be dropped to give us the desired formula.

Remarks

1. The distance formula is easily proved to be true through the use of the Pythagorean Theorem, which states that for a right triangle, the hypotenuse c and two sides a and b are related by the formula $a^2+b^2=c^2$, as shown in Figure 9.1.4.
2. Pythagorean Theorem is named after Pythagoras (570 - 495 B.C., Greek), who by tradition is credited with its proof, although it is often argued that knowledge of the theorem predates him. There is evidence that Babylonian mathematicians understood the formula, although there is little surviving evidence that they used in a mathematical framework. Incidentally, Mesopotamian, Indian, and Chinese mathematicians have all been known for discovering the result independently, some even providing proofs for special cases (Reference 3).
3. Based upon a generalization of Pythagorean theorem, P. de Fermat (1601-1665, French) conjectured in 1637 his famous last theorem, "There are no triplets (a, b, c) of positive integers that satisfy the equation $a^n+b^n=c^n$, where n is a positive integer greater than 2". For a further discussion, see Example 18.4.1(c) (Reference 4).

Example 9.1.2
Find the distance between $P(-1, 3)$ and $Q(-2, 5)$.
Solution
By substituting $x_1 = -1$, $y_1 = 3$, $x_2 = -2$, and $y_2 = 5$ into Eq. (9.1.1), we obtain

$$d(P,Q) = \sqrt{[-1-(-2)]^2 + (3-5)^2}$$
$$= \sqrt{(-1+2)^2 + (-2)^2}$$
$$= \sqrt{1+4} = \sqrt{5}.$$

Theorem 9.1.2 (The Midpoint Formula)
The midpoint R of the line segment joining the points $P(x_1, y_1)$ and $Q(x_2, y_2)$ in the Cartesian plane is given by

$$R\left(\frac{x_1 + x_2}{2}, \frac{y_1 + y_2}{2}\right). \qquad (9.1.2)$$

Remark
It is easily verified that $d(P, R) = d(R, Q)$; Hence, R is indeed a midpoint between P and Q.

Example 9.1.3
Find the midpoint of the line segment joining the points $P(-1, 3)$ and $Q(5, -7)$.
Solution
By substituting $x_1 = -1$, $y_1 = 3$, $x_2 = 5$, and $y_2 = -7$ into Eq. (9.1.2), the midpoint formula, we obtain

$$R\left(\frac{-1+5}{2}, \frac{3+(-7)}{2}\right) = R(2, -2).$$

Exercises

In each of the following problems, (a) plot the points, (b) find the distance between the points, and (c) find the midpoint of the line segment joining the points.

1. $P(0, 6)$, $Q(4, 2)$. **2.** $P(2, 5)$, $Q(-3, 7)$. **3.** $P(-1, 2)$, $Q(7, 10)$.

4. $P(\frac{1}{2},\frac{1}{4})$, $Q(\frac{1}{3},\frac{1}{5})$. **5.** $P(2.3, 5.4)$, $Q(-1.7, -2.6)$.

9.2 The Line

Definition 9.2.1 (The Slope of a Line)
Given that $P(x_1, y_1)$ and $Q(x_2, y_2)$ be two distinct points sitting on a line L, where $x_1 \neq x_2$. Then the slope, denoted by m, of this line L is defined by

$$m = \frac{y_2 - y_1}{x_2 - x_1}, \quad x_1 \neq x_2. \tag{9.2.1}$$

If $x_1 = x_2$, L is a vertical line and the slope m of L is undefined (or $\pm\infty$).

Remarks

1. The slope of a line may be computed from $P(x_1, y_1)$ to $Q(x_2, y_2)$, or from Q to P, since
$$\frac{y_1 - y_2}{x_1 - x_2} = \frac{-(y_2 - y_1)}{-(x_2 - x_1)} = \frac{y_2 - y_1}{x_2 - x_1}.$$
2. Since a line is determined by two points uniquely, any two distinct points sitting on the line can be used to compute the slope of the line.
3. If the slope m of a line L is positive, then the larger the number of m is, the steeper the line is.

Example 9.2.1
Given that the lines L_1, L_2, L_3, and L_4 containing the following pairs of points:

$L_1 : P_1(3,5), Q_1(1,2)$.

$L_2 : P_2(2,4), Q_2(3,6)$.

$L_3 : P_3(1,3), Q_3(5,3)$.

$L_4 : P_4(4,1), Q_4(4,2)$.

Find the slopes of L_1, L_2, L_3, and L_4.

Solution

Let m_1, m_2, m_3, and m_4 denote the slopes of L_1, L_2, L_3, and L_4, respectively. Then, by using Eq. (9.2.1), we obtain

$$m_1 = \frac{2-5}{1-3} = \frac{-3}{-2} = \frac{3}{2},$$

$$m_2 = \frac{6-4}{3-2} = \frac{2}{1} = 2,$$

$$m_3 = \frac{3-3}{5-1} = \frac{0}{4} = 0,$$

$$m_4 = \frac{2-1}{4-4} = \frac{1}{0} = \textit{undefined}.$$

Remark

L_3 is a horizontal line, whereas L_4 is a vertical line. Between L_1 and L_2, the line L_2 is steeper than the line L_1.

Definition 9.2.2 (An Equation of the Line)

There are three ways to find an equation of the line depending on the given information about the line:

1. Point-Slope Form

Given that both the slope m of the line L and a point $P(x_1, y_1)$ sitting on the line are known. Then an equation of the line L is given by

$$y - y_1 = m(x - x_1). \qquad (9.2.2)$$

2. Two-Points Form

Given that the line L is known to pass through two points $P(x_1, y_1)$ and $Q(x_2, y_2)$. Then an equation of the line L is given by

$$y - y_1 = \frac{y_2 - y_1}{x_2 - x_1} \cdot (x - x_1). \qquad (9.2.3)$$

3. Slope-Intercept Form

Given that the slope m of the line L is known and L is known to pass through a point $Q(0, b)$, called the y-intercept of the line L, where b is a real number. Then an equation of the line L is given by

$$y = mx + b. \qquad (9.2.4)$$

Remark

Conventionally, an equation of the line L is preferred to be simplified into the slope-intercept form as a final answer.

Example 9.2.2
Find an equation of the line written in the slope-intercept form for each of the following questions:
(a) The line L has a slope 3 and passes through the point $P(2, 1)$.
(b) The line L passes through two points $P(1, 2)$ and $Q(3, 5)$.
(c) The line L has a slope $\frac{3}{4}$ and passes through the y-intercept $P(0, 2)$.

Solution
(a) By substituting $m = 3$, $x_1 = 2$, and $y_1 = 1$ into Eq. (9.2.2), we have

$$y - 1 = 3(x - 2), \Rightarrow \; y - 1 = 3x - 6 \; y - 1 = 3x - 6, \Rightarrow$$
$$y = 3x - 6 + 1, \Rightarrow \; y = 3x - 5.$$

Hence, the equation of the line L is $y = 3x - 5$.
(b) By substituting $x_1 = 1$, $y_1 = 2$, $x_2 = 3$, and $y_2 = 5$ into Eq. (9.2.3), we have

$$y - 2 = \frac{5 - 2}{3 - 1} \cdot (x - 1) = \frac{3}{2}(x - 1),$$
$$y - 2 = \frac{3}{2}x - \frac{3}{2},$$
$$y = \frac{3}{2}x - \frac{3}{2} + 2,$$
$$y = \frac{3}{2}x + \frac{1}{2}.$$

Hence, the equation of the line L is $y = \frac{3}{2}x + \frac{1}{2}$.

(c) By substituting $m = \frac{3}{4}$ and $b = 2$ into Eq. (9.2.4), we have

$$y = \frac{3}{4}x + 2.$$

Hence, the equation of the line is $y = \frac{3}{4}x + 2$.

Definition 9.2.3 (Parallel and Perpendicular lines)

Given that the slopes of two distinct non-vertical lines L_1 and L_2 are m_1 and m_2, respectively.

1. The Parallel-line Rule

If two lines L_1 and L_2 are parallel, then $m_1 = m_2$. The converse statement is also true, namely, if $m_1 = m_2$, then two lines L_1 and L_2 are parallel. (9.2.5)

2. The Perpendicular-line Rule

If two lines L_1 and L_2 are perpendicular, then $m_1 \cdot m_2 = -1$, or equivalently, $m_2 = -\dfrac{1}{m_1}$. The converse statement is also true, namely, if $m_1 \cdot m_2 = -1$, then two lines L_1 and L_2 are perpendicular. (9.2.6)

Example 9.2.3 (Find a line that is parallel to a given line)

Find an equation of the line passing through the point $P(1, 3)$ and parallel to the line $3x + 2y - 4 = 0$.

Solution

Let the slope of the desired line be m. Since the desired line is known to be parallel to the given line $3x + 2y - 4 = 0$, we know that m equals the slope of the given line by Eq. (9.2.5). However, to figure out the slope of the given line $3x + 2y - 4 = 0$, it has to be written in the slope-intercept form, i.e.,

$$3x + 2y - 4 = 0,$$

$$3x + 2y - 4 - 3x + 4 = 0 - 3x + 4,$$

$$2y = -3x + 4,$$

$$y = \frac{-3x + 4}{2} = -\frac{3}{2}x + 2.$$

Therefore, $m = -\dfrac{3}{2}$. By substituting $x_1 = 1$, $y_1 = 3$, and $m = -\dfrac{3}{2}$ into the slope-point form of the line, we have

$$y - 3 = -\frac{3}{2}(x - 1) = -\frac{3}{2}x + \frac{3}{2},$$

$$y = -\frac{3}{2}x + \frac{3}{2} + 3 = -\frac{3}{2}x + \frac{9}{2}.$$

Hence, the equation of the desired line is $y = -\dfrac{3}{2}x + \dfrac{9}{2}$.

Example 9.2.4 (Finding a line that is perpendicular to a given line)
Find an equation of the line passing through the point $P(3, 1)$ and perpendicular to the line $2x + 3y + 5 = 0$.
Solution
Step 1. Find the slope m_1 of the given line as follows:

$$2x+3y+5=0,$$
$$2x+3y+5-2x-5=0-2x-5,$$
$$3y=-2x-5, \Rightarrow \frac{3y}{3}=\frac{-2x-5}{3}, \Rightarrow y=-\frac{2}{3}x-\frac{5}{3}.$$

Therefore, $m_1 = -\frac{2}{3}$.

Step 2. Let the slope of the desired line be m_2. Then, by using Eq. (9.2.6), we have

$$m_1 \cdot m_2 = -1, \text{ or } -\frac{2}{3}\cdot m_2 = -1, \Rightarrow m_2 = \frac{-1}{-\frac{2}{3}}, \Rightarrow m_2 = \frac{3}{2}.$$

Step 3. By using Eq. (9.2.4),

$$y-1=\frac{3}{2}\cdot(x-3)=\frac{3}{2}x-\frac{9}{2},$$
$$y=\frac{3}{2}x-\frac{9}{2}+1=\frac{3}{2}x-\frac{7}{2}.$$

Hence, the equation of the desired line is $y=\frac{3}{2}x-\frac{7}{2}$.

Exercises

Find the equation of the line with the given information for each of the following questions:

1. Slope = – 3; passing through $P(-1, 2)$.
2. Slope = – 2; passing through $P(0, 5)$.
3. Slope = undefined; passing through $P(3, 2)$.
4. Slope = 0; passing through $P(2, -1)$.
5. Passing through $P(5, 2)$ and $Q(1, 3)$.
6. Passing through $P(-1, 4)$ and $Q(-1, 7)$.

7. Parallel to the line $4x+5y-1=0$ and passing through $O(0, 0)$.
8. Parallel to the line $x = 5$ and passing through $P(-3, 4)$.
9. Perpendicular to the line $2x+y+3=0$ and passing through $P(2, 0)$.
10. Perpendicular to the line $y = 2$ and passing through $P(-2, 5)$.

9.3 Solve a Linear Equation in Two Variables

Definition 9.3.1
A polynomial equation of degree one in x and y given by $ax+by+c=0$ is called a linear equation in two variables x and y, where a, b, and c are real constants, and at least one of a and b is not zero.

Remark
A single linear equation in two variables x and y is the first prototypical example of "too few equations in too many variables". In such a case, there exist infinitely many solutions to the given equation because one of the variables can serve as an independent variable taking real numbers freely, while the remaining variable becomes a dependent variable.

Definition 9.3.2
The solution set to a linear equation in two variables x and y is the set of all ordered pairs of numbers (x, y) that satisfy the given equation.

Definition 9.3.3
In a linear equation in two variables x and y, a variable is called a dependent variable if it can be expressed in terms of the other variable, called an independent variable.

Remarks
1. In the solution set of a linear equation in two variables x and y, an independent variable can take any real number freely. Hence, in

the context of solving a linear programming problem, an independent variable is also called a "non-basic" variable, while a dependent variable is called a "basic" variable (Section 10.3).

2. In a linear equation in two variables, there are exactly one independent variable and one dependent variable. As a general rule, a variable having a nonzero coefficient can always be chosen as a dependent variable.

Example 9.3.1 (Both of the coefficients of x and y are nonzero)
Solve $2x+3y-5=0$ for the unknown (x, y).

Solution
Since both of the coefficients of x and y in the given equation are nonzero, either variable can be chosen as a dependent variable, and the other variable then serves as an independent variable.

Case 1. If y is chosen as a dependent variable, then we solve for y in terms of x, i.e.,

$$2x+3y-5-2x+5=0-2x+5,$$

$$3y=-2x+5, \Rightarrow y=\frac{-2x+5}{3}, \Rightarrow y=-\frac{2}{3}x+\frac{5}{3}.$$

Case 2. If x is chosen as a dependent variable, then we solve for x in terms of y as follows:

$$2x+3y-5-3y+5=0-3y+5,$$

$$2x=-3y+5, \Rightarrow x=\frac{-3y+5}{2}, \Rightarrow x=-\frac{3}{2}y+\frac{5}{2}.$$

Hence, the solution set is $\{(x,-\frac{2}{3}x+\frac{5}{3}) \mid x$: any real number$\}$ or $\{(-\frac{3}{2}y+\frac{5}{2}, y) \mid y$: any real number$\}$. Note that the graph of the solution set is a straight line.

Remark
To plot a straight line is equivalent to graphing the solution set to the given linear equation in x and y. In fact, graphing the solution set is a more effective way to express the infinitely many solutions.

Example 9.3.2 (The coefficient of x is zero)
Solve $2y+7=0$ for the unknown (x, y).

Solution

Since only the coefficient of y is nonzero, we solve for y as follows:

$$2y + 7 - 7 = 0 - 7, \Rightarrow 2y = -7, \Rightarrow y = -\frac{7}{2}.$$

Since the x term is missing in the given equation, it implies that the coefficient of x is zero. Hence, x is an independent variable and the solution set is $\{ (x, -\frac{7}{2}) \mid x\text{: any real number}\}$. Note that the graph of this solution set is a horizontal line $y = -\frac{7}{2}$.

Example 9.3.3 (The coefficient of *y* is zero)

Solve $3x - 5 = 0$ for the unknown (x, y).

Solution

Since only the coefficient of x is nonzero, we solve for x as follows:

$$3x - 5 + 5 = 0 + 5, \Rightarrow 3x = 5, \Rightarrow x = \frac{5}{3}.$$

Since the y term is missing in the given equation y is an independent variable. Hence, the solution set is $\{ (\frac{5}{3}, y) \mid y\text{: any real number}\}$.

Notice that the graph of the solution set is a vertical line $x = \frac{5}{3}$.

Exercises

Solve each of the following linear equations in two variables x and y for the unknown (x, y):

1. $x - y = 0$. **2.** $x - y = 2$. **3.** $x + y - 3 = 0$.
4. $3x - 4y = 12$. **5.** $2x - y + 4 = 0$. **6.** $2y - x = 0$.
7. $2x - 5 = 0$. **8.** $3y = 6$. **9.** $3x = 0$.
10. $5y = 0$.

9.4 Two Linear Equations in Two Variables

In this section we shall consider how to solve the following system of two linear equations in two variables x and y:

$$a_1 x + b_1 y = c_1, \tag{9.4.1}$$

$$a_2 x + b_2 y = c_2, \tag{9.4.2}$$

where at least one of a_1 and b_1 is not zero, so is a_2 and b_2.

Definition 9.4.1

The solution set to a system of two equations in two variables given by Eqs. (9.4.1) and (9.4.2) is the set of all ordered pairs of numbers (x, y) that satisfy both of Eqs. (9.4.1) and (9.4.2) simultaneously.

Remarks

1. Unlike a single linear equation in two variables that always has infinitely many solutions, the solution set to a system of two linear equations in two variables has three possible cases: (i) a unique solution, (ii) no solution, and (iii) infinitely many solutions. All three cases can be visualized from a geometric standpoint. Since the solution set to a single linear equation in two variables is a line, finding a solution to a system of Eqs. (9.4.1) and (9.4.2) is equivalent to finding the intersection point between two lines.
2. Babylonians already knew how to solve a system of linear equations (Reference 1, pp. 51-52).

Definition 9.4.2

A system of two linear equations in two variables x and y given by Eqs. (9.4.1) and (9.4.2) is said to be consistent if the solution set is not empty. Otherwise, the system is said to be inconsistent.

Definition 9.4.3

Two systems of two linear equations in two variables are said to be equivalent If their solution sets are the same.

Remark

Basically, the idea of solving a system of Eqs. (9.4.1) and (9.4.2) is to reduce the given system to a simpler system through the use of the appropriate operational rules for obtaining the equivalent new system of two linear equations in two variables.

Operational Rules for Obtaining Equivalent Systems of Linear Equations

1. Scalar Multiplication Rule
Both sides of an equation are multiplied by a nonzero number.
2. Switching Rule
Exchange the order of one equation with another equation.
3. Linear Combination Rule
Adds one equation multiplied by a nonzero number on both of its sides to another equation.

Remark
The above operational rules hold for any system of n linear equations in n variables, where $n \geq 2$, which is covered in Chapter 17 in Part II.

Methods for Solving a System of Linear Equations in Two Variables

1. The method of elimination (9.4.3)
To reduce the given system into a simpler system that can be easily solved through eliminating one of the two variables by using the appropriate operational rule.
2. The method of substitution (9.4.4)
To express one variable in terms of the other variable through one equation and then substitute the expression of this variable into another equation to reduce that equation to become in one variable only.

Example 9.4.1 (A Unique Solution)

(a) Solve $\begin{cases} x + 2y = 3, \\ 2x + y = 0. \end{cases}$ for the unknown (x, y).

(b) Determine if the given system is consistent.

Solution
We are going to show either one of the two methods mentioned above could be used to solve the given system.
Method I (Eq. (9.4.3)): By labeling the two given equations with (1) and (2), we have

$$x + 2y = 3, \qquad (1)$$
$$2x + y = 0. \qquad (2)$$

To eliminate the variable x, we multiply both sides of Eq. (1) by a number -2 and then add accordingly to both sides of Eq. (2) and obtain

$$-2x - 4y + 2x + y = -6 + 0, \text{ (Operational Rule 3)}$$
$$-3y = -6, \Rightarrow y = \frac{-6}{-3}, \Rightarrow y = 2. \qquad (3)$$

After substituting Eq. (3) into Eq. (1), we have

$$x + 2 \cdot 2 = 3, \Rightarrow x + 4 = 3, \Rightarrow x = 3 - 4, \Rightarrow x = -1.$$

Hence, the solution set is $\{(-1, 2)\}$.

Method II (Eq. (9.4.4)): To express x in terms of y from Eq. (1), we have

$$x + 2y - 2y = 3 - 2y, \Rightarrow x = 3 - 2y. \qquad (4)$$

After substituting Eq. (4) into Eq. (2), we have

$$2 \cdot (3 - 2y) + y = 0,$$
$$6 - 4y + y = 0, \text{ or } 6 - 3y = 0,$$
$$\frac{6}{3} = y, \Rightarrow y = 2. \qquad (5)$$

By substituting Eq. (5) into Eq. (4), we have

$$x = 3 - 2 \cdot 2 = 3 - 4 = 1.$$

Hence, the solution set is $\{(-1, 2)\}$.

(b) Since the solution set is not empty, the given system is consistent.

Remark

The solution $(-1, 2)$ represents geometrically the intersection point of two non-parallel lines defined), respectively, by Eqs. (1) and (2).

Example 9.4.2 (No solution)

(a) Solve $\begin{cases} x + y = 2, \\ 2x + 2y = 5. \end{cases}$ for the unknown (x, y).

(b) Determine if the given system is consistent.

Solution

(a) By applying Eq. (9.4.3), we multiply both sides of the first equation by a number -2 and then add it to both sides of the second equation, and obtain

$$-2x - 2y + 2x + 2y = -4 + 5, \Rightarrow 0 = 1,$$

which is a contradiction. Hence, there is no solution to the given system of equations.
(b) Since the solution set is empty, the given system is inconsistent.

Remark
Geometrically, the given system of equations represents two parallel lines. Hence, two lines do not intersect that means there is no solution to the given system of equations.

Example 9.4.3 (Infinitely Many Solutions)
(a) Solve $\begin{cases} x+2y=3, \\ 2x+4y=6. \end{cases}$ for the unknown (x, y).
(b) Determine if the given system of equations is consistent.
Solution
(a) If we examine the second equation closely, we notice that the second equation is in fact exactly the same as the first equation after dividing both sides of the second equation by a number 2. Therefore, the given system reduces in effect to one equation in two variables. By using the result of Section 9.3, the solution set is $\{(x, \frac{3-x}{2}) \mid x$: any real number$\}$.
(b) Since the solution set is not empty, the given system is consistent.

Remark
Geometrically, the given system represents two lines coincide together to become a single line.

Exercises
Solve each of the following system of linear equations in two variables for the unknown (x, y), (y, z), or (z, w). Also, determine if the given system is consistent.

1. $\begin{cases} x+y=3, \\ 3x+2y=8. \end{cases}$ **2.** $\begin{cases} 2x+3y=1, \\ 3x-2y=2. \end{cases}$ **3.** $\begin{cases} 2y-z=3, \\ z+5y=4. \end{cases}$

4. $\begin{cases} 2z+w=4, \\ z=2w-1. \end{cases}$ **5.** $\begin{cases} x+y=2, \\ 2y=4-2x. \end{cases}$ **6.** $\begin{cases} y-z=1, \\ 3y=3z-5. \end{cases}$

7. $\begin{cases} x+3y-4=0, \\ 2x-y+3=0. \end{cases}$

8. $\begin{cases} 2x-y+4=0, \\ 3x-5=0. \end{cases}$

9. $\begin{cases} 3y-z-2=0, \\ 4z+3=0. \end{cases}$

10. $\begin{cases} \frac{1}{2}z-\frac{1}{3}w=-1, \\ \frac{1}{3}z+\frac{1}{4}w=1. \end{cases}$

9.5 The Word Problems

Procedures for Solving the Word Problem

Step 1. Use the variables to represent the desired unknown quantities.

Step 2. Translate the given information into a set of equations in terms of the variables introduced in Step 1.

Step 3. Solve the set-up system of equations for the unknown variables.

Step 4. Translate the solution obtained in Step 3 into a word answer to the given word problem.

Example. 9.5.1

The sum of the digits of a two-digit number is 9. If the digits are reversed, the new number is 45 more than the original number. What is the original number?

Solution

Step 1. Let x and y be the tens and the ones digit of the original number, respectively.

Step 2. Since the sum of the digits of this two-digit number is 9, we have

$$x+y=9. \qquad (1)$$

Also, since the new number is 45 more than the original number if the digits are reversed, we have

$$10\cdot y+x=10\cdot x+y+45,$$
$$10y+x-10x-y=45,$$
$$-9x+9y=45, \Rightarrow -x+y=5. \qquad (2)$$

Step 3. By adding Eq. (1) to Eq. (2), we obtain

$$x+y-x+y=9+5, \Rightarrow 2y=14, \Rightarrow y=7. \qquad (3)$$

After substituting Eq. (3) into Eq. (1), we obtain

$$x + 7 = 9, \Rightarrow x = 9 - 7, \Rightarrow x = 2.$$

Step 4. Hence, the original two-digit number is 27.

Example 9.5.2

A madrigal music show was attended by 520 people. Adults were charged \$10, and children \$4. If the total receipts for the show were \$4,780, how many adults and how many children attended the show?

Solution

Step 1. Let x and y be, respectively, the number of adults and children who attended the show.

Step 2. From the given information, we have

$$x + y = 520, \qquad (1)$$

$$10 \cdot x + 4 \cdot y = 4{,}780. \qquad (2)$$

Step 3. By adding to Eq. (2) both sides of Eq. (1) multiplied by – 10, we obtain

$$-10x - 10y + 10x + 4y = -5{,}200 + 4{,}780.$$

$$-6y = -420, \Rightarrow y = \frac{-420}{-6}, \Rightarrow y = 70. \qquad (3)$$

By substituting Eq. (3) into Eq. (1), we obtain

$$x + 70 = 520, \Rightarrow x = 520 - 70, \Rightarrow x = 450.$$

Step 4. Hence, there are 450 adults and 70 children who attended the show.

Example 9.5.3

Tom invests some money to buy a 12-month certificate of deposit (CD) from a bank at 5 percent annual interest; and his wife, Susan, invests some money to buy a 12-month CD from an e-bank at 6 percent annual interest. If their total investment is \$10,000 and their total return in interest after 12 months is \$560, how much money does each have invested?

Solution

Step 1. Let x and y be the amount of money invested, respectively, by Tom and Susan.

Step 2. From the given information described in the above statement, we have

$$x + y = 10{,}000, \qquad (1)$$

$$0.05 \cdot x + 0.06 \cdot y = 560. \qquad (2)$$

Step 3. By adding Eq. (1) multiplied by – 0.05 on both sides to Eq. (2), we obtain

$$-0.05x-0.05y+0.05x+0.06y=-0.05\cdot 10{,}000+560,$$

$$0.01y=60, \Rightarrow\ y=\frac{60}{0.01}=6{,}000. \qquad (3)$$

By substituting Eq. (3) into Eq. (1), we obtain

$$x+6{,}000=10{,}000, \Rightarrow x=10{,}000-6{,}000=4{,}000.$$

Step 4. Hence, Tom invested \$4,000 and Susan invested \$6,000.

Remark

This example is actually the same as Example 5.3.4. A point in case is that all word problems in Section 5.3 can be solved in terms of two variables.

Example 9.5.4

Tim and Jerry working together can paint a house in 54 hours. If Tim worked alone for 45 hours and Jerry joined him, and together they finished the rest of the painting job in 27 hours. How long would it take each working alone to do the house-painting job?

Solution

Step 1. Let x and y be the amount of time in hours, respectively, for Tim and Jerry to work alone in finishing the job. Then $\frac{1}{x}$ and $\frac{1}{y}$ would be Tim's and Jerry's house-painting speed.

Step 2. From the given information, we have

$$\frac{1}{x}+\frac{1}{y}=\frac{1}{54}, \qquad (1)$$

$$45\cdot\frac{1}{x}+27\cdot\left(\frac{1}{x}+\frac{1}{y}\right)=1,$$

$$72\cdot\frac{1}{x}+27\cdot\frac{1}{y}=1. \qquad (2)$$

Step 3. By letting $v=\frac{1}{x}$ and $w=\frac{1}{y}$, we have from Eqs. (1) and (2)

$$v+w=\frac{1}{54}, \qquad (3)$$

$$72v+27w=1. \qquad (4)$$

By adding both sides of Eq. (3) multiplied by – 72 to Eq. (4), we obtain

$$-72v - 72w + 72v + 27w = -72 \cdot \frac{1}{54} + 1 = -\frac{1}{3}, \Rightarrow$$

$$-45w = -\frac{1}{3}, \Rightarrow \ w = \frac{-\frac{1}{3}}{-45} = \frac{1}{135} \tag{5}$$

By substituting Eq. (5) into Eq. (3), we have

$$v + \frac{1}{135} = \frac{1}{54},$$

$$v + \frac{1}{135} - \frac{1}{135} = \frac{1}{54} - \frac{1}{135} = \frac{1}{2 \cdot 3^3} - \frac{1}{5 \cdot 3^3} = \frac{5-2}{2 \cdot 5 \cdot 3^3},$$

$$v = \frac{3}{2 \cdot 5 \cdot 3^3} = \frac{1}{2 \cdot 5 \cdot 3^2} = \frac{1}{90}. \tag{6}$$

By using the relationship between x, y and v, w, we have

$$\frac{1}{x} = v = \frac{1}{90}, \Rightarrow x = 90.$$

$$\frac{1}{y} = w = \frac{1}{135}, \Rightarrow y = 135.$$

Step 4. Hence, it takes 90 hours for Tim to work alone in finishing the job, while it takes 135 hours for Jerry to work alone in getting the job done.

Exercises

1. Find two numbers whose sum is 102 and whose difference is 16.

2. Two numbers differ by 8. If twice the smaller number is added to the larger number, the result is 95. Find the two numbers.

3. A two-digit number is 3 more than 7 times the sum of its digits. If the digits are reversed, the new number is 18 less than the original number. Find the original number.

4. Eight years ago, Amanda was $\frac{4}{5}$ as old as Bob, while 10 years later she will be $\frac{7}{8}$ as old. What are their ages now?

5. How many pounds of coffee worth $1.55 per pound and how many pounds of coffee worth $1.85 per pound must be mixed to get 100 pounds of coffee worth $1.64 per pound?

6. A piggy bank contains $3.20 in quarters and dimes. If there are 20 coins in the piggy bank, how many of the coins are quarters? How many of them are dimes?

7. A man invested part of his money at 7% and the rest at 8%. The income from both investments totaled $1,180. If he interchanged his investments, his income would have totaled $1,220. How much did he have in each investment?

8. Bill and John working together can do a job in 24 hours. If Bill works alone for 8 hours and John finishes the job in 32 hours, how many hours would it take each working alone to do the job?

9. Flying with the wind it takes a plane 4 hours to go 2,000 miles, while flying against the wind it takes 5 hours. What are the speed of the wind and the speed of the plane in still air?

10. A rectangular rug that is 4 feet longer than it is wide in room with a perimeter of 88 feet. The perimeter of the rug is 12 feet less than the perimeter of the room. What are the dimensions of the rug?

9.6 References

1. Bunt, L. N. H., Jones, P. S., and Bedient, J. D. (1988). The Historical Roots of Elementary Mathematics, Dover Publications, Inc., New York.
2. https://en.wikipedia.org/wiki/Analytic_geometry.

3. https://en.wikipedia.org/wiki/Pythagorean_theorem.
4. https://en.wikipedia.org/wiki/Fermat%27s_last_theorem.

Chapter 10
Linear Inequalities in Two Variables

In this chapter we're going to learn how to graph the solution set to a set of linear inequalities in two variables. Then, apply this technique to solve the linear programming problem.

10.1 Solve a Linear Inequality in Two Variables

Definition 10.1.1
An inequality of one of the following forms: $ax + by + c < 0$, $ax + by + c \leq 0$, $ax + by + c > 0$, $ax + by + c \geq 0$, is called a linear inequality in two variables x and y.

Definition 10.1.2
An ordered pair of numbers is said to be a solution to an inequality if it satisfies the linear inequality in two variables when the coordinates are substituted for the variables that makes the inequality true. The set of all the solutions is called the solution set to the given inequality.

Remarks

1. Just like a linear inequality in one variable covered in Chapter 7, a linear inequality in two variables has infinitely many solutions. However, unlike a linear inequality in one variable, we are short of algebraic notation to express these infinitely many solutions. A more effective way to express these infinitely many solutions is to graph geometrically the solution set in the Cartesian plane. It is easily shown that the graph of the solution set is a half-plane.
2. Even though a system of linear equations had been conquered thousands of years ago in Babylonian times, a system of linear inequalities remained inaccessible until modern times. By 1827 J. B. J. Fourier (1768-1830, French) generalized the elimination method to solve a system of linear inequalities. The method now

known as the Fourier or Fourier-Motzkin elimination method is one of the earliest methods proposed for solving a system of linear inequalities (Reference 1).

Procedure for Graphing a Linear Inequality in Two Variables

Step 1. Replace the inequality with an equality in the given linear inequality in two variables. Then, graph the corresponding linear equation. Its graph is a line that divides the entire Cartesian plane into two half-planes. One half-plane lies above (or to the right of) the line, while the other half-plane lies below (or to the left of) the line. One of these half-planes represents the solution set to the given inequality.

Step 2. Select any point from just one of the two half-planes. Test if this point is a solution to the given inequality. If the answer is yes, then the half-plane that contains the testing point is the solution set to the given inequality. Otherwise, the other half-plane that does not contain the testing point is the desired solution set. Conventionally, we shade the half-plane to represent the solution set.

Remark

If the given inequality also contains an equality, then the solution set is a closed half-plane, which means the line is included in the solution set. Otherwise, the solution set is an open half-plane, which means the line is not included in the solution set. If the line is included in the solution set, we use a solid line to represent it in the Cartesian plane. Otherwise, we use a dashed line to represent that the line is not included in the solution set.

Example 10.1.1

Determine whether the ordered pair (1, – 3) is a solution to the inequality: $2x-3y+4\leq 0$.

Solution

By substituting $x = 1$ and $y = -3$ into the left side of the given inequality, we have

$$2\cdot 1-3\cdot(-3)+4=2+9+4=15,$$

which is not less than or equal to zero. Hence, the ordered pair (1, –3) is not a solution to the given inequality.

Example 10.1.2

Graph the solution set to the inequality: $2x - y - 3 < 0$.

Solution

Step 1. Replacing the sign '< ' by the sign '= ', we have

$$2x - y - 3 = 0.$$

To graph $2x - y - 3 = 0$, we need two points. There are many ways to choose the two points. If the line does not pass through the origin, namely, the constant term in the equation is not zero, we usually choose the two points as the x-intercept and the y-intercept of the line. By setting $y = 0$ in the equation, we have

$$2x - y - 3 = 0, \Rightarrow 2x = 3, \Rightarrow x = \frac{3}{2}.$$

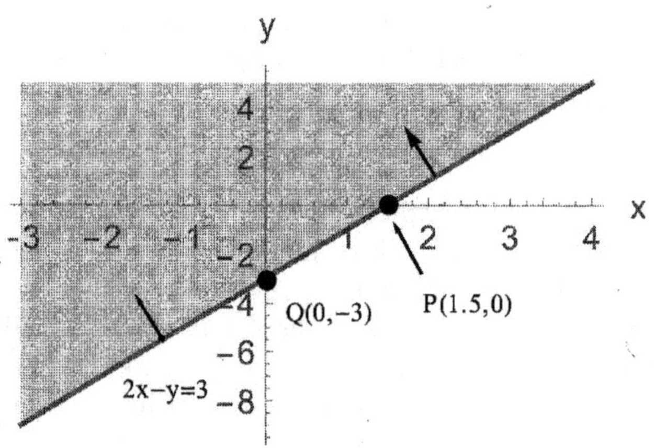

Figure 10.1.1

So, the x-intercept is $P(\frac{3}{2}, 0)$. By setting $x = 0$ in the equation, we have

$$2 \cdot 0 - y - 3 = 0, \Rightarrow -y = 3, \Rightarrow y = -3.$$

Therefore, the y-intercept is $Q(0, -3)$. After connecting the points $P(\frac{3}{2}, 0)$ and $Q(0, -3)$, we have the line L that represents the solution set to $2x-y-3=0$. Clearly, this line divides the Cartesian plane into two half-planes (Figure 10.1.1).

Step 2. Choose the origin $O(0, 0)$ as a testing point. By substituting $x=0$ and $y=0$ into the left-side of the given inequality, we have

$$2\cdot 0-0-3=-3,$$

which is less than 0. Therefore, the origin $O(0, 0)$ is a solution to the given inequality. This implies that the left open half-plane containing the origin is the solution set to the given inequality.

Example 10.1.3
Graph the solution set to the inequality: $x-2y\geq 3$.
Solution

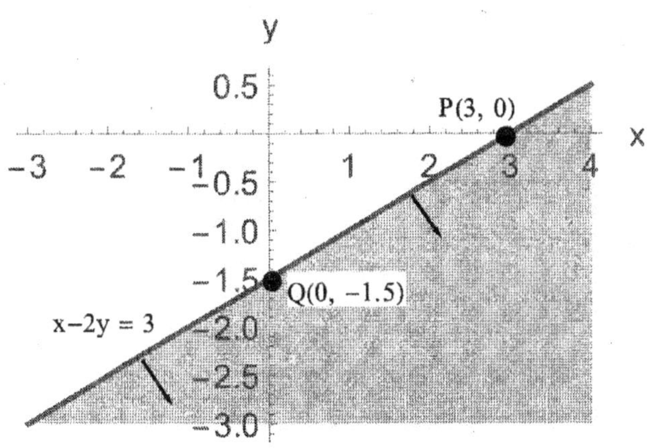

Figure 10.1.2

Step 1. Replacing the inequality sign "≥" with an equality sign "=", we have $x-2y=3$. By setting $y=0$ in the equation, we obtain

$$x-2\,0=3,\Rightarrow x=3,\Rightarrow \text{ the } x\text{-intercept is } P(3, 0).$$

By setting $x=0$ in the equation, we obtain

$$0-2y=3,\Rightarrow -2y=3,\Rightarrow$$

$$y=-\frac{3}{2}.\Rightarrow \text{ the } y\text{-intercept is } Q(0,-\frac{3}{2}).$$

After connecting the points $P(3, 0)$ and $Q(0,-\frac{3}{2})$ in the Cartesian plane, we obtain a line L that divides the plane into two half-planes.

Step 2. Choose the origin $O(0, 0)$ as a testing point. By substituting $x = 0$ and $y = 0$ into the left side of the inequality, we have $0 - 2 \cdot 0 = 0$, which is not greater than or equal to 3. This implies that the origin $O(0, 0)$ is not a solution to the given inequality. Since the given inequality contains the equality sign, the closed half-plane that does not contain the origin is the solution set to the given inequality (Figure 10.1.2).

Exercises

Graph the solution set to each of the following linear inequalities in two variables:

1. $x + y - 1 < 0$. **2.** $-2x + y \geq 2$. **3.** $x - 2y \leq 0$.

4. $3x - 2y > 6$. **5.** $5x - 2y \leq 10$.

10.2 A System of Linear Inequalities in Two Variables

Definition 10.2.1

A system of linear inequalities in two variables is that at least two linear inequalities are paired together.

Definition 10.2.2

A solution to a system of linear inequalities in two variables is an ordered pair of real numbers that satisfy all inequalities simultaneously.

Definition 10.2.3

The solution set that is the set of all the solutions to a system of linear inequalities is called the feasible set for the given system of inequalities.

Remarks

1. Unlike a linear inequality, a system of linear inequalities may not have a solution, namely, the feasible set could be empty.

2. If the feasible set is not empty, its geometric shape is usually a polygon. The vertices of the feasible set, play an important role in solving a linear programming problem in the next section.

Definition 10.2.4
The vertices of a non-empty feasible set is called the corner points of the feasible set.

Procedure for Graphing the Feasible Set to a System of Linear Inequalities

Step 1. Graph the solution set to each inequality in the given system individually.
Step 2. Find a common intersection of all the shaded half-planes that will be the desired feasible set to the given system.

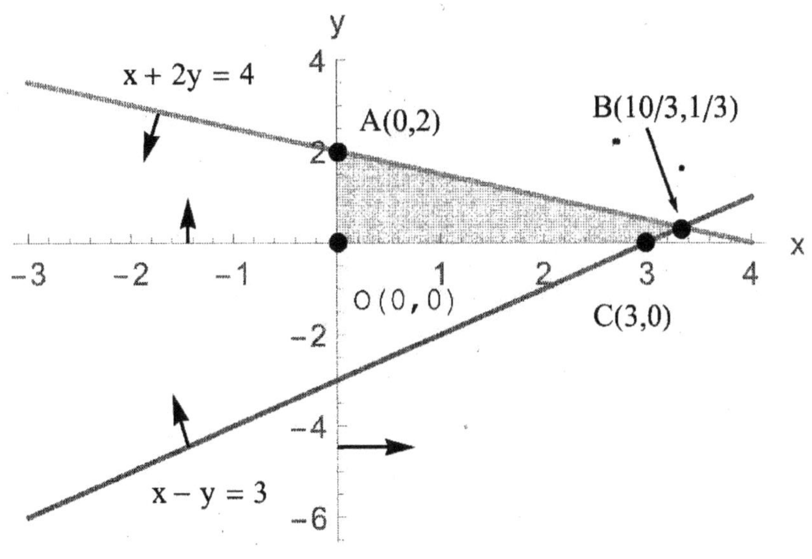

Figure 10.2.1

Example 10.2.1
Given a system of linear inequalities as follows:

$$\begin{cases} x - y \le 3, \\ x + 2y \le 4, \\ x \ge 0, \\ y \ge 0. \end{cases}$$

Graph the feasible set to the given system. Find the corner points of the feasible set, if they exist.

Solution

By graphing each linear inequality of the given system separately, the final common intersection of all the half-planes is a quadrilateral *OABC* with four corner points *O*(0, 0), *A*(0, 2), *C*(3, 0), and *B* is the intersection point of two lines $x - y = 3$ and $x + 2y = 4$ (Figure 10.2.1). By using either Eq. (9.4.1) or (9.4.2) given in Section 9.4, we obtain the solution $x = \frac{10}{3}$ and $y = \frac{1}{3}$. Hence, the corner point *B* is $B(\frac{10}{3}, \frac{1}{3})$.

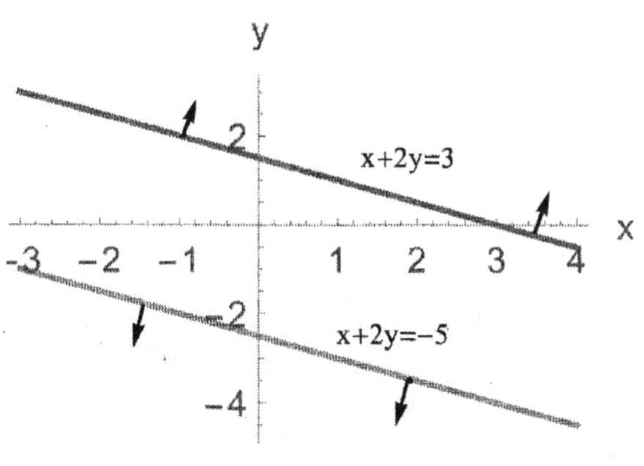

Figure 10.2.2

Example 10.2.2

Given a system of linear inequalities as follows:

$$\begin{cases} x + 2y \ge 3, \\ x + 2y \le -5. \end{cases}$$

Graph the feasible set to the given system. Find the corner points of the feasible set, if they exist.

Solution

After graphing the solution set to each linear inequality in the given system, we notice that the intersection of two closed half-planes is empty (Figure 10.2.2). Hence, the feasible set to the given system is also empty. It has no corner points either.

Exercises

Graph the feasible set to each of the following system of linear inequalities in two variables. If the feasible set is not empty, find its corner points.

1. $x + y \geq 2$, $x \geq 0$, $y \geq 0$.
2. $x - 2y > 3$, $2x - 4y < -1$, $x \geq 0$, $y \geq 0$.
3. $2x + 3y \leq 6$, $3x + 2y \leq 6$, $x \geq 0$, $y \geq 0$.
4. $x + 2y \geq 3$, $2x + y \geq 1$, $x \geq 0$, $y \geq 0$.
5. $x + y \geq 1$, $2x + 3y \leq 6$, $x \geq 0$, $y \geq 0$.

10.3 Linear Programming

Definition 10.3.1
A linear programming problem is a constrained optimization problem that either maximize or minimize a linear objective function subject to some constraints, which are constituted by a system of linear inequalities.

Remarks
1. Historically, linear programming was invented during World War II as a practical tool for solving a transportation problem of moving man and supplies for the U.S. Air Force in such a way as to minimize the cost. Today, this tool is widely used in industry to either maximize the profit or minimize the cost under the constraint that the resources are scarce.
2. In solving a practical real-world problem, a linear programming problem usually involves a system of linear inequalities in several hundred variables. However, in this section we shall limit our discussion to problems containing only two variables, namely, consider

Maximize (or Minimize) $f(x, y) = ax + by + f_0$ (10.3.1)

(x, y)

Subject to

$c_1 x + d_1 y \le e_1$,

$c_2 x + d_2 y \le e_2$, (10.3.2)

$\ldots \ldots \ldots \ldots$,

$c_m x + d_m y \le e_m$,,

$x \ge 0$, $y \ge 0$,

where a, b, f_0, c_i, d_i and e_i , $i = 1, 2, \ldots, m$ are real numbers, and m is a positive integer (Reference 2).

Definition 10.3.2

The solution set to the constraints of Eq. (10.3.2) in the above linear programming problem of Eqs. (10.3.1) and (10.3.2) is called the feasible set of the given linear programming problem.

Definition 10.3.3

An ordered pair of numbers (x_0, y_0) is called an optimal solution to the linear maximization (or minimization) problem of Eqs. (10.3.1) and (10.3.2) if (x_0, y_0) satisfies the following two conditions:

(i) (x_0, y_0) is a point in the feasible set, namely, (x_0, y_0) satisfies Eq. (10.3.2).

(ii) $f(x_0, y_0)$ is the maximum (or the minimum) of $f(x, y)$ among all ordered pairs of numbers (x, y) in the feasible set.

Theorem 10.3.1 (The Fundamental Theorem of Linear Programming)

If a linear programming problem of Eqs. (10.3.1) and (10.3.2) has an optimal solution, it will occur at one or more corner points of the feasible set to Eq. (10.3.2).

Remarks

1. A linear programming problem may not possess an optimal solution because the feasible set could be empty.
2. If an optimal solution occurs at two corner points of the feasible set, then every point sitting on the line segment that connects

these two corner points is also an optimal solution to the given linear programming problem.

3. The proof of the theorem that is beyond the scope of this book can be found in Reference 3, pp.133-134.
4. Since the objective function that is a linear function in x and y is a plane that is monotonically increasing/decreasing over the feasible set, it can be visualized geometrically that the maximum/minimum of the objective function surely occurs at one of the corner points of the feasible set.
5. If the objective function in Eq. (10.3.1) has more than two variables, the geometric method presented in this chapter is not going to work, and has to be replaced by the algebraic one, called the simplex method invented by G. B. Dantzig (1914 - 2005, American) (Reference 1).

Procedure for Solving a Linear Programming Word Problem

Step 1. Introduce the variables to represent the desired unknown quantities.

Step 2. Translate the given information in the statement of the word problem into setting up the appropriate linear programming problem, either a maximization or a minimization problem.

Step 3. Graph the feasible set of the linear programming problem obtained in Step 2 to determine if the feasible set is empty or not. If the feasible set is empty, then stop and there is no optimal solution to the given problem. Otherwise, go to Step 4.

Step 4. Find the corner points of the feasible set. Then, evaluate the objective function value of $f(x, y)$ at the corner points of the feasible set. The corner point that has the maximum (or the minimum) function value is an optimal solution to the given maximization (or minimization) problem.

Step 5. Translate the optimal solution obtained in Step 4 back as an answer to the word problem.

Example 10.3.1 (A Maximization Problem)
A farmer has 40 acres of land available for planting corn or soybeans. Profit per acre in corn production is \$40 and that in soybeans production is \$50. The cost of preparing the soil for growing corn is \$30

and that for growing soybeans is \$60. Each acre of land in corn production requires 3 hours of labor, whereas production of soybeans requires 1 hour per acre. The total labor available during the production period is 100 hours and the farmer cannot spend more than \$1,800 in preparation costs. Determine how the farmer should divide his land between growing corns and soybeans in order to gain the maximum profit from planting his land.

Solution

Step 1. Let x and y be the acres of the farmland planned by a farmer to grow corn and soybeans, respectively, and P be the profit of growing x acres of corn and y acres of soybeans.

Step 2. Since the goal of this farmer is to maximize his profit, this is a constrained maximization problem. From the given information stated in the problem, a linear programming problem can be set up as follows:

$$\underset{(x,\,y)}{\text{Maximize}} \qquad P = 40x + 50y$$

Subject to

$$x + y \leq 40, \qquad (1)$$

$$30 \cdot x + 60 \cdot y \leq 1{,}800, \qquad (2)$$

$$3 \cdot x + 1 \cdot y \leq 100, \qquad (3)$$

$$x \geq 0\,,\, y \geq 0.$$

Step 3. The feasible set that is not empty is graphed as given in Figure 10.3.1. There are five corner points $O(0, 0)$, $A(0, 30)$, $B(?, ?)$, $C(?, ?)$, and $D(\frac{100}{3}, 0)$, where B is the intersection point between two lines $x + y = 40$ and $30x + 60y = 1{,}800$, whereas C is the intersection point between two lines $x + y = 40$ and $4x + y = 120$.

To find the point B, we need to solve the following system of equations:

$$x + y = 40, \qquad (4)$$

$$30x + 60y = 1{,}800. \qquad (5)$$

By adding both sides of Eq. (4) multiplied by -30 to Eq. (5), we have

$$-30x - 30y + 30x + 60y = -30 \cdot 40 + 1{,}800 = 600\,,$$

$$30y = 600\,, \Rightarrow\ y = \frac{600}{30} = 20\,. \qquad (6)$$

By substituting Eq. (6) into Eq. (4), we have

$$x + 20 = 40, \Rightarrow x = 40 - 20 = 20.$$

Hence, the point B is $B(20, 20)$.

To find the point C, we need to solve the following system of equations:

$$x + y = 40, \quad (7)$$

$$3x + y = 100. \quad (8)$$

By adding both sides of Eq. (7) multiplied by – 3 to Eq. (8), we have

$$-3x - 3y + 3x + y = -3 \cdot 40 + 100 = -20, \Rightarrow$$

$$-2y = -20, \Rightarrow y = \frac{-20}{-2} = 10. \quad (9)$$

After substituting Eq. (9) into Eq. (7), we have

$$x + 10 = 40,$$

$$x = 40 - 10 = 30.$$

Hence, the point C is $C(30, 10)$.

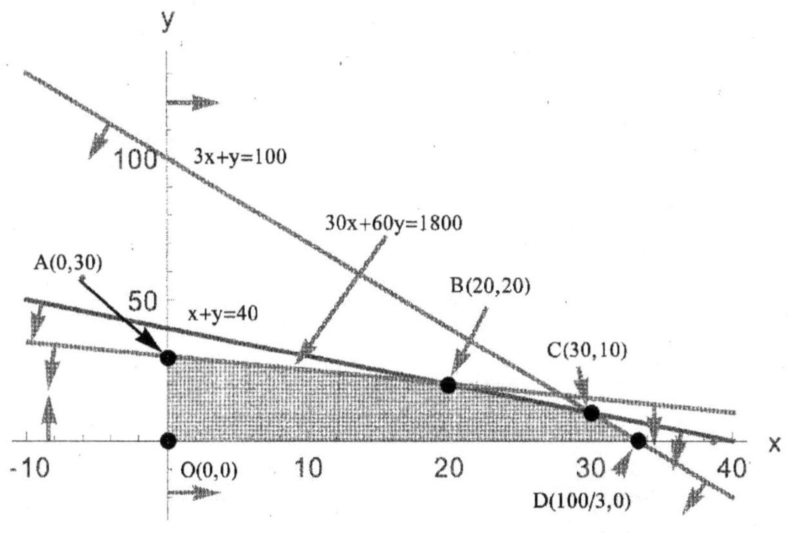

Figure 10.3.1

Step 4. Evaluate the profit at the five corner points as follows:

Profit at $O(0, 0) = 40 \cdot 0 + 50 \cdot 0 = 0$,

Profit at $A(0, 30) = 40 \cdot 0 + 50 \cdot 30 = 1{,}500$.

Profit at $B(20, 20) = 40 \cdot 20 + 50 \cdot 20 = 1{,}800$.

Profit at $C(30, 10) = 40 \cdot 30 + 50 \cdot 10 = 1{,}700$.

Profit at $D(\frac{100}{3},0)=40\cdot\frac{100}{3}+50\cdot 0=\frac{4{,}000}{3}=1{,}333.3$.

Therefore, the maximum profit occurs at the corner point $B(20,20)$.

Step 5. This farmer should grow 20 acres of corn and 20 acres of soybean with the maximum profit of $1,800.

Example 10.3.2 (A Minimization Problem)

Suppose that it takes 12 units of carbohydrates and 30 units of protein to satisfy the minimum daily nutrition requirements for a certain person. A certain meat per pound contains 8 units of protein and 2 units of carbohydrates. A certain cheese per pound contains 3 units of protein and 3 units of carbohydrates. The meat costs $3.00 per pound and the cheese costs $2.00 per pound. How many pounds of each are needed in order to minimize the cost and still meet the minimum daily nutrition requirements?

Solution

Step 1. Let x and y be the needed amount of meat and cheese in pounds, and C be the cost of buying x pounds of meat and y pounds of cheese.

Step 2. Since the goal of this person is to minimize his/her cost, this is a constrained minimization problem. From the given information stated in the problem, a linear programming problem is set up as follows:

Minimize $\quad C = 3x + 2y$

(x, y)

Subject to

$2x + 3y \geq 12,$ (carbohydrates)

$8x + 3y \geq 30,$ (protein)

$x \geq 0,\ y \geq 0.$

Step 3. The feasible set is graphed and is not empty (Figure 10.3.2). The feasible set has three corner points $A(0, 10)$, $B(?, ?)$, and $C(6, 0)$, where B is the intersection point between two lines $2x + 3y = 12$ and $8x + 3y = 30$.

To find the point B, we need to solve the following system of equations:

$$2x + 3y = 12, \qquad (1)$$

$$8x + 3y = 30. \qquad (2)$$

By adding both sides of Eq. (1) multiplied by -1 to Eq. (2), we have

$$-2x-3y+8x+3y=-12+30=18\,,\Rightarrow$$

$$6x = 18, \Rightarrow\ x=\frac{18}{6}=3\,, \qquad (3)$$

After substituting Eq. (3) into Eq. (1), we have

$$2\cdot 3 + 3y = 12, \Rightarrow\ 3y = 12 - 6 = 6, \Rightarrow\ y=\frac{6}{3}=2\,.$$

Hence, the point B is $B(3, 2)$.

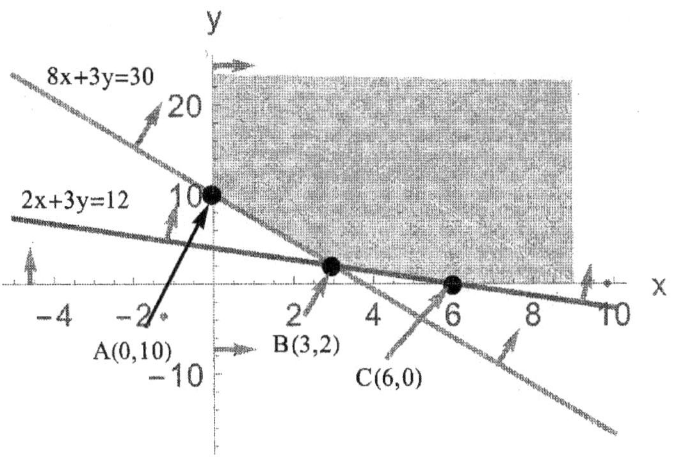

Figure 10.3.2

Step 4. Evaluate the cost at the three corner points:
Cost at $A(0, 10)$ = \$3·0 + \$2·10 = \$20,
Cost at $B(3, 2)$ = \$3·3 + \$2·2 = \$13,
Cost at $C(6, 0)$ = \$3·6 + \$2·0 = \$18.
Therefore, the least cost occurs at the point $B(3, 2)$.
Step 5. This person who should buy three pounds of meat and two pounds of cheese has the least cost of \$13.

Exercises

1. A retired couple has up to \$30,000 to invest. Their financial planner recommends that they place at most \$10,000 in corporate bonds yielding 8.5% and at least \$20,000 in Treasury bills yielding 5.8%.

How should this couple divide their money into investing in corporate bonds and Treasury bills in order to gain the maximum yearly return from their investments?

2. A domestic car dealer with warehouses in Chicago and Indianapolis receives orders from dealers in Springfield, Illinois and Louisville, Kentucky. The dealer in Springfield needs 5 cars and the dealer in Louisville needs 8. The Chicago warehouse has 7 cars and the Indianapolis warehouse has 10. The cost of shipping from Chicago to Springfield is $150 per car, from Chicago to Louisville $200 per car, from Indianapolis to Springfield $120 per car, from Indianapolis to Louisville $100 per car. Find the number of cars to be shipped from each warehouse to each dealer in order to minimize the shipping cost.

3. A student is working part time on campus to pay off college expenses. The student may work no more than 20 hours per week. She is paid $8 per hour for working as a tutor in the mathematics help center and $6 per hour for working in the library. Suppose that she can work at most 12 hours in either the mathematics help center or the library. What combination of working as a tutor and/or in the library will yield her maximum weekly income?

4. A furniture maker plans to make two products: chairs and desks. Each chair requires $10 capital and 2 hours of labor. Each desk requires $20 capital and 6 hours of labor. The profit per chair is $30 and the profit per desk $70. The furniture maker has available each day 18 hours of labor and $80 of capital. How many chairs and how many desks should be produced each day in order to maximize the profit?

5. A diet is to contain at least 12 ounces of vitamin C and 20 ounces vitamin E. These requirements are to be obtained from two types of foods - type I and type II. Each pound of type I food contains 2 ounces of vitamin C and 5 ounces of vitamin E, whereas each pound of type II food contains 3 ounces of vitamin C and 4 ounces of vitamin E. If type I food costs $6 per pound, and type II food costs $8 per pound, how many pounds of each type of foods should be purchased at the minimum cost and still meet the minimum daily requirements in vitamin C and E?

10.4 References

1. Chvatal, V. (1983). Linear Programming, W. H. Freeman and Company, New York.
2. Murty, K. G. (2006). Linear equations, inequalities, linear programs (LP), and a new efficient algorithm, http://www-personal.engin.umich.edu/~ murty/.
3. http://en.wikipedia.org/wiki/Linear_programming.

Chapter 11
Ratio, Interpolation, and Variation

In this chapter we're going to learn the concepts of ratio, proportion, interpolation, extrapolation, and, variation. Then, we will show how to apply them to solve some word problems.

11.1 Ratio and Proportion

Definition 11.1.1
The ratio of one quantity to another quantity is the quotient of one quantity divided by another. The two quantities must be measured in the same physical units.

Remarks

1. Conventionally, we denote the ratio as a:b that is equivalent to $\frac{a}{b}$, $b \neq 0$. Also, a ratio is always simplified so that no common factor exists between a and b.
2. It is impossible to trace the origin of the concept of ratio, because the ideas from that it developed would have been familiar to pre-literate culture. In Book 5 of Euclid's "Elements" gave 18 definitions, all of that are related to ratio. In modern day mathematics, the concept of ratio is also extended to the concept of odds in probability (Chapter 19 in Part II) or percentage (or relative frequency) of the frequency distribution in statistics (Reference 6).

Example 11.1.1
If a father's weight is 185 pounds and a son's weight is 225 pounds, what is the ratio of a father's weight to a son's weight?
Solution
The ratio of a father's weight to a son's weight is

$$185:225 = \frac{185}{225} = \frac{37}{45} = 37:45.$$

Definition 11.1.2
If two ratios equal each other, it is called a proportion.

Remarks
1. If two ratios *a*:*b* and *c*:*d* are equal, this implies
$$\frac{a}{b} = \frac{c}{d}, \ (b \neq 0, \ d \neq 0), \text{ or } \quad a \cdot d = b \cdot c.$$
2. Historically, the notion of proportion was already used by Eudoxus (408-355 B.C., Greek) as presented in Book Five of Euclid's book "Elements". But it was used to express the ratio between two similar quantities rather than a numerical value as we think of it today (Reference 4, p. 52). In China, it was mentioned in "Zhou-Bi Suan-Jing (周髀算經)" ("The Mathematical Manuals of the Gnomon and the Circular Paths of Heaven"). However, neither the exact date when this book was written nor its author were known. The written date was estimated to lie between 644 B.C. and 402 B.C. (Reference 2, p.59).

Example 11.1.2
If a variable x satisfies the proportion of $x:2 = 4:8$, find the value of x.
Solution
Since x satisfies the given proportion, we have
$$x:2 = 4:8, \Rightarrow \frac{x}{2} = \frac{4}{8}. \tag{1}$$
By multiplying both sides of Eq. (1) by a number 2, we obtain
$$x = 2 \cdot \frac{4}{8} = \frac{8}{8} = 1.$$
Hence, the value of x that satisfies the given proportion is 1.

Example 11.1.3
A private college claims that the faculty-student ratio is $1:15$. If this private college has a student body of 2,010, how many faculty members does this private college have?
Solution

Let x be the number of faculty members that this private college has. Then, since the faculty-student ratio is 1:15, we have

$$1 : 15 = x : 2{,}010, \Rightarrow \frac{1}{15} = \frac{x}{2{,}010} \qquad (1)$$

By multiplying both sides of Eq. (1) by 2,010, we obtain

$$2{,}010 \cdot \frac{1}{15} = 2{,}010 \cdot \frac{x}{2{,}010}, \Rightarrow x = 134.$$

Hence, this private college has 134 faculty members.

Definition 11.1.3

Two triangles are said to be similar if the lengths of corresponding sides are proportional.

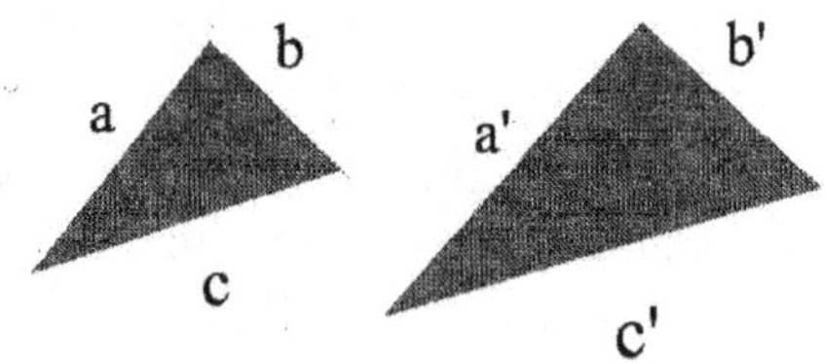

Figure 11.1.1

Remark

If two triangles are similar, then the corresponding angles are equal. The converse statement is also true, namely if the corresponding angles of two triangles are equal, then two triangles are similar (Figure 11.1.1).

Example 11.1.4

If in the similar triangles of Figure 11.1.1, $a = 5$, $b = 6$, and $a' = 10$, find $b' = ?$.

Solution

Since the triangles are similar, we have

$$\frac{a}{a'} = \frac{b}{b'}, \Rightarrow \frac{5}{10} = \frac{6}{b'}, \Rightarrow 5 \cdot b' = 6 \cdot 10 = 60, \Rightarrow$$

$$\frac{5\cdot b'}{5}=\frac{60}{5}, \Rightarrow b'=12.$$

Hence, the length of b' is 12.

Definition 11.1.4

A rectangle is called a golden rectangle if the ratio of its length to its width is $\frac{1+\sqrt{5}}{2}:1$.

Remark

The construction of the golden rectangle is given in Reference 3, p. 61.

Example 11.1.5

If the length of a golden rectangle is 12 inches, find its width.

Solution

Let the width of this golden rectangle is x inches. Then, we have

$$12:x=\frac{1+\sqrt{5}}{2}:1, \Rightarrow \frac{12}{x}=\frac{1+\sqrt{5}}{2}, \Rightarrow$$

$$24=(1+\sqrt{5})x \Rightarrow \frac{24}{1+\sqrt{5}}=x, \Rightarrow$$

$$\frac{(\sqrt{5}-1)\cdot 24}{(\sqrt{5}-1)(\sqrt{5}+1)}=x, \Rightarrow \frac{24(\sqrt{5}-1)}{(\sqrt{5})^2-1^2}=x,$$

$$\frac{24(\sqrt{5}-1)}{4}=x, \Rightarrow x=6(\sqrt{5}-1).$$

Hence, the width of this golden rectangle is $6(\sqrt{5}-1)$ inches.

Definition 11.1.5

The odds on the occurrence of a certain event is the ratio of the number of times that this event will occur to the number of times that this event will not occur.

Remark

The odds is readily translated into the language of probability. If the odds on the occurrence of an event are a to b, then the probability that this event will occur is $\frac{a}{a+b}$ (Chapter 19 in Part II).

Example 11.1.6
The odds is 2 to 100,000 that a certain tire manufacturer will produce a defective tire. Out of the production of 6 million tires, how many tires are expected to be defective?
Solution
Let x be the expected number of defective tires out of 6 million tires produced. Then, since the odds in producing a defective tier is 2 to 100,000, we have

$$2:100{,}000 = x:6{,}000{,}000,$$

$$\frac{2}{100{,}000} = \frac{x}{6{,}000{,}000}. \qquad (1)$$

By multiplying both sides of Eq. (1) by a number 6,000,000, we have

$$6{,}000{,}000 \cdot \frac{2}{100{,}000} = x, \Rightarrow x = 120.$$

Hence, the expected number of defective tires is 120 tires.

Exercises

1. If a father's age is 52 and a son's age is 16, what is the age ratio of father to son?

2. The triangles are similar with the length of corresponding sides are denoted by $\{a, b, c\}$ and $\{a', b', c'\}$, respectively. If it is known that $a = 4$, $b = 8$, $c = 10$, and $a' = 12$. Find $b' = ?$, and $c' = ?$.

3. If the width of a golden rectangle is 8 inches, what is the length of this rectangle?

4. In a certain state university there are 600 faculty members and the faculty-student ratio is 1 to 30. How many students are there?

5. A bag of six grapefruits costs \$1.98. How much does three grape-fruits cost?

6. On a road atlas, an inch represents 200 miles. How many miles are represented by two and a half inches?

7. The odds on twin births is 1 to 105. If Dr. Lin has delivered 2,100 births, how many pairs of twins would he expect to deliver?

8. A 60-foot rope is cut into two pieces whose lengths are in the ratio of 17 to 13. How long is the longer piece of rope?

9. A mutual fund invests in bonds and stocks in the ratio of 3 to 5. How much of $240,000 invested will go into stocks?

10. Jack earns $30.75 for 5 hours of work. At this rate how much will he earn for 12 hours of work?

11.2 Linear Interpolation/Extrapolation

A Method of Linear Interpolation or Extrapolation

Given two variables x and y. Suppose that the relationship between x and y is linear, i.e., a straight line. Furthermore, if the corresponding y-values for two gives distinct x-values, x_1 and x_2 $(x_1 < x_2)$ are known and given as y_1 and y_2, respectively, then

Case 1. Linear interpolation:

The unknown y-value of a given x_0, where $x_1 < x_0 < x_2$ is denoted by y_0, can be obtained by a linear interpolation from the following equation of proportion:

$$\frac{y_0 - y_1}{x_0 - x_1} = \frac{y_2 - y_1}{x_2 - x_1},$$

$$y_0 = y_1 + \frac{y_2 - y_1}{x_2 - x_1} \cdot (x_0 - x_1) \qquad (11.2.1)$$

Case 2. Linear extrapolation:

The unknown y-value of a given x_3, where $x_1 < x_2 < x_3$, denoted by y_3, can be obtained by a linear extrapolation from the following equation of proportion:

$$\frac{y_3 - y_2}{x_3 - x_2} = \frac{y_2 - y_1}{x_2 - x_1},$$

$$y_3 = y_2 + \frac{y_2 - y_1}{x_2 - x_1} \cdot (x_3 - x_2) \qquad (11.2.2)$$

Remarks

1. The derivation of Eqs. (11.2.1) or (11.2.2) is based on the fact that the slope of a line is the same between any two points on the line.
2. This method is very useful in finding the desired y-value that is not available in the table when we are looking it up in a certain table, for example, the logarithmic table or the probability table of a standard normal distribution.
3. Linear interpolation has been used since antiquity for filling the gaps in tables, often with astronomical data. It was certainly used by Babylonian astronomers and mathematicians in Seleucid Mesopotamia (last three centuries B.C.), and by the Greek astronomer and mathematician, Hipparchus (180-125 B.C.) (Reference 5).

Example 11.2.1

The following is a table of numbers between two variables x and y:

x	y
0.4	1.4918
0.5	1.6487
0.6	1.8221

Find: (a) y = ? when x = 0.44, (b) y = ? when x = 0.62.

Solution

(a) Since $x_0 = 0.44$ lies between $x_1 = 0.4$ and $x_2 = 0.5$, we obtain y_0 by using Eq. (11.2.1) of linear interpolation as follows:

$$y_0 = y_1 + \frac{y_2 - y_1}{x_2 - x_1} \cdot (x_0 - x_1),$$

$$= 1.4918 + \frac{1.6487 - 1.4918}{0.5 - 0.4} \cdot (0.44 - 0.4),$$

$$= 1.4918 + \frac{0.1569}{0.1} \cdot 0.04 = 1.4918 + 1.569 \cdot 0.04,$$

$$= 1.4918 + 0.06276 = 1.55456 \approx 1.5546.$$

Hence, the value of y is 1.5546 when x = 0.44.

(b) Since $x_3 = 0.62$ lies outside $x_1 = 0.5$ and $x_2 = 0.6$, we obtain y_3 by using Eq. (11.2.2) of linear extrapolation as follows:

$$y_3 = 1.8221 + \frac{1.8221 - 1.6487}{0.6 - 0.5} \cdot (0.62 - 0.6),$$
$$= 1.8221 + \frac{0.1734}{0.1} \cdot 0.02 = 1.8221 + 1.734 \cdot 0.02,$$
$$= 1.8221 + 0.03468 = 1.85678 \approx 1.8568.$$

Hence, the value of y is 1.8568 when $x = 0.62$.

Remark

Even though the relationship between x and y in this example is not linear (in fact, the table was taken from a table of the natural exponential function e^x), the y-value obtained through either an interpolation or an extrapolation still provides a good approximation to the exact value. Compare $y_0 = 1.5546$ with the exact value $e^{0.44} = 1.5527$ and $y_3 = 1.8568$ with the exact value $e^{0.62} = 1.8589$.

Exercises

1. From the following natural logarithmic table of values between x and y, find: (i) $y = ?$ when $x = 2.23$, (ii) $y = ?$ when $x = 2.46$.

x	y
2.2	0.7885
2.3	0.8329
2.4	0.8755

Remark

The function of natural logarithm is covered in Chapter 16 in Part II.

2. From the following probability table of a standard normal random variables, find (i) $p = ?$ when $z = -1.76$, (ii) $p = ?$ when $z = -1.94$.

z	p
−1.9	0.0287
−1.8	0.0359
-1.7	0.0446

11.3 Variation

Definition 11.3.1
A variable y is said to be directly proportional to another variable x (or to vary directly as x) if there exists a proportional constant $k \neq 0$ such that

$$y = k \cdot x. \tag{11.3.1}$$

Remarks
1. If y is directly proportional to x, then so is x directly proportional to y. In fact, $x = \frac{1}{k} \cdot y$.
2. If y is directly proportional to x, then the ratio of y to x is a constant.

Example 11.3.1
Assume that y is directly proportional to x.
(a) Determine the proportional constant k if $y = 5$ when $x = 10$.
(b) Using the relationship between y and x obtained in part (a), what is the value of y when $x = 25$?
Solution
(a) Since y is directly proportional to x, there exists by using Eq. (11.3.1) a proportional constant k such that $y = k \cdot x$.
To determine the value of k, we substitute $x = 10$ and $y = 5$ into Eq. (11.3.1) and obtain

$$5 = k \cdot 10 = 10k\ , \Rightarrow\ \frac{5}{10} = k\ , \Rightarrow\ k = \frac{1}{2}.$$

Hence, the relationship between x and y is $y = \frac{1}{2}x$.

(b) When $x = 25$, we obtain the corresponding value of y from the relationship between x and y obtained in part (a) as follows:

$$y = \frac{1}{2} \cdot 25 = \frac{25}{2} = 12.5.$$

Hence, the value of y is 12.5 when $x = 25$.

Remark
Many physical laws are set up through the proportional relationship between two physical variables as shown in the next example.

Example 11.3.2

Hooke's law states that within the elastic range of loads the elongation ΔL of a spring subjected to a stretching force by hanging a weight on it is directly proportional to the applied load *F*, namely, *F* and ΔL satisfy

$$F = k \cdot \Delta L \tag{11.3.2}$$

where *k* is the proportional constant, also known as the spring constant, to be determined (Figure 11.3.1). Assume the elongation of the spring is 2 inches when the applied load is 10 pounds. What is the elongation of this spring when the applied load is 25 pounds?

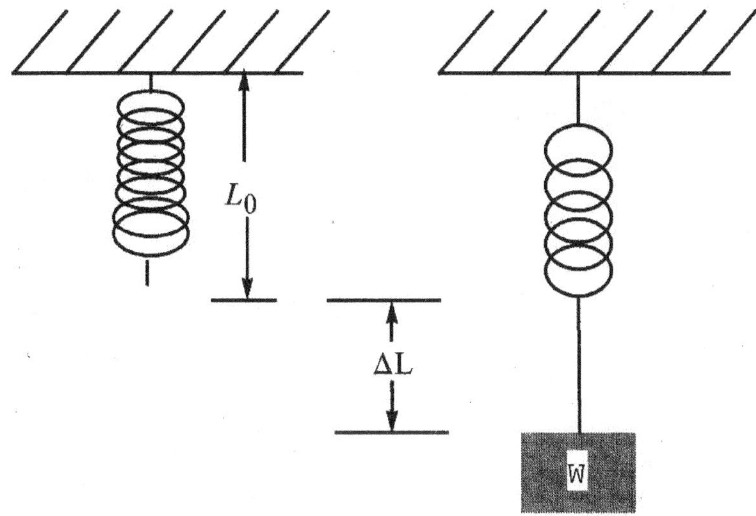

Figure 11.3.1

Solution

We first determine the unknown proportional constant *k*. By substituting $\Delta L = 2$ inches and $F = 10$ pounds into Eq. (1), we have

$$10 = k \cdot 2, \Rightarrow k = 5 \text{ (pound/inch)}.$$

Hence, the relationship between ΔL and *F* for this spring is

$$F = 5 \cdot \Delta L \tag{11.3.3}$$

Then, by substituting $F = 25$ pounds into Eq. (11.3.3), we obtain

$$25 = 5 \cdot \Delta L \Rightarrow \frac{25}{5} = \Delta L\,, \Rightarrow \Delta L = 5 \text{ (inches)}.$$

Hence, the elongation of this spring is 5 inches when the applied load is 25 pounds.

Remark

ΔL is a standard symbol in calculus. It represents a small increment in the length *L*. Also, see problems 8-10 in Exercises for Section 13.1 in Part II.

Definition 11.3.2

A variable y is said to be directly proportional to (or to vary directly as) the nth power of another variable x if there exists a proportional constant $k \neq 0$ such that

$$y = kx^n \text{ , where } n \text{ is a positive integer.} \tag{11.3.4}$$

Example 11.3.3

Suppose that y is directly proportional to x^2. Also, it is know that y = 10 when $x = 2$. What is the value of y when $x = 5$?

Solution

Because y is directly proportional to x^2, there exists a proportional constant $k \neq 0$ such that

$$y = kx^2 . \tag{11.3.5}$$

To determine the value of *k*, we obtain by substituting the given information that $y = 10$ when $x = 2$ into Eq. (11.3.3) as follows:

$$10 = k \cdot 2^2 = 4k, \Rightarrow \frac{10}{4} = \frac{4k}{4} \Rightarrow k = 2.5.$$

Therefore, the relationship between x and y is given by

$$y = 2.5x^2 . \tag{11.3.6}$$

After substituting $x = 5$ into Eq. (2), we have

$$y = 2.5 \cdot 5^2 = 2.5 \cdot 25 = 62.5.$$

Hence, the value of y is 62.5 when $x = 5$.

Example 11.3.4

In geometry, it is found that

(a) the area A of a circle with a radius r is directly proportional to r^2 and the proportional constant has been shown to be π. Thus, $A = \pi r^2$.

(b) the surface area S of a sphere with a radius r is directly proportional to r^2 and the proportional constant has been shown to be 4π. Thus, $S = 4\pi r^2$.

(c) the volume V of a sphere with a radius r is directly proportional to r^3 and the proportional constant has been shown to be $\frac{4}{3}\pi$.

Thus, $V = \frac{4}{3}\pi r^3$.

Definition 11.3.3

A variable y is said to be inversely proportional to x (or to vary inversely as x) if there exists a proportional constant $k \neq 0$ such that

$$y = k \cdot \frac{1}{x} = \frac{k}{x}. \ (x \neq 0) \tag{11.3.7}$$

Remarks

1. If y is inversely proportional to x, so is x is inversely proportional to y. In fact, $x = \frac{k}{y}$, $y \neq 0$.
2. The product of x and y remains constant because $x \cdot y = k$.
3. This definition is also equivalent to saying that y is directly proportional to $\frac{1}{x}$.

Example 11.3.5

Suppose that y is inversely proportional to x, and it is known that $y = \frac{1}{2}$ when $x = 4$. What is the value of y when $x = 7$?

Solution

Since y is inversely proportional to x, there exists by using Eq. (11.3.7) a proportional constant $k \neq 0$ such that

$$y = \frac{k}{x},$$

where k is to be determined. To determine the value of k, we obtain by substituting the given information $x = 4$ and $y = \frac{1}{2}$ into Eq. (11.3.7) as follows:

$$\frac{1}{2} = \frac{k}{4}, \Rightarrow 4 \cdot \frac{1}{2} = k, \Rightarrow k = 2.$$

Therefore, the relationship between x and y is given by

$$y = \frac{2}{x}. \qquad (1)$$

After substituting $x = 7$ into Eq. (1), we have

$$y = \frac{2}{7}.$$

Hence, the value of y is $\frac{2}{7}$ when $x = 7$.

Example 11.3.6

In modern physics, it is known that the momentum p of the photon is inversely proportional to the wavelength λ of the scattered photon and the proportional constant is the Planck's constant h (= 6.626 $\cdot 10^{-34}$ J·sec). Thus,

$$p = \frac{h}{\lambda} = \frac{6.626 \cdot 10^{-34}}{\lambda} \qquad (11.3.8)$$

(Reference 1, p. 619).

Definition 11.3.4

A variable z is said to be jointly direct-proportional to x and y (or to vary jointly as x and y) if there exists a proportional constant $k \neq 0$ such that

$$z = k \cdot x \cdot y. \qquad (11.3.9)$$

Remark

To extend this definition of a dependent variable z to be proportional either directly or inversely to more than two independent variables x and y is straightforward.

Example 11.3.7

Suppose that a variable z is jointly direct-proportional to x and inverse-proportional to y. It is known that $z = 5$ when $x = 2$ and $y = 4$. What is the value of z when $x = 4$ and $y = 2$.

Solution

Since z is jointly direct-proportional to x and inverse-proportional to y, there exists a proportional constant $k \neq 0$ such that

$$z = k \cdot \frac{x}{y}, \tag{1}$$

where k is to be determined. To determine the value of k, we obtain by substituting the given information $x = 2$, $y = 4$, and $z = 5$ into Eq. (1) as follows:

$$5 = k \cdot \frac{2}{4} = \frac{1}{2}k, \Rightarrow 2 \cdot 5 = k, \Rightarrow k = 10.$$

Therefore, the relationship between x, y, and z is given by

$$z = 10 \cdot \frac{x}{y} = \frac{10x}{y}. \tag{2}$$

By substituting $x = 4$ and $y = 2$ into Eq. (2), we have

$$z = \frac{10 \cdot 4}{2} = 20.$$

Hence, the value of z is 20 when $x = 4$ and $y = 2$.

Example 11.3.8

Newton's Law of Gravitation says that the attractive force F between two spherical objects is jointly directly-proportion to their individual mass m_1 and m_2, and inverse-proportion to the distance r between the centers of two spherical bodies, and the proportional constant is the universal gravitational constant G ($= 6.673 \cdot 10^{-11}$ N$\cdot m^2$/kg^2). Thus, we have

$$F = G \cdot \frac{m_1 m_2}{r^2} \tag{11.3.10}$$

(Reference 1, pp. 65-67).

Exercises

1. Suppose that y is directly proportional to x. If $y = 18$ when $x = 2$, what is the value of y when $x = 3$?

2. Suppose that y is directly proportional to x^3. If $y = 24$ when $x = 2$, what is the value of y when $x = 4$?

3. Suppose that y is inversely proportional to x. If $y = 9$ when $x = 3$, what is the value of y when $x = 5$?

4. Suppose that w is jointly direct-proportional to x^2 and y^3 and inverse-proportional to z^4. If $w = 72$ when $x = 3$, $y = 4$, and $z = 2$, what is the value of w when $x = 2$, $y = 3$, and $z = 4$?

5. The perimeter P of a circle is directly proportional to its radius r. If the perimeter is 4π cm when the radius is 2 cm, what is the value of the perimeter when the radius is 5 cm?

6. A load of 30 pounds stretches a spring 3 inches. How far will a load of 40 pounds stretch the spring?

7. A freely falling body falls 64 feet in 2 seconds. The distance (d) fallen is directly proportional to the square of the time elapsed (t). How far will it fall in10 seconds?

8. The resistance of an electrical wire of fixed length is inversely proportional to the square of the radius of the wire. If the resistance is 50 ohms when the radius of the wire is 0.001 inches, what is the resistance of a wire of radius 0.05 inch?

9. The cost of building a concrete driveway that leads to the garage of a house is directly proportional to the square of its length. If a 5-feet long driveway costs $250 to build, how much does a 30-feet driveway cost to build?

10. A company has found that the demand for its product is inversely proportional to the price of the product. If the demand is1,000 units when the price is $4, what is the demand when the price is $5?

11.4 References

1. Bueche, F. (1972). Principles of Physics, second edition, McGraw-Hill Book Company, New York.
2. Ho, P. Y. (1985). Li, Qi and Shu: An Introduction to Science and Civilization in China, Dover Publications, Inc., New York.
3. Huntley, H. E. (1970). The Divine Proportion: A Study in Mathematical Beauty, Dover Publications, Inc., New York.
4. Struik, D. J. (1948). A concise History of Mathematics, Dover Publications, Inc., New York.
5. http://en.wikipedia.org/wiki/Linear_interpolation.
6. http://en.wikipedia.org/wiki/Ratio.

Chapter 12
A Quadratic Equation in Two Variables

In this chapter we're going to learn how to graph the solution set to a quadratic equation in two variables. They are the well-known conic sections including the parabola, circle, ellipse, and hyperbola.

12.1 Solving Quadratic Equations in Two Variables

Definition 12.1.1
A polynomial equation of degree two in x and y given by

$$Ax^2 + Bxy + Cy^2 + Dx + Ey + F = 0, \tag{12.1.1}$$

where A, B, C, D, E, and F are real constants, and at least one of A and C is not zero, is called a quadratic equation in two variables x and y.

Remarks

1. Since Eq. (12.1.1) has more variables (two variables) than the number of equations (one equation), a quadratic equation in two variables must have infinitely many solutions if it has a solution. In analytic geometry, a conic is defined as a plane algebraic curve of degree two.

2. Greek mathematicians such as Menaechmus, Aristaeus the Elder and Euclid had studied the conics in the 4th century B.C. The greatest progress in the study of conics by the ancient Greeks is due to Apollonius of Perga (262-190 B.C.), whose eight-part "Conic Sections" summarized the existing knowledge at the time and greatly extended it. Apollonius's major innovation was to characterize a conic using properties within the plane and intrinsic to the curve; this greatly simplified analysis. With this tool, it was now possible to show that any plane cutting the cone, regard-

less of its angle, will produce a conic according to the earlier definition, leading to the definition commonly used today (Reference 4).

3. A study on the solution set to a polynomial equation of degree greater than two in two or more variables leads to an interesting family of algebraic curves (Reference 2).
4. If the coefficients of a polynomial equation and its solution set are limited to integers/rational numbers, this leads to the theory of Diophantine equation, which is a branch of number theory (Reference 3).

Definition 12.1.2
The solution set to a quadratic equation in two variables x and y is the set of all ordered pairs of numbers (x, y) that satisfy the given equation.

Remarks
1. An effective way to represent a solution set of infinitely many solutions is to graph the solution set. This is what we are going to do in this chapter. It turns out that the graph of a solution set to a quadratic equation in two variables x and y can be classified into one of the four categories: (i) parabola, (ii) circle, (iii) ellipse, and (iv) hyperbola, depending on the form of the given quadratic equation.
2. All these four graphs can be obtained by the intersection of a plane and a double-napped cone. Hence, they are also called conic sections.

Definition 12.1.3 (Classification for the Solution Set to a Quadratic Equation in x and y)
Assume that the solution set to a quadratic equation in x and y is not empty, the graph of the solution set to Eq. (12.1.1) is
1. A parabola if either $A = B = 0$ or $B = C = 0$. (12.1.2)
2. A circle if $A = C$ and $B = 0$. (12.1.3)
3. An ellipse if $A \neq C$, $B = 0$, and A and C are of the same sign. (12.1.4)

4. A hyperbola if $A \neq C$, $B = 0$, and A and C are of the opposite sign. (12.1.5)

Remarks

In this chapter we only deal with the case when $B = 0$. When $B \neq 0$, the graph of the solution set to the given quadratic equation is beyond the scope of this book (Reference 1, pp. 731-738).

Example 12.1.1

Solve $2x^2 - 3y - 4x + 5 = 0$ for the unknown (x, y).

Solution

From the given equation, we solve for y in terms of x as follows:

$$2x^2 - 3y - 4x + 5 + 3y = 0 + 3y, \Rightarrow$$

$$2x^2 - 4x + 5 = 3y, \Rightarrow \frac{2x^2 - 4x + 5}{3} = y, \Rightarrow$$

$$y = \frac{2}{3}x^2 - \frac{4}{3}x + \frac{5}{3}.$$

Hence, the solution set is $\{(x, \frac{2}{3}x^2 - \frac{4}{3}x + \frac{5}{3}) \mid x: \text{any real number}\}$.

Remark

The graph of this solution set is a parabola (Figure 12.2.2).

Example 12.1.2

Solve $3y^2 - 2x - y + 4 = 0$ for the unknown (x, y).

Solution

From the given equation, we solve for x in terms of y as follows:

$$3y^2 - 2x - y + 4 + 2x = 0 + 2x = 2x,$$

$$3y^2 - y + 4 = 2x, \Rightarrow \frac{3y^2 - y + 4}{2} = x, \Rightarrow$$

$$\frac{3}{2}y^2 - \frac{1}{2}y + 2 = x.$$

Hence, the solution set is $\{(\frac{3}{2}y^2 - \frac{1}{2}y + 2, y) \mid y: \text{ any real number}\}$.

Example 12.1.3

Solve $x^2 + y^2 - 2x + 4y - 4 = 0$ for the unknown (x, y).

Solution

We first complete the square in both x and y as follows:

$$x^2 - 2x + (\frac{-2}{2})^2 + y^2 + 4y + (\frac{4}{2})^2 - (\frac{-2}{2})^2 - (\frac{4}{2})^2 - 4 = 0,$$

$$(x - 1)^2 + (y + 2)^2 - 9 = 0. \qquad (1)$$

Then we can solve either for y in terms of x or for x in terms of y. Let us say that we decide to solve for y in terms of x. Thus, from Eq. (1), we have

$(x - 1)^2 + (y + 2)^2 - 9 - (x - 1)^2 + 9 = 0 - (x - 1)^2 + 9 = 9 - (x - 1)^2, \Rightarrow$

$(y + 2)^2 = 9 - (x - 1)^2 = (3 + x - 1)(3 - (x - 1)) = (2 + x)(4 - x), \Rightarrow$

$$y + 2 = \pm\sqrt{(2 + x)(4 - x)}, \Rightarrow y = -2 \pm \sqrt{(2 + x)(4 - x)}.$$

Since the expression under the symbol of the square root is required to be nonnegative, we have

$$(2 + x)(4 - x) \geq 0. \qquad (2)$$

The solution set to the quadratic inequality of Eq. (2) is $\{x \in R \mid -2 \leq x \leq 4\}$. Hence, the solution set is

$$\{ (x, -2 \pm \sqrt{(2 + x)(4 - x)} \mid -2 \leq x \leq 4 \}.$$

Remarks

1. If we decide to solve for x in terms of y, then the solution set is $\{(1 \pm \sqrt{(1 - y)(5 + y)}, y) \mid -5 \leq y \leq 1\}$.
2. The graph of the solution set is a circle with the center located at $Q(1, -2)$ with a radius of 3 (Figure 12.2.3).

Example 12.1.4

Solve $2x^2 + 3y^2 - 4x - 6y - 1 = 0$ for the unknown (x, y).

Solution

We first complete the square in both x and y as follows:

$2x^2 - 4x + 3y^2 - 6y - 1 = 0, \Rightarrow$

$2[x^2 - 2x + (-1)^2 - (-1)^2] + 3[y^2 - 2y + (-1)^2 - (-1)^2] - 1 = 0, \Rightarrow$

$2[(x - 1)^2 - 1] + 3[(y - 1)^2 - 1] - 1 = 0, \Rightarrow$

$2(x - 1)^2 - 2 + 3(y - 1)^2 - 3 - 1 = 0, \Rightarrow$

$2(x - 1)^2 + 3(y - 1)^2 - 6 = 0. \qquad (1)$

From Eq. (1), we decide to solve for y in terms of x as follows:

$2(x - 1)^2 + 3(y - 1)^2 - 6 - 2(x - 1)^2 + 6 = 0 - 2(x - 1)^2 + 6,$

$3(y-1)^2 = 6 - 2(x-1)^2$, $\Rightarrow$ $\dfrac{3(y-1)^2}{3} = \dfrac{6-2(x-1)^2}{3}$, $\Rightarrow$

$(y-1)^2 = \frac{2}{3}(\sqrt{3}+x-1)(\sqrt{3}-x+1)$, $\Rightarrow$

$y-1 = \pm\sqrt{\frac{2}{3}(\sqrt{3}-1+x)(\sqrt{3}+1-x)}$,

$y = 1 \pm \sqrt{\frac{2}{3}(\sqrt{3}-1+x)(\sqrt{3}+1-x)}$.

Since the expression under the symbol of the square root is required to be nonnegative, we have

$$(\sqrt{3}-1+x)(\sqrt{3}+1-x) \ge 0. \qquad (2)$$

The solution set to the quadratic inequality of Eq. (2) is $\{x \in R \mid 1-\sqrt{3} \le x \le 1+\sqrt{3}\}$. Hence, the solution set is

$$\{(x, 1 \pm \sqrt{\tfrac{2}{3}(\sqrt{3}-1+x)(\sqrt{3}+1-x)}) \mid 1-\sqrt{3} \le x \le 1+\sqrt{3}\}.$$

Remarks

1. If we decide to solve for x in terms of y, then the solution set is

$$\{(1 \pm \sqrt{\tfrac{3}{2}(\sqrt{2}-1+y)(\sqrt{2}+1-y)}, y) \mid 1-\sqrt{2} \le y \le 1+\sqrt{2}\}.$$

2. The graph of the solution set is an ellipse (Figure 12.3.5).

Example 12.1.5

Solve $x^2 - y^2 - 4x - 6y - 6 = 0$ for the unknown (x, y).

Solution

We first complete the square in both x and y as follows:

$x^2 - 4x - (y^2 + 6y) - 6 = 0$, $\Rightarrow$

$x^2 - 4x + (-2)^2 - (-2)^2 - (y^2 + 6y + 3^2 - 3^2) - 6 = 0$, $\Rightarrow$

$(x-2)^2 - 4 - [(y+3)^2 - 9] - 6 = 0$, $\Rightarrow$

$(x-2)^2 - (y+3)^2 - 4 + 9 - 6 = 0$, $\Rightarrow$

$$(x-2)^2 - (y+3)^2 - 1 = 0. \qquad (1)$$

If we solve for y in terms of x, then we have from Eq. (1)

$(x-2)^2 - 1 = (y+3)^2$, $\Rightarrow$

$(x-2+1)(x-2-1) = (y+3)^2$, $\Rightarrow$

$(x-1)(x-3) = (y+3)^2$, $\Rightarrow$

$\pm\sqrt{(x-1)(x-3)} = y+3$, $\Rightarrow$

$$y = -3 \pm \sqrt{(x-1)(x-3)}\,.$$

Since the expression under the symbol of the square root is required to be nonnegative, we have

$$(x-1)(x-3) \geq 0. \tag{2}$$

The solution set to the quadratic inequality of Eq. (2) is $\{x \in R \mid x \lesssim 1 \text{ or } x \geq 3\}$. Hence, the solution set is

$$\{(x, -3 \pm \sqrt{(x-1)(x-3)} \mid x \leq 1,\ x \geq 3\}\,.$$

Remarks

1. If we solve for x in terms of y, then the solution set is

$$\{\,(2 \pm \sqrt{1+(y+3)^2}\,, y \mid y\text{: any real number}\}.$$

2. The graph of the solution set is a hyperbola (Figure 12.4.5).

Exercises

Solve each of the following quadratic equations for the unknown (x, y):

1. $2x^2 - 4y - x - 5 = 0.$
2. $3y^2 - 2y - 5x - 1 = 0.$
3. $2x^2 + 2y^2 - 4x - 6y - 1 = 0.$
4. $2x^2 + 4y^2 - 8x - 8y - 3 = 0.$
5. $2x^2 - y^2 - 6x - 4y - 1 = 0.$

12.2 Parabolas

Definition 12.2.1

A parabola is the set of all points $P(x, y)$ in the plane that have the same distance from a given point F and a given line L where F is not on the line L. The point F is called the focus and the line L is called the directrix.

Remark

Many applications of parabolas occur in science and engineering. For example, the paths of comets are parabolic. The cable of a suspension bridge is parabolic. An automobile headlight is designed in the shape of parabolic reflector so that if the bulb is put at the focus, then the light is reflected off the parabola along parallel rays.

Theorem 12.2.1 (A Standard Equation of a Parabola with Vertex at $Q(h, k)$)
The standard form of the equation of a parabola with vertex at $Q(h, k)$ and directrix $y = -b$ is given by

$$(x-h)^2 = 4(b+k)(y-k)\ ,\ b \neq 0. \tag{12.2.1}$$

If the directrix is $x = -b$, then the equation is given by

$$(y-k)^2 = 4(b+h)(x-h),\ b \neq 0. \tag{12.2.2}$$

The focus is on the line of symmetry (or the axis of a parabola) that is perpendicular to the directrix (either $x = h$ or $y = k$) with a directed distance that is b units away from the vertex $Q(h, k)$.

Remarks

1. To derive Eq. (12.2.1), we use the fact that the vertex $Q(h, k)$ has the same distance to the focus F and any point $R(x, -b)$ on the line L (Figure 12.2.1). This implies that the focus should be given by

$$F(h, 2k+b). \tag{12.2.3}$$

Let $P(x, y)$ be an arbitrary point lying on the parabola. Then, according to the definition of a parabola, we have

$$d(P, F) = d(P, R). \tag{12.2.4}$$

By using the distance formula between two points given in Section 9.1, we have

$$d(P,F) = \sqrt{(x-h)^2 + [y-(2k+b)]^2}\ , \tag{12.2.5}$$

and

$$d(P,R) = \sqrt{(x-x)^2 + [y-(-b)]^2} = |\,y+b\,|. \tag{12.2.6}$$

By substituting Eqs. (12.2.5) and (12.2.6) into Eq. (12.2.4), we have

$$\sqrt{(x-h)^2 + [y-(2k+b)]^2} = |\,y+b\,|,\ \Rightarrow$$

$$(\sqrt{(x-h)^2 + [y-(2k+b)]^2}\,)^2 = (|\,y+b\,|)^2,\ \Rightarrow$$

$$(x-h)^2 + [y-(2k+b)]^2 = (y+b)^2,\ \Rightarrow$$

$$(x-h)^2 + y^2 - 2(2k+b)y + (2k+b)^2 = y^2 + 2by + b^2,\ \Rightarrow$$
$$(x-h)^2 = 2by + b^2 + 2(2k+b)y - (2k+b)^2$$
$$= 2by + b^2 + 2(2k+b)y - (4k^2 + 4bk + b^2)$$
$$= 2by + 2(2k+b)y - 4k^2 - 4bk$$
$$= 2y(b + 2k + b) - 4k(b+k)$$
$$= 2y(2b + 2k) - 4k(b+k)$$

$$= 4y(b + k) - 4k(b + k)$$
$$= 4(b + k)(y - k),$$

which is Eq. (12.2.1). Eq. (12.2.2) can similarly be derived with the focus given by

$$F(2h + b, k). \qquad (12.2.7)$$

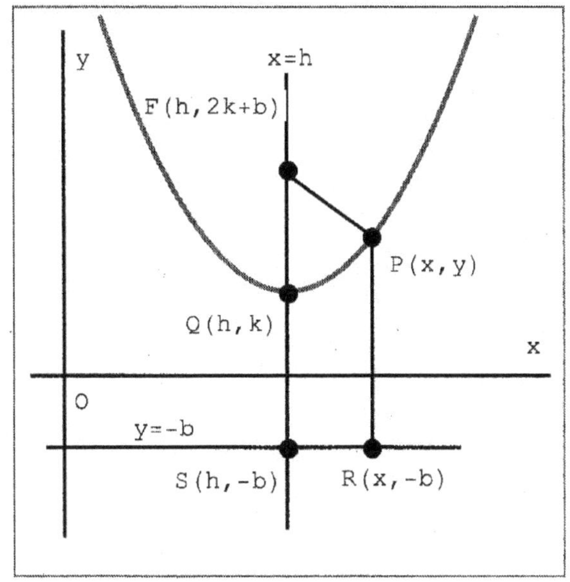

Figure 12.2.1

2. To graph a parabola we need a minimum number of three points lying on the parabola. These three points are the vertex $Q(h, k)$ and two other points $P_1(x_1, y_1)$ and $P_2(x_2, y_2)$ with their x-coordinates being chosen as one unit on either side of the x-coordinate of the vertex $Q(h, k)$, namely $x_1 = h + 1$ and $x_2 = h - 1$, and then using Eq. (12.2.1) (or Eq. (12.2.2)) to figure out their corresponding y-coordinates y_1 and y_2 of P_1 and. P_2 Note that due to a symmetry we have $y_1 = y_2$. Connect these three points smoothly $Q(h, k)$, $P_1(h+1, y_1)$, and $P_2(h-1, y_2)$, and then extend the graph out accordingly.

Example 12.2.1

Given a quadratic equation in two variables x and y as follows:

$$2x^2 - 3y - 4x + 5 = 0.$$

(a) Show that the graph of the solution set is a parabola. Also, determine the vertex, the focus, the directrix, and the line of symmetry of the parabola.

(b) Graph the parabola.

Solution

(a) Since the coefficient of y^2 in the given equation is zero, the graph of the solution set to the given equation is a parabola. Thus, we shall write the given equation to conform Eq. (12.2.1) through completing the square in x as follows:

$$2x^2 - 4x = 3y - 5, \Rightarrow$$

$$2x^2 - 4x + 2 = 3y - 5 + 2 = 3y - 3 = 3(y - 1), \Rightarrow$$

$$2(x^2 - 2x + 1) = 3(y - 1), \Rightarrow$$

$$2(x - 1)^2 = 3(y - 1), \Rightarrow$$

$$(x-1)^2 = \frac{3(y-1)}{2}, \Rightarrow$$

$$(x-1)^2 = \frac{3}{2}(y-1). \qquad (1)$$

By comparing Eq. (1) with Eq. (12.2.1), we have

$$h = 1,\ k = 1,$$

$$4(b + k) = \frac{3}{2}. \qquad (2)$$

By substituting $k = 1$ into Eq. (2), we have

$$4(b + 1) = \frac{3}{2}, \Rightarrow 4b + 4 = \frac{3}{2}, \Rightarrow$$

$$4b = \frac{3}{2} - 4 = -\frac{5}{2}, \Rightarrow b = -\frac{5}{8}.$$

By using Eq. (12.2.3), the focus is found to be $F(1, 2 \cdot 1 - \frac{5}{8}) = F(1, \frac{11}{8})$. From Eq. (1), the line of symmetry is obtained by setting $x - 1 = 0$, or $x = 1$. Hence, the vertex is $Q(1, 1)$, the focus is $F(1, \frac{11}{8})$, the directrix is $y = -(-\frac{5}{8}) = \frac{5}{8}$, and the line of symmetry is $x = 1$.

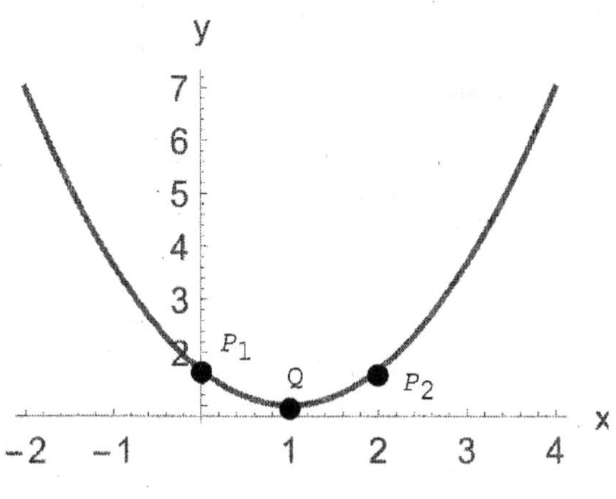

Figure 12.2.2

(b) To graph the parabola, we have already the vertex $Q(1,\ 1)$. All we need is two extra points, $P_1(x_1, y_1)$ and $P_2(x_2, y_2)$. By substituting $x_1 = h - 1 = 1 - 1 = 0$ and $x_2 = h + 1 = 1 + 1 = 2$ into Eq. (1), we have

$$(0-1)^2 = \frac{3}{2}(y-1)\,,\ \Rightarrow\ (-1)^2 = \frac{3}{2}y - \frac{3}{2}\,,$$

$$\frac{5}{2} = \frac{3}{2}y\,,\ \Rightarrow\ \frac{\frac{5}{2}}{\frac{3}{2}} = \frac{\frac{3}{2}y}{\frac{3}{2}}\,,\ \Rightarrow\ y = \frac{5}{3}\,.$$

Therefore, $y_1 = y_2 = \frac{5}{3}$, i.e., $P_1(0, \frac{5}{3})$ and $P_2(2, \frac{5}{3})$.

By connecting three points $Q(1,\ 1)$, $P_1(0, \frac{5}{3})$, and $P_2(2, \frac{5}{3})$ smoothly, the graph of the parabola is given in Figure 12.2.2.

Remark
Figure 12.2.2 is also the graph of the solution set in Example 12.1.1.

Example 12.2.2
Given a quadratic equation in two variables x and y as follows:

$$3y^2 - 2x - y + 4 = 0.$$

(a) Show that the graph of the solution set to the given equation is a parabola. Also, determine the vertex, the focus, the directrix, and the line of symmetry of the parabola.
(b) Graph the parabola.
Solution
(a) Since the coefficient of x^2 in the given equation is zero, the graph of the solution set is a parabola. Thus, we shall write the given equation to conform Eq. (12.2.2) through completing the square in y as follows:

$$3y^2 - y = 2x - 4, \Rightarrow$$

$$3[y^2 - \frac{1}{3}y + (-\frac{1}{6})^2 - (-\frac{1}{6})^2] = 2x - 4, \Rightarrow$$

$$3[(y - \frac{1}{6})^2 - \frac{1}{36}] = 2x - 4, \Rightarrow$$

$$3(y - \frac{1}{6})^2 - \frac{1}{12} = 2x - 4, \Rightarrow$$

$$3(y - \frac{1}{6})^2 = 2x - 4 + \frac{1}{12} = 2x - \frac{47}{12} = 2(x - \frac{47}{24}), \Rightarrow$$

$$(y - \frac{1}{6})^2 = \frac{2}{3}(x - \frac{47}{24}). \qquad (1)$$

By comparing Eq. (1) with Eq. (12.2.2), we have

$$k = \frac{1}{6},\ h = \frac{47}{24},$$

$$4(b + h) = \frac{2}{3}. \qquad (2)$$

By substituting $h = \frac{47}{24}$ into Eq. (2), we have

$$4 \cdot (b + \frac{47}{24}) = \frac{2}{3}, \Rightarrow 4b + \frac{47}{6} = \frac{2}{3}, \Rightarrow$$

$$4b = \frac{2}{3} - \frac{47}{6} = \frac{-43}{6}, \Rightarrow b = -\frac{43}{24}.$$

By using Eq. (12.2.7), the focus is found to be $F(2 \cdot \frac{47}{24} - \frac{43}{24}, \frac{1}{6})$ $= F(\frac{51}{24}, \frac{1}{6})$. From Eq. (1), the line of symmetry is obtained by setting $y - \frac{1}{6} = 0$, or, $y = \frac{1}{6}$. Hence, the vertex is $Q(\frac{47}{24}, \frac{1}{6})$, the focus is $F(\frac{51}{24}, \frac{1}{6})$, the directrix is $x = -(-\frac{43}{24}) = \frac{43}{24}$, and the line of symmetry of the parabola is $y = \frac{1}{6}$.

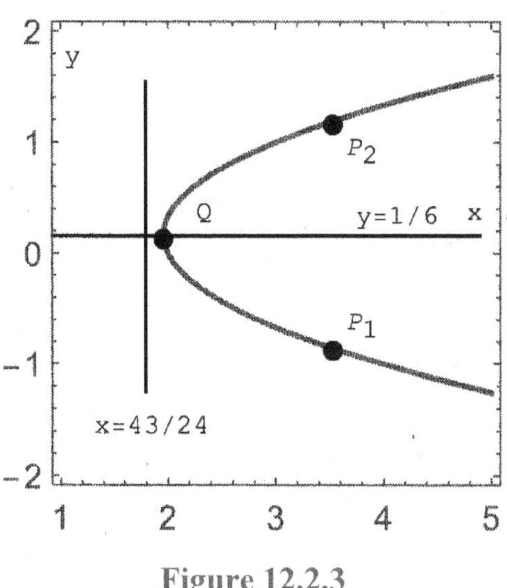

Figure 12.2.3

(b) To graph the parabola, we have already the vertex $Q(\frac{47}{24}, \frac{1}{6})$. All we need is two extra points $P_1(x_1, y_1)$ and $P_2(x_2, y_2)$. By substituting $y_1 = \frac{1}{6} - 1 = -\frac{5}{6}$ (or $y_2 = \frac{1}{6} + 1 = \frac{7}{6}$) into Eq. (1), we have

$$(-\frac{5}{6} - \frac{1}{6})^2 = \frac{2}{3}(x - \frac{47}{24}),$$

$$1 = (-1)^2 = \frac{2}{3}x - \frac{47}{36},$$

$$\frac{83}{36} = 1 + \frac{47}{36} = \frac{2}{3}x, \Rightarrow \frac{83}{24} = \frac{3}{2} \cdot \frac{83}{36} = x.$$

Therefore, $x_1 = x_2 = \frac{83}{24}$, i.e., $P_1(\frac{83}{24}, -\frac{5}{6})$, and $P_2(\frac{83}{24}, \frac{7}{6})$. By connecting $Q(\frac{47}{24}, \frac{1}{6})$, $P_1(\frac{83}{24}, -\frac{5}{6})$, and $P_2(\frac{83}{24}, \frac{7}{6})$ smoothly, the graph of the parabola is given in Figure 12.2.3.

Remark
Figure 12.2.3 is also the graph of the solution set in Example 12.1.2.

Exercises

For each of the following problems, (a) show that the graph of the solution set to the given equation is a parabola. Also, determine the vertex, the focus, the directrix, and the line of symmetry of the parabola, (b) graph the parabola:

1. $2x^2 - 4y - 5 = 0$. **2.** $3y^2 - 2x + 6 = 0$.
3. $6x^2 - 12x - y + 8 = 0$. **4.** $2y^2 - 6y - 3x = 1$.
5. $y = 4x^2 + 8x - 3$.

12.3 Circles and Ellipses

Definition 12.3.1
A circle is the set of all points of the plane that has a fixed distance from a given point. The given point is called the center of the circle. This fixed distance from the center to a point on the circle is called a radius of the circle.

Remark
The lengths of all radiuses are the same.

Theorem 12.3.1 (The Standard Equation of a Circle with its Center at $Q(h, k)$ and a Radius r)
The standard form of the equation of a circle with a radius r and a center at $Q(h, k)$ is given by

$$(x-h)^2 + (y-k)^2 = r^2, \; (r > 0) \qquad (12.3.1)$$

Remarks

1. To prove Eq. (12.3.1), we use the fact that the distance between an arbitrary point $P(x, y)$ on the circle to its center $Q(h, k)$ equals its radius r, namely,

$$d(P, Q) = r. \qquad (12.3.2)$$

On the other hand,

$$d(P, Q) = \sqrt{(x-h)^2 + (y-k)^2}\,. \qquad (12.3.3)$$

By substituting Eq. (12.3.3) into Eq. (12.3.2) and then squaring both sides, Eq. (12.3.1) follows.

2. To graph a circle, we need a minimum number of four points. These four points are $P_1(h + r, k)$, $P_2(h - r, k)$, $P_3(h, k + r)$, and $P_4(h, k - r)$. Connect these four points P_1, P_2, P_3, and P_4 smoothly, and we get the graph of a circle accordingly.

Example 12.3.1

Given a quadratic equation as follows:

$$x^2 + y^2 - 2x - 4y - 4 = 0.$$

(a) Show that the solution set to the above equation is a circle. Also, determine its center $Q(h, k)$ and its radius r.

(b) Graph the circle.

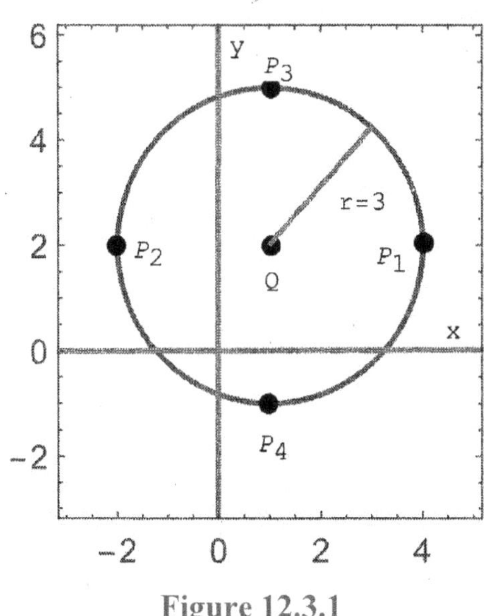

Figure 12.3.1

Solution

(a) Since the coefficients of both x^2 and y^2 are the same, the solution set to the given equation is, therefore, a circle. Since the given equation is the same as the one in Example 12.1.3, we have after completing the square in both variables x and y (See Example 12.1.3 for details)

$$(x-1)^2+(y-2)^2-9=0, \Rightarrow$$
$$(x-1)^2+(y-2)^2=9. \qquad (1)$$

By comparing Eq. (1) with Eq. (12.3.1), we obtain

$$h=1,\ k=2,\ r^2=9\ , \text{ or } r=3.$$

Hence, the center of the circle is located at $Q(1, 2)$ and its radius is 3.
(b) To graph a circle, we need a minimum number of four points that are chosen as $P_1(h+r, k)=P_1(1+3, 2)=P_1(4,2)$, $P_2(h-r, k)=P_2(1-3, 2)=P_2(-2, 2)$, $P_3(h, k+r)=P_3(1, 2+3)=P_3(1, 5)$, and $P_4(h, k-r)=P_4(1, 2-3)=P_4(1, -1)$. Then we connect these four points smoothly. The graph of the circle is shown in Figure 12.3.1.

Definition 12.3.2

An ellipse is the set of all points in the plane so that the sum of whose distances from two fixed points is a given positive constant. The two fixed points are called the foci of the ellipse, (each fixed point is a focus) and the midpoint of the line segment connecting the foci is called the center of the ellipse (Figure 12.3.2).

Remarks

1. Most of the planets' orbits in the solar system are ellipses. For example, the earth's orbit around the sun is approximately an ellipse with the sun at one of its foci.
2. The line segment through the foci and across the ellipse is called the major axis, while the line segment through the center across the ellipse, and perpendicular to the major axis is called the minor axis. Conventionally, the length of the major axis is denoted by $2a$, the length of the minor axis by $2b$, and the distance between the foci by $2c$. The number a and b are called the length of semi-major axis and semi-minor axis, respectively. The relationship between three numbers a, b, and c can be shown to satisfy (Figure 12.3.3)

$$a=\sqrt{b^2+c^2}\ . \qquad (12.3.4)$$

3. If the foci coincide, an ellipse reduces to a circle.

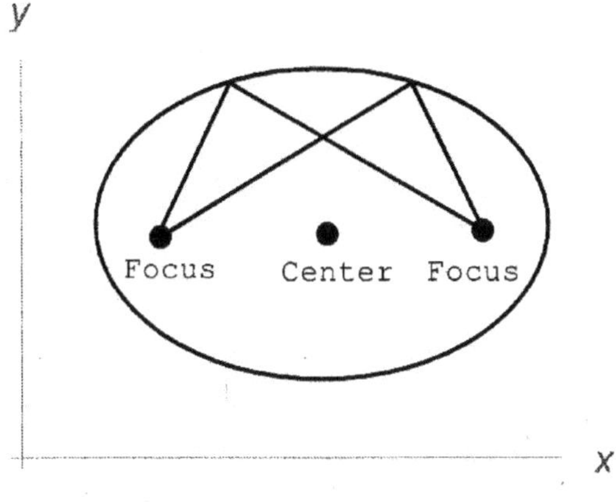

Figure 12.3.2

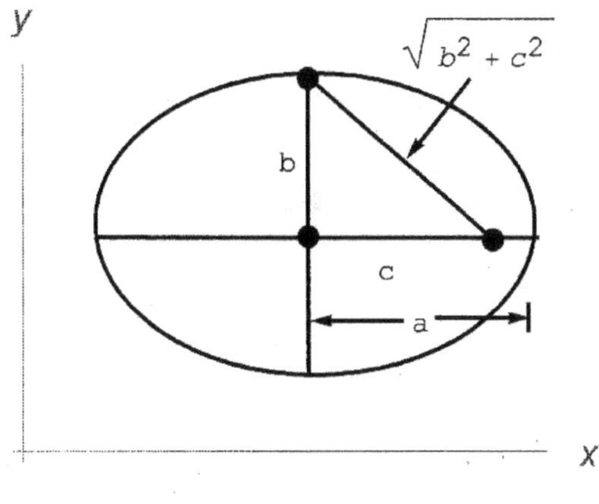

Figure 12.3.3

Theorem 12.3.2 (The standard equation of an ellipse with center at *Q*(*h*, *k*))

Case 1. If the major axis is parallel to the x-axis, then it is given by

$$\frac{(x-h)^2}{a^2}+\frac{(y-k)^2}{b^2}=1. \qquad (12.3.5)$$

Case 2. If the major axis is parallel to the y-axis, then it is given by

$$\frac{(x-h)^2}{b^2}+\frac{(y-k)^2}{a^2}=1. \qquad (12.3.6)$$

Remarks

1. To prove Eq. (12.3.5), we use the fact that for all points on the ellipse the sum of the distances to the foci is $2a$ by realizing that if we measure the distance from one of the end points on the minor axis to the foci and using Eq. (12.3.4), then the sum of the distances is indeed $2a$. Let $P(x, y)$ be an arbitrary point on the ellipse (Figure 12.3.4a). Then, by the definition of an ellipse,

$$d(P, F_1) + d(P, F_2) = 2a, \tag{12.3.7}$$

However,

$$d(P, F_1) = \sqrt{[x-(h-c)]^2 + (y-k)^2}\ , \tag{12.3.8}$$

and

$$d(P, F_2) = \sqrt{[x-(h+c)]^2 + (y-k)^2}\ . \tag{12.3.9}$$

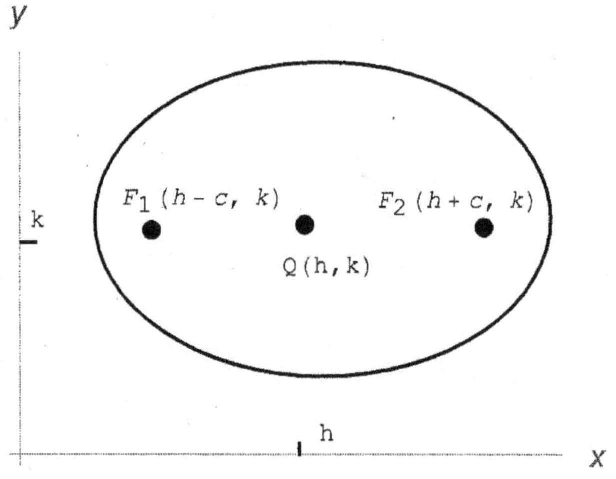

Figure 12.3.4a

By substituting Eqs. (12.3.8) and (12.3.9) into Eq. (12.3.7), we have

$$\sqrt{[x-(h-c)]^2 + (y-k)^2} + \sqrt{[x-(h+c)]^2 + (y-k)^2} = 2a\,.$$

By moving the second radical of the above equation to its right side and then squaring both sides of the equation, we have

$$[x-(h-c)]^2 + (y-k)^2 = 4a^2 - 4a\cdot\sqrt{[x-(h+c)]^2 + (y-k)^2}$$
$$+[x-(h+c)]^2 + (y-k)^2\,,$$

After simplifying the above equation, we have

$$\sqrt{[x-(h+c)]^2+(y-k)^2} = a - \frac{c}{a}(x-h).$$

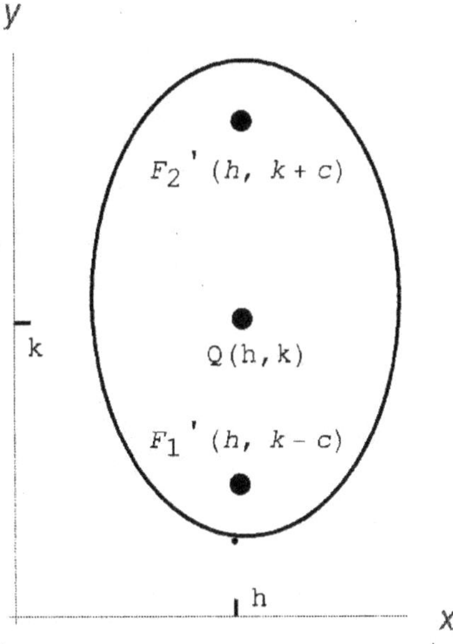

Figure 12.3.4b

By squaring again both sides of the above equation, we have

$$[x-(h+c)]^2+(y-k)^2 = [a - \frac{c}{a}(x-h)]^2, \Rightarrow$$

$$(x-h)^2 - 2c(x-h) + c^2 + (y-k)^2 = a^2 - 2c(x-h) + \frac{c^2}{a^2}(x-h)^2, \Rightarrow$$

$$(x-h)^2 - \frac{c^2}{a^2}(x-h)^2 + (y-k)^2 = a^2 - c^2 = b^2, \Rightarrow$$

$$(x-h)^2(1-\frac{c^2}{a^2}) + (y-k)^2 = b^2, \Rightarrow$$

$$(x-h)^2(\frac{a^2-c^2}{a^2}) + (y-k)^2 = b^2, \Rightarrow$$

$$(x-h)^2 \cdot \frac{b^2}{a^2} + (y-k)^2 = b^2, \Rightarrow$$

$$\frac{b^2(x-h)^2+a^2(y-k)^2}{a^2b^2}=1, \Rightarrow$$

$$\frac{(x-h)^2}{a^2}+\frac{(y-k)^2}{b^2}=1,$$

where the last equation is the desired Eq. (12.3.5). Eq. (12.3.6) can be similarly derived with its foci located at $F_1'(h,k-c)$ and $F_2'(h,k+c)$ (Figure 12.3.4b).

2. To graph an ellipse, we need a minimum of four points. These four points are chosen as the four vertices, P_1 $(h-a, k)$, P_2 $(h+a, k)$, P_3 $(h, k-b)$, and P_4 $(h, k+b)$. Then, connect these four vertices accordingly. We will have the graph of an ellipse.

Example 12.3.2

Given a quadratic equation in x and y as follows:

$$2x^2+3y^2-4x-6y-1=0.$$

(a) Show that the graph of the solution set to the above equation is an ellipse with the major axis parallel to the x-axis. Also, locate the center of the ellipse and its foci.

(b) Graph the ellipse.

Solution

The given equation is exactly the same as that in Example 12.1.4. Hence, after completing the square in both x and y, we have (See Example 12.1.4 for details)

$$2(x-1)^2+3(y-1)^2-6=0, \Rightarrow$$

$$2(x-1)^2+3(y-1)^2=6, \Rightarrow$$

$$\frac{2(x-1)^2+3(y-1)^2}{6}=1, \Rightarrow$$

$$\frac{(x-1)^2}{3}+\frac{(y-1)^2}{2}=1. \qquad (1)$$

Since Eq. (1) conforms Eq. (12.3.5), the graph of its solution set is an ellipse with the major axis parallel to the x-axis. Furthermore, by comparing Eq. (1) with Eq. (12.3.5), we have

$$h=1, k=1,$$

and

$$a^2=3, b^2=2, \Rightarrow a=\sqrt{3}, b=\sqrt{2}.$$

Therefore, $c^2 = a^2 - b^2 = 3 - 2 = 1, \Rightarrow c = 1$. Hence, the center of the ellipse is located at $Q(1, 1)$ and its foci (Figure 12.3.4a) are $F_1(h - c, k) = F_1(1 - 1, 1) = F_1(0, 1)$ and $F_2(h + c, k) = F_2(1 + 1, 1) = F_2(2, 1)$.
(b) To graph the ellipse, we use the four vertices $P_1(h - a, k) = P_1(1 - \sqrt{3}, 1)$, $P_2(h + a, k) = P_2(1 + \sqrt{3}, 1)$, $P_3(h, k - b) = P_3(1, 1 - \sqrt{2})$, $P_4(h, k + b) = P_4(1, 1 + \sqrt{2})$. Then, connect P_1, P_2, P_3, and P_4 smoothly. The graph of the ellipse is shown in Figure 12.3.5.

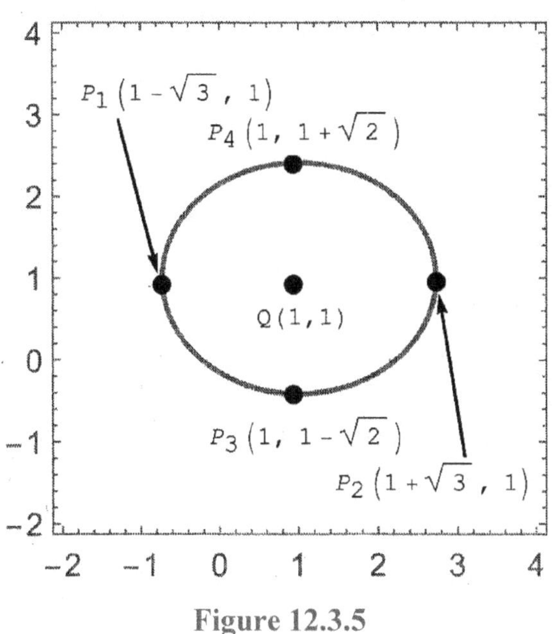

Figure 12.3.5

Example 12.3.3
Given a quadratic equation in x and y as follows:

$$9x^2 + 4y^2 + 18x - 16y - 11 = 0.$$

(a) Show that the graph of the solution set to the above equation is an ellipse with the major axis parallel to the y-axis. Also, locate the center of the ellipse and its foci.
(b) Graph the ellipse.
Solution
(a) Since the coefficients of x^2 and y^2 in the given quadratic equation are of the same sign with unequal magnitude, the graph of the solution set to the given equation is an ellipse. To show that the major

axis of the ellipse is parallel to the y-axis, we need to complete the square in both x and y of the given equation as follows:

$$9x^2 + 18x + 4y^2 - 16y = 11, \Rightarrow$$
$$9(x^2 + 2x + 1^2 - 1^2) + 4[y^2 - 4y + (-2)^2 - (-2)^2] = 11, \Rightarrow$$
$$9[(x + 1)^2 - 1] + 4[(y - 2)^2 - 4] = 11, \Rightarrow$$
$$9(x + 1)^2 - 9 + 4(y - 2)^2 - 16 = 11, \Rightarrow$$
$$9(x + 1)^2 + 4(y - 2)^2 = 11 + 9 + 16 = 36, \Rightarrow$$
$$\frac{9(x+1)^2 + 4(y-2)^2}{36} = 1, \Rightarrow$$
$$\frac{(x+1)^2}{4} + \frac{(y-2)^2}{9} = 1. \qquad (1)$$

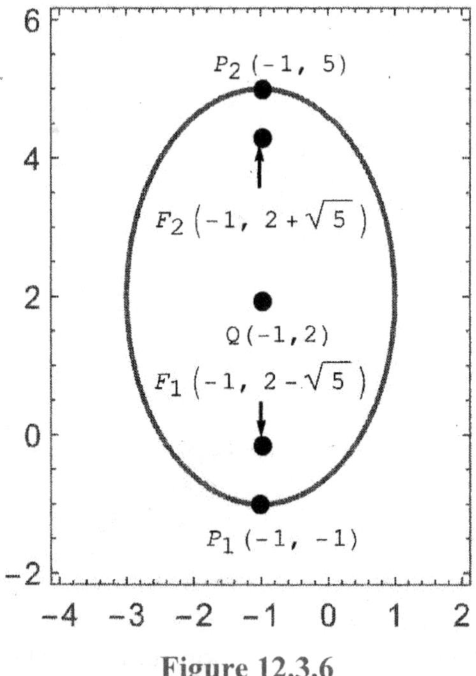

Figure 12.3.6

Since Eq. (1) conforms to Eq. (12.3.6), the major axis of the ellipse is parallel to the y-axis. Also, by comparing Eq. (1) with Eq. (12.3.6), we have

$$h = -1,\ k = 2,$$

and

$a^2 = 9$, $b^2 = 4$, $\Rightarrow$ $a = 3$, $b = 2$.
Therefore, $c^2 = a^2 - b^2 = 9 - 4 = 5$, $\Rightarrow$ $c = \sqrt{5}$. Hence, the center of the ellipse is located at $Q(-1, 2)$ and its foci (Figure 12.3.4b) are located at $F_1(h, k - c)$=$F_1(-1, 2 - \sqrt{5})$ and $F_2(h, k + c) = F_2(-1, 2 + \sqrt{5})$.
(b) To graph the ellipse, we plot the four vertices $P_1(h, k - a) = P_1(-1, 2 - 3) = P_1(-1. -1)$, $P_2(h, k + a) = P_2(-1, 2 + 3) = P_2(-1, 5)$, $P_3(h - b, k) = P_3(-1 - 2, 2) = P_3(-3, 2)$, and $P_4(h + b, k) = P_4(-1 + 2, 2) = P_4(1, 2)$. Then, connect P_1, P_2, P_3, and P_4 smoothly. The graph of the ellipse is given in Figure 12.3.6.

Exercises

For each of the following problems, show that whether the graph of the solution set to the given equation is a circle or an ellipse. If it is a circle, find its center and its radius and then graph it. If it is an ellipse, determine its center, to which coordinate axis is the major axis parallel, the length of semi-major axis (a) and semi-minor axis (b), the foci of the ellipse, and then graph it.

1. $x^2 + y^2 + 4x + 6y + 12 = 0$.
2. $2x^2 + 2y^2 - 8x - 12y = 6$.
3. $25x^2 + 16y^2 = 400$.
4. $9x^2 + 16y^2 + 18x - 64y - 71 = 0$.
5. $9x^2 + 4y^2 - 9x + 12y - \dfrac{99}{4} = 0$.

12.4 Hyperbolas

Definition 12.4.1

A hyperbola is the set of all points in the plane such that the difference of the distances from two fixed points is a given positive constant. The two fixed points are called the foci, and the midpoint of the line segment connecting the foci is called the center of the hyperbola. The line through the foci is called the focal axis (or transverse axis) and the line through the center and perpendicular to the focal axis is called the conjugate axis. The hyperbola intersects the focal axis at two points called vertices. The two separate parts of a hyperbola are called the branches of the hyperbola (Figure 12.4.1).

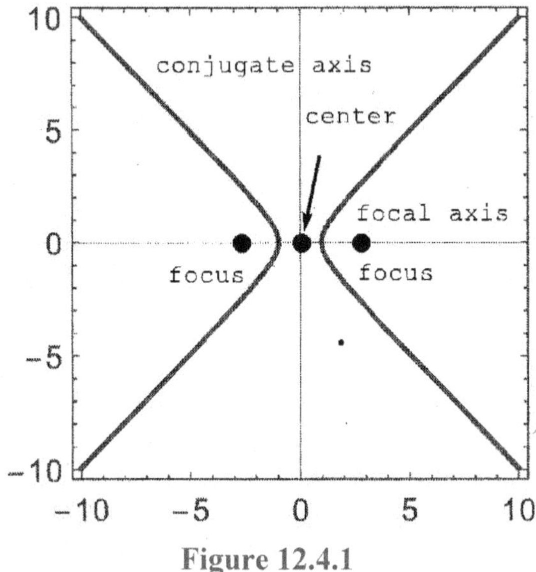

Figure 12.4.1

Remarks

1. The difference of the distances has to be a positive number. Hence, the difference means that the longer distance from a point on the hyperbola to one fixed point minus the shorter distance from the same point on the hyperbola to the other fixed point.
2. A hyperbola possesses a geometric property that a ray of light emanating from one focus of a hyperbola will be reflected back along the line from the opposite focus (Figure 12.4.2). This light reflection property of a hyperbola is used in the design of high-quality telescopes.
3. Conventionally, the distance between the vertex and the center is denoted by a, and the distance between the focus and the center is denoted by c, where $c > a$. A positive number b is defined as $b = \sqrt{c^2 - a^2}$ (Figure 12.4.3).

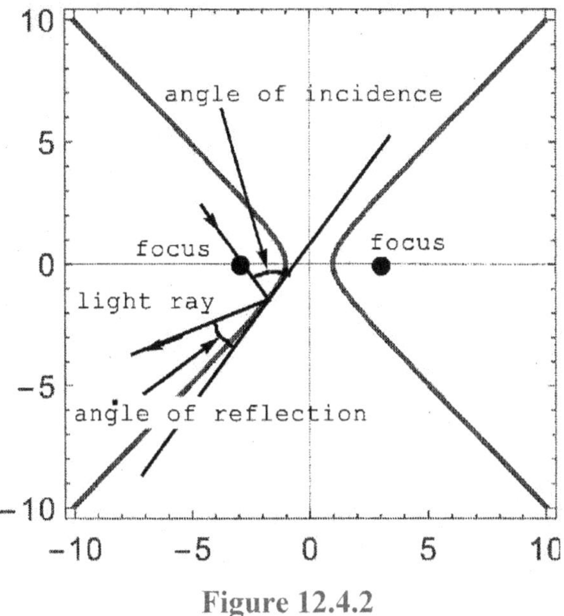

Figure 12.4.2

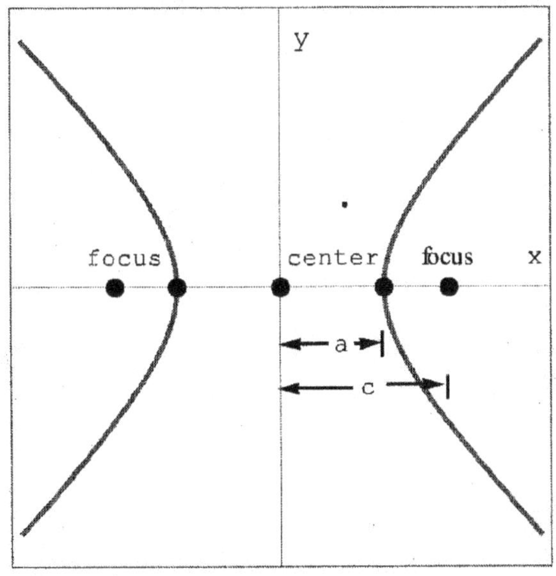

Figure 12.4.3

Theorem 12.4.1 (The Standard Equation of a Hyperbola with Center at $Q(h, k)$)

Case 1. If the focal axis is parallel to the x-axis, it is given by

$$\frac{(x-h)^2}{a^2}-\frac{(y-k)^2}{b^2}=1, \tag{12.4.1}$$

where the vertices are located at $V_1(h-a, k)$ and $V_2(h+a, k)$.

Case 2. If the focal axis is parallel to the y-axis, it is given by

$$\frac{(y-k)^2}{a^2}-\frac{(x-h)^2}{b^2}=1, \tag{12.4.2}$$

where the vertices are located at $V_1^{'}(h, k-a)$ and $V_2^{'}(h, k+a)$.

Remarks

1. To prove Eq. (12.4.1), we use the fact that for all points on the hyperbola, the longer distance to the farther focus minus the shorter distance to the closer focus is always $2a$. This fact can be realized by putting the point on the hyperbola right at one of the vertices. Let $P(x, y)$ be an arbitrary point on the hyperbola. Depending on whether $P(x, y)$ lies on the right branch or the left branch of the hyperbola, the set-up equations are different (Figure 12.4.4). If $P(x, y)$ lies on the right branch of the hyperbola, then we have

   $$d(P, F_1)-d(P, F_2)=2a,$$

   or

   $$\sqrt{[x-(h-c)]^2+(y-k)^2}-\sqrt{[x-(h+c)]^2+(y-k)^2}=2a. \tag{12.4.3}$$

   If $P(x, y)$ lies on the left branch of the hyperbola, then we have

   $$d(P, F_2)-d(P, F_1)=2a,$$

   or

   $$\sqrt{[x-(h+c)]^2+(y-k)^2}-\sqrt{[x-(h-c)]^2+(y-k)^2}=2a. \tag{12.4.4}$$

   To solve Eq. (12.4.3), we move the second radical to the right side of the equation and then squaring both sides of the equation as follows:

   $$\{\sqrt{[x-(h-c)]^2+(y-k)^2}\}^2=\{2a+\sqrt{[x-(h+c)]^2+(y-k)^2}\}^2,$$

   $$[x-(h-c)]^2+(y-k)^2=4a^2+4a\cdot\sqrt{[x-(h+c)]^2+(y-k)^2}$$

$$+[x-(h+c)]^2+(y-k)^2.$$

By subtracting $(y-k)^2$ from both sides of the above equation and moving the term of $[x-(h+c)]^2$ from the right side to the left side of the equation, we obtain after simplification

$$\frac{c}{a}(x-h)-a=\sqrt{[x-(h+c)]^2+(y-k)^2}.$$

By squaring both sides of the above equation and then simplifying, we have

$$[\frac{c}{a}(x-h)-a]^2=(\sqrt{[x-(h+c)]^2+(y-k)^2})^2, \Rightarrow$$

$$\frac{c^2}{a^2}(x-h)^2-2c(x-h)+a^2=[x-(h+c)]^2+(y-k)^2$$

$$=(x-h)^2-2c(x-h)+c^2+(y-k)^2, \Rightarrow$$

$$(\frac{c^2}{a^2}-1)(x-h)^2-(y-k)^2=c^2-a^2, \Rightarrow$$

$$\frac{b^2}{a^2}(x-h)^2-(y-k)^2=b^2, (\text{ Since } b=\sqrt{c^2-a^2}),$$

which is the desired Eq. (12.4.1). It can be shown that Eq. (12.4.4) leads to Eq. (12.4.1) after simplification. Eq. (12.4.2) can be similarly derived (Figure 12.4.4b).

2. To graph a hyperbola we need a minimum of six points, two vertices and four other points. These four points are chosen in such a way that two points on each branch of a hyperbola are symmetric with respect to the focal axis. By using Figure 12.4.4a as an illustration, we solve for y by substituting $x=h-c$ into Eq. (12.4.1) as follows:

$$\frac{(h-c-h)^2}{a^2}-\frac{(y-k)^2}{b^2}=1, \Rightarrow \frac{c^2}{a^2}-1=\frac{(y-k)^2}{b^2}, \Rightarrow$$

$$\frac{c^2-a^2}{a^2}=\frac{(y-k)^2}{b^2}, \Rightarrow \frac{b^2}{a^2}=\frac{(y-k)^2}{b^2}, \Rightarrow$$

$$\frac{b^4}{a^2}=(y-k)^2, \Rightarrow \pm\sqrt{\frac{b^4}{a^2}}=y-k, \Rightarrow y=k\pm\frac{b^2}{a}.$$

Therefore, the two points on the left branch of a hyperbola are $P_1(h-c, k-\frac{b^2}{a})$ and $P_2(h-c, k+\frac{b^2}{a})$. Similarly, by substituting $x=h+c$ into Eq. (12.4.1) we obtain the two points on the right branch of a hyperbola that are $P_3(h + c, k - \frac{b^2}{a})$ and $P_4(h + c, k + \frac{b^2}{a})$.

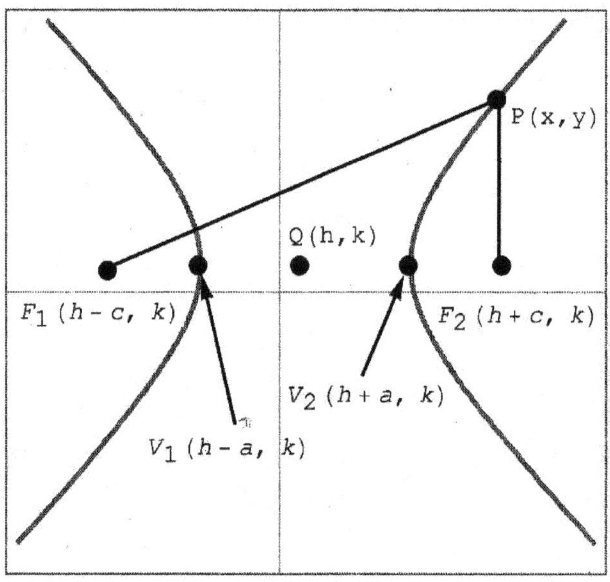

Figure 12.4.4a

3. For any hyperbola there exists a pair of asymptotes that can be obtained by setting the left side of either Eq. (12.4.1) or Eq. (12.4.2) equal to zero.
4. The graph of a hyperbola is symmetric with respect to both its focal axis and its conjugate axis.
5. The focal axis is determined by setting the term with a negative sign equal to zero after writing the given equation into the standard form of either Eq. (12.4.1) or Eq. (12.4.2), and solve the resulting equation. The result is summarized in the following theorem.

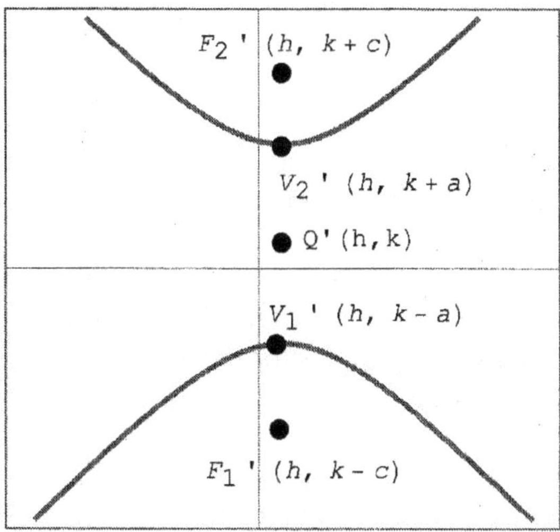

Figure 12.4.4b

Theorem 12.4.2 (Existence of Asymptotes for Hyperbolas)
Case 1. A hyperbola of Eq. (12.4.1) has a pair of oblique asymptotes given, respectively, by

$$y-k=\frac{b}{a}(x-h), \qquad (12.4.5a)$$

and

$$y-k=\frac{b}{a}(x-h). \qquad (12.4.5b)$$

Case 2. A hyperbola of Eq. (12.4.2) has a pair of oblique asymptotes given, respectively, by

$$y-k=\frac{a}{b}(x-h), \qquad (12.4.6a)$$

and

$$y-k=-\frac{a}{b}(x-h). \qquad (12.4.6b)$$

Remarks

1. The concept of an asymptote is that the distance between the points on a hyperbola and the corresponding point on the asymptote is getting closer and closer to zero as x approaches $\pm\infty$.

2. A proof of this theorem requires the concept of the limit that is covered in calculus and is beyond the scope of this level (Reference 3).

Example 12.4.1

Given a quadratic equation in x and y as follows:

$$x^2 - y^2 - 4x - 6y - 6 = 0.$$

(a) Show that the graph of the solution set to the above equation is a hyperbola. Also, determine the center, the focal axis, the vertices, the foci, and a pair of asymptotes for the hyperbola.

(b) Graph the hyperbola.

Solution

The given equation is exactly the same as that in Example 12.1.5. By completing the square in both x and y, we have (See Example 12.1.5 for details)

$$(x-2)^2 - (y+3)^2 - 1 = 0, \Rightarrow$$

$$\frac{(x-2)^2}{1} - \frac{(y+3)^2}{1} = 1. \qquad (1)$$

Since Eq. (1) conforms Eq. (12.4.1), the focal axis is parallel to the x-axis, or $y = -3$. Also, by comparing Eq. (1) with Eq. (12.4.1), we have

$$h = 2, k = -3,$$

$$a^2 = 1, b^2 = 1, \Rightarrow a = 1, b = 1.$$

Therefore, $c^2 = a^2 + b^2 = 1 + 1 = 2, \Rightarrow c = \sqrt{2}$.

Hence, the vertices are $V_1(h-a, k) = V_1(2-1, -3) = V_1(1, -3)$ and $V_2(h+a, k) = V_2(2+1, -3) = V_2(3, -3)$, the foci are $F_1(h-c, k) = F_1(2 - \sqrt{2}, -3)$ and $F_2(h+c, k) = F_2(2+\sqrt{2}, -3)$. By using Eqs. (12.4.5a-b), the asymptotes are given, respectively, by

$$y-(-3) = \frac{1}{1}(x-2), \Rightarrow y+3 = x-2, \Rightarrow y = x-5,$$

and

$$y-(-3) = -\frac{1}{1}(x-2), \Rightarrow y+3 = -(x-2) = -x+2, \Rightarrow$$

$$y = -x-1.$$

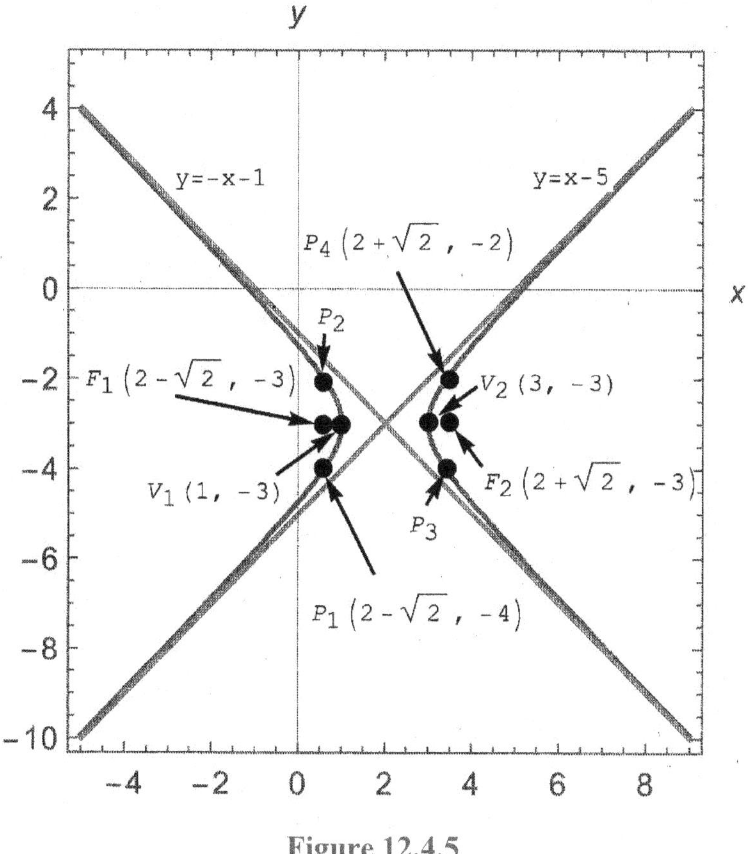

Figure 12.4.5

(b) To graph the hyperbola, we need a minimum of six points. We already have two vertices $V_1(1, -3)$ and $V_2(3, -3)$ obtained in part (a). By using the idea mentioned in Remarks #2 of Theorem 12.4.1, we solve for y by substituting $x = h - c = 2 - \sqrt{2}$ into Eq. (1) as follows:

$$\frac{(2-\sqrt{2}-2)^2}{1} - \frac{(y+3)^2}{1} = 1, \Rightarrow$$

$$1 = 2 - 1 = (-\sqrt{2})^2 - 1 = (y+3)^2, \Rightarrow$$

$$\pm 1 = \sqrt{(y+3)^2} = y+3, \Rightarrow \; -3 \pm 1 = y, \Rightarrow$$

$$y = -3 + 1 = -2, \text{ or } y = -3 - 1 = -4.$$

Therefore, the two points on the left branch of the hyperbola are $P_1(2 - \sqrt{2}, -4)$ and $P_2(2 - \sqrt{2}, -2)$. Due to a symmetry of the hyperbola with respect to the conjugate axis, the two other points on the right branch of the hyperbola are $P_3(2+\sqrt{2}, -4)$ and $P_4(2+\sqrt{2}, -2)$. By connecting three points on each branch of the hyperbola smoothly, the graph of the hyperbola is given in Figure 12.4.5. A pair of asymptotes is also plotted in Figure 12.4.5.

Example 12.4.2
Given a quadratic equation in x and y as follows:

$$9x^2 - 4y^2 + 9x + 8y + \frac{137}{4} = 0.$$

(a) Show that the graph of the solution set to the above equation is a hyperbola. Also, determine the center, the vertices, the focal axis, the foci, and a pair of asymptotes for the hyperbola.
(b) Graph of the hyperbola.
Solution
(a) We note that because the coefficients of x^2 and y^2 in the given equation are of the opposite sign, the graph of its solution set is, therefore, a hyperbola. By completing the square in both x and y, we have

$$9x^2 + 9x - 4y^2 + 8y = -\frac{137}{4}, \Rightarrow$$

$$9[(x+\frac{1}{2})^2 - \frac{1}{4}] - 4[(y-1)^2 - 1] = -\frac{137}{4}, \Rightarrow$$

$$9(x+\frac{1}{2})^2 - \frac{9}{4} - 4(y-1)^2 + 4 = -\frac{137}{4}, \Rightarrow$$

$$9(x+\frac{1}{2})^2 - 4(y-1)^2 = -\frac{137}{4} + \frac{9}{4} - 4 = -\frac{144}{4} = -36 \quad ,$$

$\Rightarrow$

$$\frac{9(x+\frac{1}{2})^2 - 4(y-1)^2}{-36} = 1, \Rightarrow$$

$$\frac{(y-1)^2}{9} - \frac{(x+\frac{1}{2})^2}{4} = 1. \qquad (1)$$

By comparing Eq. (1) with Eq. (12.4.2), we have

$$h = -\frac{1}{2}, k = 1,$$

$a^2 = 9$, $b^2 = 4$, $\Rightarrow$ $a = 3$, $b = 2$.

Therefore, $c^2 = a^2 + b^2 = 9 + 4 = 13$, $\Rightarrow$ $c = \sqrt{13}$.

Hence, the center is $Q(-\frac{1}{2}, 1)$, the vertices are $V_1^{'}(h, k - a) = V_1^{'}(-\frac{1}{2}, 1 - 3) = V_1^{'}(-\frac{1}{2}, -2)$ and $V_2^{'}(h, k + a) = V_2^{'}(-\frac{1}{2}, 1 + 3) = V_2^{'}(-\frac{1}{2}, 4)$, the focal axis is $x = -\frac{1}{2}$, the foci are $F_1^{'}(h, k - c) = F_1^{'}(-\frac{1}{2}, 1 - \sqrt{13})$ and $F_2^{'}(h, k + c) = F_2^{'}(-\frac{1}{2}, 1 + \sqrt{13})$, and a pair of asymptotes are given by using Eq. (12.4.6a-b) as follows:

$$y - 1 = \frac{3}{2}(x + \frac{1}{2}), \Rightarrow y - 1 = \frac{3}{2}x + \frac{3}{4}, \Rightarrow$$

$$y = \frac{3}{2}x + \frac{7}{4},$$

and

$$y - 1 = -\frac{3}{2}(x + \frac{1}{2}), \Rightarrow y - 1 = -\frac{3}{2}x - \frac{3}{4}, \Rightarrow$$

$$y = -\frac{3}{2}x + \frac{1}{4}.$$

(b) To graph the hyperbola we need a minimum of six points. We already have two vertices $V_1^{'}(-\frac{1}{2}, -2)$ and $V_2^{'}(-\frac{1}{2}, 4)$ obtained in part (a). To obtain four other points, we solve for y by substituting $y = k - c = 1 - \sqrt{13}$ into Eq. (1) as follows:

$$\frac{(1 - \sqrt{13} - 1)^2}{9} - \frac{(x + \frac{1}{2})^2}{4} = 1, \Rightarrow$$

$$\frac{16}{9} = (x + \frac{1}{2})^2, \Rightarrow \pm\frac{4}{3} = x + \frac{1}{2}, \Rightarrow$$

$$x = -\frac{1}{2} \pm \frac{4}{3}, \Rightarrow x = -\frac{1}{2} + \frac{4}{3} = \frac{5}{6}, \text{ or } x = -\frac{1}{2} - \frac{4}{3} = -\frac{11}{6}.$$

Therefore, two points on the lower branch of the hyperbola are $P_1^{'}(-\frac{11}{6}, 1 - \sqrt{13})$ and $P_2^{'}(\frac{5}{6}, 1 - \sqrt{13})$. Due to a symmetry of the hyperbola with respect to the conjugate axis, two other points on the upper branch of the hyperbola are $P_3'(-\frac{11}{6}, 1+\sqrt{13})$ and $P_4'(\frac{5}{6}, 1+\sqrt{13})$. By connecting three points on each branch of the hyperbola smoothly, the graph of the hyperbola is given in Figure 12.4.6.

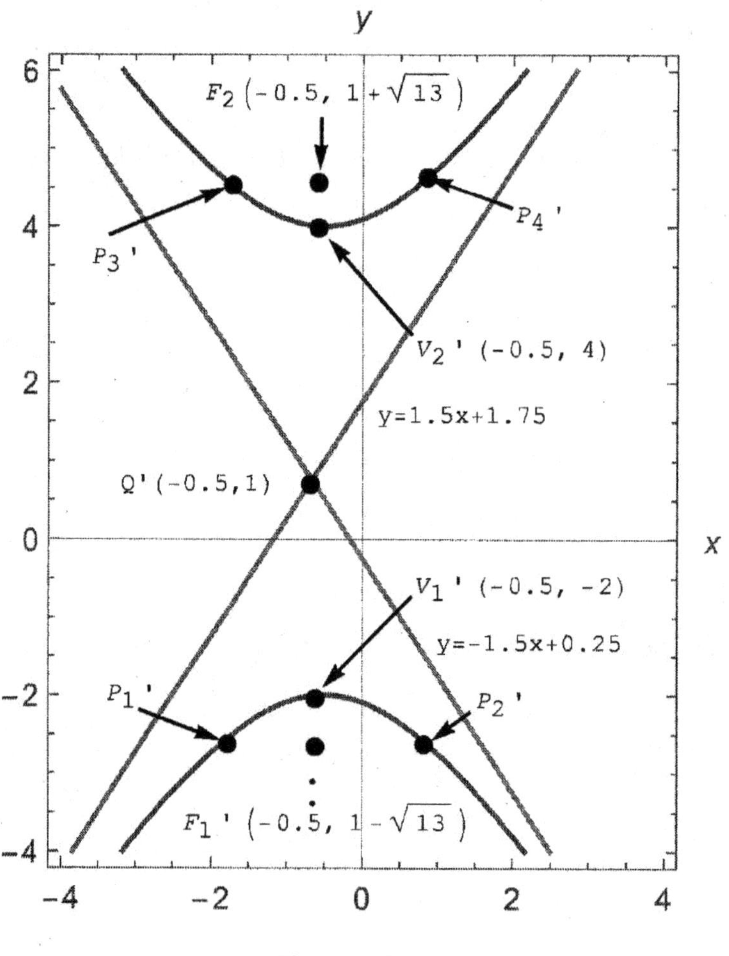

Figure 12.4.6

Exercises

For each of the following problems, (a) show that the graph of the solution set to the given equation is a hyperbola. Also, determine the center, the vertices, the focal axis, the foci, and a pair of asymptotes for the hyperbola, (b) graph the hyperbola:

1. $9x^2 - 16y^2 + 144 = 0.$
2. $2x^2 - 3y^2 - 6 = 0.$
3. $9(x + 2)^2 - 4(y - 1)^2 + 36 = 0.$
4. $4x^2 - 5y^2 - 8x - 10y - 9 = 0.$
5. $5x^2 - 6y^2 + 10x + 12y - 31 = 0.$

12.5 References

1. Anton, H. (1988). Calculus with Analytic Geometry, 3rd edition, John Wiley & Sons, New York.
2. Bix, R. (1998). Conics and Cubics: A Concrete Introduction to Algebraic Curves, Springer-Verlag Inc., New York.
3. Silverman, J. H. and Tate, J. (1992). Rational Points on Elliptic Curves, Springer Science+Business Media, LLC, New York.
4. http://en.wikipedia.org/wiki/Conic_section.

Part II

Functions, Mathematical Induction, Matrices, Probability, Sequences, and Series

Chapter 13
Functions

In this chapter we're going to learn the concept of functions. Please pay attention to that function is very different from that of equation. In particular, an equation of degree two in two variables can define more than one function (Example 13.1.2). By the way, all graphs in this book is plotted by using the software "Mathematica" (Reference 5).

13.1 Dependent Variable and Function

Definition 13.1.1
Given two variables x and y. The letter y is called a dependent variable and the letter x is called an independent variable if y varies as x varies.

Remarks
1. Since the variation of y depends on the variation of x, there must exist some kind of (mathematical) relationship between x and y.
2. The variation of a dependent variable may depends on more than the variation of one independent variable, namely, a dependent variable could depend on several independent variables.

Example 13.1.1
(a) The formula for the area A of a circle with a radius r is given by $A = \pi r^2$. Since A also varies as r varies, A is a dependent variable and r is an independent variable.
(b) The formula for the volume V of a circular cylinder with a base radius r and a height h is given by $V = \pi r^2 h$. In this equation, V is a dependent variable, which varies as two independent variables r and h vary.
(c) Newton's law of gravitation is given by

$$F = G \cdot \frac{m_1 \cdot m_2}{r^2},$$

where F denotes the attractive force between two spherical objects with masses given by m_1 and m_2, respectively, r is the distance between the centers of two objects, and G is the gravitational constant. In this equation, F is a dependent variable that varies as three independent variables m_1, m_2, and r vary.

Remarks

1. The role of a dependent variable and an independent variable can be interchangeable. For example, in the area formula of a circle, if we solve for r in terms of A, namely,

   $$r = \sqrt{\frac{A}{\pi}},$$

   where r is a dependent variable that varies as an independent variable A varies.
2. The notion of dependent variable is intimately related to that of a function as shown in the following definition. Using the notion of a function, the area A of a circle is a function of one independent variable r. The volume V of a circular cylinder is a function of two independent variables r and h. The attractive force F of Newton's gravitational law is a function of three independent variables of m_1, m_2, and r.

Definition 13.1.2

A function f of a real variable x is a rule of correspondence between a dependent variable y and an independent variable x such that for every value of x there corresponds exactly one value of y. The set of all values taken by an independent variable x is called the domain of f, denoted by Domain(f) and the set of all corresponding values of y is called the range of f, denoted by Range(f).

Remarks

1. A function f, denoted by $y = f(x)$ (read "f of x"), is a many-to-one, but not a one-to-many mapping that maps the points in Domain(f) into the points in Range(f) (Figure 13.1.1).

2. A function $y = f(x)$ could, in fact, be considered as a special case of an equation in two variables x and y with a restriction on the variable of y such that both the degree and the coefficient of y are one if it is rewritten as $y - f(x) = 0$. Hence, the graph of a function $y = f(x)$ can be viewed as the graph of the solution set to an equation in two variables x and y of the form $y - f(x) = 0$ that has infinitely many solutions. Consequently, a function is always an equation. However, an equation in x and y does not necessarily define their relationship as a function (see, e.g., Example 13.1.2).
3. If the domain of a function $f(x)$ is not specified in advance, then it is taken as the largest possible set of all (real number) x-values such that the corresponding y-values given by $y = f(x)$ is well defined.
4. A function $y = f(x_1, x_2, \ldots, x_n)$ of n independent variables $x_1, x_2, \ldots, x_n$ can be similarly defined. However, functions of both the multivariate and/or a complex variable are much advanced, which are beyond the scope of this course (References 3 and 4).
5. The function concept is one of the most fundamental concepts in modern mathematics. It did not arise suddenly. Historically, it may not be too generous to credit the Babylonians with an instinct for functionality, because they were among the most indefatigable compilers of arithmetic tables in history by noting that a function is also succinctly defined as a table or a correspondence (Reference 1, pp. 31-32). As a mathematical term, it was coined by G. W. von Leibniz (1641-1716, Germany). For more details, see Reference 5.

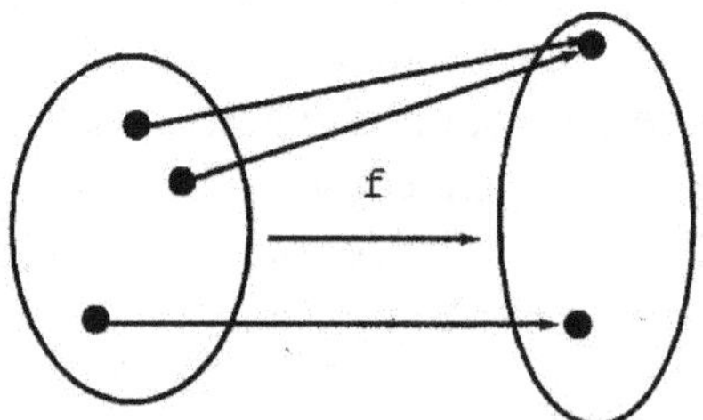

Figure 13.1.1

Example 13.1.2

Given a quadratic equation in x and y as follows:

$$x^2 + y^2 - 2x + 4y - 4 = 0.$$

(a) Show that this equation is not a function with y as a dependent variable and x as an independent variable.
(b) Show that the given equation actually defines two functions.

Proof

It suffices to show that for every x-values, there corresponds two distinct y-values. Hence, it violates the definition of a function. By completing the square in both x and y, we have

$$[x^2 - 2x + (\frac{-2}{2})^2 - (\frac{-2}{2})^2] + [y^2 + 4y + (\frac{4}{2})^2 - (\frac{4}{2})^2] - 4 = 0,$$

$$(x-1)^2 - 1 + (y+2)^2 - 4 - 4 = 0, \Rightarrow$$

$$(x-1)^2 + (y+2)^2 = 9, \Rightarrow$$

$$(y+2)^2 = 9 - (x-1)^2 = 3^2 - (x-1)^2 = (3 + x - 1)(3 - (x-1))$$

$$= (2+x)(4-x), \Rightarrow$$

$$y + 2 = \pm\sqrt{(2+x)(4-x)}, \Rightarrow$$

$$y = -2 \pm\sqrt{(2+x)(4-x)}. \qquad (1)$$

Since the expression under the square root is required to be nonnegative, we have

$$(2+x)(4-x) \geq 0. \Rightarrow -2 \leq x \leq 4, \Rightarrow$$

$$x \in [-2, 4]. \qquad (2)$$

Therefore, for every x in the closed interval $[-2, 4]$, there corresponds two distinct y-values obtained from Eq. (1) as follows:

$$y = -2 + \sqrt{(2+x)(4-x)}, \qquad (3a)$$

and

$$y = -2 - \sqrt{(2+x)(4-x)}. \qquad (3b)$$

This violates the characteristic feature of a function that is a many-to-one, but not a one-to-many mapping from its domain to its range.
(b) Obviously, the given equation defines two distinct functions that can be obtained from Eqs. (3a-b) as follows:

$$f_1(x) = -2 + \sqrt{(2+x)(4-x)},$$

and

$$f_2(x) = -2 - \sqrt{(2+x)(4-x)}.$$

The domain of $f_1(x)$ and $f_2(x)$ is obtained from Eq. (2) and given as follows:

$$\text{Domain}(f_1) = \text{Domain}(f_2) = \{x \in R | -2 \leq x \leq 4\}$$

$= [-2, 4]$

The graphs of $f_1(x)$ and $f_2(x)$ are represented by the upper and lower semi-circle, respectively (Figure 13.1.2).

Remark

Please see Section 12.2 in Part I for graphing a circle shown in Figure 13.1.2.

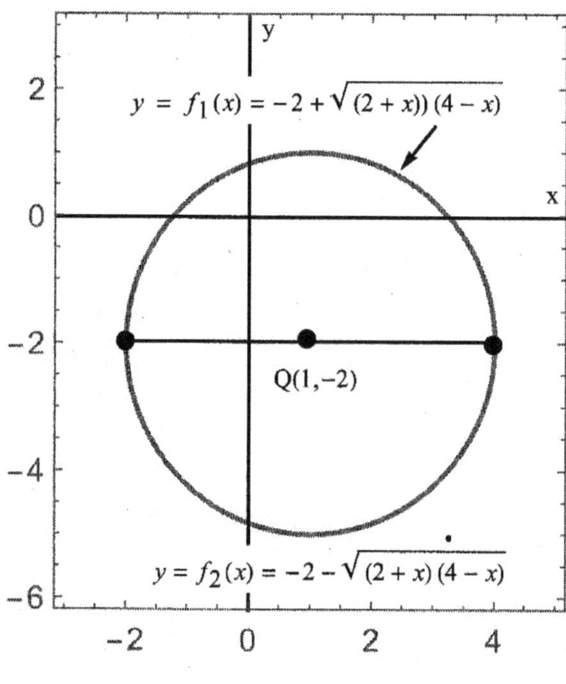

Figure 13.1.2

Example 13.1.3

Given: $f(x) = -x^2 + x - 3$. Evaluate each of the following function values:

(a) $f(4) = ?$, (b) $f(s + 1) = ?$, (c) $\dfrac{f(2+\Delta x) - f(2)}{\Delta x} = ?$, $\Delta x \neq 0$.

Solution

(a) By substituting $x = 4$ into the right side of the given function, we have

$$f(4) = -4^2 + 4 - 3 = -16 + 4 - 3 = -15.$$

(b) By substituting $x = s + 1$ into the right side of the given function, we have

$$\begin{aligned} f(s+1) &= -(s+1)^2 + (s+1) - 3 \\ &= -(s^2 + 2s + 1) + s + 1 - 3 \\ &= -s^2 - 2s - 1 + s - 2 \\ &= -s^2 - s - 3. \end{aligned}$$

(c) By substituting $x = 2 + \Delta x$ into the right side of the given function, we have

$$\begin{aligned} f(2+\Delta x) &= -(2+\Delta x)^2 + (2+\Delta x) - 3 \\ &= -[4 + 4\Delta x + (\Delta x)^2] + 2 + \Delta x - 3 \\ &= -4 - 4\Delta x - (\Delta x)^2 + 2 + \Delta x - 3 \\ &= -(\Delta x)^2 - 3\cdot\Delta x - 5, \end{aligned}$$

and

$$f(2) = -2^2 + 2 - 3 = -4 + 2 - 3 = -5.$$

By substituting the above two function values into the desired evaluation, we have

$$\begin{aligned} \frac{f(2+\Delta x) - f(2)}{\Delta x} &= \frac{-(\Delta x)^2 - 3\cdot\Delta x - 5 - (-5)}{\Delta x} \\ &= \frac{-(\Delta x)^2 - 3\cdot\Delta x - 5 + 5}{\Delta x} = \frac{-(\Delta x)^2 - 3\cdot\Delta x}{\Delta x} \\ &= \frac{-\Delta x(\Delta x + 3)}{\Delta x} = -(\Delta x + 3). \end{aligned}$$

Hence, $\dfrac{f(2+\Delta x) - f(2)}{\Delta x} = -(\Delta x + 3)$.

Remark

In part (c), Δx is viewed as a single variable, rather than two variables, that simply means a very small increment in x. This variable of Δx will be commonly used in calculus to define the first derivative of $f(x)$ (Reference 1, pp. 139-142).

Example 13.1.4

Find the domain from each of the following functions:

(a) $f(x) = x^2 - 2x + 2$.

(b) $g(x) = \sqrt{x^2 - x - 2}$.

(c) $h(x) = \dfrac{2x+3}{x^2 - x - 6}$.

Solution

(a) Since $f(x)$ is well-defined for all real numbers, the domain of $f(x)$ is therefore, given by

$$\text{Domain}(f) = \{x \mid x \in R\} = (-\infty, +\infty)$$
$$= \text{the set of all real numbers.}$$

(b) Since $g(x)$ is well-defined only for those real numbers such that the expression under the square root is nonnegative, it is, therefore, required for x to satisfy the following inequality:

$$x^2 - x - 2 \geq 0, \Rightarrow (x-2)(x+1) \geq 0, \Rightarrow$$
$$x \leq -1 \text{ or } x \geq 2.$$

Hence, $\text{Domain}(g) = \{x \in R \mid x \leq -1 \text{ or } x \geq 2\} = (-\infty, -1] \cup [2, +\infty)$.

(c) Since $h(x)$ is well-defined only for those real numbers such that its denominator of $x^2 - x - 6$ does not equal zero, we deliberately set its denominator equal to zero and solve for x as follows:

$$x^2 - x - 6 = 0, \Rightarrow (x-3)(x+2) = 0, \Rightarrow$$
$$x - 3 = 0 \text{ or } x + 2 = 0, \Rightarrow x = 3 \text{ or } x = -2.$$

Hence, $\text{Domain}(h) = \{x \in R \mid x \neq -2 \text{ and } x \neq 3\} = (-\infty, -2) \cup (-2, 3) \cup (3, +\infty)$.

Remark

Please see Chapter 6 and Section 8.1 in Part I on how to solve a quadratic equation and a quadratic inequality, respectively.

Exercises

In problems 1–7, determine if the given equation defines y as function of x:

1. $2x - 3y + 4 = 0$.
2. $x^2 - 3x + 4y - 5 = 0$.
3. $y^2 + 2x - 4y - 1 = 0$.
4. $x^2 + y^2 + 4x + 6y + 12 = 0$.
5. $2x^2 + 3y^2 - 4x - 6y - 1 = 0$.
6. $x^2 - y^2 - 4x - 6y - 6 = 0$.
7. $x^2y - x + 4y = 5$.

In problems 8–12, evaluate the function at the specified value of the independent variable and simplify the result.

8. $f(x) = 3x - 5$; (a) $f(2) = ?$ (b) $f(2 + \Delta x) = ?$
(c) $\dfrac{f(2+\Delta x) - f(2)}{\Delta x} = ?$, $\Delta x \neq 0$.

9. $g(t) = \sqrt{t+1}$; (a) $g(3) = ?$ (b) $g(3 + \Delta t) = ?$
(c) $\dfrac{g(3+\Delta t) - g(3)}{\Delta t} = ?$, $\Delta t \neq 0$.

10. $h(s) = \dfrac{1}{s^2+1}$; (a) $h(1) = ?$ (b) $h(1 + \Delta s) = ?$
(c) $\dfrac{h(1+\Delta s) - h(1)}{\Delta s} = ?$, $\Delta s \neq 0$.

11. $k(u) = |u| - 2$; (a) $k(-3) = ?$ (b) $k(3) = ?$ (c) $k(-5) = ?$
(d) $k(-1) = ?$

12. $p(x) = \begin{cases} x+3, x \leq 1, \\ -2x+1, x > 1. \end{cases}$ (a) $p(0) = ?$ (b) $p(1) = ?$ (c) $p(3) = ?$
(d) $p(-2) = ?$

In problems 13–15, find the domain of the given function.

13. $f(x) = \sqrt{x-3}$. **14.** $g(t) = \dfrac{2t}{t+1}$.

15. $h(s) = \dfrac{\sqrt{s-2}}{s-5}$.

13.2 The Graph of a Function

A General Procedure for Graphing a Function

To graph $y = f(x)$ is equivalent to graph the solution set to the given equation $y - f(x) = 0$. Since $y - f(x) = 0$ is a single equation in two variables x and y, it has infinitely many solutions. Each solution that is represented by an ordered pair (x, y) can be plotted as a point in the Cartesian plane. Then, by connecting all these points smoothly, we get the graph of $f(x)$.

Remarks

1. Please see Chapters 9 and 12 for a review on graphing the solution set of a linear and quadratic equation in two variables x and y, respectively.
2. An effective way to graph a function is to locate its relative maximum and minimum points. However, this topic is beyond the scope of this course and is only covered in calculus.

Definition 13.2.1
A function f is said to be increasing (or decreasing) on an open interval (a, b) if for any two points x_1 and x_2 in (a, b), $f(x_1) < f(x_2)$ (or $f(x_1) > f(x_2)$) whenever $x_1 < x_2$.

Remarks
1. For the time being, the only way to determine if a function is increasing or decreasing on a certain interval is to graph the function and discern the pattern of variation of its graph. A function is judged to be increasing (a, b) if as the x-coordinate of the points on the graph increases, the corresponding y-coordinate increases too. A function is judged to be decreasing on (a, b) if as the x-coordinate of the point on the graph increases, the corresponding y-coordinate decreases instead.
2. A more effective way to judge whether a function is increasing or decreasing on an interval is through the use of the first derivative of a function that is only covered in calculus.

Definition 13.2.2
A function $f(x) = x^n$ is called a power function of x, where n is a positive integer.

Remarks
1. A power function is also a polynomial function of degree n that is covered in Chapter 14.
2. When $n = 1$, $f(x) = x$ is called an identity function that serves as an identity element for the operation of function-composition (see Remark 6 after operational rules for functions in Section 13.3). The identity function is oftenimes denoted by $i(x) = x$.

Example 13.2.1
Given: $i(x) = x$, the identity function.

(a) Graph $i(x)$.
(b) Find the intervals on that i(x) is increasing or decreasing.

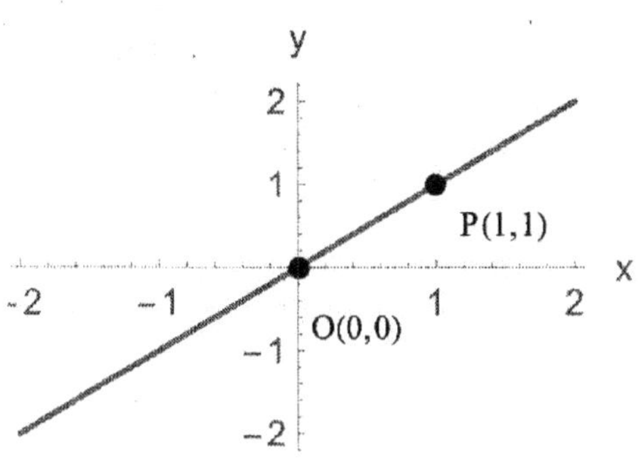

Figure 13.2.1

Solution
(a) First, note that $i(x)$ is a linear function whose graph is a straight line (Section 14.1). Since we just need a minimum of two points to graph a line, we evaluate $i(x) = x$ at $x = 0$ and $x = 1$ as follows:

x	$y = i(x) = x$
0	$i(0) = 0 \Rightarrow O(0, 0)$
1	$i(1) = 1 \Rightarrow P(1, 1)$

By connecting $O(0, 0)$ and $P(1, 1)$, we get the 45-degree line as shown in Figure 13.2.1.
(b) Since the y-coordinate of the points on the line increases too as the x-coordinate of the points increases from $-\infty$ to $+\infty$, $i(x) = x$ is increasing on $(-\infty, +\infty)$, the entire real line.

Example 13.2.2
Given $f(x) = |x|$, that is called the absolute value function of x.
(a) Graph $f(x)$.
(b) Find the intervals on that $f(x)$ is increasing or decreasing.
Solution
(a) By the definition of the absolute value of x, the given function is actually defined by two pieces of different functions on different intervals as follows:

$$f(x)=\begin{cases} x, x\geq 0, \\ -x, x<0. \end{cases}$$

For $x \geq 0$, $f(x) = x$ is an identity function whose graph is a 45-degree line in the first quadrant as shown in Example 13.2.1. Similarly for $x < 0$, the graph of $f(x) = -x$ will be another 45-degree line in the second quadrant. By composing these two pieces of the line segments, the graph of $f(x) = |x|$ is shown in Figure 13.2.2.

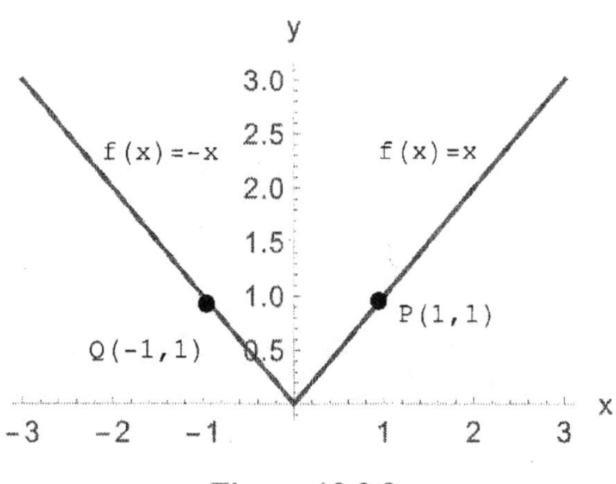

Figure 13.2.2

(b) As it can be seen from Figure 13.2.2, $f(x) = |x|$ is decreasing on the interval $(-\infty, 0]$ since the corresponding y-coordinate decreases from $+\infty$ to 0 as the x-coordinate of the points on the graph of $f(x) = |x|$ increases from $-\infty$ to 0. In contrast, as the x-coordinate of the points on the graph of $f(x) = |x|$ increases from 0 to $+\infty$, the corresponding y-coordinate also increases from 0 to $+\infty$, $f(x) = |x|$ is increasing on the interval $[0, +\infty)$.

Example 13.2.3

Given $f(x)=\sqrt{x}$, the square root function of x.

(a) Graph $f(x)$.

(b) Find the intervals on that $f(x)$ is increasing or decreasing.

Solution

(a) Note that the domain of $f(x)=\sqrt{x}$ is $\{x \in R| \, x \geq 0\} = [0, +\infty)$. To graph the given function, we evaluate $f(x)=\sqrt{x}$ at $x = 0, 1, 4$ as follows:

x	$y = f(x) = \sqrt{x}$
0	$f(0) = \sqrt{0} = 0 \Rightarrow O(0, 0)$.
1	$f(1) = \sqrt{1} = 1 \Rightarrow P_1(1, 1)$.
4	$f(4) = \sqrt{4} = 2 \Rightarrow P_2(4, 2)$.

By connecting the three points $O(0, 0)$, $P_1(1, 1)$, and $P_2(4, 2)$ smoothly, we get the graph of $f(x)=\sqrt{x}$ as shown in Figure 13.2.3.
(b) From the graph of $f(x)=\sqrt{x}$ as shown in Figure 13.2.3, we conclude that $f(x)$ is increasing on the interval $[0, +\infty)$ since as the x-coordinate of the points on the graph of $f(x)=\sqrt{x}$ increases from 0 to $+\infty$, the corresponding y-coordinate also increases from 0 to $+\infty$.

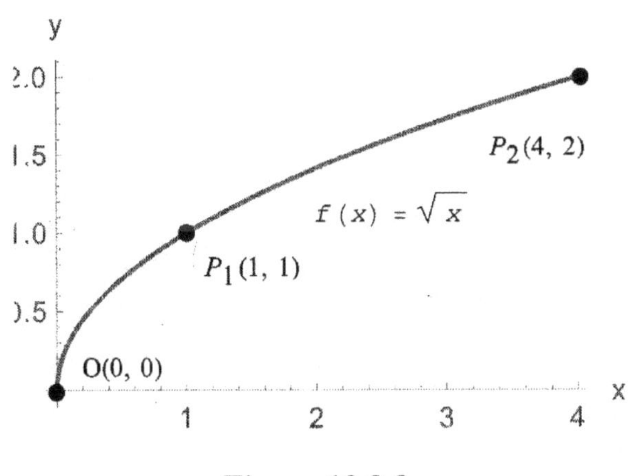

Figure 13.2.3

Definition 13.2.3
A function $y = f(x)$ is called an even function if for every x in the domain of f, the function values of $f(x)$ and $f(-x)$ are the same, i.e., $f(x) = f(-x)$.

Definition 13.2.4

A function $y = f(x)$ is called an odd function if for every x in the domain of f, the function value of $f(-x)$ equals the negative of the function value of $f(x)$, i.e., $f(-x) = -f(x)$.

Remark
The graph of an even function is symmetric with respect to the y-axis, whereas the graph of an odd function is symmetric with respect to the origin.

Example 13.2.4
Given: $f(x) = x^2$.
(a) Show that the given function is an even function.
(b) Graph $f(x)$.
(c) Find the intervals on that $f(x)$ is increasing or decreasing.
Solution
(a) Since $f(-x) = (-x)^2 = x^2 = f(x)$, $f(x) = x^2$ is, therefore, an even function by Definition 13.2.3.
(b) To graph the given function, we evaluate $f(x) = x^2$ at $x = 0, \pm 1, \pm 2$ as follows:

x	$y = f(x) = x^2$
-2	$f(-2) = (-2)^2 = 4 \Rightarrow P_1(-2, 4)$.
-1	$f(-1) = (-1)^2 = 1 \Rightarrow P_2(-1, 1)$.
0	$f(0) = 0^2 = 0 \Rightarrow O(0, 0)$.
1	$f(1) = 1^2 = 1 \Rightarrow P_3(1, 1)$.
2	$f(2) = 2^2 = 4 \Rightarrow P_4(2, 4)$.

By connecting the five points $P_1(-2, 4)$, $P_2(-1, 1)$, $O(0, 0)$, $P_3(1, 1)$, and $P_4(2, 4)$ smoothly, we get the graph of $f(x) = x^2$ as shown in Figure 13.2.4.
(c) From the graph of $f(x) = x^2$ (Figure 13.2.4), we conclude that $f(x)$ is decreasing on the interval $(-\infty, 0]$ and increasing on $[0, +\infty)$.

Remarks
1. The graph of $f(x) = x^2$ is clearly symmetric with respect to the y-axis.
2. $f(x) = x^2$ is a member of the family of quadratic function. A more general way to graph any quadratic function is covered in Section 14.2.

3. It is clear that any power function of $f(x) = x^n$ with n being an even positive integer is an even function. For other specific case, see problem 2 in Exercises for Section 13.2.

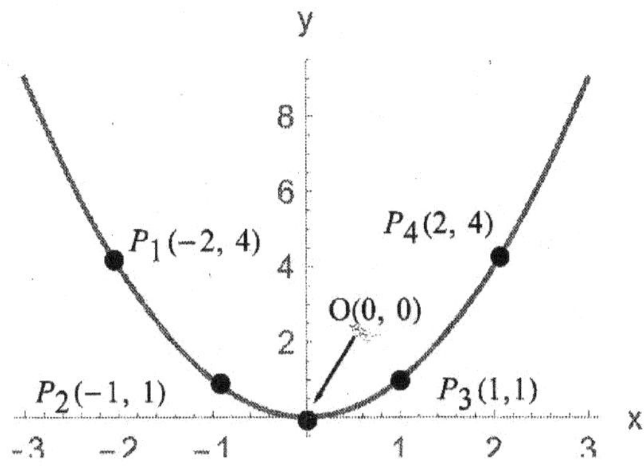

Figure 13.2.4

Example 13.2.5
Given: $f(x) = x^3$.
(a) Show that the given function is an odd function.
(b) Graph $f(x)$.
(c) Find the intervals on that $f(x)$ is increasing or decreasing.
Solution
(a) Since $f(-x) = (-x)^3 = -x^3 = -f(x)$, $f(x) = x^3$ is an odd function by Definition 13.2.4.
(b) To graph the given function, we evaluate $f(x) = x^3$ at $x = 0, \pm 1, \pm 2$ as follows:

x	$y = f(x) = x^3$
-2	$f(-2) = (-2)^3 = -8 \Rightarrow P_1(-2, -8)$.
-1	$f(-1) = (-1)^3 = -1 \Rightarrow P_2(-1, -1)$.
0	$f(0) = 0^3 = 0 \Rightarrow O(0, 0)$.
1	$f(1) = 1^3 = 1 \Rightarrow P_3(1, 1)$.
2	$f(2) = 2^3 = 8 \Rightarrow P_4(2, 8)$.

By connecting the five points $P_1(-2,-8)$, $P_2(-1,-1)$, $O(0,0)$, $P_3(1,1)$, and $P_4(2,8)$ smoothly, we get the graph of $f(x)=x^3$ as shown in Figure 13.2.5.
(c) From the graph of $f(x)=x^3$ as shown in Figure 13.2.5, we conclude that $f(x)=x^3$ is increasing on $(-\infty,+\infty)$, i.e., on the entire real line.

Remarks

1. The graph of $f(x)=x^3$ is clearly symmetric with respect to the origin.
2. It is clear that any power function of $f(x)=x^n$ with n being an odd positive integer is an odd function. For specific case, see problem 3 in Exercises for Section 13.2.

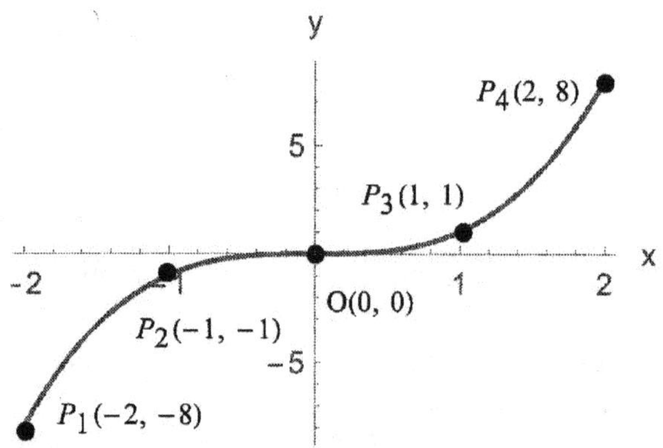

Figure 13.2.5

Exercises

1. Show that the absolute value function of $f(x)=|x|$ is an even function.

In problems 2–10, (a) determine if the given function is an even function, or an odd function, or neither. (b) graph the given function, (c) find the intervals on that the given function is increasing or decreasing.

2. $f(x) = x^4$. **3.** $f(x) = x^5$. **4.** $f(x) = 2x + 3$.
5. $f(x) = -2x + 1$. **6.** $f(x) = x^2 - 1$. **7.** $f(x) = -2x^2 + 1$.
8. $f(x) = \sqrt{1-x}$. **9.** $f(x) = \sqrt{1-x^2}$.

10. $f(x) = \begin{cases} x+2, x \geq -1, \\ -x+1, x < -1. \end{cases}$.

13.3 Operational Rules for Functions

When we would like to combine two functions in order to get a new function, the following rules governs the operations between functions.

Operational Rules for Functions
Given two functions f and g.
1. Function Addition Rule

$$(f+g)(x) = f(x) + g(x). \quad (13.3.1)$$

2. Function Subtraction Rule

$$(f-g)(x) = f(x) - g(x). \quad (13.3.2)$$

3. Scalar Multiplication Rule (or Constant Multiplies a Function Rule)

$(c \cdot f)(x) = c \cdot f(x)$, where c is a real constant. (13.3.3)

4. Function Multiplication Rule

$$(f \cdot g)(x) = f(x) \cdot g(x). \quad (13.3.4)$$

5. Function Division Rule

$(\frac{f}{g})(x) = \frac{f(x)}{g(x)}$, if $g(x) \neq 0$. (13.3.5)

6. Function Composition Rule

$$(f \circ g)(x) = f(g(x)) . \quad (13.3.6)$$

Remarks
1. All operations between functions are defined pointwisely.
2. It is understood that we will get a new function after applying the operational rules to combine two functions.
3. Associative law holds for function -addition, -multiplication, and -composition, namely,

$$(f+g) + h = f + (g+h),$$

$(f \cdot g) \cdot h = f \cdot (g \cdot h)$,

$(f \circ g) \circ h = f \circ (g \circ h)$,

where h is another function, but does not hold for function subtraction and division.

4. Commutative law holds for function- addition and multiplication, but does not hold for function-subtraction, and composition, namely,

$f + g = g + f$,

$f \cdot g = g \cdot f$,

but

$f - g \neq g - f$,

$\frac{f}{g} \neq \frac{g}{f}$,

$f \circ g \neq g \circ f$.

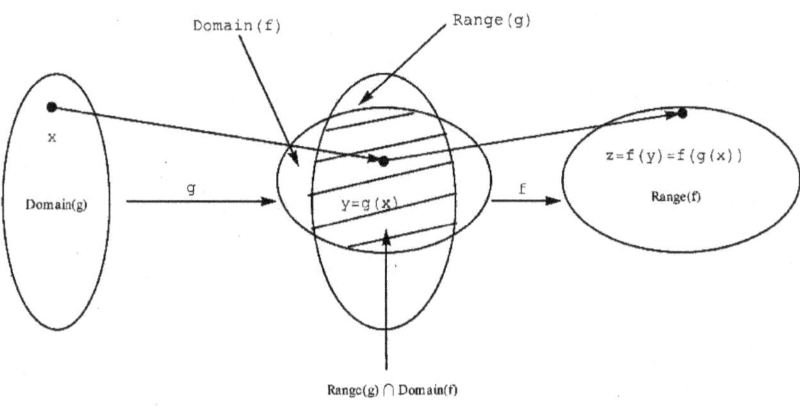

Figure 13.3.1

5. Usually, the domain of the new function obtained through applying the operational rules to any two functions is the intersection of the domains of those two functions. For function division, the domain of the new function must exclude those x-values such that $g(x) = 0$. For function composition, the domain of the new function h (= $f \circ g$) is the set of all x in the domain of g such that $g(x)$ must be in the domain of f (see Figure 13.3.1).
6. Let $i(x) = x$. Now, for any function f, we have

$$(f \circ i)(x) = f(i(x)) = f(x) \Rightarrow f \circ i = f,$$

and

$$(i \circ f)(x) = i(f(x)) = f(x) \Rightarrow i \circ f = f.$$

Since $f \circ i = f = i \circ f$, the function $i(x)$ is the identity function for the function composition. Note that the graph of $i(x)$ is the 45-degree line in the first and third quadrants passing through the origin (Figure 13.2.1).

Example 13.3.1

Given: $f(x) = \sqrt{x}$, where Domain(f) = $[0, +\infty)$, and $g(x) = x^2 - 1$, where Domain(g) = $(-\infty, +\infty)$.

(a) Find $h(x) = ?$, where $h = 2{\cdot}f - 3{\cdot}g$. Also, find the domain of h.

(b) Find $k(x) = ?$, where $k = f \cdot g$. Also, find the domain of k.

(c) Find $\ell(x) = ?$, where $\ell = \dfrac{f}{g}$. Also, find the domain of ℓ.

(d) Find $p(x) = ?$, where $p = f \circ g$. Also, find the domain of p.

(e) Find $q(x) = ?$, where $q = g \circ f$. Also, find the domain of q.

Solution

(a) By using Eqs. (13.3.2) and (13.3.3), we have

$$h(x) = (2{\cdot}f - 3{\cdot}g)(x) = 2{\cdot}f(x) - 3{\cdot}g(x)$$
$$= 2\sqrt{x} - 3{\cdot}(x^2 - 1) = 3 + 2\sqrt{x} - 3x^2. \quad (1)$$

From Eq. (1), Domain(h) = $\{x \in R | \, x \geq 0\} = [0, +\infty)$ = Domain(f) $\cap$ Domain(g).

(b) By using Eq. (13.3.4), we have

$$k(x) = (f{\cdot}g)(x) = f(x){\cdot}g(x)$$
$$= \sqrt{x} \cdot (x^2 - 1) = \sqrt{x} \cdot x^2 - \sqrt{x}$$
$$= x^{\frac{1}{2}} \cdot x^2 - \sqrt{x} = x^{\frac{1}{2}+2} - \sqrt{x} = x^{\frac{5}{2}} - \sqrt{x}$$
$$= \sqrt{x^5} - \sqrt{x}. \quad (2)$$

From Eq. (2), Domain(k) = $\{x \in R | \, x \geq 0\} = [0, +\infty)$ = Domain(f) $\cap$ Domain(g).

(c) By using Eq. (13.3.5), we have

$$\ell(x) = \left(\frac{f}{g}\right)(x) = \frac{f(x)}{g(x)} = \frac{\sqrt{x}}{x^2 - 1}. \quad (3)$$

From Eq. (3), Domain(ℓ) = $\{x \in R | \, x \geq 0, x \neq 1\} = [0, 1) \cup (1, +\infty)$.

(d) By using Eq. (13.3.6), we have

$$p(x) = (f \circ g)(x) = f(g(x)) = \sqrt{g(x)} = \sqrt{x^2 - 1}. \qquad (4)$$

From Eq. (4), Domain(p) = $\{x \in R | x \leq -1, \text{or } x \geq 1\} = (-\infty, -1] \cup [1, +\infty)$.

(e) By using Eq. (13.3.6), we have

$$q(x) = (g \circ f)(x) = g(f(x)) = (f(x))^2 - 1 = (\sqrt{x})^2 - 1. \qquad (5)$$

From Eq. (5), Domain(q) = $\{x \in R | x \geq 0\} = [0, +\infty)$. Under this domain of Domain(q), Eq. (5) can be further simplified to $q(x) = x - 1$.

Remark

Clearly, $p(x) \neq q(x)$. Consequently, $f \circ g \neq g \circ f$.

Example 13.3.2

Given: $f(x) = \sqrt{x}$ and $g(x) = x^2 - 1$. Find:

(a) $(2 \cdot f - 3 \cdot g)(4) = ?$

(b) $(f \cdot g)(9) = ?$

(c) $(\frac{f}{g})(9) = ?$

(d) $(f \circ g)(16) = ?$

(e) $(g \circ f)(16) = ?$

Solution

By using the results of Example 13.3.1, we have

(a) $(2 \cdot f - 3 \cdot g)(4) = h(4) = 3 + 2\sqrt{4} - 3 \cdot 4^2 = 3 + 2 \cdot 2 - 3 \cdot 16$
$= 3 + 4 - 48 = -41.$

(b) $(f \cdot g)(9) = k(9) = \sqrt{9^5} - \sqrt{9} = 3^5 - 3 = 243 - 3 = 240.$

(c) $(\frac{f}{g})(9) = \ell(9) = \frac{\sqrt{9}}{9^2 - 1} = \frac{3}{80}.$

(d) $(f \circ g)(16) = p(16) = \sqrt{16^2 - 1} = \sqrt{255}.$

(e) $(g \circ f)(16) = q(16) = 16 - 1 = 15.$

Example 13.3.3

Given: $h(x) = \sqrt{x^2 + 1}$. Find two functions f and g such that $h = f \circ g$.

Solution

To express h as a composition of f and g, the idea is to simplify h to an elementary function that is the square root function in this case. So, let $g(x) = x^2 + 1$. Then $h(x)$ can be re-written as $h(x) = \sqrt{g(x)} = \sqrt{y} = f(y)$, where $y = g(x)$. Hence, two functions f and g are given by $f(x)=\sqrt{x}$ and $g(x)=x^2+1$.

Remark

Many complicated functions are in fact obtained through using the function composition on some basic elementary functions. Recognition of expressing a complicated function as a composition of two simpler elementary functions is vital in understanding the chain rule in calculus.

Exercises

In problems 1–3, find: (a) $(3{\cdot}f - 4{\cdot}g)(x) = ?$, (b) $(f{\cdot}g)(x) = ?$, (c) $(\frac{f}{g})(x)=?$, (d) $(f\circ g)(x)=?$, (e) $(g\circ f)(x)=?$. Also, find the domain of the new function resulting from the indicated operation.

1. $f(x) = x - 3$ and $g(x) = x^2 + 1$.
2. $f(x)=\sqrt{x-1}$ and $g(x) = x^2 + 3$.
3. $f(x)=\dfrac{x}{x-6}$ and $g(x) = x^2 + 2$.

In problems 4–8, evaluate the function value at the indicated x-value for $f(x)=\sqrt{x-1}$ and $g(x) = x^2 + 3$.

4. $(3{\cdot}f - 4{\cdot}g)(2) = ?$ **5.** $(f{\cdot}g)(10) = ?$ **6.** $(\frac{f}{g})(5) = ?$

7. $(f\circ g)(5)=?$ **8.** $(g\circ f)(5)=?$

In problems 9–15, find two functions g and h such that $(g\circ h)(x)=f(x)$.

9. $f(x) = (x + 5)^3$. **10.** $f(x) = (x^2 + 1)^{13}$. **11.** $f(x) = |x^2 + 5|$.
12. $f(x)=\sqrt{x^2+2}$. **13.** $f(x)=\sqrt[3]{x-2}$. **14.** $f(x)=(x^2+3)^{\frac{3}{2}}$.
15. $f(x)=\dfrac{1}{x-5}$.

13.4 The Inverse of a Function

In Section 13.1, we note that a function is a many-to-one, but not one-to-many mapping between its domain and its range. It is evident that if an inverse mapping that maps its range back to its domain by a reverse of the rule of correspondence, then this inverse mapping could become a one-to-many mapping that violates the definition of a function. Therefore, in order for the inverse mapping to be a function, the original function is required to be a one-to-one mapping.

Definition 13.4.1
A function f is called a one-to-one function if for any two distinct values $x_1 \neq x_2$ in its domain, their corresponding y-values are also distinct, i.e., $f(x_1) \neq f(x_2)$.

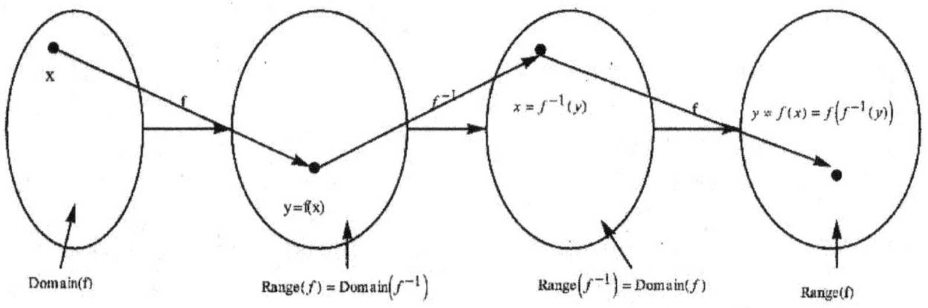

Figure 13.4.1

Remarks

1. Equivalently, a function f is one-to-one if "$f(x_1) = f(x_2)$" implies "$x_1 = x_2$".
2. If the graph of $f(x)$ is available, you draw many horizontal lines to see if the horizontal lines intersect the graph of $f(x)$ at a single point. If the answer is yes, then $f(x)$ is one-to-one. This is called the horizontal line test for testing the one-to-one property of a function.

3. For any one-to-one function f, its inverse is a function that is denoted by f^{-1} (read "f inverse"). For a function $y = f(x)$, y is a dependent variable and x is an independent variable. For the inverse function $x = f^{-1}(y)$, x is a dependent variable and y is an independent variable. Note that Domain(f^{-1}) = Range(f) and Range(f^{-1}) = Domain(f) (Figure 13.4.1). Also, it follows that a function f and its inverse f^{-1} satisfies the following relationship:

$$(f^{-1} \circ f)(x) = x\,, \tag{13.4.1}$$

and

$$(f \circ f^{-1})(y) = y \tag{13.4.2}$$

This leads to the following definition of inverse functions.

Definition 13.4.2

Two functions f and g are inverses of each other if for every $x \in$ Domain(g),

$$(f \circ g)(x) = x, \tag{13.4.3}$$

and for every $x \in$ Domain(f),

$$(g \circ f)(x) = x. \tag{13.4.4}$$

Thus, we denote g by f^{-1} or f by g^{-1}.

Remarks

1. Strictly speaking, y should be viewed as an independent variable for f^{-1} as shown in Eq. (13.4.2). However, as a convention, we ordinarily use x to denote an independent variable. This is the reason why x is used in Eq. (13.4.3) in contrast to y used in Eq. (13.4.2). If we use $i(x) = x$ as an identity function, then we can suppress the use of an independent variable by writing Eq. (13.4.3) and (13.4.4) as follows:

$$f \circ g = i = g \circ f\,,$$

or

$$f \circ f^{-1} = i = f^{-1} \circ f\,.$$

2. The inverse function f^{-1} is unique if it exists.
3. If an arbitrary point on the graph of $y = f(x)$ is denoted by $P(x, y) = P(x, f(x))$, then $Q(y, x) = Q(y, f^{-1}(y))$ must be a point on the graph of $x = f^{-1}(y)$. Notice that two points $P(x, y)$ and $Q(y, x)$ are

symmetric with respect to the 45-degree line of $y = i(x) = x$ (Figure 13.4.2). Therefore, a function f and its inverse function f^{-1} possess such a property that the graph of f and f^{-1} are symmetric with respect to the 45-degree line.

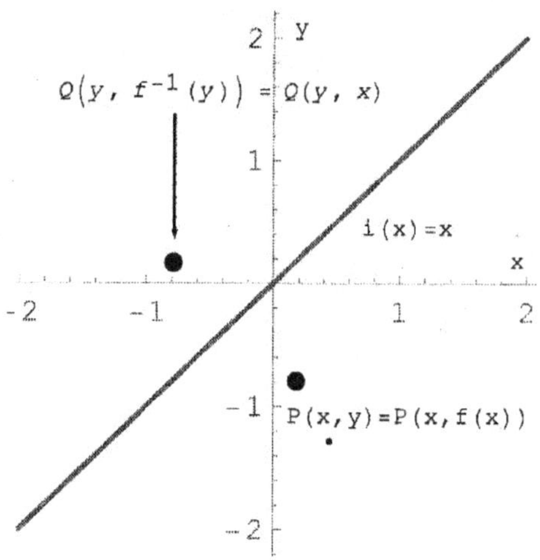

Figure 13.4.2

A question arises: How to find the inverse function if it exists? An answer is provided in the following procedure.

A Procedure for Finding the Inverse Function

Given that $f(x)$ is an one-to-one function. To find the inverse function f^{-1}, we proceed as follows:

Step 1. Let $y = f(x)$.

Step 2. View $y = f(x)$ as an equation $y - f(x) = 0$. Solve this equation for x in terms of y. The solution then defines x as a function of y, i.e., $x = f^{-1}(y)$ that is the desired inverse function.

Step 3. Conventionally, we prefer to use y as a symbol for a dependent variable and x for an independent variable. Therefore, we replace y by x and x by y in the inverse function of $x = f^{-1}(y)$ obtained in Step 2 to obtain $y = f^{-1}(x)$.

Example 13.4.1

Given: $f(x) = x^2$, where Domain(f) = $(-\infty, \infty)$. Show that the given function is not a one-to-one function on $(-\infty, \infty)$.

Proof

Suppose that for any $x_1, x_2 \in$ Domain(f),

$$f(x_1) = f(x_2). \qquad (1)$$

By solving Eq. (1), we have

$$x_1^2 = x_2^2, \Rightarrow \; x_1^2 - x_2^2 = 0, \Rightarrow$$

$$(x_1 + x_2)(x_1 - x_2) = 0,$$

$$\Rightarrow \quad x_1 + x_2 = 0, \text{ or } x_1 - x_2 = 0,$$

$$\Rightarrow \quad x_1 = -x_2 \text{ or } x_1 = x_2. \qquad (2)$$

From Eq. (2), this does not lead to a unique solution of $x_1 = x_2$. Hence, the given function is not one-to-one by definition.

Remark

If the domain of the given function is restricted to Domain(f) = $[0, +\infty)$, it can be shown that the given function is one-to-one on $[0, +\infty)$. Also, the inverse function f^{-1} is given by $y = f^{-1}(x) = \sqrt{x}$, where Domain(f^{-1}) = $[0, +\infty)$.

Example 13.4.2

Given: $f(x) = 3x + 4$, where Domain(f) = $(-\infty, +\infty)$.

(a) Show that the given function is one-to-one.

(b) Find $f^{-1}(x) = ?$.

Solution

(a) Suppose that for any $x_1, x_2 \in$ Domain(f),

$$f(x_1) = f(x_2). \qquad (1)$$

By solving Eq. (1), we have

$$3x_1 + 4 = 3x_2 + 4, \Rightarrow \; 3x_1 + 4 - 4 = 3x_2 + 4 - 4,$$

$$3x_1 = 3x_2, \Rightarrow \; x_1 = x_2. \qquad (2)$$

We begin with Eq. (1). It leads to Eq. (2). Hence, the given function is one-to-one by Definition 13.4.1.

(b) To find the inverse function f^{-1}, we let

$$y = f(x), \text{ or } y = 3x + 4. \qquad (3)$$

By viewing Eq. (3) as an equation in two variables x and y, we solve for x in terms of y as follows:

$$y - 4 = 3x + 4 - 4 = 3x, \Rightarrow \; \frac{y-4}{3} = x, \Rightarrow$$

$$x = f^{-1}(y) = \frac{1}{3}y - \frac{4}{3}. \qquad (4)$$

Finally, replacing x by y and y by x in Eq. (4), we have

$$y = f^{-1}(x) = \frac{1}{3}x - \frac{4}{3}, \qquad (5)$$

which is the desired inverse function.

Remark

We can always check if the inverse function $f^{-1}(x)$ given by Eq. (5) is correct by checking if Eqs. (13.4.1 –2) hold or not. Usually, just checking one of the equations (or Eq. (13.4.1)) is sufficient. We proceed our double check as follows:

$$(f \circ f^{-1})(x) = f(f^{-1}(x)) = 3 \cdot f^{-1}(x) + 4$$
$$= 3 \cdot (\frac{1}{3}x - \frac{4}{3}) + 4 = 3 \cdot \frac{1}{3}x - 3 \cdot \frac{4}{3} + 4 = x - 4 + 4 = x.$$

Since Eq. (13.4.1) is satisfied, we are assured that Eq. (5) is the correct inverse function.

Exercises

1. Show that the inverse function of the identity function $i(x) = x$ is itself. Hint: Show that $i \circ i = i = i \circ i$.

In problems 2–8, (a) determine whether the given function is one-to-one. (b) If yes, find its inverse function with its domain.

2. $f(x) = 2x + 5$, where Domain(f) = $(-\infty, +\infty)$.
3. $f(x) = x^2 - 3$, where Domain(f) = $(-\infty, +\infty)$.
4. $f(x) = x^2 - 3$, where Domain(f) = $[0, +\infty)$.
5. $f(x) = x^2 - 3$, where Domain(f) = $(-\infty, 0]$.
6. $f(x) = x^3 + 1$, where Domain(f) = $(-\infty, +\infty)$.
7. $f(x) = \sqrt{x-2}$, where Domain(f) = $[2, +\infty)$.
8. $f(x) = \dfrac{1}{x+3}$, where Domain(f) = $(-\infty, -3) \cup (-3, +\infty)$.

In problems 9–10, use the functions $f(x) = \sqrt{x-2}$ and $g(x) = 2x + 5$ to find the indicated value.

9. $(f^{-1} \circ g^{-1})(3) = ?$

10. $(g^{-1} \circ f^{-1})(-2) = ?$

13.5 References

1. Anton, H. (1988). Calculus with Analytic Geometry, 3rd edition, Wiley, New York.
2. Bell, E. T. (1940). The Development of Mathematics, Dover Publications, Inc., New York.
3. Flanigan, F. (1972). Complex Variables, Dover Publications, Inc., New York.
4. Stewart, J. (2007). Multivariable Calculus, 6th edition, Cengage Learning, Boston, MA.
5. http://en.wikipedia.org/wiki/History_of_the_function_concept.
6. www.wolfram.com/Mathematica10.

Chapter 14
Polynomial Functions

Without the notion of function, we can merely explore slightly the polynomial expression in Chapter 2 in Part I. Now, we're ready to study in depth the polynomial expression from the concept of the function. First, the graph of a polynomial function is just to graph the solution set for the equation of $y - p_n(x) = 0$. Second, to find the zeroes of a polynomial function is equivalent to solving for the roots of a polynomial equation.

Historically, mathematicians began to focus their attention on solving a polynomial equation of degree greater than two, once the quadratic equation was more or less solved. Beginning from the 13th century and ending in the 16th century, two different approaches were taken respectively by Italian and Chinese mathematicians (Reference 4). On the one hand, Italian schools mainly concentrate on finding solution by radical, that is, the closed-form formula expressed in terms of the radical root of the coefficients in the polynomial equations. It led to two important formula: one is Tartaglia-Cardano's formula (Theorem 14.5.4) for cubic equations and Ferrari's formula (Theorem 14.5.5) for quartic equations. On the other hand, Chinese schools proposed the Tien-Yuan Shu (天元术) in the 13th century to get the solution by approximation (Reference 8, pp. 258-271).

It is worthy to note that before N. H. Abel (1802-1829, Norwegian) proved in 1824 that it is impossible to find solution by radical for a polynomial equation of degree five or higher, P. Ruffini (1765-1822, Italian) had already attempted to prove the same proposition. But his proof was found to be incomplete. Also, E. Galois (1811-1832, French) proved independently the same thing that there is no quintic formula. Their proof demonstrates that the approach employed by Italian school is the dead end as far as finding a closed-form formula for quintic or higher equations is concerned (Reference 16). Yet, the Chinese approach was reinvented by W. G. Horner (1773-1827, English) as the Horner's method published in

1819 (Reference 17). In this respect, Chinese mathematics was ahead of the Western by more than half a millennium.

14.1 Linear Functions

Definition 14.1.1

A function $p_n(x) = a_n x^n + a_{n-1} x^{n-1} + ... + a_1 x + a_0$ is called a polynomial function of x with degree n, where the leading coefficient $a_n \neq 0$.

Remarks

1. The graph of a polynomial function is just to graph the solution set for he equation of $y - p_n(x) = 0$.
2. Polynomial functions play a very important role in approximating a general nonlinear function through the technique of a Taylor's series expansion in calculus. Some special (orthogonal) polynomials like Legendre-, Hermite-, and Laguerre-polynomials are fundamental tools in solving differential equations arising from modeling practical problems in physics (Reference 2).

Definition 14.1.2

A polynomial function of x with degree 0, $p_0(x) = a_0$, is called a constant function.

Remark

The graph of a constant function is a horizontal line parallel to the x-axis.

Definition 14.1.3

A polynomial function of x with degree one, $p_1(x) = a_1 x + a_0$, is called a linear function of x, where $a_1 \neq 0$.

Remarks

1. Let $y = p_1(x) = a_1 x + a_0$. By viewing it as an equation, it is a linear equation in two variables x and y. From Chapter 9 in Part I, we learn that the graph of the solution set to a linear equation

in two variables x and y is a straight line. Hence, the graph of a linear function is also a line that usually intersects with the x-axis and the y-axis, respectively.

2. A linear function is widely used as a model for some economic laws (Reference 7). See Example 14.1.2 and problems 7 and 8 in Exercises for Section 14.1.

Definition 14.1.4

The point that is the intersection of a line and the x-axis (or the y-axis) is called the x-intercept (or the y-intercept) of the line.

Remark

To graph a linear function is simply to find the x- and y-intercept of the line and then connect these two points to get the desired line. To find the x-intercept, we need to set $y = 0$ in the equation of $y - p_1(x) = 0$ and solve for x. So, the x-intercept is a root of a linear equation. To find the y-intercept, we need to set $x = 0$ in the equation of $y - p_1(x) = 0$ and solve for y.

Example 14.1.1

Given: $p_1(x) = 2x + 5$.

(a) Find the x-intercept and the y-intercept for the graph of $p_1(x)$.

(b) Graph $p_1(x)$.

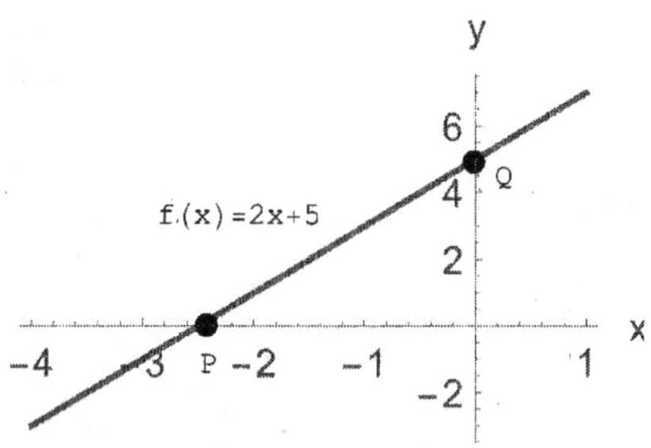

Figure 14.1.1

Solution

(a) Let

$$y = p_1(x) = 2x + 5. \qquad (1)$$

By setting $y = 0$ in Eq. (1), we have

$$0 = 2x + 5, \Rightarrow -2x = 5, \Rightarrow \frac{-2x}{-2} = \frac{5}{-2}, \Rightarrow x = -\frac{5}{2}.$$

Therefore, the x-intercept is $P(-\frac{5}{2}, 0)$.

By setting $x = 0$ in Eq, (1), we have $y = 2 \cdot 0 + 5 = 5$. Therefore, the y-intercept is $Q(0,5)$.

(b) By connecting the two points $P(-\frac{5}{2}, 0)$ and $Q(0, 5)$, the graph of $p_1(x)$ is given in Figure 14.1.1.

Example 14.1.2 (The Demand Curve)

In economics, it is commonly observed that the consumer's buying a good depends on its price. The higher the price, the less the quantity is demanded. On the contrary, given other things being equal, the lower the price, the greater the demand. Evidently, there exists at any one time a definite relation, called the demand curve between the market price of a good and the quantity demanded of that good. The simplest model for the demand curve is to assume that the relation between the market price and the quantity demanded is a linear function. Suppose that when the market price for wheat is \$5 per bushel, the quantity demanded in the market for wheat is 9 million bushels per month. If the market price for wheat drops to \$3 per bushel, the quantity demanded for wheat is 12 million bushels per month. Assume that the relation between the market price and the quantity demanded for wheat is linear.

(i) Find linear model for the demand curve.

(ii) Using the linear model obtained in part (i), predict the quantity demanded per month for wheat when the market price drops to \$2 per bushel.

Solution

(i) Let p denote the market price and q the quantity demanded for wheat. Since the relation between the market price and the quantity demanded is linear, there exists two unknown numbers a and b such that

$$q = ap + b, \qquad (1)$$

where a and b are to be determined. By using the given information, we have

$$q = 9 \text{ million bushels when } p = \$5, \qquad (2)$$

and

$$q = 12 \text{ million bushels when } p = \$3, \qquad (3)$$

After substituting both of Eqs. (2) and (3) into Eq. (1), we have

$$9 = a \cdot 5 + b = 5a + b, \qquad (4)$$

$$12 = a \cdot 3 + b = 3a + b, \qquad (5)$$

Eqs. (4) and (5) constitute two linear equations in two unknowns a and b. By using the method of elimination described in Section 9.4 of Part I, we obtain by subtracting Eq. (4) from Eq. (5)

$$3 = -2a, \Rightarrow \frac{3}{-2} = \frac{-2a}{-2}, \Rightarrow a = -\frac{3}{2} = -1.5. \qquad (6)$$

By substituting Eq. (6) back into Eq. (4), we have

$$9 = 5 \cdot (-1.5) + b = -7.5 + b, \Rightarrow$$

$$b = 9 + 7.5 = 16.5 \qquad (7)$$

From Eqs. (6) and (7), the linear model for the demand curve is then given by

$$q = -1.5p + 16.5. \qquad (8)$$

(ii) By substituting $p = 2$ into Eq. (8), we have

$$q = -1.5 \cdot 2 + 16.5 = -3 + 16.5 = 13.5.$$

Hence, the quantity demanded for wheat is 13.5 million bushels per month when the market price for wheat is \$2 per bushel.

Remark

Eq. (8) is a decreasing function of p that is true in general for a demand curve. The demand curve is examined from the perspective of the consumer. However, from the perspective of the producer, the relation between the market price and the quantity supplied, called the supply curve, is in general an increasing function of p (See problem 7 in Exercises for Section 14.1).

Exercises

In problems 1–6, (a) find the x-intercept and the y-intercept for the graph of the given function and (b) graph the given function.

1. $f(x) = 3$. **2.** $f(x) = 2x$.
3. $f(x) = 3x + 4$. **4.** $f(x) = 3x - 4$.
5. $f(x) = -4x + 3$. **6.** $f(x) = -4x - 3$.

7. (The Supply Curve) In contrast to the consumer, the producer of a good would be happy to produce more goods if the price of a good is high. Hence, from the producer's perspective, the relation between the market price and the quantity produced, called the supply curve, is an increasing function of the price. Suppose that when the market price for wheat is \$5 per bushel, the quantity supplied in the market is 12 million bushels per month. If the market price for wheat drops to \$3 per bushel, the quantity supplied in 9 million bushels per month. Assume that the supply curve is linear.
(i) Find a linear model for the supply curve.
(ii) Using the linear model obtained in part (i), predict the quantity supplied for wheat when the market price is raised up to \$6 per bushel.

8. (The Equilibrium Price) When the quantity demand equals the quantity supplied, the corresponding price is called the equilibrium price for a good. The supply and the demand for a good in the perfectly competitive market is not stabilized until the market price reaches its equilibrium one. By using the demand curve of Example 14.1.2 and the supply curve of problem 7 in this exercise, find its equilibrium market price for wheat.

14.2 Quadratic Functions

Definition 14.2.1
A polynomial function of x with degree two, $p_2(x) = a_2x^2 + a_1x + a_0$, is called a quadratic function of x, when $a_2 \neq 0$.

Remarks
1. If we let $y = p_2(x) = a_2x^2 + a_1x + a_0$ and viewing it as an equation, then it is a quadratic equation in two variables x and y. Since the term of y^2 is absent in the equation, we learn from Section 12.2 that the graph of the solution set to this type of quadratic equation in two

variables x and y is a parabola. Hence, the graph of any quadratic function is a parabola.

2. Instead of writing a quadratic function as $p_2(x) = a_2x^2 + a_1x + a_0$, we often use $f(x) = ax^2 + bx + c$, where $a \neq 0$, to represent a general quadratic function.

3. A quadratic function always has either an absolute maximum or an absolute minimum over its entire domain. This property is useful in practical applications.

Definition 14.2.2

A function $f(x)$ is said to have an absolute (or global) maximum (or minimum) at $x = x_0$ (or $x = x_1$) if for all $x \in$ Domain(f) $f(x_0) \geq f(x)$ (or $f(x_1) \leq f(x)$).

Remarks

1. This definition applies to any nonlinear function, not limited to a quadratic function of interest to us in this section.
2. Whether a quadratic function has an absolute maximum or minimum depends on the sign of the leading coefficient a of $f(x) = ax^2 + bx + c$. It is easily shown that $f(x)$ has an absolute minimum at $x = -\dfrac{b}{2a}$ if $a > 0$, whereas $f(x)$ has an absolute maximum at $x = -\dfrac{b}{2a}$ if $a < 0$.
3. For quadratic functions, the absolute extrema coincides with the local (or relative) ones (Reference 1). For other polynomial functions of degree three of greater, local extrema exists, whereas absolute extrema may not exist (see Figure 14.3.1).

A General Procedure for Graphing a Quadratic Function

Given: $f(x) = ax^2 + bx + c$, where $a \neq 0$. To graph $f(x)$, we proceed as follows:

Step 1. By completing the square in x, we have

$$f(x) = a(x^2 + \frac{b}{a}x + \frac{c}{a})$$

$$= a[x^2 + \frac{b}{a}x + (\frac{b}{2a})^2 - (\frac{b}{2a})^2 + \frac{c}{a}]$$

$$= a[(x+\frac{b}{2a})^2 - \frac{b^2}{4a^2} + \frac{c}{a}]$$

$$= a(x+\frac{b}{2a})^2 + a\cdot\frac{4ac-b^2}{4a^2}.$$

$$= a(x+\frac{b}{2a})^2 + \frac{4ac-b^2}{4a}. \quad (14.2.1)$$

Sep 2. By setting the expression inside the square term equal to zero, we have

$$x+\frac{b}{2a}=0, \Rightarrow x=-\frac{b}{2a}, \quad (14.2.2)$$

which is the line of symmetry (or the axis of the parabola) in the sense that the parabola is symmetric with respect to the line of Eq. (14.2.2). By substituting Eq. (14.2.2) back into Eq. (14.2.1), we have

$$f(-\frac{b}{2a}) = a\cdot(-\frac{b}{2a}+\frac{b}{2a})^2 + \frac{4ac-b^2}{4a} = \frac{4ac-b^2}{4a}$$

Therefore, the vertex of the parabola is

$$V(-\frac{b}{2a}, \frac{4ac-b^2}{4a}). \quad (14.2.3)$$

Step 3. To plot a parabola, we need a minimum number of three points. We already have the vertex V in Step 2. All we need are two extra points. By taking advantage of the symmetric property of a parabola with respect to the line of symmetry, we substitute $x=-\frac{b}{2a}\pm 1$ into Eq. (14.2.1) and obtain

$$f(-\frac{b}{2a}\pm 1) = a\cdot(-\frac{b}{2a}\pm 1+\frac{b}{2a}) + \frac{4ac-b^2}{4a}$$

$$= a\cdot(\pm 1)^2 + \frac{4ac-b^2}{4a} = a + \frac{4ac-b^2}{4a}.$$

Therefore, we have two other points $P(-\frac{b}{2a}+1, a+\frac{4ac-b^2}{4a})$ and $Q(-\frac{b}{2a}-1, a+\frac{4ac-b^2}{4a})$.

Step 4. By connecting the three points $V(-\frac{b}{2a}, \frac{4ac-b^2}{4a})$,

$P(-\frac{b}{2a}+1, a+\frac{4ac-b^2}{4a})$ and $Q(-\frac{b}{2a}-1, a+\frac{4ac-b^2}{4a})$

smoothly, we get the desired graph of $f(x) = ax^2 + bx + c$.

Remarks

1. We do not need to memorize the formula given in Eqs. (14.2.1-2). We just do it again each time by completing the square in x, and then follow the ideas described in Steps 2 and 3.
2. If $b^2 - 4ac > 0$, then sometimes we use two x-intercepts $P(\frac{-b+\sqrt{b^2-4ac}}{2a},0)$ and $Q(\frac{-b-\sqrt{b^2-4ac}}{2a},0)$ plus the vertex to plot the graph of a parabola.
3. If we let $h = -\frac{b}{2a}$ and $k = \frac{4ac-b^2}{4a}$, the vertex of Eq. (14.2.3) reduces to $V(h, k)$. Further, Eq. (14.2.1) is simplified as
$$f(x) = a \cdot (x-h)^2 + k, \qquad (14.2.4)$$
which is called the standard form of a quadratic function.

Example 14.2.1

Given: $f(x) = 2x^2 + 4x + 1$.

(a) Find the vertex and the line of symmetry for the given function.

(b) Graph $f(x)$.

Solution

(a) By completing the square in x, we have
$$f(x) = 2(x^2 + 2x) + 1$$
$$= 2[x^2 + 2x + (\frac{2}{2})^2 - (\frac{2}{2})^2] + 1 = 2 \cdot [(x+1)^2 - 1] + 1$$
$$= 2(x+1)^2 - 2 + 1 = 2(x+1)^2 - 1. \qquad (1)$$
By setting $x + 1 = 0$ in Eq. (1), $x = -1$ is the line of symmetry. By evaluating $f(x)$ of Eq. (1) at $x = -1$, we have
$$f(-1) = 2 \cdot (-1+1)^2 - 1 = 2 \cdot 0^2 - 1 = -1.$$
Therefore, the vertex of the parabola is $V(-1, -1)$.

(b) Since we already have the vertex $V(-1, -1)$ of the parabola from part (a), all we need is two other points. Let us evaluate $f(x)$ at $x = 0$ and $x = -2$ that are one unit to the left and one unit to the right from the line of symmetry as follows:

x	$y = f(x) = 2(x+1)^2 - 1$
-2	$y = 2 \cdot (-2+1)^2 - 1 = 2 \cdot 1 - 1 = 1 \Rightarrow P(-2, 1)$.
0	$y = 2 \cdot (0+1)^2 - 1 = 2 \cdot 1 - 1 = 1 \Rightarrow Q(0, 1)$.

Also, from the given function of $f(x) = 2x^2 + 4x + 1$, we have that $a = 2$, $b = 4$, and $c = 1$. Since $b^2 - 4ac = 4^2 - 4 \cdot 2 \cdot 1 = 16 - 8 = 8 > 0$, the two x-intercepts are given by $R(\frac{-4-\sqrt{8}}{2\cdot 2},0) = R(-1-\frac{\sqrt{2}}{2},0)$ and $S(\frac{-4+\sqrt{8}}{2\cdot 2},0) = S(-1+\frac{\sqrt{2}}{2},0)$. By connecting the five points $V(-1,-1)$, $P(-2, 1)$, $Q(0, 1)$, $R(-1-\frac{\sqrt{2}}{2},0)$ and $S(-1+\frac{\sqrt{2}}{2},0)$ smoothly, the graph of $f(x)$ is given in Figure 14.2.1.

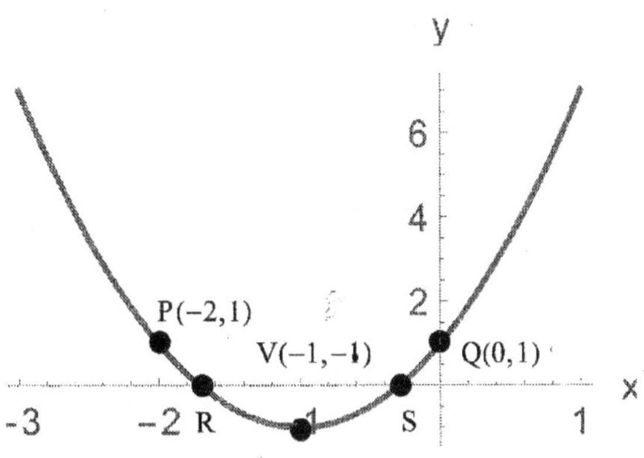

Figure 14.2.1

Remarks

1. From Figure 14.2.1, it is seen that $f(x)$ has an (absolute) minimum at $x = -1$, i.e., the function value of $f(x)$ at $x = -1$ is the smallest one.
2. This example also confirms what was observed in Remark 2 of Definition 14.2.2, because $a = 2 > 0$.

Example 14.2.2

Given: $f(x) = -2x^2 + 4x + 1$.

(a) Find the vertex and the line of symmetry.

(b) Graph $f(x)$.

Solution

(a) By completing the square in x, we have

$$f(x) = -2(x^2 - 2x) + 1$$

$$= -2[x^2 - 2x + (\frac{-2}{2})^2 - (\frac{-2}{2})^2] + 1$$
$$= -2[(x-1)^2 - 1] + 1 = -2(x-1)^2 + 2 + 1$$
$$= -2(x-1)^2 + 3. \qquad (1)$$

By setting $x - 1 = 0$ in Eq. (1), we obtain that $x = 1$ is the line of symmetry. By evaluating $f(x)$ of Eq. (1) at $x = 1$, we have

$$f(1) = -2(1-1)^2 + 3 = -2 \cdot 0^2 + 3 = 3.$$

Therefore, the vertex of the parabola is $V(1, 3)$.

(b) Since we already have the vertex $V(1, 3)$ of the parabola from part (a), all we need is two other points. Let us evaluate $f(x)$ at $x = 0$ and $x = 2$ that are one unit to the left and one unit to the right from the line of the symmetry as follows:

x	$y = f(x) = -2(x-1)^2 + 3$
0	$f(0) = -2(0-1)^2 + 3 = -2 + 3 = 1 \Rightarrow P(0, 1).$
2	$f(2) = -2(2-1)^2 + 3 = -2 + 3 = 1 \Rightarrow Q(2, 1).$

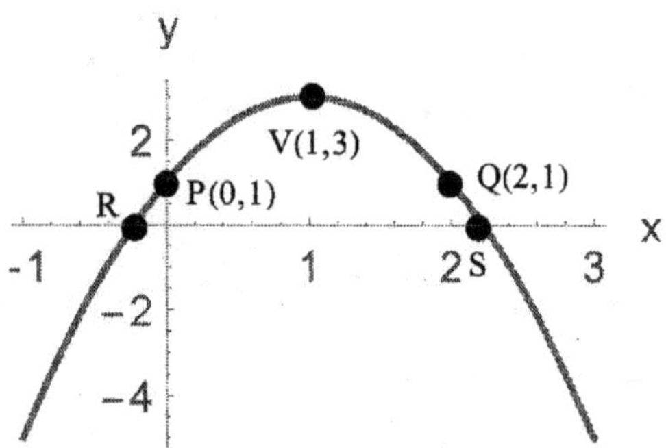

Figure 14.2.2

Also, from the given function of $f(x) = -2x^2 + 4x + 1$, we have that $a = -2$, $b = 4$, and $c = 1$. Since $b^2 - 4ac = 4^2 - 4 \cdot (-2) \cdot 1 = 16 + 8 = 24 > 0$, the two x-intercepts are given by $R(\frac{-4+\sqrt{24}}{2\cdot(-2)},0) = R(1-\frac{\sqrt{6}}{2},0)$ and $S(\frac{-4-\sqrt{24}}{2\cdot(-2)},0) = S(1+\frac{\sqrt{6}}{2},0)$. By connecting the five points $V(1, 3)$, $P(0, 1)$, $Q(2, 1)$, $R(1-\frac{\sqrt{6}}{2},0)$, and $S(1+\frac{\sqrt{6}}{2},0)$ smoothly, the graph of $f(x)$ is given in Figure 14.2.2.

Remarks

1. From Figure 14.2.2, it is seen that $f(x)$ has an (absolute) maximum at $x = 1$, i.e., the function value of $f(x)$ at $x = 1$ is the largest one.
2. This example also confirms what was observed in remark 2 of Definition 14.2.2, that is, the function has an absolute maximum at $x = 1$ because $a = -2 < 0$.

Example 14.2.3

Suppose that the average cost (AC) of producing x units daily of a good is given by a quadratic function $AC = f(x) = 1000 - 20x + 0.5x^2$, where $x \geq 0$. How many units should be produced each day so that the average cost is at a minimum? What is the minimum average cost?

Solution

To find the minimum average cost at a certain production level is by finding the absolute minimum of the given quadratic function. By completing the square in x, we have

$$\begin{aligned} f(x) &= 0.5(x^2 - 40x) + 1000 \\ &= 0.5[x^2 - 40x + (\frac{-40}{2})^2 - (\frac{-40}{2})^2] + 1000 \\ &= 0.5[(x-20)^2 - 400] + 1000 \\ &= 0.5(x-20)^2 - 0.5 \cdot 400 + 1000 \\ &= 0.5(x-20)^2 - 200 + 1000 \\ &= 0.5(x-20)^2 + 800. \end{aligned} \qquad (1)$$

From Eq. (1), it is easily seen that $f(x)$ has an absolute minimum of \$800 at $x = 20$ since $f(x) \geq 800$ for all $x \geq 0$. Hence, the production level of minimum average cost is to produce 20 units each day and the minimum average cost at this production level is \$800 per day.

Exercises

In problems 1–4, (a) determine if the given quadratic function has an absolute maximum or minimum, (b) find the vertex and the line of symmetry, and (c) graph of the given function.

1. $f(x) = -x^2 + 3.$ **2.** $f(x) = 2x^2 + 8x + 9.$

3. $f(x) = -2x^2 + 4x.$ **4.** $f(x) = (x + 3)^2 - 2.$

In problems 5–6, find the quadratic function that has the given vertex and whose graph passes through the given point.

5. Vertex: $V(-1, 3)$; point: $P(1, 1)$.
6. Vertex: $V(2, -1)$; point: $P(1, 3)$.

In problems 7–8, find the quadratic function that has the given x-intercepts and whose graph passes through the given point.

7. x-intercepts: $P_1(-2, 0)$ and $P_2(0, 0)$; point: $P_3(-1, -1)$.
8. x-intercepts: $P_1(2, 0)$ and $P_2(6,0)$; point: $P_3(4, -1)$.

9. Find two positive numbers such that the sum of the first and three times the second is 12 and their product is a maximum.

10. Suppose that the furniture manufacturer has found that the profit (P) from sales of x sofas per month (in dollars) is $P = -2x^2 + 200x$. How many sofas need to be produced per month to maximize the profit? What is the maximum profit per month?

14.3 Polynomial Functions of Degree Three or Higher

To graph effectively a polynomial function of degree three or higher requires to identify the local maximum or minimum (Figure 14.3.1), which needs the knowledge of calculus. At this level, we can only use the x-intercepts in aiding the graph of the given polynomial function.

Definition 14.3.1
A real number x_0 is called a zero of a polynomial function $p_n(x)$ of degree n in x if $p_n(x_0) = 0$.

Remarks

1. To find the zeroes of a polynomial function $p_n(x)$ is equivalent to solving for the roots of a polynomial equation $p_n(x) = 0$; hence, the words "zero" and "root" are synonym.

2. A point $P(x_0, 0)$ is exactly a x-intercept of the graph of $p_n(x)$, provided that x_0 is a zero of $p_n(x)$ (Figure 14.3.1).
3. If x_0 is a zero of a polynomial function $p_n(x)$, then $(x - x_0)$ is a factor of $p_n(x)$, namely, there exists a polynomial function $q_{n-1}(x)$ of degree $n - 1$ such that $p_n(x) = (x - x_0) \cdot q_{n-1}(x)$.

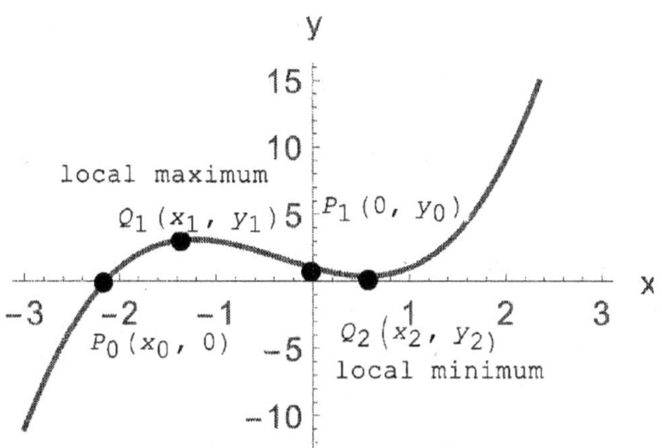

Figure 14.3.1

Definition 14.3.2
A function $f(x)$ is said to have a local (or relative) maximum at $x = x_0$ if $f(x)$ is increasing on its left-half neighborhood $(x_0 - \delta, x_0)$ and decreasing on its right-half neighborhood $(x_0, x_0 + \delta)$, where $\delta > 0$ is a small positive real number.

Definition 14.3.3
A function $f(x)$ is said to have a local minimum at $x = x_1$ if $f(x)$ is decreasing on $(x_1 - \delta, x_1)$ and increasing on $(x_1, x_1 + \delta)$, where $\delta > 0$ is a small positive real number.

Definition 14.3.4
A function $f(x)$ is said to have a turning point $P(x_2, f(x_2))$ at $x = x_2$ if $f(x)$ has either a local maximum or minimum at $x = x_2$.

Definition 14.3.5
A real number x_0 is called a zero of multiplicity r of $p_n(x)$ if $(x-x_0)^r$ is a factor of $p_n(x)$, but $(x-x_0)^{r+1}$ is not a factor of $p_n(x)$, where r is a positive integer

Remarks
1. A zero of multiplicity 1 is also called a simple root for a polynomial equation $p_n(x) = 0$. A zero of multiplicity 2 is also called a double root for the given polynomial equation $p_n(x) = 0$.
2. The x-intercept corresponding to a zero of even multiplicity will be a turning point for the graph of the given polynomial function. Otherwise, the graph of $p_n(x)$ simply crosses the x-axis through this x-intercept.

A General Procedure for Graphing a Polynomial Function of Degree Three or Higher
Given a polynomial function $p_n(x)$, where $n \geq 3$.

Step 1. Solve the polynomial equation $p_n(x) = 0$ for the roots.

Step 2. Use the multiplicity of the root x_0 to decide whether $p_n(x)$ has a turning point at $x = x_0$.

Step 3. Select the midpoints between two successive x-intercepts and evaluate the function at these midpoints.

Step 4. Find the y-intercept and evaluate the function at two extra values of x in each direction of the x-axis beyond the right-most and the left-most x-intercepts.

Step 5. Connecting all these points smoothly gives us the desired graph of the given polynomial function.

Remark
More precisely, we need techniques in calculus to identify the location of local extrema (maximum/minimum) (Reference 1).

Example 14.3.1
Given: $p_3(x) = x^3 + x^2 - 2x$.

(a) Find all the x-intercepts of $p_3(x)$.

(b) Graph $p_3(x)$.

Solution

(a) To find the x-intercepts of $p_3(x)$, we solve the following equation for the roots:

$$p_3(x) = 0, \Rightarrow x^3 + x^2 - 2x = 0, \Rightarrow x\cdot(x^2 + x - 2) = 0,$$
$$x\cdot(x - 1)\cdot(x + 2) = 0, \Rightarrow x = 0,\ x - 1 = 0, \text{ or } x + 2 = 0, \Rightarrow$$
$$x = 0, x = 1, \text{ or } x = -2.$$

Hence, the x-intercepts are $P_1(-2, 0)$, $P_2(0, 0)$, and $P_3(1, 0)$.

(b) To graph the given function, we evaluate the function values at the midpoints between the successive x-intercepts, the y-intercept, and two extra values of x, i.e., $x = -3$ and $x = 2$:

x	$y = p_3(x) = x^3 + x^2 - 2x$
-3	$p_3(-3) = (-3)^3 + (-3)^2 - 2\cdot(-3) = -27 + 9 + 6 = -12 \Rightarrow Q_1(-3, -12)$.
-1	$p_3(-1) = (-1)^3 + (-1)^2 - 2\cdot(-1) = -1 + 1 + 2 \Rightarrow Q_2(-1, 2)$.
$\frac{1}{2}$	$p_3(\frac{1}{2}) = (\frac{1}{2})^3 + (\frac{1}{2})^2 - 2\cdot\frac{1}{2} = \frac{1}{8} + \frac{1}{4} - 1 = -\frac{5}{8} \Rightarrow Q_3(\frac{1}{2}, -\frac{5}{8})$.
2	$p_3(2) = 2^3 + 2^2 - 2\cdot 2 = 8 + 4 - 4 = 8 \Rightarrow Q_4(2, 8)$.

By connecting the seven points $Q_1(-3, -12)$, $P_1(-2, 0)$, $Q_2(-1, 2)$, $P_2(0, 0)$, $Q_3(\frac{1}{2}, -\frac{5}{8})$, $P_3(1, 0)$, and $Q_4(2, 8)$, the graph of $p_3(x)$ is given in Figure 14.3.2.

Remark

The exact location of the local maximum and local minimum of $p_3(x)$ can be shown to be located at $x = \frac{-1-\sqrt{7}}{3}$ and $x = \frac{-1+\sqrt{7}}{3}$, respectively by using the technique in calculus.

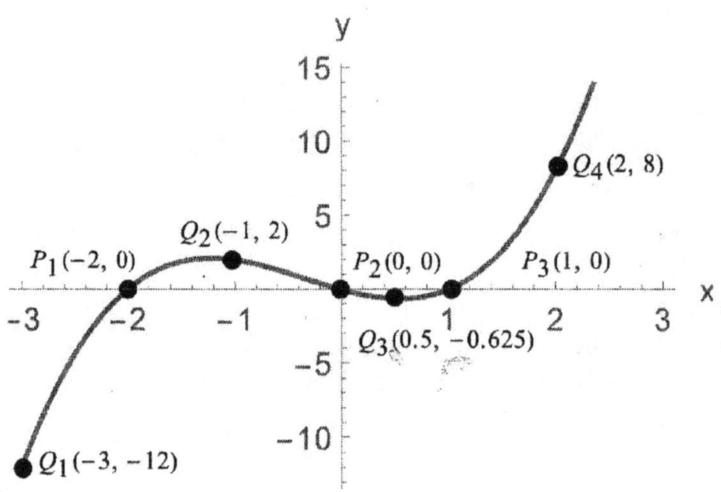

Figure 14.3.2

Example 14.3.2

Given: $p_4(x) = x^2(x-2)(x+2)$.

(a) Find the zeroes with its multiplicity of $p_4(x)$.

(b) Graph $p_4(x)$.

Solution

(a) By setting $p_4(x) = 0$, we have

$$x^2 \cdot (x-2) \cdot (x+2) = 0, \Rightarrow$$
$$x^2 = 0,\ x-2 = 0, \text{ or } x+2 = 0, \Rightarrow$$
$$x = 0, 0,\ x = 2, \text{ or } x = -2.$$

Hence, the zeroes of $p_4(x)$ are {0 with a multiplicity 2 (a double root), both – 2 and 2 with a multiplicity 1 (a simple root)}.

(b) From the result of part (a), the x-intercepts of the graph of $p_4(x)$ are $P_1(-2, 0)$, $P_2(0, 0)$, and $P_3(2, 0)$, where the graph has a turning point at $P_2(0,0)$. To graph $p_4(x)$, we evaluate $p_4(x)$ at the midpoints between the successive x-intercepts and two x-values beyond the x-intercepts in each direction of the x-axis as follows:

x	$y = p_4(x) = x^2 \cdot (x-2) \cdot (x+2)$
–3	$p_4(-3) = (-3)^2 \cdot (-3-2) \cdot (-3+2) = 9 \cdot (-5) \cdot (-1) = 45$ $\Rightarrow Q_1(-3, 45)$.

−1	$p_4(-1) = (-1)^2 \cdot (-1-2) \cdot (-1+2) = 1 \cdot (-3) \cdot 1 = -3 \Rightarrow Q_2(-1, -3)$.
1	$p_4(1) = 1^2 \cdot (1-2) \cdot (1+2) = 1 \cdot (-1) \cdot 3 = -3 \Rightarrow Q_3(1, -3)$.
3	$p_4(3) = 3^2 \cdot (3-2) \cdot (3+2) = 9 \cdot 1 \cdot 5 = 45 \Rightarrow Q_4(3, 45)$.

By connecting the seven points $Q_1(-3, 45)$, $P_1(-2, 0)$, $Q_2(-1, -3)$, $P_2(0, 0)$, $Q_3(1, -3)$, $P_3(2, 0)$, and $Q_4(3, 45)$ smoothly, the graph of $p_4(x)$ is given in Figure 14.3.3.

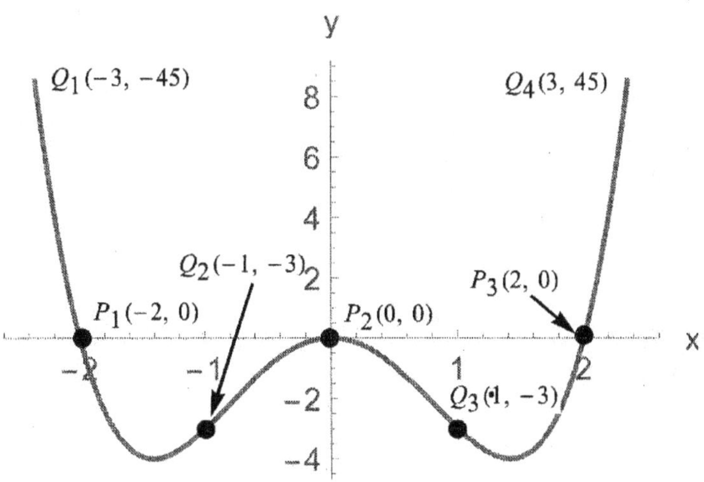

Figure 14.3.3

Remark

Since $p_4(x) = 0$ has a double root at $x = 0$, $p_4(x)$ has a turning point at $x = 0$, which turns out to be a local maximum. It can also be shown by using a technique from calculus that $p_4(x)$ has a local minimum at $x = \pm\sqrt{2}$.

Exercises

In problems 1–5, (a) find the real zeroes of the given function with its multiplicity, (b) find the x-intercepts and the y-intercept for the graph, and (c) graph the given function.

1. $p_3(x) = x^3 - 4x$.
2. $p_3(x) = x^3 - 8$.
3. $p_4(x) = x^4 + 8x$.
4. $p_5(x) = (x + 2)^2 \cdot (x - 1)^3$.
5. $p_6(x) = (x + 1)(x - 1)^2(x - 2)^3$.

14.4 The Remainder and Factor Theorems with Synthetic Division

To graph a polynomial function $p_n(x)$, where $n \geq 3$. At this level, we rely on finding the x-intercepts of the graph of $p_n(x)$ that is equivalent to solving $p_n(x) = 0$ for the roots. In fact, this is the central issue in this chapter: "How to solve $p_n(x) = 0$ for the unknown x, where $n \geq 3$?" We have to be aware of that it is not trivial to address this question.

In Section 14.3, we just gave some simple examples in the sense that these equations are easily being factored or already in the factor forms. In general, it is not easy to factor a polynomial function of degree three or greater into the product of several linear factors. We need some tools to aid us in accomplishing the factoring task. These tools are covered in both of this section and the next section. In this section, a primary tool is the factor theorem. Before we can describe what the factor theorem is, we need to know the remainder theorem.

Just like in the number system, any integer that is divided by a nonzero integer will end up with a quotient and a remainder. For example, 53 divided by 8 equals 6 plus a remainder of 5, i.e., $\frac{53}{8} = 6 + \frac{5}{8}$, or $53 = 6 \cdot 8 + 5$. The remainder theorem describes a similar property for polynomial functions.

Theorem 14.4.1 (The Quotient and Remainder for Polynomial Division)

For any polynomial function $p_n(x)$ of degree $n \geq 1$, and each real number x_0, there exists a polynomial function $q_{n-1}(x)$, called the quotient, of degree $n - 1$ such that

$$p_n(x) = (x - x_0) \cdot q_{n-1}(x) + R, \qquad (14.4.1a)$$

where R, called the remainder, is given by

$$R = p_n(x_0). \qquad (14.4.1b)$$

Remarks

1. The proof of (14.4.1a) is given as follows. By applying the division rule for polynomials in Section 14.2, there exists a quotient polynomial $q_{n-1}(x)$ of degree $n - 1$ and a remainder R such that

$$\frac{p_n(x)}{x - x_0} = q_{n-1}(x) + \frac{R}{x - x_0}. \qquad (14.4.2)$$

By multiplying both sides of Eq. (14.4.2) with a factor of $x - x_0$, this leads to Eq. (14.4.1a).

2. The proof of Eq. (14.4.1b) follows by evaluating $p_n(x)$ of Eq. (14.4.1a) at $x = x_0$.
3. If it happens that $R = 0$, then Eq. (14.4.1a) reduces to

$$p_n(x) = (x - x_0) \cdot q_{n-1}(x). \qquad (14.4.3)$$

From Eq. (14.4.3), $p_n(x)$ is the product of a linear factor $x - x_0$ and a quotient of $q_{n-1}(x)$. This observation leads to the following factor theorem.

Theorem 14.4.2 (Finding a Linear Factor)

Given any polynomial function $p_n(x)$, where $n \geq 2$, and each real number x_0, $x - x_0$ is a factor of $p_n(x)$ if and only if the remainder R of Eq. (14.4.1b) is zero.

Proof

To prove the factor theorem, the "if" part follows from remark 3 of Theorem 14.4.1. So, it suffices to prove the "only if" part, i.e., to prove that if $x - x_0$ is a factor of $p_n(x)$, the remainder R of Eq. (14.4.1b) is zero. Since it is known that $x - x_0$ is a factor of $p_n(x)$, this implies that there must exist another polynomial function $Q(x)$ such that

$$p_n(x) = (x - x_0) \cdot Q(x) . \qquad (14.4.4)$$

By evaluating $p_n(x)$ of Eq. (14.4.4) at $x = x_0$, we have

$$p_n(x_0) = (x_0 - x_0) \cdot Q(x_0) = 0 \cdot Q(x_0) = 0 . \qquad (14.4.5)$$

By combining Eqs. (14.4.1b) and (14.4.5), we have $R = p_n(x_0) = 0$. Hence, the proof is complete.

Remark

From this theorem, we realize that in order to find the zero of a polynomial function $p_n(x)$, we need an effective algorithm in evaluating $p_n(x)$ at $x = x_0$ to see if $p_n(x_0) = 0$. An effective function evaluation is accomplished by the following synthetic division.

A Procedure for Implementing the Synthetic Division

Consider a polynomial function of degree n given by $p_n(x) = a_n x^n + a_{n-1} x^{n-1} + ... + a_1 x + a_0$, where $a_n \neq 0$. Let x_0 be any real number. To evaluate $p_n(x)$ at $x = x_0$, we proceed as follows:

Step 1. List x_0 and the coefficients $a_n, a_{n-1}, ..., a_1, a_0$ in the following format:

x_0	a_n	a_{n-1}	......	a_1	a_0

Step 2. Bring a_n down to the third row. Then multiply a_n by x_0 and add it to a_{n-1} as shown below:

x_0	a_n	a_{n-1}	......	a_1	a_0
		$a_n x_0$			
	a_n	$a_n x_0 + a_{n-1}$	......		

Step 3. Multiply the second number in the third row, $a_n x_0 + a_{n-1}$, by x_0 again and then add it to a_{n-2} as shown below:

x_0	a_n	a_{n-1}	a_{n-2}	$\cdots$	a_0
		$a_n x_0$	$(a_n x_0 + a_{n-1})x_0$		
	a_n	$a_n x_0 + a_{n-1}$	$(a_n x_0 + a_{n-1})x_0 + a_{n-2}$		

Step 4. Repeat a similar calculation as shown in Step 3 again and again until the last coefficient a_0 as shown below:

x_0	a_n	a_{n-1}	$\cdots$	a_0
		$a_n x_0$		$(...((a_n x_0 + a_{n-1})x_0 + ... + a_1)x_0$
	a_n	$a_n x_0 + a_{n-1}$		$(...((a_n x_0 + a_{n-1})x_0 + ... + a_1)x_0 + a_0$

Step 5. The last number in the third row is the remainder R of Eq. (14.4.1b) or the function value of $p_n(x_0)$.

Remarks

1. The synthetic division can be viewed as putting the given polynomial function $p_n(x)$ in the nested form, i.e.,

$$p_n(x) = (a_n x^{n-1} + a_{n-1}x^{n-2} + ... + a_2 x + a_1)\cdot x + a_0$$
$$= [(a_n x^{n-2} + a_{n-1}x^{n-3} + ... + a_3 x + a_2)\cdot x + a_1]\cdot x + a_0$$
$$= \ldots$$
$$= (((((...(a_n x + a_{n-1})\cdot x + ... + a_2)\cdot x + a_1)\cdot x + a_0, \quad (14.4.6)$$

and then evaluate $p_n(x)$ of Eq. (14.4.5) at $x = x_0$ as follows:

$$p_n(x_0) = (((((...(a_n x_0 + a_{n-1})\cdot x_0 + a_{n-2})\cdot x_0 + ... + a_2)\cdot x_0 + a_1)\cdot x_0 + a_0. \quad (14.4.7)$$

To reorganize Eq. (14.4.7) into the tabular form, it ends up as shown in the form of synthetic division.

2. All the numbers before the last one in the third row of Step 4 are the coefficients of the quotient polynomial $q_{n-1}(x)$ of Eq. (14.4.1a) as can be shown below:

$$p_n(x) - p_n(x_0) = a_n x^n + a_{n-1}x^{n-1} + ... + a_1 x + a_0 - (a_n x_0^n + a_{n-1}x_0^{n-1} + ... + a_1 x_0 + a_0)$$
$$= a_n(x^n - x_0^n) + a_{n-1}(x^{n-1} - x_0^{n-1}) + ... + a_1(x - x_0)$$
$$= a_n(x - x_0)(x^{n-1} + x^{n-2}x_0 + ... + xx_0^{n-2} + x_0^{n-1}) + a_{n-1}(x - x_0)(x^{n-2} + x^{n-3}x_0 + ... + xx_0^{n-3} + x_0^{n-2}) + ... + a_1(x - x_0)$$
$$= (x - x_0)[a_n(x^{n-1} + x^{n-2}x_0 + ... + xx_0^{n-2} + x_0^{n-1}) + a_{n-1}(x^{n-2} + x^{n-3}x_0$$

$$+\ldots+xx_0^{n-3}+x_0^{n-2})+\ldots+a_1]$$
$$=(x-x_0)[a_nx^{n-1}+(a_nx_0+a_{n-1})x^{n-2}+(a_nx_0^2+a_{n-1}x_0+a_{n-2})x^{n-3}$$
$$+\ldots+a_nx_0^{n-1}+a_{n-1}x_0^{n-2}+\ldots+a_2x_0+a_1]$$
$$=(x-x_0)\cdot q_{n-1}(x). \qquad (14.4.8)$$

where $q_{n-1}(x)$ is given by

$$q_{n-1}(x)=a_nx^{n-1}+(a_nx_0+a_{n-1})x^{n-2}+\ldots+a_nx_0^{n-1}+\ldots+a_2x_0+a_1. \qquad (14.4.9)$$

By comparing Eq. (14.4.9) with the third row of the synthetic table in Step 4, the coefficients of $q_{n-1}(x)$ in Eq. (14.4.9) and the third row of the synthetic table in Step 4 are indeed the same.

3. In forming the table for an implementation of the synthetic division, the coefficients must be listed in the descending order of the degree in x. The coefficient of the missing term has to be listed as zero.

Example 14.4.1

Given: $p_3(x)=x^3+x^2-x-4$. Find $p_3(2)=?$ by using the synthetic division.

Solution

We form the table of the synthetic division as follows:

2	1	1	−1	−4
		2	6	10
	1	3	5	6

From the last entry of the third row in the above table, we have $p_3(2)=6$.

Example 14.4.2

Given that $p_4(x)=2x^4+5x^2-3x+8$ and $p_1(x)=x-3$. Use the synthetic division to find the quotient and the remainder when $p_4(x)$ is divided by $p_1(x)$.

Solution

In order to employ the synthetic division, we need to find the value of x_0. By setting $p_1(x)=0$, we have

$$x-3=0, \Rightarrow x=3.$$

Hence, $x_0 = 3$. Notice that the x^3-term is missing in $p_4(x)$. Thus, $p_4(x)$ is re-written as follows:

$$p_4(x) = 2x^4 + 5x^2 - 3x + 8.$$

Now, form the table of the synthetic division as follows:

3	2	0	5	–3	8
		6	18	69	198
	2	6	23	66	206

From the first four numbers of the third row in the above table, the quotient is given by

$$q_3(x) = 2x^3 + 6x^2 + 23x + 66,$$

From the last number of the third row in the above table, the remainder is given by

$$R = p_4(3) = 206.$$

Example 14.4.3

Given: $p_3(x) = x^3 - 6x^2 + 11x - 6$. Determine if $x = 2$ is a zero of $p_3(x)$.

Solution

By using the factor theorem, the function value of $p_3(2)$ will be zero if $x = 2$ is a zero of $p_3(x)$. We use the synthetic division to evaluate $p_3(2)$ as follows:

2	1	–6	11	–6
		2	–8	6
	1	–4	3	0

Since the last number of the third row in the above table is zero, we have $p_3(2) = 0$. This shows that $x = 2$ is a zero of $p_3(x)$.

Exercises

In problems 1–5, use the synthetic division to evaluate the indicated function value.

1. Given: $p_3(x) = -2x^3 - x^2 + x + 1$. Find: $p_3(-4) = ?$

2. Given: $p_4(x)=x^4+x^2-2$. Find: $p_4(3)=?$

3. Given: $p_5(x)=3x^5-x^3+x^2+2x$. Find: $p_5(-1)=?$

4. Given: $p_6(x)=-x^6+3x^5-2x^4+x^3-2x^2+1$. Find: $p_6(2)=?$

5. Given: $p_7(x)=x^7-x^6+x^5-2x^4+3x^2-5x+1$. Find: $p_7(-2)=?$

In problems 6–10, use the synthetic division to find the quotient and the remainder.

6. $p_3(x)=-2x^3-x^2+x+1$. is divided by $p_1(x)=x+4$.

7. $p_4(x)=x^4+x^2-2$ is divided by $p_1(x)=x-3$.

8. $p_5(x)=3x^5-x^3+x^2+2x$ is divided by $p_1(x)=x+1$.

9. $p_6(x)=-x^6+3x^5-2x^4+x^3-2x^2+1$ is divided by $p_1(x)=x-2$.

10. $p_7(x)=x^7-x^6+x^5-2x^4+3x^2-5x+1$ is divided by $p_1(x)=x+2$.

In problems 11–15, determine if the given x-value is a zero of the given function.

11. $p_3(x)=2x^3+3x^2-8x+3$ and $x_0=-3$.

12. $p_3(x)=x^3+x^2-3x-4$ and $x_0=1$.

13. $p_4(x)=x^4-5x^2+4$ and $x_0=-2$.

14. $p_4(x)=2x^4-8x^2+x+5$ and $x_0=2$.

15. $p_5(x)=x^5-5x^4+3x^3-15x^2-4x+20$ and $x_0=5$.

14.5 Real Zeroes of a Polynomial Function of Degree at Least Three

In this section we are ready to tackle the problem of finding all real zeroes of a polynomial function of degree at least three. It depends whether a polynomial function has any rational zero or not. If a polynomial function has a rational zero, a general strategy is to find this rational zero first, and use the synthetic division to find the quotient polynomial function having its degree one less than that of the original polynomial function. Then, find the zeroes of the quotient polynomial function. Repeat the process until the quotient polynomial function reduces to a quadratic function that is easily solved by using a quadratic formula in Section 6.3 in Part I. If a polynomial function does not have any rational zero, then we need to employ either specific closed form formula, Tartaglia-Cardano's formula for cubic equations (Theorem 14.5.4) and Ferrari's formula for quartic equations (Theorem 14.5.5), to solve the corresponding polynomial equation for the real (irrational) roots or a numerical bisection method to find an approximate real root (Example 14.5.9).

Now, let us suppose that a polynomial function has a rational zero. A question arises: How to find the first rational zero of a polynomial function? The answer relies on the following rational zero test.

Theorem 14.5.1 (The Rational Zero Test for Polynomial Functions with Integer Coefficients)
Given that a polynomial function $p_n(x)=a_nx^n+a_{n-1}x^{n-1}+...+a_1x+a_0$ of degree n in x with integer coefficients has a rational zero of the form $\frac{r}{s}$, where r and $s \neq 0$ are integers, and have no common factors except 1. Then r must be a factor of a_0 and s be a factor of a_n.

Proof

Since $\frac{r}{s}$ is a zero of $p_n(x)$, we have

$$p_n(\frac{r}{s})=0,$$

$$a_n(\frac{r}{s})^n+a_{n-1}(\frac{r}{s})^{n-1}+...+a_1(\frac{r}{s})+a_0=0,$$

$$a_n\cdot\frac{r^n}{s^n}+a_{n-1}\cdot\frac{r^{n-1}}{s^{n-1}}+...+a_1\cdot\frac{r}{s}+a_0=0,$$

$$a_nr^n+a_{n-1}r^{n-1}s+...+a_1rs^{n-1}+a_0s^n=0. \tag{14.5.1}$$

By moving either a_nr^n or a_0s^n to the right-side of the equation, we have from Eq. (14.5.1) the following:

$$s\cdot(a_{n-1}r^{n-1}+...+a_1rs^{n-2}+a_0s^{n-1})=-a_nr^n, \tag{14.5.2}$$

and

$$r\cdot(a_nr^{n-1}+a_{n-1}r^{n-2}s+...+a_1s^{n-1})=-a_0s^n, \tag{14.5.3}$$

From the left-side of Eq. (14.5.2), it implies that s is a factor of $a_n r^n$. However, since s ad r have no common factors, s must be a factor of a_n. By a similar argument applied to Eq. (14.5.3), r is indeed a factor of a_0. The proof is, therefore, complete.

Remarks

1. The rational zero test does not imply that every polynomial function with integer coefficients have the rational zeroes. All it says is that if a polynomial function with integer coefficients has a rational zero, this rational zero must have the stated property.
2. This theorem provides a useful tool to search for the rational zeroes. The idea is that we first form all possible rational zeroes as follows:

   Possible rational zeroes = (factors of constant term (a_0)) ÷ (factors of leading coefficient (a_n))

   Next, we use the synthetic division to test if a possible rational zero is indeed a zero of the given polynomial function by computing its remainder at this possible rational zero. If the remainder is zero, then this possible rational zero is proved to be a zero of the given polynomial function. Otherwise, this possible rational zero is not a zero of the given polynomial function.

Example 14.5.1

Given : $p_3(x) = 3x^3 - 4x^2 - 5x + 2$. Find them if $p_3(x)$ has any rational zeroes.

Solution

Since $p_3(x)$ has integer coefficients, the rational zero test (Theorem 14.5.1) applies to $p_3(x)$. First, we list all possible integer factors for $a_0 = 2$ and $a_3 = 3$ as follows:

$$r : \pm 1, \pm 2. \text{ and } s : \pm 1, \pm 3.$$

Next, we form all possible rational zeroes $\dfrac{r}{s}$ as follows:

$$\frac{r}{s} : \frac{\pm 1}{\pm 1}, \frac{\pm 2}{\pm 1}, \frac{\pm 1}{\pm 3}, \frac{\pm 2}{\pm 3},$$

or

$$\frac{r}{s}: \pm 1, \pm 2, \pm\frac{1}{3}, \pm\frac{2}{3}.$$

By using the synthetic division, we test the possible rational zeros one-by-one as follows:

1	3	– 4	– 5	2
		3	– 1	– 6
	3	– 1	– 6	– 4

Since the remainder is $-4 \neq 0$ that is the last entry in the third row of the above table, $x = 1$ is not a zero of $p_3(x)$.

– 1	3	– 4	– 5	2
		– 3	7	– 2
	3	– 7	2	0

Since the remainder is 0 that is the last entry in the third row of the above table, $x = -1$ is a zero of $p_3(x)$. Once we obtain one of the zeroes of $p_3(x)$, we are going to find the remaining zeroes of $p_3(x)$ by working with the quotient polynomial $q_2(x)$ that can be obtained from the first three entries in the third row of the above table, namely, $q_2(x) = 3x^2 - 7x + 2$. By setting $q_2(x) = 0$, we have

$$3x^2 - 7x + 2 = 0, \Rightarrow (3x - 1)(x - 2) = 0, \Rightarrow$$

$$3x - 1 = 0 \text{ or } x - 2 = 0, \Rightarrow x = \frac{1}{3} \text{ or } x = 2.$$

Hence, all three zeroes of $p_3(x)$ are rational zeroes given as $\{-1, 2, \frac{1}{3}\}$.

Remarks

1. Conventionally, we begin with testing the integer zeroes from the smallest one to the largest one and from the positive one to the negative one, and then testing the fractional numbers. Again, it is done from the smallest one to the largest and also from the positive one to the negative one. Once a zero is found, we immediately work with the corresponding quotient polynomial function

that has one degree less than the original polynomial function; Hence, it is easier to solve for the remaining zeroes.

2. In general, a polynomial function of degree n in x can have at most n rational zeroes. In fact, a polynomial function of degree n with real coefficients can have exactly n zeroes if complex zeroes are allowed. This is the "Fundamental Theorem of Algebra" that will be covered in Section 14.6.

Example 14.5.2

Given: $p_3(x)=x^3-x^2+x+1$. Find them if $p_3(x)$ has rational zeroes.

Solution

Since $p_3(x)$ has integer coefficients, the rational zero test of Theorem 14.5.1 applies to $p_3(x)$. First, we list all possible integer factors for $a_0=1$ and $a_3=1$ as follows:

$r:\pm 1$ and $s:\pm 1$.

Next, we form all possible rational zeroes $\frac{r}{s}$ as follows:

$$\frac{r}{s}:\frac{\pm 1}{\pm 1}=\pm 1$$

By using the synthetic division, we test the two possible rational zeroes one-by-one as follows:

$$\begin{array}{r|rrrr} 1 & 1 & -1 & 1 & 1 \\ & & 1 & 0 & 1 \\ \hline & 1 & 0 & 1 & 2 \end{array}$$

Since the remainder is 2 ($\neq 0$), $x = 1$ is not a zero of $p_3(x)$.

$$\begin{array}{r|rrrr} -1 & 1 & -1 & 1 & 1 \\ & & -1 & 2 & -3 \\ \hline & 1 & -2 & 3 & -2 \end{array}$$

Again, because the remainder is -2 ($\neq 0$), $x = -1$ is not a zero of $p_3(x)$. Since both possible rational zeroes $x = \pm 1$ are not zeroes of $p_3(x)$, $p_3(x)$ has no rational zeroes.

Remarks

1. Although the given polynomial function $p_3(x)$ has no rational zeroes, it does not imply that $p_3(x)$ has no irrational zeroes. In fact, $p_3(x)$ has a real zero between $x = -1$ and $x = 0$.
2. Before we learn how to find a real (irrational) zero of a polynomial function of degree at least 3, we introduce Descartes' rule of signs that enables us to determine the number of real zeroes theoretically.

Theorem 14.5.2 (Descartes' Rule of Signs)
Given that $p_n(x)=a_nx^n+a_{n-1}x^{n-1}+...+a_2x^2+a_1x+a_0$ is a polynomial function of degree n with real coefficients, where $a_n \neq 0$, and $a_0 \neq 0$. Then the following statement holds:
(i) The number of positive real zeroes of $p_n(x)$ equals at most the number of variations in sign of the coefficients of $p_n(x)$ or could equal that number less by an even integer.
(ii) The number of negative real zeroes of $p_n(x)$ equals at most the number of variations in sign of the coefficients of $p_n(-x)$ or equals that number less by an even integer.

Remarks

1. $p_n(x)$ must be written in descending powers of x in that the coefficients of the missing terms are regarded as zero. When counting the variations in sign of the coefficients of $p_n(x)$, the missing terms should be skipped since their coefficients are zero that are considered as having neither positive nor negative signs.
2. Since the complex zeroes always occur in pairs for a polynomial function with real coefficients (Theorem 14.6.2), this is the reason why the number of positive or negative real roots either exactly equals the number of variations in sign or equals that number less by an even integer.
3. The proof of Descartes' rule of signs can be found in Reference 15, pp. 121-124.

Example 14.5.3

Given : $p_5(x)=x^5+x^4+2x^3+2x^2-8x-8$. By using Descartes' rule of signs (Theorem 14.5.2), determine the number of positive and negative real zeroes of $p_5(x)$.

Solution

Since $p_5(x)$ has only one variation in sign of the coefficients of $p_5(x)$, i.e., from $2x^2$ to $-8x$, $p_5(x)$ has exactly one positive real zero. To determine the number of negative real zeroes of $p_5(x)$, we need to count the number of variations in sign of the coefficients of $p_5(-x)$. Now,

$$p_5(-x)=(-x)^5+(-x)^4+2(-x)^3+2(-x)^2-8(-x)-8$$
$$=-x^5+x^4-2x^3+2x^2+8x-8.$$

The number of variations in sign of the coefficients of $p_5(-x)$ is 4, i.e., from $-x^5$ to x^4, from x^4 to $-2x^3$, from $-2x^3$ to $2x^2$, and from $8x$ to -8. Hence, the number of negative real zeroes is at most 4. It could be exactly four negative real zeroes, or two negative real zeroes, or no negative real zeroes.

Remarks

1. $p_5(x)$ is able to be factored as follows:

   $$p_5(x)=(x+1)(x^2-2)(x^2+4).$$

   By setting $p_5(x)=0$ and solve for the unknown x, we have that the zeroes of $p_5(x)$ are $\{-1, -\sqrt{2}, \sqrt{2}, -2i, 2i\}$. Therefore, $p_5(x)$ has exactly one positive real zero that is $\sqrt{2}$, two negative real zeroes that are -1 and $-\sqrt{2}$, and two complex roots that are $-2i$, and $2i$.
2. Another useful qualitative criterion on the real zeroes of a polynomial function with real coefficients is the upper and the lower bounds for the real zeroes as shown below.

Theorem 14.5.3 (Upper and Lower Bounds for the Real Zeroes of a Polynomial Function)

Given that $p_n^*(x)=x^n+b_{n-1}x^{n-1}+...+b_1x+b_0$ be a polynomial with real coefficients and a leading coefficient being one. Then the most negative coefficient, increased by one is an upper bound for the positive real zeroes of $p_n^*(x)$. A lower bound for the negative real zeroes of $p_n^*(x)$ is the negative of an upper bound for the positive real zeroes of $p_n^*(-x)$.

Proof

Let b^* be the absolute value of the most negative coefficient among b_{n-1}, b_{n-2}, ..., b_1, b_0, i.e., the most negative coefficient is $-b^*$. Then, observe that for any $x > 0$,

$$p_n^*(x) = x^n + b_{n-1}x^{n-1} + ... + b_1 x + b_0$$
$$\geq x^n - b^* x^{n-1} - b^* x^{n-2} - ... - b^* x - b^*,$$
(Since $b_i \geq -b^*$ for all i)
$$= x^n - b^*(x^{n-1} + x^{n-2} + ... + x + 1)$$
$$= x^n - b^* \cdot \frac{x^n - 1}{x - 1} \geq x^n - 1 - b^* \frac{x^n - 1}{x - 1},$$
$$= (x^n - 1)(1 - \frac{b^*}{x - 1}) = (x^n - 1)(\frac{x - 1 - b^*}{x - 1}), \tag{14.5.4}$$

Note that the right side of Eq. (14.5.4) is greater than zero whenever $x > 1 + b^*$. Since $p_n^*(x)$ is no less than the right side of Eq. (14.5.4), $p_n^*(x) > 0$ for any $x > 1 + b^*$. Hence, $1 + b^*$ is an upper bound for the positive real zeroes of $p_n^*(x)$. By noting that the negative real zeroes of $p_n^*(x)$ are exactly the positive real zeroes of $p_n^*(-x)$, a lower bound for the negative real zeroes is clearly the negative of an upper bound for the positive real zeroes of $p_n^*(-x)$. The proof is, therefore, complete.

Remark

The upper and lower bounds given here are not the sharpest one. There are some bounds that are sharper than the ones given here (Reference 5, pp. 44-46).

Example 14.5.4

Given: $p_5(x) = 2x^5 + 8x^4 - 36x^3 + 7x^2 + 9x - 19$. Find an upper and lower bound for the real zeroes of $p_5(x)$.

Solution

In order to apply Theorem 14.5.3, the leading coefficient of the given polynomial function has to be one. Thus, let

$$p_5^*(x) = \frac{1}{2} \cdot p_5(x) = x^5 + 4x^4 - 18x^3 + \frac{7}{2}x^2 + \frac{9}{2}x - \frac{19}{2}.$$

Note that $p_5^*(x)$ and $p_5(x)$ have exactly the same real zeroes. To find an upper and lower bound for the real zeroes of $p_5(x)$ is equivalent to finding the bounds for that of $p_5^*(x)$. Since the most negative coefficient of $p_5^*(x)$ is – 18, the coefficient of x^3, this implies that $b^* = |-18| = 18$. Hence, an upper bound for the positive real zeroes of $p_5(x)$ is 19 (= 1 + b^* = 1 + 18). Now, to find a lower bound for the negative real zeroes of $p_5(x)$, we consider

$$p_5^*(-x) = (-x)^5 + 4(-x)^4 - 18(-x)^3 + \frac{7}{2}(-x)^2 + \frac{9}{2}(-x) - \frac{19}{2}$$

$$= -x^5 + 4x^4 + 18x^3 + \frac{7}{2}x^2 - \frac{9}{2}x - \frac{19}{2}.$$

Since the most negative coefficient of $p_5^*(-x)$ is $-\frac{19}{2}$, the constant term, this implies that $b^* = |-\frac{19}{2}| = \frac{19}{2}$. Hence, a lower bound for the negative real zeroes of $p_5(x)$ is given by

$-(b^* + 1) = -(\frac{19}{2} + 1) = -\frac{21}{2}$. Therefore, the real zeroes of $p_5(x)$ lies between $-\frac{21}{2}$ and 19.

So far, we have not learned any algorithm to find an irrational zero of a polynomial function of degree at least three. Since finding the zeroes of a polynomial function is equivalent to solving the corresponding polynomial equation for the roots, we shall pose our question as follows: How to solve for the irrational roots of a polynomial equation of degree at least three? There are two approaches to address this question. One approach is analytical, while the other approach is numerical. An analytical approach asks the following question: Does there exist any closed form formula for the roots of a polynomial equation of higher degree, in terms of the coefficients of the given polynomial equation, just like the quadratic formula for a quadratic equation? An answer to this question is affirmative for a polynomial equation of degree three (a cubic equation) and four (a quartic equation). However, Galois's theory states that for any polynomial equation of degree five or higher, it is impossible to find a close form formula in terms of the coefficients of the given polynomial equation (Reference 14). A numerical approach is based on the intermediate value theorem for a continuous function. First, we shall describe Tartaglia-Cardano's formula for solving any cubic equation.

Theorem 14.5.4 (Tartaglia-Cardano's Formula for Cubic Equations)

Given a cubic equation with real coefficients as follows:

$$x^3 + b_2^* x^2 + b_1^* x + b_0^* = 0,\ b_0^* \neq 0. \tag{14.5.5}$$

Let c_0^*, c_1^*, and Δ^* be defined respectively, by

$$c_0^* = b_0^* - \frac{1}{3} b_1^* b_2^* + \frac{2}{27} b_2^{*3}, \tag{14.5.6a}$$

$$c_1^* = b_1^* - \frac{1}{3} b_2^{*2}, \tag{14.5.6b}$$

$$\Delta^* = 4c_1^{*3} + 27c_0^{*2}. \tag{14.5.6c}$$

Then three cases of the roots of Eq. (14.5.5) are given as follows:

Case 1. If $\Delta^* > 0$, there are one real root and two complex roots given, respectively, by

$$x_1 = -\frac{1}{3} b_2^* + \sqrt[3]{A} + \sqrt[3]{B}, \tag{14.5.7a}$$

$$x_2 = -\frac{1}{3} b_2^* - \frac{1}{2}(\sqrt[3]{A} + \sqrt[3]{B}) + i \cdot \frac{\sqrt{3}}{2}(\sqrt[3]{A} - \sqrt[3]{B}), \tag{14.5.7b}$$

and

$$x_3 = -\frac{1}{3} b_2^* - \frac{1}{2}(\sqrt[3]{A} + \sqrt[3]{B}) - i \cdot \frac{\sqrt{3}}{2}(\sqrt[3]{A} - \sqrt[3]{B}), \tag{14.5.7c}$$

where A and B are given, respectively, by

$$A = -\frac{1}{2} c_0^* + \frac{1}{6}\sqrt{\frac{\Delta^*}{3}}, \tag{14.5.7d}$$

and

$$B = -\frac{1}{2} c_0^* - \frac{1}{6}\sqrt{\frac{\Delta^*}{3}}. \tag{14.5.7e}$$

Case 2. If $\Delta^* = 0$, there are three real roots, one simple root and one double root, given respectively, by

$$x_1 = -\frac{1}{3}b_2^* + 2\sqrt{-\frac{c_0^*}{2}}, \tag{14.5.8a}$$

and

$$x_2 = x_3 = -\frac{1}{3}b_2^* + \sqrt[3]{\frac{c_0^*}{2}}. \tag{14.5.8b}$$

Case 3. If $\Delta^* < 0$, three are three distinct real roots given, respectively, by

$$x_1 = -\frac{1}{3}b_2^* + 2\sqrt{\frac{-c_1^*}{3}} \cdot \cos(\frac{\theta}{3}), \tag{14.5.9a}$$

$$x_2 = -\frac{1}{3}b_2^* + 2\sqrt{\frac{-c_1^*}{3}} \cdot \cos(\frac{2\pi + \theta}{3}), \tag{14.5.9b}$$

and

$$x_3 = -\frac{1}{3}b_2^* + 2\sqrt{\frac{-c_1^*}{3}} \cdot \cos(\frac{4\pi + \theta}{3}), \tag{14.5.9c}$$

where θ given by

$$\theta = \tan^{-1}(\frac{-\sqrt{-\Delta^*}}{3\sqrt{3} \cdot c_0^*}), \tag{14.5.9d}$$

is taken in the first or second quadrant according as c_0^* is negative or positive.

Remarks

1. The proof of Theorem 14.5.4 can be found in Reference 15, pp. 84-93. By using the binomial theorem in Section 18.5, Theorem 14.5.4 is derived in Example 18.5.4 too.
2. Δ^* of Eq. (14.5.6c) can be called the cubic discriminant because it plays a role similar that of the quadratic discriminant in Chapter 6 in Part I.
3. With a scientific calculator Tartaglia-Cardano's formula becomes very practical. Neither Eqs. (14.5.7a-e) nor Eqs. (14.5.9a-d) pose any difficulty. In fact, all the calculations in the examples below are done with a TI-35 calculator.

4. It requires knowledge of trigonometry in order to use Eqs. (14.5.9a-d) if $\Delta^* < 0$.
5. A feud between two Italian mathematicians was associated with Theorem 14.5.4. After winning a mathematical contest in 1535, N. Tartaglia (1500-1557) asked G. Cardano (1501-1576) not to reveal it to the public his newly discovered formula after he showed the formula to Cardano in private. Yet, Cardano didn't keep his promise. He included this theorem into his book "Ars Magna (the Rules of Algebra)" that was published in 1545. He did give a credit at three places in his book to Tartaglia. Tartaglia was furious after learning that Theorem 14.5.4 was fully disclosed in the book. Even though it was with due acknowledgement, it was still viewed the same as stolen. A bitter quarrel ensued with insults hurled both ways, in that Cardano was defended by his student, L. Ferrari (1522-1565). Out of this war came some interesting documents from that the whole history about this spectacular discovery became public knowledge (Reference 13, pp. 91-92).
6. An interesting geometric connection between trisecting an angle and the solution of cubic equation is pointed out in Reference 9.

Example 14.5.5

Solve $x^3 - 3x + 2 = 0$ for the unknown roots of x.

Solution

Before applying Theorem 14.5.4, we first obtain $b_j^*, j = 0, 1, 2$, from the given cubic equation as follows:

$$b_2^* = 0,\ b_1^* = -3,\ b_0^* = 2.$$

We then proceed to find the three roots as follows:

Step 1. Calculate c_0^* of Eq. (14.5.6a) and c_1^* of Eq. (14.5.6b):

$$c_0^* = 2 - \frac{0\cdot(-3)}{3} + \frac{2\cdot 0^3}{27} = 2 - 0 + 0 = 2\ ,$$

And

$$c_1^* = -3 - \frac{0^2}{3} = -3\ .$$

Step 2. Calculate Δ^* of Eq. (14.5.6c), as follows:

$$\Delta^* = 4\cdot(-3)^3 + 27\cdot 2^2 = 108 - 108 = 0.$$

Step 3. Since the cubic discriminant $\Delta^* = 0$, calculate x_1 of Eq. (14.5.8a), $x_2 = x_3$ of Eq. (14.5.8b) as follows:

$$x_1 = -\frac{0}{3} + 2\cdot\sqrt{-\frac{2}{2}} = 2\cdot(-1) = -2,$$

and

$$x_2 = x_3 = -\frac{0}{3} + \sqrt[3]{\frac{2}{2}} = \sqrt[3]{1} = 1.$$

Hence, the three roots of the given cubic equation are $\{1, 1, -2\}$.

Remark

The given cubic equation can be solved by using the rational root test (Theorem 14.5.1) that gives the same solution

Example 14.5.6

Solve $x^3 - x^2 + 2 = 0$ for the unknown roots of x.

Solution

Before applying Theorem 14.5.4, we first obtain b_j^*, j = 0, 1, 2, from the given cubic equation as follows:

$$b_2^* = -1,\ b_1^* = 0,\ b_0^* = 2.$$

We then proceed to find the three roots as follows:

Step 1. Calculate c_0^* of Eq. (14.5.6a) and c_1^* of Eq. (14.5.6b) as follows:

$$c_0^* = 2 - \frac{0\cdot(-1)}{3} + \frac{2\cdot(-1)^3}{27} = 2 - 0 - \frac{2}{27} = \frac{52}{27},$$

and

$$c_1^* = 0 - \frac{(-1)^2}{3} = -\frac{1}{3}.$$

Step 2. Calculate Δ^* of Eq. (14.5.6c):

$$\Delta^* = 4\cdot(-\frac{1}{3})^3 + 27\cdot(\frac{52}{27})^2 = -\frac{4}{27} + \frac{52^2}{27} = 100.$$

Step 3. Since the cubic discriminant $\Delta^* = 100 > 0$, calculate A of Eq. (14.5.7d), B of Eq. (14.5.7e), x_1 of Eq. (14.5.7a), x_2 of Eq. (14.5.7b) and x_3 of Eq. (14.5.7c) as follows:

$$A = -\frac{\frac{52}{27}}{2} + \frac{1}{6}\sqrt{\frac{100}{3}} = -\frac{26}{27} + \frac{5\sqrt{3}}{9} = -0.00071251,$$

and

$$B = -\frac{\frac{52}{27}}{2} - \frac{1}{6}\sqrt{\frac{100}{3}} = -\frac{26}{27} - \frac{5\sqrt{3}}{9} = -1.92521341 .$$

Consequently,

$$\sqrt[3]{A} = -0.089316384 , \ \sqrt[3]{B} = -1.244016936 ,$$

$$\sqrt[3]{A} + \sqrt[3]{B} = -1.33333332 ,$$

$$\frac{1}{2}(\sqrt[3]{A} + \sqrt[3]{B}) = -0.666666666 ,$$

$$\frac{\sqrt{3}}{2}(\sqrt[3]{A} - \sqrt[3]{B} = 1.0000000012 \cong 1.0 ,$$

$$x_1 = -\frac{1}{3}(-1) + (-1.33333332) = -1.0 ,$$

$$x_2 = -\frac{1}{3}(-1) + 0.666666666 + i \cdot 1.0 = 1 + i ,$$

and

$$x_3 = 1 - i .$$

Hence, the given cubic equation has three roots, one real root x_1 and two complex roots: $\{-1, 1 \pm i\}$.

Remark

The three roots of the given cubic equation obtained here are the same as those obtained by using the rational root test of Theorem 14.5.1.

Example 14.5.7

Solve $x^3 - 3x - 1 = 0$ for the unknown roots of x.

Solution

Before applying Theorem 14.5.4, we first obtain b_j^*, j = 0, 1, 2, from the given cubic equation as follows:

$$b_2^* = 0 , \ b_1^* = -3 , \ b_0^* = -1 .$$

Then we proceed to find the three roots as follows:

Step 1. Calculate c_0^* of Eq. (14.5.6a) and c_1^* of Eq. (14.5.6b) as follows:

$$c_0^* = -1 - \frac{1}{3} \cdot (-3) \cdot 0 + \frac{2}{27} \cdot 0^3 = -1 - 0 + 0 = -1 ,$$

and

$$c_1^* = -3 - \frac{1}{3} \cdot 0^2 = -3 .$$

Step 2. Calculate Δ^* of Eq. (2.5.6c) as follows:

$$\Delta^* = 4 \cdot (-3)^3 + 27 \cdot (-1)^2 = -108 + 27 = -81 .$$

Step 3. Since the cubic discriminant $\Delta^* = -81 < 0$, we need to calculate the angle θ of Eq. (14.5.9d) as follows:

$$\theta = \tan^{-1}\left(\frac{-\sqrt{-(-81)}}{\sqrt{27} \cdot (-1)}\right) = \tan^{-1}\left(\frac{-9}{-3\sqrt{3}}\right) = \tan^{-1}(\sqrt{3}) = \frac{\pi}{3} .$$

Thus, the three real roots, x_1 of Eq. (14.5.9a), x_2 of Eq. (14.5.9b), and x_3 of Eq. of (14.5.9c) are calculated as follows:

$$x_1 = -\frac{1}{3} \cdot 0 + 2 \cdot \sqrt{\frac{-(-3)}{3}} \cdot \cos\left(\frac{\frac{\pi}{3}}{3}\right) = 2 \cdot \cos\left(\frac{\pi}{9}\right) = 1.879385242,$$

$$x_2 = -\frac{1}{3} \cdot 0 + 2 \cdot \sqrt{\frac{-(-3)}{3}} \cdot \cos\left(\frac{2\pi + \frac{\pi}{3}}{3}\right) = 2 \cdot \cos\left(\frac{7\pi}{9}\right)$$

$$= -1.5320889,$$

and

$$x_3 = -\frac{1}{3} \cdot 0 + 2 \cdot \sqrt{\frac{-(-3)}{3}} \cdot \cos\left(\frac{4\pi + \frac{\pi}{3}}{3}\right) = 2 \cdot \cos\left(\frac{13\pi}{9}\right)$$

$$= -0.347296355.$$

Therefore, the three real roots of the given cubic equation are $\{1.879385242, -1.5320889, -0.347296355\}$.

Remark

All three real roots can be checked for its validity by using the synthetic division as follows:

1.879385242	1	0	− 3	− 1
		1.879385242	3.532088888	1.0
	1	1.879385242	0.532088888	0.0

-1.5320889	1	0	− 3	− 1
		-15320889	2.3472964	1.0
	1	-1.5320889	-0.6527036	0.0

-0.347963552	1	0	-3	-1
		-0.347963552	0.120614758	1.0
	1	-0.347296355	-2.879385242	0.0

Theorem 14.5.5 (Ferrari's Formula for Quartic Equation)
Given a quartic equation with real coefficients as follows:

$$x^4 + b_3x^3 + b_2x^2 + b_1x + b_0 = 0,\ b_0 \neq 0. \tag{14.5.10}$$

Let c_2, c_1, and c_0 be defined respectively, by

$$c_2 = b_2 - \frac{3}{8}b_3^2, \tag{14.5.11a}$$

$$c_1 = b_1 - \frac{1}{2}b_2b_3 + \frac{1}{8}b_3^3, \tag{14.5.11b}$$

and

$$c_0 = b_0 - \frac{1}{4}b_1b_3 + \frac{1}{16}b_2b_3^2 - \frac{3}{256}b_3^4. \tag{14.5.11c}$$

Let y_0 be a root of the following cubic equation given by

$$y^3 + c_2y^2 + (\frac{1}{4}c_2^2 - c_0)y - \frac{1}{8}c_1^2 = 0. \tag{14.5.12}$$

Then two cases for the roots of Eq. (14.5.10) are given by
Case 1. If $c_1 \neq 0$, the four roots of Eq. (14.5.10) are given, respectively, by

$$x_1 = -\frac{1}{4}b_3 + \sqrt{\frac{y_0}{2}} + d_1, \tag{14.5.13a}$$

$$x_2 = -\frac{1}{4}b_3 + \sqrt{\frac{y_0}{2}} - d_1, \tag{14.5.13b}$$

$$x_3 = -\frac{1}{4}b_3 - \sqrt{\frac{y_0}{2}} + d_2, \tag{14.5.13c}$$

and

$$x_4 = -\frac{1}{4}b_3 - \sqrt{\frac{y_0}{2}} - d_2, \tag{14.5.13d}$$

where d_1 and d_2 are given, respectively, by

$$d_1 = \sqrt{-\frac{y_0}{2} - \frac{c_2}{2} - \frac{c_1}{2\sqrt{2y_0}}}, \tag{14.5.13e}$$

and

$$d_2 = \sqrt{-\frac{y_0}{2} - \frac{c_2}{2} + \frac{c_1}{2\sqrt{2y_0}}}. \tag{14.5.13f}$$

Case 2. If $c_1 = 0$, the four roots of Eq. (14.5.10) are given, respectively, by

$$x_1 = -\frac{1}{4}b_3 + e_1, \tag{14.5.14a}$$

$$x_2 = -\frac{1}{4}b_3 - e_1, \tag{14.5.14b}$$

$$x_3 = -\frac{1}{4}b_3 + e_2, \tag{14.5.14c}$$

and

$$x_4 = -\frac{1}{4}b_3 - e_2, \tag{14.5.14d}$$

where e_1 and e_2 are given, respectively, by

$$e_1 = \sqrt{-\frac{c_2}{2} + \sqrt{\frac{1}{4}c_2^2 - c_0}}, \tag{14.5.14e}$$

and

$$e_2 = \sqrt{-\frac{c_2}{2} - \sqrt{\frac{1}{4}c_2^2 - c_0}}, \tag{14.5.14f}$$

Remarks

1. The proof of Theorem 14.5.5 can be found in Reference 15, pp. 94-96. By using the binomial theorem in Section 18.5, it is also derived in Example 18.5.5.
2. Eq. (14.5.12) is called the resolvant cubic equation associated with Eq. (14.5.10). In case Eq. (14.5.12) has a rational root, this rational root will be chosen as y_0. As a consequence, the roots of Eq. (14.5.10) are expressible through the square root symbol.
3. From the viewpoint of algebraic geometry the quartic equation of Eq. (14.5.10) can be solved by using the concept of "pencil" (Reference 3).
4. There are five other algorithms for solving quartic equations: (i) Descartes-Euler-Cardano's, (ii) Ferrari-Lagrange's, (iii) Neumark's, (iv) Christianson-Brown's, and (v) Yacoub-Fraidenriach-Brown's (Reference 12).

Example 14.5.8

Solve $x^4 - 4x^2 + x + 2 = 0$ for the unknown roots of x.

Solution

Before applying Theorem 14.5.5, we first obtain b_3, b_2, b_1, b_0 from the given quartic equation:

$$b_3 = 0,\ b_2 = -4,\ b_1 = 1,\ b_0 = 2.$$

Then we proceed to find the roots of the given quartic equation as follows:

Step 1. Calculate c_2 of Eq. (14.5.11a), c_1 of Eq. (14.5.11b), c_0 of Eq. (14.5.11c), and the resolvent cubic equation of Eq. (14.5.12):

$$c_2 = -4 - \frac{3}{8}\cdot 0^2 = -4 - 0 = -4,$$

$$c_1 = 1 - \frac{1}{2}\cdot(-4)\cdot 0 + \frac{1}{8}\cdot 0^3 = 1 - 0 + 0 = 1,$$

$$c_0 = 2 - \frac{1}{4}\cdot 1\cdot 0 + \frac{1}{16}\cdot(-4)\cdot 0^2 - \frac{3}{256}\cdot 0^4 = 2,$$

and

$$y^3 - 4y^2 + 2y - \frac{1}{8} = 0. \qquad (1)$$

Step 2. Solve the above resolvent cubic equation of Eq. (1) for a root y_0 as follows. Before applying Theorem 14.5.4 to solve Eq. (1), we first obtain b_j^*, j = 0, 1, 2,:

$$b_2^* = -4,\ b_1^* = 2,\ b_0^* = -\frac{1}{8}.$$

Step 2.1. Calculate c_0^* of Eq. (14.5.6a) and c_1^* of Eq. (14.5.6b) as follows:

$$c_0^* = -\frac{1}{8} - \frac{1}{3}\cdot 2\cdot(-4) + \frac{2}{27}\cdot(-4)^3 = -\frac{1}{8} + \frac{8}{3} - \frac{128}{27} = -\frac{475}{216},$$

and

$$c_1^* = 2 - \frac{1}{3}\cdot(-4)^2 = 2 - \frac{16}{3} = -\frac{10}{3}.$$

Step 2.2. Calculate Δ^* of Eq. (14.5.6c) as follows:

$$\Delta^* = 4\cdot(-\frac{10}{3})^3 + 27\cdot(-\frac{475}{216})^2 = -\frac{4000}{27} + \frac{475^2}{8\cdot 216} = -\frac{30375}{1728}.$$

Step 2.3. Since the cubic discriminant $\Delta^* = -\frac{30375}{1728} < 0$, we need to calculate the angle of θ of Eq. (14.5.9d):

$$\theta = \tan^{-1}\left(\frac{-\sqrt{-(-\frac{30375}{1728})}}{3\sqrt{3}\cdot(-\frac{475}{216})}\right) = \tan^{-1}\left(\frac{216\cdot\sqrt{30375}}{3\sqrt{3}\cdot 475\cdot\sqrt{1728}}\right)$$

$$= \tan^{-1}\left(\frac{216\cdot 45\cdot\sqrt{5}\cdot\sqrt{3}}{3\sqrt{3}\cdot 475\cdot 24\cdot\sqrt{3}}\right) = \tan^{-1}\left(\frac{9\sqrt{15}}{95}\right) = 0.351662984 \ . \qquad (2)$$

Therefore, the three distinct real roots of Eq. (1) are given, respectively, by

$$y_1 = -\frac{1}{3}(-4) + 2\sqrt{\frac{-\frac{10}{3}}{3}}\cdot\cos(\frac{\theta}{3}) = \frac{4}{3} + \frac{2\sqrt{10}}{3}\cdot\cos(\frac{\theta}{3})$$

$$= \frac{4}{3} + \frac{2\sqrt{10}}{3}\cdot 0.993137842 = 3.427050090 \ , \qquad (3a)$$

$$y_2 = \frac{4}{3} + \frac{2\sqrt{10}}{3}\cdot\cos(\frac{2\pi+\theta}{3}) = \frac{4}{3} + \frac{2\sqrt{10}}{3}\cdot 2.21161609$$

$$= 5.995829451, \qquad (3b)$$

and

$$y_3 = \frac{4}{3} + \frac{2\sqrt{10}}{3}\cdot\cos(\frac{4\pi+\theta}{3}) = \frac{4}{3} + \frac{2\sqrt{10}}{3}\cdot(-0.3952847\ )$$

$$= 0.5, \qquad (3c)$$

where θ is given by Eq. (2).

Step 3. Let y_0 be chosen as y_3 of Eq. (3c) i.e., $y_0 = 0.5$. Since $c_1 = 1 \neq 0$, we first calculate d_1 (Eq. (14.5.13e)) and d_2 (Eq. (14.5.13f)):

$$d_1 = \sqrt{-\frac{0.5}{2} - \frac{-4}{2} - \frac{1}{2\sqrt{2\cdot 0.5}}} = \sqrt{1.25} = 1.118033989 \ ,$$

and

$$d_2 = \sqrt{-\frac{0.5}{2} - \frac{-4}{2} + \frac{1}{2\sqrt{2\cdot 0.5}}} = \sqrt{2.25} = 1.5 \, .$$

By using Eq. (14.5.13a-d), the four real roots of the given quartic equation are given, respectively, by

$$x_1 = -\frac{1}{4}\cdot 0 + \sqrt{\frac{0.5}{2}} + 1.118033989 = 1.618033989 \ ,$$

$$x_2 = -\frac{1}{4}\cdot 0 + \sqrt{\frac{0.5}{2}} - 1.118033989 = -0.618033989 \ ,$$

$$x_3 = -\frac{1}{4}\cdot 0 - \sqrt{\frac{0.5}{2}} + 1.5 = 1.0 \, ,$$

and

$$x_4 = -\frac{1}{4}\cdot 0 - \sqrt{\frac{0.5}{2}} - 1.5 = -2.0\,.$$

Hence, the four real roots of the given quartic equation are {− 2, − 0.618033989, 1, 1.618033989}.

Remark

By observation, the given quartic equation has a rational root 1. After using synthetic division, we have

1	1	0	− 4	1	2
		1	1	− 3	− 2
	1	1	− 3	− 2	0

The quotient polynomial is $q_3(x) = x^3 + x^2 - 3x - 2$. Since $q_3(-2) = (-2)^3 + (2)^2 - 3\cdot(-2) - 2 = -8 + 4 + 6 - 2 = 0$, $x = -2$ is a zero of $q_3(x)$. Again, using synthetic division on $q_3(x)$, we have

− 2	1	1	− 3	− 2
		− 2	2	2
	1	− 1	− 1	0

So, the quotient polynomial becomes $q_2(x) = x^2 - x - 1$. By setting $q_2(x) = 0$, we solve for the roots by using a quadratic formula:

$$x = \frac{-(-1) \pm \sqrt{(-1)^2 - 4\cdot 1\cdot(-1)}}{2\cdot 1} = \frac{1 \pm \sqrt{5}}{2}, \Rightarrow$$

$$x = \frac{1+\sqrt{5}}{2} = 1.618033989 \quad \text{or } x = \frac{1-\sqrt{5}}{2} = -0.618033989\ ,$$

which are exactly the same as obtained by using Theorem 14.5.5. We learn from this example a lesson that the rational root test (Theorem 14.5.1) should always be applied first. Only when it fails to find any rational root, then we apply Theorem 14.5.5.

Now, let us take a numerical approach to find an approximate real root of a polynomial equation. In this age of computer, a numerical approach is more effective than an analytical approach because it has no restriction on its degree as far as finding a real root of a polynomial equation is concerned. There are many numerical methods available in solving a polynomial equation. However, most of them require the knowledge of calculus. Here we are going to describe a bisection method that does not require techniques of calculus. Still, the idea is based upon the following zero bracketing theorem, that is a special case of the Intermediate Value Theorem in calculus (Reference 1).

Theorem 14.5.6 (The Zero Bracketing Theorem)
Given that a function $f(x)$ is continuous on the closed interval $[a, b]$. If $f(a)\cdot f(b) < 0$, then there exists at least one real number x_0 with $a < x_0 < b$ such that $f(x_0) = 0$.

Remarks

1. Informally, a continuous function is a function whose graph has no break. A formal definition of a continuous function requires a concept of the limit of a function to be covered in calculus. We will cover a little bit of it when we come to graph a rational function in Section 15.2. The proof of this theorem is not easy that requires some knowledge of advanced calculus. Yet, even without a formal proof, this theorem is intuitively appealing through seeing the graph of a continuous function (Figure 14.5.1).
2. It can be shown that a polynomial function of degree n is continuous on the entire real line. Therefore, this theorem provides a theoretical base to find a real root of a polynomial equation.
3. A numerical method, called the bisection method that is based upon the zero bracketing theorem (Theorem 14.5.6), will be constructed below to find a real root of a polynomial equation (Reference 11, pp. 243-247).

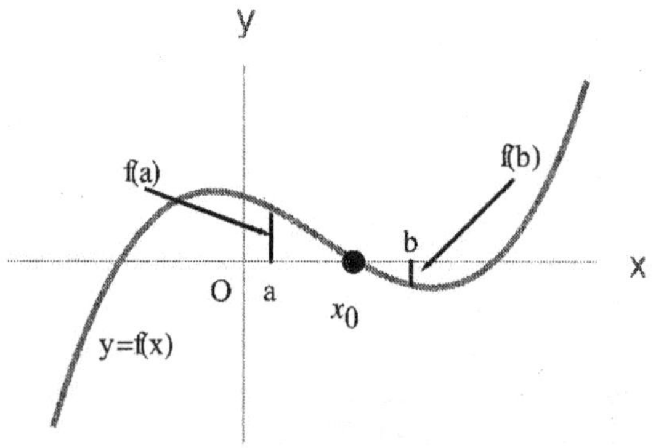

Figure 14.5.1

The Bisection Method

Given a polynomial equation $p_n(x) = 0$. We proceed to solve the given equation for a real root of x as follows:

Step 1. Use Theorem 14.5.2 (the Descartes' rule of signs) to determine the number of positive and negative real roots.

Step 2. Use Theorem 14.5.3 (An upper and lower bound for real zeroes of a polynomial function) to find an upper bound x_U for the positive real roots and a lower bound x_L for the negative real roots.

Step 3. To find an approximation to a positive real root, evaluate $p_n(x)$ at $x = 0$ and x_U. If $p_n(0) \cdot p_n(x_U) > 0$, there exists likely no positive real root. To make sure, we need to take a further action to evaluate $p_n(x)$ at every integer between 0 and x_U. Verify that all the function values are of the same sign. If $p_n(0) \cdot p_n(x_U) < 0$, it implies that there is at least one real root between 0 and x_U. Thus, using the synthetic division to evaluate $p_n(x)$ at $x = \frac{1}{2}x_U$. Determine which of the two inequalities

$$p_n(0) \cdot p_n(\frac{1}{2}x_U) < 0 \text{ and } p_n(\frac{1}{2}x_U) \cdot p_n(x_U) < 0$$

is valid. Let us say that $p_n(\frac{1}{2}x_U) \cdot p_n(x_U) < 0$. Then, evaluate $p_n(x)$ at the midpoint between $\frac{1}{2}x_U$ and x_U, which is

$$x = \frac{1}{2}(\frac{1}{2}x_U + x_U) = \frac{3}{4}x_U.$$

Again, determine which of the two inequalities

$$p_n(\frac{1}{2}x_U) \cdot p_n(\frac{3}{4}x_U) < 0 \text{ and } p_n(\frac{3}{4}x_U) \cdot p_n(x_U) < 0$$

is valid. Repeat this process until a half of the length of the interval that bracket the root of $p_n(x)$ is less than the desired accuracy. Similarly, the same process can be applied to find a negative real root.

Remarks

1. The idea of the bisection method is to repeatedly apply the zero bracketing theorem (Theorem 14.5.6) to bisect the interval that bracket the root until the desired accuracy is achieved.
2. The method provides an approximate real root with a maximum error to be less than one half of the length of the final interval that brackets the root.
3. An advantage of the bisection method is that it can be programmed and let the computer do the tedious work of calculation. Two FORTRAN programs that implement the above bisection method for finding a positive and a negative real root are given, respectively, in Appendices 14.5.1-2 at the end of Exercises for Section 14.5.

Example 14.5.9

Given: $p_3(x) = x^3 - 2x^2 - 1$.

Use the bisection method to find an approximation to all the real zeroes within an accuracy of 0.02.

Solution

We proceed to solve the given cubic equation by using the bisection method as follows:

Step 1. After applying Theorem 14.5.2 (the Descartes' rule of signs), the given equation has exactly one positive real root and no negative real root.

Step 2. By applying Theorem 14.5.3 (An upper and lower bound for real zeroes), an upper bound for the positive real root is 3 $(= |-2| + 1)$.

Step 3. Evaluate $p_3(x)$ at $x = 0$ and 3 as follows:

x	$p_3(x) = x^3 - 2x^2 - 1$
0	$p_3(0) = 0^3 - 2\cdot 0^2 - 1 = -1$.
3	$p_3(3) = 3^3 - 2\cdot 3^2 - 1 = 8$.

Step 4. Since $p_3(0)\cdot p_3(3) = (-1)\cdot 8 = -8 < 0$, there is a root between $x = 0$ and $x = 3$. Evaluate $p_3(x)$ at their midpoint $x = \frac{0+3}{2} = \frac{3}{2} = 1.5$, i.e., $p_3(1.5) = 1.5^3 - 2\cdot 1.5^2 - 1 = -2.215$.

Since $p_3(1.5)\cdot p_3(3) = -2.215\cdot 8 = -17 < 0$, the positive real root lies between $x = 1.5$ and $x = 3$. Again, evaluate $p_3(x)$ at the midpoint $x = \frac{1.5+3}{2} = \frac{4.5}{2} = 2.25$, i.e., $p_3(2.25) = 2.25^3 - 2\cdot 2.25^2 - 1$ = 11.390625 – 10.125 – 1 = 0.265625 ≈ 0.267. Since $p_3(1.5)\cdot p_3(2.25)$ = (–2.125)·0.267 = – 0.567375 < 0, the positive real root lies between x = 1.5 and x = 2.25. Again, evaluate $p_3(x)$ at the midpoint $x = \frac{1.5+2.25}{2} = 1.875$, i.e.,

$$p_3(1.875) = 1.875^3 - 2\cdot 1.875^2 - 1 = 1.439453125 \approx -1.44.$$

Since $p_3(1.875)\cdot p_3(2.25)$ = (– 1.44)·0.267 = – 0.38448 < 0, the positive real root lies between x = 1.875 and x = 2.25. Again, evaluate $p_3(x)$ at the midpoint $x = \frac{1.875+2.25}{2} = 2.0625$, i.e.,

$$p_3(2.0625) = 2.0625^3 - 2\cdot 2.0625^2 - 1 = -0.7341.$$

Since $p_3(2.0625)\cdot p_3(2.25)$ = (– 0.7341)·0.267 = – 0.196 < 0, the positive real root lies between x = 2.0625 and x = 2.25. Again, evaluate $p_3(x)$ at midpoint $x = \frac{2.0625+2.25}{2} = 2.15625$, i.e.,

$p_3(2.15625) = 2.15625^3 - 2\cdot 2.15625^2 - 1$ = 10.0253 – 9.2988 –1 = – 0.2735. Since $p_3(2.15625)\cdot p_3(2.25)$ = (– 0.2735)·0.267 = – 0.0730245 < 0, the positive real root lies between x = 2.15625 and x = 2.25. Hence, evaluate $p_3(x)$ at the midpoint

$x = \frac{2.15625+2.25}{2} = 2.203125$, i.e.,

$$p_3(2.203125) = 2.203125^3 - 2\cdot 2.203125 - 1 = -0.0141.$$

Since $p_3(2.203125)\cdot p_3(2.25)$ = (– 0.0141)·0.267 = – 0.0037647 < 0, the positive real root lies between x = 2.203125 and x = 2.25. Again, evaluate $p_3(x)$ at the midpoint $x = \frac{2.203125+2.25}{2} = 2.2265625$, i.e.,

$$p_3(2.2265625) = 2.2265625^3 - 2\cdot 2.2265625^2 - 1 = 0.0232.$$

Since $p_3(2.203125)\cdot p_3(2.2265625)$ = (–0.0141)·0.0232 = – 0.0003243 < 0, the positive real root lies between x = 2.203125 and x = 2.2265625. Since a half of the length of the interval [2.203125,

2.2265625] is 0.01171875 that is less than 0.02. Therefore, we stop and take the midpoint of [2.203125, 2.2265625], i.e.,
$x = \dfrac{2.203125 + 2.226565}{2} = 2.21484375$ as an approximate positive real root for the given cubic equation.

Remarks

1. The reason why the desired accuracy was chosen as 0.02 was that we didn't want to make you bored with tedious calculations. The bisection method does seem tedious with a pencil-and-paper calculation. However, it is very suitable to be implemented on the computer that can perform tedious calculations without being tired or bored. In fact, two FORTRAN programs that implement the bisection method for calculating a positive and a negative real zero are given in Appendices 14.5.1 and 14.5.2, respectively.
2. By using the program given in Appendix 14.5.1, we obtain the following approximate real root with an accuracy set to be within 0.001:

a	b	$c = \dfrac{a+b}{2}$	$ME = \dfrac{b-a}{2}$
0	3	1.5	1.5
1.5	3	2.25	0.75
1.5	2.25	1.875	0.375
1.875	2.25	2.0625	0.1875
2.0625	2.25	2.15625	0.09375
2.15625	2.25	2.20312	0.046875
2.20312	2.25	2.22656	0.023438
2.20312	2.22656	2.21484	0.017719
2.20312	2.21484	2.20898	0.005860
2.20312	2.20898	2.20605	0.002930
2.20312	2.20605	2.20459	0.001465
2.20459	2.20605	2.20532	0.000732

$f(a)$	$f(b)$	$f(c)$
– 1	8	– 2.2125
– 2.125	8	0.265625
– 2.125	0.265625	– 1.43945
– 1.43945	0.265625	– 0.734131

– 0.734131	0.265625	– 0.273529
– 0.273529	0.265625	– 0.01408
– 0.01408	0.265625	0.123201
– 0.01408	0.123201	0.053923
– 0.01408	0.053923	0.019763
– 0.01408	0.019763	0.002802
– 0.01408	0.002802	– 0.00565

The approximate real root is 2.20532 with an accuracy within 0.001.

Exercises

In problems 1–5, (a) determine the number of positive and negative real zeroes of the given polynomial function and (b) find an upper and lower bound for real zeroes of the given polynomial function.

1. $p_3(x) = 2x^3 + 3x^2 - 4x + 5$.
2. $p_4(x) = x^4 - 4x^3 + 5x^2 - 8x - 7$.
3. $p_5(x) = x^5 - 3x^4 + 5x^3 - 6x^2 + 2x - 9$.
4. $p_6(x) = x^6 - 3x^5 - 5x^2 - x - 6$.
5. $p_7(x) = -9x^7 + 5x^6 + 4x^3 - 6x^2 + x + 11$.

In problems 6–10, use the rational zero test (Theorem 14.5.2) to find all zeroes of the given function.

6. $p_3(x) = 2x^3 - x^2 + 2x - 1$.
7. $p_4(x) = 6x^4 + x^3 - 7x^2 - x + 1$.
8. $p_5(x) = 2x^5 - 6x^4 + 12x^3 - 12x^2 + 11x - 6$.
9. $p_6(x) = 6x^6 + x^5 + 5x^4 + x^3 - 13x^2 - 2x - 2$.
10. $p_7(x) = 2x^7 + 11x^6 - 7x^5 - 6x^4 - 8x^3 - 44x^2 + 28x + 24$.

In problems 11–15, Use Tartaglia-Cardano's formula (Theorem 14.5.4) to solve the given cubic equation for the roots.

11. $x^3 - x - 2 = 0$.
12. $2x^3 + 6x + 4 = 0$.
13. $x^3 + 6x^2 - 2 = 0$.
14. $x^3 - x^2 - x - 1 = 0$.
15. $x^3 + 3x^2 - 3x - 1 = 0$.

In problems 16–20, use Ferrari's formula (Theorem 14.5.5) to solve the given quartic equation.

16. $x^4 - 4x - 1 = 0.$
17. $x^4 - x^2 + x + 3 = 0.$
18. $x^4 - 3x^3 + 1 = 0.$
19. $2x^4 + 4x^2 - 4x - 6 = 0.$
20. $x^4 + 3x^3 - 5x^2 + x - 1 = 0.$

In problems 21–25, use the bisection method to find an approximation to the real zero of the given polynomial function.

21. $p_3(x) = x^3 - x - 2.$
22. $p_3(x) = x^3 - 2x^2 + 3x - 4.$
23. $p_4(x) = x^4 - x - 3.$
24. $p_4(x) = x^4 + x^3 - 4.$
25. $p_5(x) = x^5 + x - 5.$

Appendix 14.5.1 A FORTRAN program for calculating a positive real zero

```
C THIS PROGRAM USES THE BISECITON METHOD FOR SOLVING A
C POLYNOMIAL EQUAITON FOR AN APPROXIMATE POSITIVE REAL ROOT
C BASED ON THE UPPER BOUND FOR REAL ZEROS [THEOREM 14.5.3]

   REAL A, B, C, EPSILON, FNA, FNB, FNC, ME
   INTERGER NMAX

C DEFINE THE GIVEN POLYNOMIAL FUNCTION
   FN (X) =X**3 –X –2.

C THE UPPER BOUND B IS OBTAINED FROM THEOREM 14.5.3
   A = 0.
   B = 3.
```

```
C DEFINE THE DESIRED ACCURACY FOR AN APPROXI-
MATE REAL ROOT
   EPSILON=0.001

C SET UP THE MAXIMUM NUMBER OF ITERATIONS FOR
INVOKING THE
C BISECITON METHOD
   NMAX =20
   PRINT 1
1  FORMAT (5X, 'SOLVE X**3 -X -2 = 0 FOR POSITIVE
REAL ROOT')
   DO 50 I =1, NMAX

C THE MIDPOINT BETWEEN A AND B IS DEFIED AS FOL-
LOWS:
   C = (A+B) /2.

C THE MAXIMUM ERROR, DENOTED BY ME, IS DEFINED
AS FOLLOWS:
   ME = (B-A)/2.
   PRINT 100
   PRINT 110, A, B, C, ME
   IF (ME.LE.EPSILON) THEN
   GO TO 150
   ELSE
   FNA = FN(A)
   FNB = FN(B)
   FNC = FN(C)
   ENDIF
   PRINT 120
   PRINT 130, FNA, FNB, FNC
   IF (FNA*FNC.GT.0. .AND. FNB*FNC.GT. 0.) GO TO 60
   IF (FNA*FNC.LT. 0.) THEN
   B = C
   ENDIF
   IF (FNB*FNC.LT. 0.) THEN
   A = C
   ENDIF
```

```
50 CONTINUE
60 PRINT 70
70 FORMAT (2X, 'THE EQUATION LIKELY HAS NO POSI-
TIVE REAL ROOT')
100 FORMAT (8X, 'A', 16X, 'B', 14X, 'C', 16X, 'ME')
110 FORMAT (4(4X, E12.6))
120 FORMAT (8X, 'FNA', 13X, 'FNB', 12X, 'FNC')
130 FORMAT (3 (4X, E12. 6))
150 STOP
   END
```

Remarks

1. To use this program you should have basic knowledge about the computer programming language of FORTRAN. For example, in any FORTRAN program, the first five columns are reserved for the numbering of the statement. Column six is used for a continuation of the previous statement. All executable statements must start from the seventh column. Also, a FORTRAN compiler is required to compile the program.

2. If you would like to use the above program to solve your equation, then three places need to change accordingly: (i) The function in the seventh row needs to be replaced by yours; (ii) The upper bound B in the tenth row needs to be replaced by an appropriated number; (iii) Format Statement #1 in the 17th row needs to be changed.

Appendix 14.5.2 A FORTRAN program for calculating a negative real zero

```
C THIS PROGRAM USES THE BISECTION METHOS FOR SOLVING A
C POLYNOMIAL EQUATION FOR AN APPROXIMATE NEG-
ATIVE REAL ROOT
C BASED ON THE LOWER BOUND FOR REAL ZEROS [THE-
OREM 14.5.3]

   REAL A, B, C, EPSILON, FNA, FNB, FNC, ME
   INTEGER NMAX
```

```
C DEFINE THE GIVEN POLYNOMIAL FUNCTION
  FN (X) = X**3 –X –2.

C THE LOWER BOUND A IS OBTAINED FROM THEOREM
14.5.3
  A = -3
  B = 0

C DEFINE THE DESIRED ACCURACY FOR AN APPROXI-
MATE REAL ROOT
  EPSILON = 0.001

C SET UP THE MAXIMUM NUMBER OF ITERATIONS FOR
INVOKING THE BISECTION METHOD
  NMAX = 20
  PRINT 1
1  FORMAT (5X, 'SOLVE X** 3 –X –2 = 0 FOR NEGATIVE
REAL ROOT')
10 DO 50 I=1, NMAX
C THE MIDPOINT BETWEEN A AND B IS DEFINED AS FOL-
LOWS:
  C = (A+B) /2.
C THE MAXIMUM ERROR, DENOTED BY ME, IS DEFINED
AS FOLLOWS:
  ME = (B–A) /2
  PRINT 100
  PRINT 110, A, B, C, ME
  IF (ME.LE.EPSILON) THEN
  GO TO 150
  ELSE
  FNA = FN(A)
  FNB = FN(B)
  FNC = FN(C)
  ENDIF
  PRINT 120
  PRINT 130, FNA, FNB, FNC
  IF (FNA*FNC.GT.0. .AND. FNB*FNC.GT. 0.) GO TO 60
  IF (FNA*FNC.LT. 0.) THEN
```

```
   B= C
   ENDIF
   IF (FNB*FNC.LT. 0.) THEN
   A= C
   ENDIF
50 CONTINUE
60 PRINT 70
70 FORMAT (2X, 'THE EQUATION LIKELY HAS NO NEGA-
TIVE REAL ROOT')
100 FORMAT (8X, 'A', 16X, 'B', 14X, 'C', 16X, 'ME')
110 FORMAT (4(4X, E12.6))
120 FORMAT (8X, 'FNA', 13X, 'FNB', 12X, 'FNC')
130 FORMAT (3 (4X, E12. 6))
150 STOP
   END
```

Remark
The same remarks for Appendix 14.5.1 are applied here too.

14.6 Complex Zeroes and the Fundamental Theorem of Algebra

So far, we have presented some computational methods such as the rational root test, Tartaglia-Cardano's formula for cubic equations, Ferrari's formula for quartic equations, and a bisection method for finding numerically an approximate real root. In this section we shall describe some theoretical properties about the zeroes of a polynomial function.

Theorem 14.6.1 (The Fundamental Theorem of Algebra)
A polynomial function $p_n(x)$ of degree $n \geq 1$ with real coefficients has at least one complex zero.

Remarks
1. Although the coefficients of $p_n(x)$ in the above theorem does not have to be limited to the set of real numbers (they can be complex numbers), we are only concerned with a polynomial function

with real coefficients in this book. The proof of this theorem is beyond the scope of this book, but can be found in Reference 6, which contains 11 distinct proofs.

2. This theorem is only a type of the (non-constructive) existence theorem in the sense that it only tells us that there exists at least one complex zero; however, it does not tell us how to find such a complex zero. A review on algorithms finding complex roots of a polynomial equation is given in Reference 10.
3. A polynomial function with real coefficients possesses another important theoretical property, that is, its complex zeroes must occur in pairs as shown below.

Theorem 14.6.2 (Pairs of Complex Zeroes for a Real Polynomial Function)

Given a polynomial function $p_n(x)$ of degree $n \geq 2$ with real coefficients. If a complex number $z = a + ib$ $(b \neq 0)$ is a zero of $p_n(x)$, then its conjugate $\bar{z} = a - ib$ is also a zero of $p_n(x)$.

Proof

Suppose that $p_n(x) = a_n x^n + a_{n-1}x^{n-1} + \ldots + a_1 x + a_0$, where $a_n \neq 0$ and all a_i's are real numbers. Since z is a zero of $p_n(x)$, we have $p_n(z) = 0$, namely,

$$a_n z^n + a_{n-1} z^{n-1} + \ldots + a_1 z + a_0 = 0. \tag{1}$$

By taking the conjugate number of both sides of Eq. (1), we have

$$\overline{a_n z^n + a_{n-1} z^{n-1} + \ldots + a_1 z + a_0} = \bar{0},$$

$$\overline{a_n z^n} + \overline{a_{n-1} z^{n-1}} + \ldots + \overline{a_1 z} + \overline{a_0} = 0,$$

$$\overline{a_n} \cdot \overline{z^n} + \overline{a_{n-1}} \cdot \overline{z^{n-1}} + \ldots + \overline{a_1} \cdot \bar{z} + a_0 = 0,$$

$$a_n (\bar{z})^n + a_{n-1} (\bar{z})^{n-1} + \ldots + a_1 \bar{z} + a_0 = 0,$$

(Since a_i is real, $\overline{a_i} = a_i$ for all i),

or, equivalently,

$$p_n(\bar{z}) = 0. \tag{2}$$

From Eq. (2), $\bar{z}$ is shown to be a zero of $p_n(x)$. The proof is, therefore, complete.

Theorem 14.6.3 (Factors of a Real Polynomial Function)

A real polynomial function $p_n(x)$ of degree $n \ge 2$ can be written as the product of factors of linear and quadratic polynomials with real coefficients, where the factors of quadratic polynomials have no real zeroes.

Proof

By applying Theorem 14.6.1 (The Fundamental Theorem of Algebra), $p_n(x)$ has at least one complex zero, say z_1. Consequently, $x - z_1$ is a factor of $p_n(x)$. By Theorem 14.4.2 (Finding a Linear Factor), there exists a quotient polynomial function $q_{n-1}(x)$ of degree $n - 1$ such that $p_n(x) = (x - z_1) \cdot q_{n-1}(x)$. By applying Theorem 14.6.1 on $q_{n-1}(x)$, $q_{n-1}(x)$ has at least one complex zero, say z_2. By Theorem 14.4.2, there exists another quotient polynomial function $q_{n-2}(x)$ of degree $n - 2$ such that $q_{n-1}(x) = (x - z_2) \cdot q_{n-2}(x)$. By substituting this expression for $q_{n-1}(x)$ into the right side of the expression for $p_n(x)$, we have

$$p_n(x) = (x - z_1)(x - z_2) \cdot q_{n-2}(x).$$

Obviously, we can apply again Theorem 14.6.1 to $q_{n-2}(x)$. If we repeatedly apply Theorem 14.6.1 n times, then we obtain $p_n(x) = a_n(x - z_1)(x - z_2) \cdot \ldots \cdot (x - z_n)$, where $a_n \ne 0$ is the leading coefficient of $p_n(x)$. If all z_j's are real numbers, then there is nothing more to prove. However, if any of the z_j's is a complex number, say, $z_k = a + ib$, where a and b are real numbers, $b \ne 0$, then we know that its conjugate $\bar{z}_k = a - ib$ must also be one of the z_j's, say $z_\ell = \bar{z}_k$, since $p_n(x)$ has real coefficients and the result of Theorem 14.6.2. By multiplying the corresponding factors, we have

$$(x - z_k)(x - z_\ell) = [x - (a + ib)][x - (a - ib)] = x^2 - 2ax + a^2 + b^2,$$

where each coefficient is a real number. Hence, the proof is complete.

Remark

The factor of quadratic polynomials is said to be irreducible over the set of real numbers since it cannot be factored into the product of two linear factors with real zeroes.

Example 14.6.1

Given: $p_3(x) = 2x^3 - 3x^2 + 8x + 5$.

(a) Show that $1 - 2i$ is a zero of $p_3(x)$.

(b) Find all zeroes of $p_3(x)$.

Solution

(a) By using the synthetic division, we have

$1-2i$	2	-3	8	5
	0	$2(1-2i)$	$(1-2i)(-1-4i)$	$(1-2i)(-1-2i)$
	2	$-1-4i$	$-1-2i$	0

Since the remainder, that is the last number in the third row of the above table is 0, $1 - 2i$ is shown to be a zero of $p_3(x)$ by Theorem 14.4.2.

(b) Since $z = 1 - 2i$ is a zero of $p_3(x)$ and $p_3(x)$ is a polynomial function with real coefficients, its conjugate $\bar{z} = \overline{1-2i} = 1 + 2i$ is also a zero of $p_3(x)$. By Theorem 14.6.3, $p_3(x)$ has a quadratic factor given by

$$d(x) = [x - (1 - 2i)]\cdot[x - (1+2i)] = x^2 - 2x + 5 \qquad (1)$$

By using the long division to divide $p_3(x)$ by $d(x)$ of Eq. (1), we have

$$\begin{array}{r} 2x+1 \\ x^2-2x+5 \overline{)\, 2x^3 - 3x^2 + 8x + 5} \\ \underline{2x^3 - 4x^2 + 10x} \\ x^2 - 2x + 5 \\ \underline{x^2 - 2x + 5} \\ 0 \end{array}$$

Therefore, $2x + 1$ is a linear factor of $p_3(x)$ since the remainder is zero. By setting $2x + 1 = 0$, we have that $x = -\frac{1}{2}$ is another zero of $p_3(x)$. Hence, the three zeroes of $p_3(x)$ are $\{-\frac{1}{2}, 1 \pm 2i\}$.

Example 14.6.2

Find a quartic polynomial $p_4(x)$ with real coefficients that have –2, $\frac{1}{3}$, and $1 + i \cdot \sqrt{2}$ as zeroes and such that $p_4(0) = -6$.

Solution

Since $p_4(x)$ is required to be a polynomial function with real coefficients and has a complex zero $z = 1 + i \cdot \sqrt{2}$, $\bar{z} = 1 - i \cdot \sqrt{2}$ must be another complex zero of $p_4(x)$ by Theorem 14.6.2. Let a_4 be the leading coefficient of $p_4(x)$. Then, by Theorem 14.6.3 (Factors of a Real Polynomial Function), we have

$$p_4(x) = a_4[x - (-2)](x - \frac{1}{3})[x - (1 + i\sqrt{2})][x - (1 - i\sqrt{2})]$$

$$= a_4(x + 2)(x - \frac{1}{3})(x^2 - 2x + 3). \qquad (1)$$

By using the given information of $p_4(0) = -6$, we have from Eq. (1)

$$p_4(0) = a_4(0 + 2)(0 - \frac{1}{3})(0^2 - 2 \cdot 0 + 3) = -6, \Rightarrow$$

$$-a_4 \cdot 2 \cdot \frac{1}{3} \cdot 3 = -6, \Rightarrow -2a_4 = -6, \Rightarrow a_4 = \frac{-6}{-2} = 3.$$

Hence, the desired $p_4(x)$ is given by

$$p_4(x) = 3(x + 2)(x - \frac{1}{3})(x^2 - 2x + 3)$$

$$= 3(x^2 + \frac{5}{3}x - \frac{2}{3})(x^2 - 2x + 3)$$

$$= (3x^2 + 5x - 2)(x^2 - 2x + 3)$$

$$= 3x^4 - x^3 - 3x^2 + 19x - 6.$$

Exercises

In problems 1–5, (a) show that the given polynomial function has the given complex number z as a zero, (b) find all zeroes of the given polynomial function.

1. $p_3(x) = 2x^3 - 3x^2 + 8x - 12$ and $z = -2i$, where $i = \sqrt{-1}$.

2. $p_3(x) = 3x^3 - 13x^2 + 31x - 9$ and $z = 2 + i \cdot \sqrt{5}$, where $i = \sqrt{-1}$.

3. $p_4(x) = 4x^4 + 20x^3 + 65x^2 + 40x - 39$ and $z = -2 - 3i$, where $i = \sqrt{-1}$.

4. $p_4(x) = 3x^4 + 6x^3 + 10x^2 + 2x + 3$ and $z = 3 + i \cdot \sqrt{3}$, where $i = \sqrt{-1}$.

5. $p_5(x) = x^5 - x^4 + 6x^3 - 2x^2 + 5x + 15$ and $z = 1 - i \cdot \sqrt{2}$, where $i = \sqrt{-1}$.

In problems 6–10, find a polynomial function of a specific degree with real coefficients that have the given zeroes and the given function value.

6. $p_3(0) = -11$ and has zeroes: $3 - i \cdot \sqrt{2}$ and $-\frac{1}{2}$, where $i = \sqrt{-1}$.

7. $p_3(1) = -12$ and has zeroes : $2 - i$ and 3, where $i = \sqrt{-1}$.

8. $p_4(0) = 10$ and has zeroes: $\frac{1}{4}$, -2, and $1 - 2i$, where $i = \sqrt{-1}$.

9. $p_4(1) = 39$ and has zeroes: $2 + i \cdot \sqrt{2}$, $-5i$, where $i = \sqrt{-1}$.

10. $p_5(1) = 240$ and has zeroes: -3, -2, $\frac{1}{3}$, and $2 - 3i$, where $i = \sqrt{-1}$.

14.7 References

1. Anton, H. (1988). Calculus with Analytic Geometry, 3rd edition, John Wiley & Sons, New York.

2. Askey, R. (1975). Orthogonal Polynomials and Special Functions, Regional Conference Series in Applied Mathematics, #21, Society for Industry and Applied Mathematics, Philadelphia.
3. Auckly, D. (2007). Solving the quartic with a pencil, American Mathematical Monthly, 114, 29-39.
4. Bell, E. T. (1940). The Development of Mathematics, Dover Publications, Inc., New York.
5. Cajori, F. (1969). An Introduction to the Theory of Equations, Section 39, pp. 44 – 45, Dover Publications, Inc., New York.
6. Fine, B. and Rosenberger, G. (1997). The Fundamental Theorem of Algebra, Springer-Verlag, Inc., New York.
7. Gale, D. (1960). The Theory of Linear Economic Models, The University of Chicago Press, Chicago, Illinois.
8. Libbrecht, U. (1987). Chinese Mathematics in the Thirteen Century, Dover Publications, Inc., New York.
9. Nickalls, W. R. D. (2006). Viete, Descarte and the cubic equation, The Mathematics Gazette, 90, 203-208.
10. Pan, V. Y. (1997). Solving a polynomial equation: Some history and recent progress, SIAM Review, 39, 187-220.
11. Press, W. H., Flannery, B. P., Teukolsky, S. A., and Vetterling, W. T. (1986). Numerical Recipes: The Art of Scientific Computing, Cambridge University Press, Cambridge, England.
12. Shmakov, S. (2011). A universal method of solving quartic equations, International Journal of Pure and Applied mathematics, 71, 251-259.
13. Struik, D. J. (1948). A Concise History of Mathematics, Dover Publications, Inc., New York.
14. Tignol, J. P. (1980). Galois' Theory of Algebraic Equations, Longman Scientific & Technical, Essex, England.
15. Uspensky, J. V. (1948). Theory of Equations, McGraw-Hill Book Company, Inc., New York.
16. https://en.wikipedia.org/wiki/Galois_theory.
17. https://en.wikipedia.org/wiki/Horner%27s_method.

Chapter 15
Rational Functions

In this chapter we're going to learn how to graph a rational function with special features including the determination of the (horizontal, oblique, and vertical) asymptotes. In addition, a technique of partial fraction is presented that is very useful to calculate the indefinite integral of rational functions in calculus.

15.1 Rational Functions

In the number system, ratios of two integers are called rational numbers. In the function space, ratios of two polynomial functions are called rational functions.

Definition15.1.1
A function of the form

$$r(x) = \frac{p_n(x)}{q_m(x)}, \tag{15.1.1}$$

is called a rational function of x, where $p_n(x)$ and $q_m(x)$ ($m \geq 1$) are polynomial functions of degrees n and m, respectively, and have no common (nontrivial) factors.

Remarks

1. Rational expressions in Example 4.1.1 (Chapter 4 in Part I) can be viewed as rational functions of several independent variables. Here we only consider a rational function of one independent (real) variable x.
2. The domain of a rational function $r(x)$ is restricted to the set of those real numbers such that $q_m(x) \neq 0$ since the function value of $r(x)$ becomes undefined at $x = x_0$ if $q_m(x_0) = 0$. In fact, those x-values such that $q_m(x) = 0$ serve as the vertical asymptotes for the graph of $r(x)$.

3. The zeroes of a rational function $r(x)$ are the same as that of the numerator $p_n(x)$.
4. Elliptic rational functions are extensively used in the design of elliptic electronic filters (Reference 3).
5. Many physical laws are expressed in the form of rational function. For example, Newton's Law of Gravitation of Example 13.1.1 (c) states that the attractive force F is a rational function of the distance r given that the masses of two objects m_1 and m_2 are fixed. The momentum p of a photon (Example 11.3.6 in Part I) is a rational function of the wavelength λ of the scattered photon. In economics, the average cost is a rational function of the production level, i.e., the quantities produced. In advanced mathematics, rational functions are used to approximate an exponential function in the theory of Pade approximation (Reference 4). Mobius transformation that is a rational function of a complex variable maps a circle in one plane to a circle in another plane (Reference 2, p. 304). In calculus, the average rate of change of y for a unit change in x, denoted by $\frac{\Delta y}{\Delta x}$, is a rational function of Δx.

Example 15.1.1 (The Law of Ideal Gas)

The law of ideal gas states that the pressure P of a gas is directly proportional to its temperature T and inversely proportional to its volume V with the universal gas constant R, called molar gas constant, given by

$$P = R \cdot \frac{n \cdot T}{V}, \tag{1}$$

where n is the number of *kg* mol of the gas in the volume V, T is measured in Kelvin degree (K), and R is given by

$$R = 8.3144\ J/(\text{g}\cdot\text{mol}\cdot\text{K}) = 8{,}314\ J/(\text{kg}\cdot\text{mol}\cdot\text{K}). \tag{2}$$

Find the volume of 1 kg mol of gas would occupy at 0°C and a pressure of 1 *atm* (or atmosphere), where 1 *atm* is given by

$$1\ atm = 101{,}325\ N/m^3. \tag{3}$$

Solution

We solve for V from Eq. (1) to obtain

$$V = R \cdot \frac{n \cdot T}{P}. \tag{4}$$

By substituting $n = 1$ kg mol, $T = 273$°K (since 0°C is equivalent to 273°K), R of Eq. (2), and $P = 1$ *atm* of Eq. (3) into the right side of Eq. (4), we have

$$V = 8{,}314\ J/(\text{kg}\cdot\text{mol}\cdot\text{K})\cdot(1\ \text{kg mol}\cdot 273°\text{K})\ /\ (101325\ N/\text{m}^2)$$
$$= 22.4\ \text{m}^3/\text{kg mol}.$$

Hence, 1 kg mol, that contains $6.02\cdot10^{26}$ molecules of a gas, occupies a volume of 22.4 m^3 at the standard temperature and pressure, i.e., $T = 0°C$ and $P = 1$ *atm*.

Remarks

1. Note that P of Eq. (1) can be viewed as a rational function of three independent variables n, T, and V, i.e., $P = f(n, T, V)$.
2. The universal gas constant R is defined as the product of Avogadro's number ($N_A = 6.02\cdot10^{26}$ molecules/kg·mol), and the Boltzmann's constant ($k = 1.38\cdot10^{-23}$ J/K). For a discussion of Eq. (1), see Reference 5.
3. If liter (L) is used for measuring the volume, the universal gas constant R is given by
   $$R = 0.08206\ L\cdot atm/(\text{g}\cdot\text{mol}\cdot\text{K}).$$

Definition 15.1.2

A function $r(x)$ of Eq. (15.1.1) is called a proper rational function if the degree of $p_n(x)$ is less than the degree of $q_m(x)$, i.e., $n < m$. Otherwise, $r(x)$ of Eq. (15.1.1) is called an improper rational function.

Remark

By using the long division, an improper rational function can be written as the sum of a quotient polynomial function and a remainder that is a proper rational function.

Example 15.1.2

Find the domain of each of the following rational functions:

(a) $r_1(x) = \dfrac{x^2 - 1}{x + 5}$. (b) $r_2(x) = \dfrac{5}{x^2 - 1}$.

(c) $r_3(x) = \dfrac{x + 1}{x^2 - 5x + 6}$. (d) $r_4(x) = \dfrac{x - 1}{x^2 + 2}$.

Solution

(a) By setting the denominator of $r_1(x)$ equal to zero, we have

$$x + 5 = 0, \Rightarrow x = -5.$$

Hence, Domain(r_1) = $\{x \in R | x \neq -5\} = (-\infty, -5) \cup (-5, +\infty)$.

(b) By setting the denominator of $r_2(x)$ equal to zero, we have

$$x^2 - 1 = 0, \Rightarrow (x + 1)(x - 1) = 0, \Rightarrow$$

$$x + 1 = 0 \text{ or } x - 1 = 0, \Rightarrow x = -1 \text{ or } x = 1.$$

Since $r_2(x)$ is undefined at $x = \pm 1$, the domain of $r_2(x)$ is given by

Domain(r_2) = $\{x \in R | x \neq \pm 1\} = (-\infty, -1) \cup (-1, 1) \cup (1, +\infty)$.

(c) By setting the denominator of $r_3(x)$ equal to zero, we have

$$x^2 - 5x + 6 = 0, \Rightarrow (x - 2)(x - 3) = 0, \Rightarrow$$

$$x - 2 = 0 \text{ or } x - 3 = 0, \Rightarrow x = 2 \text{ or } x = 3.$$

Since $r_3(x)$ is undefined at $x = 2$ and 3, the domain of $r_3(x)$ is given by

Domain(r_3) = $\{x \in R | x \neq 2 \text{ and } x \neq 3\} = (-\infty, 2) \cup (2, 3) \cup (3, +\infty)$.

(d) By setting the denominator of $r_4(x)$ equal to zero, we have

$$x^2 + 2 = 0.$$

However, the above equation has no real roots. Hence, the domain of $r_4(x)$ is given by

Domain(r_4) = $(-\infty, +\infty)$.

Remark

In calculus, it can be shown that the x-values at which a rational function is undefined are the points of discontinuity for the graph of a rational function. Moreover, the points of discontinuity of a rational function are actually the vertical asymptotes that will be discussed in the next section.

Example 15.1.3

Determine if each of the following functions is a proper rational function or not. If not, use a long division to simplify it as the sum of a polynomial function and a proper rational function.

(a) $s_1(x) = \dfrac{2x+1}{x^2+3}$. (b) $s_2(x) = \dfrac{x-2}{3x+1}$. (c) $s_3(x) = \dfrac{3x^2-2x+1}{4x^2+x-3}$.

(d) $s_4(x)=\dfrac{x^2+3}{2x+1}$. (e) $s_5(x)=\dfrac{2x^3-3x^2+4x-5}{4x-1}$.

Solution

(a) Since the degree of the numerator of $s_1(x)$ is less than the degree of its denominator, $s_1(x)$ is a proper rational function.

(b) Since the degree of the numerator of $s_2(x)$ equals the degree of its denominator, $s_2(x)$ is an improper rational function. By using the long division, we have

$$\begin{array}{r|l} & \frac{1}{3} \\ \hline 3x+1 & x-2 \\ & x+\frac{1}{3} \\ \hline & -\frac{7}{3} \end{array}$$

Hence, $s_2(x) = \dfrac{1}{3}-\dfrac{\frac{7}{3}}{3x+1}=\dfrac{1}{3}-\dfrac{7}{3}\cdot\dfrac{1}{3x+1}=\dfrac{1}{3}-\dfrac{7}{3\cdot(3x+1)}$.

(c) Since the degree of the numerator of $s_3(x)$ equals the degree of its denominator, $s_3(x)$ is an improper rational function. By using the long division, we have

$$\begin{array}{r|l} & \frac{3}{4} \\ \hline 4x^2+x-3 & 3x^2-2x+1 \\ & 3x^2+\frac{3}{4}x-\frac{9}{4} \\ \hline & -\frac{11}{4}x+\frac{13}{4} \end{array}$$

Hence, $s_3(x) = \dfrac{3}{4}+\dfrac{-\frac{11}{4}x+\frac{13}{4}}{4x^2+x-3}=\dfrac{3}{4}+\dfrac{-11x+13}{4(4x^2+x-3)}$.

(d) Since the degree of the numerator of $s_4(x)$ is greater than the degree of its denominator, $s_4(x)$ is an improper rational function. By using the long division, we have

$$
\begin{array}{r}
\frac{1}{2}x-\frac{1}{4} \\
2x+1\overline{\big)x^2+0\cdot x+3} \\
\underline{x^2+\frac{1}{2}x\quad\quad} \\
-\frac{1}{2}x+3 \\
\underline{-\frac{1}{2}x-\frac{1}{4}} \\
\frac{13}{4}
\end{array}
$$

Hence, $s_4(x)=\frac{1}{2}x-\frac{1}{4}+\frac{\frac{13}{4}}{2x+1}=\frac{1}{2}x-\frac{1}{4}+\frac{13}{4(2x+1)}$.

(e) Since the degree of the numerator of $s_5(x)$ is greater than the degree of its denominator, $s_5(x)$ is an improper rational function. By using the long division, we have

$$
\begin{array}{r}
\frac{1}{2}x^2-\frac{5}{8}x+\frac{27}{32} \\
4x-1\overline{\big)2x^3-3x^2+4x-5} \\
\underline{2x^3-\frac{1}{2}x^2\quad\quad\quad\quad} \\
-\frac{5}{2}x^2+4x\quad\; \\
\underline{-\frac{5}{2}x^2+\frac{5}{8}x\quad\;} \\
\frac{27}{8}x-5 \\
\underline{\frac{27}{8}x-\frac{37}{32}} \\
-\frac{123}{32}
\end{array}
$$

Hence,

$$s_5(x)=\frac{1}{2}x^2-\frac{5}{8}x+\frac{27}{32}-\frac{\frac{123}{32}}{4x-1}=\frac{1}{2}x^2-\frac{5}{8}x+\frac{27}{32}-\frac{123}{32(4x-1)}.$$

Remark

It is important to discern whether a given rational function is proper or improper. This information is useful in finding the horizontal/oblique asymptotes in order to graph a rational function that is covered in Section 15.2.

Example 15.1.4

Given: $r(x)=\dfrac{x-1}{x^2-x-2}$ and $s(x)=\dfrac{x+1}{x^2-3x+2}$.

Find: (a) $t(x)=(r-s)(x)=?$ (b) $u(x)=(\frac{3}{2}\cdot r)(x)=?$

(c) $v(x)=(r\cdot s)(x)=?$ (d) $w(x)=(\frac{r}{s})(x)=?$

(e) $f(x)=(r\circ s)(x)=?$ (f) $g(x)=(s\circ r)(x)=?$

Solution

First, their denominators of $r(x)$ and $s(x)$ are factored into the product of two linear factors, namely,

$$r(x)=\frac{x-1}{(x+1)(x-2)}, \tag{1}$$

and

$$s(x)=\frac{x+1}{(x-1)(x-2)}. \tag{2}$$

Second, all operational rules for functions defined in Section 13.3 are equally applicable to rational functions.

(a) By using the function subtraction rule with $r(x)$ of Eq. (1) and $s(x)$ of Eq. (2), we have

$$\begin{aligned} t(x)&=(r-s)(x)=r(x)-s(x)\\ &=\frac{x-1}{(x+1)(x-2)}-\frac{x+1}{(x-1)(x-2)}\\ &=\frac{(x-1)^2}{(x+1)(x-1)(x-2)}-\frac{(x+1)^2}{(x+1)(x-1)(x-2)}\\ &=\frac{x^2-2x+1-(x^2+2x+1)}{(x+1)(x-1)(x-2)}\\ &=\frac{-4x}{x^3-2x^2-x+2}. \end{aligned}$$

(b) By using the scalar multiplication rule for functions, we have

$$u(x) = (\frac{3}{2} \cdot r)(x) = \frac{3}{2} \cdot r(x) = \frac{3}{2} \cdot \frac{x-1}{x^2-x-2}$$
$$= \frac{3(x-1)}{2(x^2-x-2)} = \frac{3x-3}{2x^2-2x-4}.$$

(c) By using the function multiplication rule with $r(x)$ of Eq. (1) and $s(x)$ of Eq. (2), we have

$$v(x) = (r \cdot s)(x) = r(x) \cdot s(x)$$
$$= \frac{x-1}{(x+1)(x-2)} \cdot \frac{x+1}{(x-1)(x-2)} = \frac{1}{(x-2)^2} = \frac{1}{x^2-4x+4}.$$

(d) By using the function division rule with $r(x)$ of Eq. (1) and $s(x)$ of Eq. (2), we have

$$w(x) = (\frac{r}{s})(x) = \frac{r(x)}{s(x)} = \frac{\frac{x-1}{(x+1)(x-2)}}{\frac{x+1}{(x-1)(x-2)}} = \frac{(x-1)^2}{(x+1)^2}$$
$$= \frac{x^2-2x+1}{x^2+2x+1}.$$

(e) By using the function composition rule with $r(x)$ of Eq. (1) and $s(x)$ of Eq. (2), we have

$$f(x) = (r \circ s)(x) = r(s(x)) = \frac{s(x)-1}{[s(x)+1][s(x)-2]}$$
$$= \frac{\frac{x+1}{(x-1)(x-2)} - 1}{[\frac{x+1}{(x-1)(x-2)} + 1][\frac{x+1}{(x-1)(x-2)} - 2]}$$
$$= \frac{\frac{x+1-(x-1)(x-2)}{(x-1)(x-2)}}{[\frac{x+1+(x-1)(x-2)}{(x-1)(x-2)}][\frac{x+1-2(x-1)(x-2)}{(x-1)(x-2)}]}$$
$$= \frac{\frac{x+1-(x^2-3x+2)}{(x-1)(x-2)}}{\frac{[x+1+(x^2-3x+2)] \cdot [x+1-2(x^2-3x+2)]}{(x-1)^2(x-2)^2}}$$

$$= \frac{\dfrac{-x^2+4x-1}{1}}{\dfrac{(x^2-2x+3)(-2x^2+7x-3)}{(x-1)(x-2)}}$$

$$= \frac{(-x^2+4x-1)(x-1)(x-2)}{(x^2-2x+3)(-2x^2+7x-3)}$$

(Since $\dfrac{\frac{a}{b}}{\frac{c}{d}} = \dfrac{ad}{bc}$)

$$= \frac{(-x^2+4x-1)(x^2-3x+2)}{(x^2-2x+3)(-2x^2+7x-3)}$$

$$= \frac{-x^4+7x^3-15x^2+11x-2}{-2x^4+11x^3+23x^2+27x-9}$$

$$= \frac{x^4-7x^3+15x^2-11x+2}{2x^4-11x^3-23x^2-27x+9}.$$

(f) By using the function composition rule with $r(x)$ of Eq. (1) and $s(x)$ of Eq. (2), we have

$$g(x) = (s \circ r)(x) = s(r(x)) = \frac{r(x)+1}{[r(x)-1][r(x)-2]}$$

$$= \frac{\dfrac{x-1}{(x+1)(x-2)}+1}{[\dfrac{x-1}{(x+1)(x-2)}-1][\dfrac{x-1}{(x+1)(x-2)}-2]}$$

$$= \frac{\dfrac{x-1+(x+1)(x-2)}{(x+1)(x-2)}}{[\dfrac{x-1-(x+1)(x-2)}{(x+1)(x-2)}][\dfrac{x-1-2(x+1)(x-2)}{(x+1)(x-2)}]}$$

$$= \frac{\dfrac{x-1+(x^2-x-2)}{(x+1)(x-2)}}{\dfrac{[x-1-(x^2-x-2)][x-1-2(x^2-x-2)]}{(x+1)^2(x-2)^2}}$$

$$= \frac{\dfrac{x^2-3}{1}}{\dfrac{(-x^2+2x+1)(-2x^2+3x+3)}{(x+1)(x-2)}}$$

$$= \frac{(x^2-3)(x+1)(x-2)}{(-x^2+2x+1)(-2x^2+3x+3)}$$

$$= \frac{(x^2-3)(x^2-x-2)}{(-x^2+2x+1)(-2x^2+3x+3)}$$

$$= \frac{x^4-x^3-5x^2+3x+6}{2x^4-7x^3+x^2+9x+3}.$$

Remark

The reader needs to have a review over Section 4.3 in Part I if he/she is not familiar with operational rules for rational expressions.

Exercises

In problems 1–5, find the domain of the given rational function.

1. $r_1(x) = \dfrac{2x+3}{x+\sqrt{2}}$.

2. $r_2(x) = \dfrac{4}{2x^2-3}$.

3. $r_3(x) = \dfrac{x^2+x-2}{3x^2+5}$.

4. $r_4(x) = \dfrac{x^2-x-6}{x^3-x-2}$.

5. $r_5(x) = \dfrac{9}{x^4-5x^2+6}$.

In problems 6–10, determine if the given rational function is proper or not. If not, use the long division to write it as the sum of a polynomial function and another proper rational function.

6. $r_6(x) = \dfrac{3x^2+5}{x+3}$.

7. $r_7(x) = \dfrac{5x^2-2x+1}{2x^2+3x-1}$.

8. $r_8(x) = \dfrac{x^3-x-4}{x^2-x-6}$.

9. $r_9(x) = \dfrac{2x^3-5x^2+x-4}{x-3}$.

10. $r_{10}(x) = \dfrac{2x^4 - 3x^3 - 5x^2 + 6x - 1}{x^2 + x - 6}$.

In problems 11–15, given that $r(x) = \dfrac{x+3}{2x-1}$ and $s(x) = \dfrac{x-3}{2x+1}$, find each of the following new rational function:

11. $t(x) = (3 \cdot r - 4 \cdot s)(x) = ?$

12. $u(x) = (r \cdot s)(x) = ?$

13. $v(x) = (\dfrac{s}{r})(x) = ?$

14. $f(x) = (r \circ s)(x) = ?$

15. $g(x) = (s \circ r)(x) = ?$

In problems 16–18, use the law of the ideal gas described in Example 15.1.1 to answer each question:

16. A tank of a certain gas with the molecular weight of 32 kg at 0°C has a volume of 0.001 m^3 and the pressure is 10 atm. What mass of the gas is in the tank?

17. A car is filled with air to a pressure of 32 lb/in^2 on a warm day when the temperature is 27°C. Assume that the volume of the tire remains constant. What would the pressure be in the tire when the temperature drops to – 13°C?

18. The gas in the piston of a diesel engine is originally at a temperature of 32°C and a pressure of 80 cm Hg when it is suddenly compressed. If the final pressure of the gases is 3,200 cm Hg and the temperature has been increased to 407°C, what is the final volume of the gas if the initial volume of the gas is 10m^3/(kg·mol)?

15.2 The Graph of a Rational Function

To graph a rational function requires the knowledge of the vertical and horizontal asymptotes and sometimes even the oblique (or slant) asymptote. Before we are able to introduce the concept of asymptotes, we need the concept of x approaching a real number a from either side of a.

Definition 15.2.1

An independent variable x is said to approach a real number a from the right side of a, denoted by $x \to a^+$, if x is moving closer and closer to a from its right side as close to a as we wish, but never exactly reaches a. Similarly, an independent variable x is said to approach a real number a from the left side of a, denoted by $x \to a^-$, if x is moving closer and closer to a from its left side as close to a as we wish, but never exactly reaches a. Again, an independent variable x is said to approach a real number a from both sides of a, denoted by $x \to a$, if x is moving closer and closer to a from both sides of a as close to a as we wish, but never exactly reaches a.

Remark

At a first glimpse, the concept of $x \to a$ sounds somewhat puzzling and hard to understand it. This involves the concept of the limit, that serves as the fundamental building block for "Analysis", a branch of mathematics that is totally different from Algebra. Historically, the limit concept was rooted in the 4th century B.C. from an empirical experience that the area of a circle could be exhausted by inscribing in it a regular polygon and then increasing the number of sides indefinitely (Reference 1, pp. 90-92). Here we give a simpler explanation why we need such a concept. In algebra, mathematicians were bothered by the circumstance in which the function value became "undefined" at some x-values. For example, a rational function $r(x) = \dfrac{x+1}{x-1}$ is undefined at $x = 1$, since $r(1) = \dfrac{1+1}{1-1} = \dfrac{2}{0}$. Note that the concept of the limit can help us to get by this impasse. As long as x is moving to 1 from either side of 1, but never exactly reaches 1, $r(x) = \dfrac{x+1}{x-1}$ is well defined for all such x-values. Hence, there is no difficulty encountered in this new type of function evaluations. This is the motivation to learn the concept of the limit.

Definition 15.2.2

An independent variable x is said to approach $\pm\infty$, denoted by $x \to \pm\infty$, if x is moving closer and closer to $\pm\infty$ as close to $\pm\infty$ as we wish, but never exactly reaches $\pm\infty$.

Remarks

1. Note that x approaches $+\infty$ only from its let side, while x approaches $-\infty$ only from its right side.
2. A definition for a dependent variable $y = f(x)$ approaching either a real number b or $\pm\infty$ can be similarly defined.

Definition 15.2.3

The vertical line $x = a$ is said to be a vertical asymptote for the graph of a function $f(x)$ if $f(x) \to \pm\infty$ as $x \to a^+$ or $x \to a^-$.

Remarks

1. The graph of $f(x)$ never crosses its vertical asymptotes $x = a$.
2. For a rational function $r(x) = \dfrac{p_n(x)}{q_m(x)}$ of Eq. (15.1.1), the vertical asymptotes for the graph of $r(x)$ are simply those x-values such that $q_m(x) = 0$, or the points of x not in the domain of $r(x)$.

Definition 15.2.4

The horizontal line $y = b$ is said to be a horizontal asymptote for the graph of a function $f(x)$ if $f(x) \to b^+$ or $f(x) \to b^-$ as $x \to \pm\infty$.

Remarks

1. In calculus, b is called the limit of $f(x)$ as $x \to \pm\infty$ It could be denoted by $b = \lim\limits_{x \to +\infty} f(x)$ or $b = \lim\limits_{x \to -\infty} f(x)$. To lessen the burden of getting used to this symbol, we will not use it in this book.
2. For a rational function $r(x) = \dfrac{p_n(x)}{q_m(x)}$ of Eq. (15.1.1), the horizontal asymptote exists only when the degree of $p_n(x)$ is no greater than the degree of $q_m(x)$, i.e., $n \le m$.

Definition 15.2.5

The graph of a polynomial function of $Q_\ell(x)$ of degree $\ell \geq 1$ is said to be a curvilinear asymptote for the graph of a function $f(x)$ if $f(x) \to Q_\ell(x)$ as $x \to \pm\infty$. When $\ell = 1$, the graph of a linear function $Q_\ell(x)$ is called a oblique (or slant) asymptote for the graph of $f(x)$.

Remarks

1. We had already encountered the concept of the oblique asymptote in Section 12.4 in Part I. The graph of a hyperbola has a pair of oblique asymptotes.
2. For an improper rational function $r(x)$ of Eq. (15.1.1) with $n \geq m + 1$, the quotient polynomial function obtained by applying the long division on $r(x)$ is a curvilinear asymptote for the graph of $r(x)$.

A Procedure for Graphing Rational Functions

Given a rational function $r(x) = \dfrac{p_n(x)}{q_m(x)}$ of Eq. (15.1.1). To graph $r(x)$, we proceed as follows:

Step 1. Find the y-intercept for the graph of $r(x)$ by evaluating $r(0) = ?$.

Step 2. Find the x-intercepts by setting $r(x) = 0$, or equivalently, $p_n(x) = 0$, and solve $p_n(x) = 0$ for the unknown roots of x.

Step 3. Find the vertical asymptotes by setting $q_m(x) = 0$ and solve $q_m(x) = 0$ for the roots of x.

Step 4. Find the horizontal or curvilinear asymptotes by using the long division on $r(x)$ if $n \geq m+1$.

Step 5. Find a few extra points by evaluating $r(x)$ both between and beyond the x-intercepts and the vertical asymptotes.

Step 6. Connecting all the points smoothly by keeping in mind that the graph of $r(x)$ never crosses its asymptotes then provides us the desired graph of $r(x)$.

Example 15.2.1

Given: $r(x) = \dfrac{1}{x}$, where $x \neq 0$.

(a) Show that the line $x = 0$ is a vertical asymptote for the graph of $r(x)$.
(b) Show that the line $y = 0$ (or x-axis) is a horizontal asymptote for the graph of $r(x)$.
(c) Graph $r(x)$.

Solution

(a) To show that $x = 0$ is a vertical asymptote for the graph of $r(x)$, we need to show that $r(x) \to \pm\infty$ as $x \to 0^+$ or $x \to 0^-$. Since $x \to 0^+$ or $x \to 0^-$ is a dynamic process, this is best accomplished by using a tabulation as follows:

x	0.1	0.01	0.001	0.0001
$r(x) = \frac{1}{x}$	1/0.1 = 10	1/0.01 = 100	1/0.001 = 1000	1/0.0001 = 10000

...	$x \to 0^+$
...	$r(x) \to +\infty$

x	– 0.1	– 0.01	– 0.001	–0.0001
$r(x) = \frac{1}{x}$	1/(– 0.1) = – 10	1/(- 0.01) = – 100	1/(- 0.001) = – 1000	1/(-0.0001) = – 10000

...	$x \to 0^-$
...	$r(x) \to -\infty$

The above tables show that $r(x) \to +\infty$ as $x \to 0^+$, while $r(x) \to -\infty$ as $x \to 0^-$. Hence, by definition, $x = 0$ is proved to be a vertical asymptote for the graph of $r(x)$.

(b) To show that $y = 0$ is a horizontal asymptote for the graph of $r(x)$, we need to show that $r(x) \to 0$ as $x \to +\infty$ or $x \to -\infty$ Again, since $x \to +\infty$ or $x \to -\infty$ is a dynamic process, it is best accomplished through the use of a tabulation as follows:

x	10000	100000	10^6
$r(x) = \frac{1}{x}$	1/10000 = 0.0001	1/100000 = 0.00001	$1/10^6 = 10^{-6}$

...	$x \to +\infty$
...	$r(x) \to 0^+$

x	-10000	-100000	-10^6
$r(x) = \frac{1}{x}$	$1/(-10000)$ $= -0.0001$	$1/(-100000)$ $= -0.00001$	$1/(-10^6)$ $= -10^{-6}$

...	$x \to -\infty$
...	$r(x) \to 0^-$

The above tables show that $r(x) \to 0^+$ as $x \to +\infty$, while $r(x) \to 0^-$ as $x \to -\infty$. Hence, by definition, $y = 0$ is proved to be a horizontal asymptote for the graph of $r(x)$.

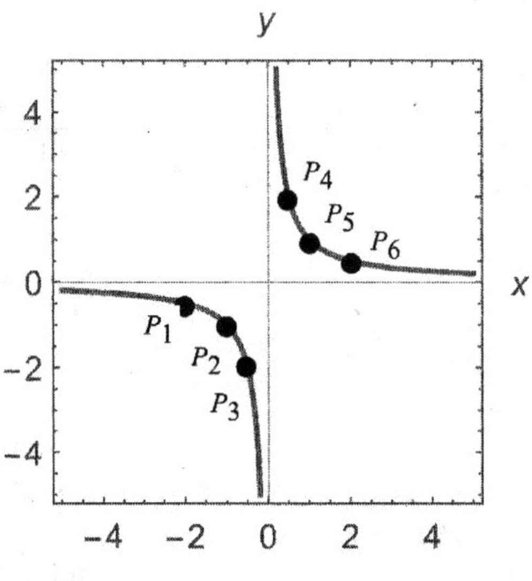

Figure 15.2.1

(c) To graph $r(x)$, we need to find the y-intercept and the x-intercepts. However, since $r(0) = \frac{1}{0} =$ undefined, there is no y-intercept. In fact, it is shown in (a) that $x = 0$ is a vertical asymptote. To find the x-intercept, we set $r(x) = 0$ and solve for the roots of x, that is equivalent to setting the numerator of $r(x)$ equal zero and solve for the roots of

x. However, the numerator of $r(x) = \dfrac{1}{x}$ is a constant 1, which has no way to be zero. Therefore, $r(x)$ has no x-intercept. We evaluate $r(x)$ at $x = \pm\dfrac{1}{2}$, ± 1, ± 2 as follows:

x	$\pm\dfrac{1}{2}$	± 1	± 2
$r(x) = \dfrac{1}{x}$	$1/(\pm½)$ $= \pm 2$	$1/(\pm 1)$ $= \pm 1$	$1/(\pm 2)$ $= \pm½$

This provides us six points $P_1(-2,-\frac{1}{2})$, $P_2(-1,-1)$, $P_3(-\frac{1}{2},-2)$, $P_4(\frac{1}{2},2)$, $P_5(1,1)$, and $P_6(2,\frac{1}{2})$. By connecting these six points smoothly, we get the graph of $r(x) = \dfrac{1}{x}$ that is a hyperbola as shown in Figure 15.2.1.

Remark

From this example, we note that $\dfrac{1}{x} \to 0^+$ as $x \to +\infty$ while $\dfrac{1}{x} \to 0^-$ as $x \to -\infty$. This observation provides a tip on how to find a horizontal asymptote for other more complex rational functions by rewriting the given rational function to become a function of $\dfrac{1}{x}$. This is done by dividing both its numerator and denominator by a power function of the highest degree between that of the numerator and denominator.

Example 15.2.2

Given: $r(x) = \dfrac{x^2 - 4}{2x^2 + 1}$.

(a) Show that there are no vertical asymptotes for $r(x)$.
(b) Find a horizontal asymptote for $r(x)$ if any exists.
(c) Graph $r(x)$.

Solution

(a) By setting the denominator of $r(x)$ equal to zero, we have

$$2x^2 + 1 = 0. \qquad (1)$$

However, Eq. (1) has no real solutions since $2x^2 + 1 > 0$ for any real number x. Hence, there are no vertical asymptotes for the graph of $r(x)$.

(b) Since the given rational function $r(x)$ is improper, we re-write $r(x)$ as the sum of a quotient polynomial function and the remainder as follows:

$$\begin{array}{r|l} & \frac{1}{2} \\ 2x^2+1 & x^2 - 4 \\ & x^2 + \frac{1}{2} \\ \hline & -\frac{9}{2} \end{array}$$

Hence, $r(x) = \frac{1}{2} - \frac{\frac{9}{2}}{2x^2+1} = \frac{1}{2} - \frac{9}{2(2x^2+1)}$. (2)

As $x \to +\infty$, the second term of Eq. (2) approaches 0 from the right side of 0. Consequently, as $x \to -\infty$, $r(x) \to 0.5^-$. Hence, $y = 0.5$ is a horizontal asymptote for the graph of $r(x)$ as $x \to \pm\infty$.

(c) To graph $r(x)$, we need to find the y-intercept and the x-intercepts. Since $r(0) = \frac{0^2 - 4}{2 \cdot 0^2 + 1} = -4$, the y-intercept is $P_0(0, -4)$. By setting the numerator of $r(x)$ equal to zero, we have

$$x^2 - 4 = 0, \Rightarrow (x+2)(x-2) = 0, \Rightarrow$$
$$x + 2 = 0, \text{ or } x - 2 = 0, \Rightarrow x = -2, \text{ or } x = 2.$$

Hence, the x-intercepts are $P_1(-2, 0)$, $P_2(2, 0)$. We evaluate $r(x)$ at $x = \pm 1$, and ± 3 as follows:

x	± 1	± 3
$r(x) = \frac{x^2-4}{2x^2+1}$	$\frac{(\pm 1)^2 - 4}{2 \cdot (\pm 1)^2 + 1} = -1$	$\frac{(\pm 3)^2 - 4}{2 \cdot (\pm 3)^2 + 1} = \frac{5}{19}$

The four extra points are $P_3(-1, -1)$, $P_4(1, -1)$, $P_5(-3, 0.263)$, and $P_6(3, 0.263)$. By plotting the vertical and horizontal asymptotes and connecting the seven points smoothly, the graph of $r(x)$ is shown in Figure 15.2.2.

Remark

$r(x)$ is an even function since $r(-x) = r(x)$. This fact is reflected by that the graph of $r(x)$ is symmetric with respect to the y-axis. Also, $r(x)$ has a local minimum at $x = 0$.

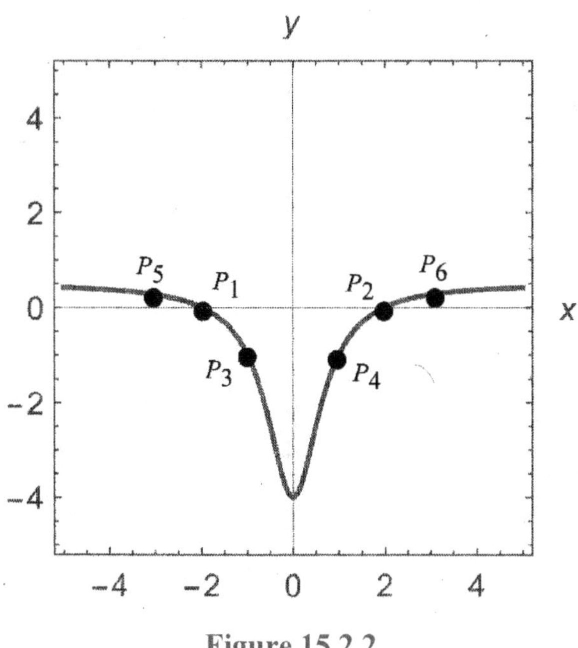

Figure 15.2.2

Example 15.2.3

Given: $r(x) = \dfrac{x^3 - 2x^2 + 1}{x^2}$, where $x \neq 0$.

(a) Find the vertical asymptotes for the graph of $r(x)$ if any exists.
(b) Find the horizontal or oblique or curvilinear asymptotes for the graph of $r(x)$ if any exists.
(c) Graph $r(x)$.

Solution

(a) By setting the denominator of $r(x)$ equal to zero, we have

$$x^2 = 0, \Rightarrow x = 0, 0 \text{ (0 is a double root).}$$

Hence, $x = 0$ (or y-axis) is a vertical asymptote for the graph of $r(x)$.

(b) Since the degree of the numerator of $r(x)$ is one degree higher than the degree of its denominator, there exists an oblique asymptote for the graph of $r(x)$. Also, since the denominator of $r(x)$ has just one term, we can use the term-by-term division to re-write $r(x)$ as follows:

$$r(x) = \frac{x^3}{x^2} - \frac{2x^2}{x^2} + \frac{1}{x^2} = x - 2 + \frac{1}{x^2}. \qquad (1)$$

As $x \to \pm\infty$, the third term of Eq. (1) approaches zero. Consequently, as $x \to \pm\infty$,

$$r(x) \to x - 2. \qquad (2)$$

From Eq. (2), $y = x - 2$ is by definition an oblique asymptote for the graph of $r(x)$.

(c) To graph $r(x)$, we need to find the y-intercept and the x-intercept. Since $r(0) = \dfrac{1}{0^2}$ = undefined, there is no y-intercept. By setting the numerator of $r(x)$ equal to zero, we have

$$x^3 - 2x^2 + 1 = 0. \qquad (3)$$

By applying the rational zeroes test (Theorem 14.5.1) on Eq. (3), we immediately note that $x = 1$ is a root of Eq. (3). Using the synthetic division with $x = 1$, we obtain the quotient polynomial as follows:

1	1	– 2	0	1
		1	– 1	– 1
	1	– 1	– 1	0

Consequently, Eq. (3) can be written as

$$(x-1)\cdot(x^2 - x - 1) = 0, \Rightarrow x - 1 = 0 \text{ or } x^2 - x - 1 = 0, \Rightarrow$$

$$x = 1 \text{ or } x = \frac{1 \pm \sqrt{5}}{2}, \Rightarrow$$

$$x = 1,\ x = \frac{1+\sqrt{5}}{2} \approx 1.618, \text{ or } x = \frac{1-\sqrt{5}}{2} \approx -0.618.$$

Hence, the x-intercepts are $P_1(-0.618, 0)$, $P_2(1, 0)$, and $P_3(1.618, 0)$. To get a few extra points, we evaluate $r(x)$ at $x = -2, -1, 1.5$, and 3 as follows:

x	-2	-1
$r(x) = x - 2 + \frac{1}{x^2}$	$-2 - 2 + \frac{1}{(-2)^2}$ $= -3.75$	$-1 - 2 + \frac{1}{(-1)^2}$ $= -2$

x	1.5	3
$r(x) = x - 2 + \frac{1}{x^2}$	$1.5 - 2 + \frac{1}{(1.5)^2}$ $= -0.56$	$3 - 2 + \frac{1}{3^2}$ $= 1.11$

Consequently, the four extra points are $P_4(-2, -3.75)$, $P_5(-1, -2)$, $P_6(1.5, -0.56)$, and $P_7(3, 1.11)$. By plotting the vertical and oblique asymptotes, and connecting the seven points smoothly, the graph of $r(x)$ is shown in Figure 15.2.3.

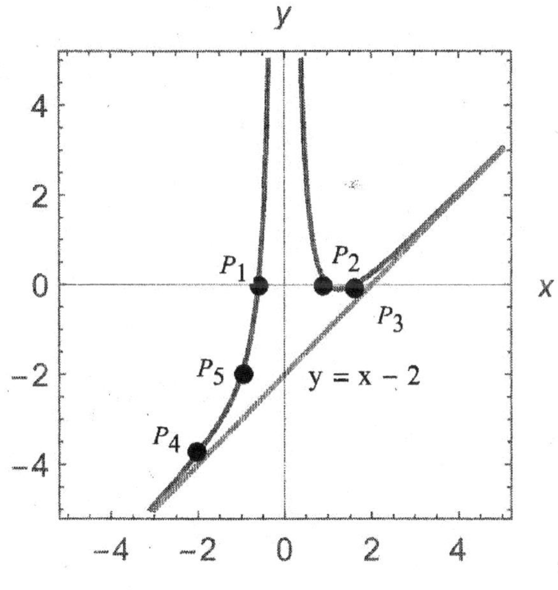

Figure 15.2.3

Remark

From Figure 15.2.3, it is seen that $r(x)$ has a local minimum between $x = 1$ and $x = 1.5$. By using a technique in calculus, it can be shown that $r(x)$ has a local minimum at $x = \sqrt[3]{2} = 1.25992105 \approx 1.26$.

Example 15.2.4

Given: $r(x) = \dfrac{x^4}{x^2 - 1}$, where $x \neq \pm 1$.

(a) Find the vertical asymptote for the graph of $r(x)$ if any exists.

(b) Find the horizontal, or oblique, or curvilinear asymptote for the graph of $r(x)$ if any exists.

(c) Graph $r(x)$.

Solution

(a) By setting the denominator of $r(x)$ equal to zero, we have

$$x^2 - 1 = 0, \Rightarrow (x + 1)\cdot(x - 1) = 0, \Rightarrow$$
$$x + 1 = 0 \text{ or } x - 1 = 0, \Rightarrow x = -1 \text{ or } x = 1.$$

Consequently, both $x = -1$ and $x = 1$ are the vertical asymptotes for the graph of $r(x)$.

(b) Since the degree of the numerator of $r(x)$ is two degrees higher than the degree of the denominator, there exists a curvilinear asymptote for the graph of $r(x)$. By applying the long division on $r(x)$, we have

$$\begin{array}{r|l} & x^2 + 1 \\ \hline x^2 - 1 & x^4 \\ & \underline{x^4 - x^2} \\ & \quad x^2 \\ & \quad \underline{x^2 - 1} \\ & \qquad 1 \end{array}$$

Consequently, $r(x)$ can be written as

$$r(x) = x^2 + 1 + \frac{1}{x^2 - 1}. \tag{1}$$

Observe that the third term $\dfrac{1}{x^2 - 1}$ of Eq. (1) approaches zero as $x \to \pm \infty$. Hence, as $x \to \pm \infty$, we have

$$r(x) \to x^2 + 1. \tag{2}$$

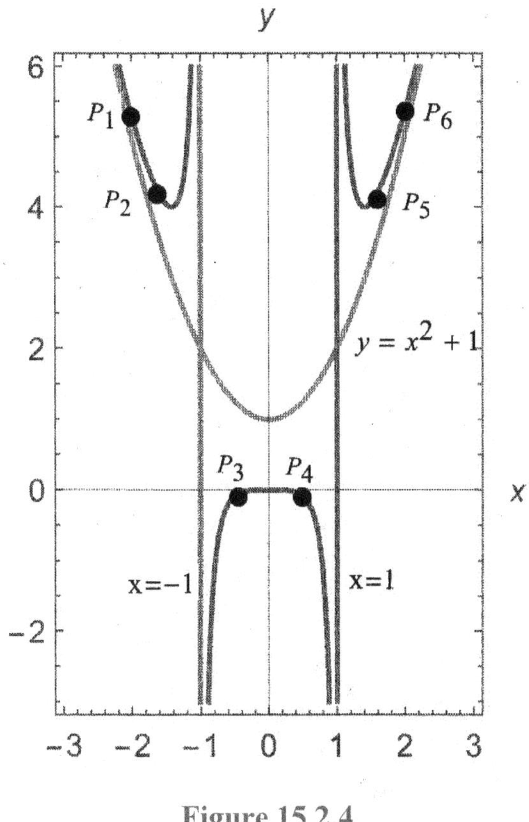

Figure 15.2.4

From Eq. (2), $y = x^2 + 1$ is by definition a curvilinear asymptote for the graph of $r(x)$.

(c) To graph $r(x)$, we need to find the y-intercept and the x-intercepts. Since $r(0) = \dfrac{0^4}{0^2 - 1} = 0$, the y-intercept is $P_0(0, 0)$. By setting the numerator of $r(x)$ equal to zero, we have $x^4 = 0, \Rightarrow x = 0$ ($x = 0$ is a root with a multiplicity 4). To get a few extra points, we evaluate $r(x)$ at $x = \pm\dfrac{1}{2}, \pm\dfrac{3}{2}$, and ± 2 as follows:

x	$\pm½$	±1.5	±2

$r(x)=\dfrac{x^4}{x^2-1}$	$\dfrac{(\pm\frac{1}{2})^4}{(\pm\frac{1}{2})^2-1}$ $=-\dfrac{1}{12}=-0.08$	$\dfrac{(\pm 1.5)^4}{(\pm 1.5)^2-1}$ $=4.05$	$\dfrac{(\pm 2)^4}{(\pm 2)^2-1}$ $=\dfrac{16}{3}=5.3$

Consequently, the six extra points are $P_1(-2, 5.3)$, $P_2(-1.5, 4.05)$, $P_3(-0.5, -0.08)$, $P_4(0.5, -0.08)$, $P_5(1.5, 4.05)$, and $P_6(2, 5.3)$. By plotting the vertical and the curvilinear asymptotes, and connecting the seven points smoothly, the graph of $r(x)=\dfrac{x^4}{x^2-1}$ is shown in Figure 15.2.4.

Remark

From Figure 15.2.4, $r(x)$ is seen to have a local maximum at $x = 0$ and two local minima either between $x = -2$ and $x = -1$ or between $x = 1$ and $x = 2$. By using a technique in calculus, it can be shown that $r(x)$ has local minima at $x = \pm\sqrt{2}$.

Exercises

In problems 1–8, (a) find the vertical asymptote (if any), (b) find the horizontal, or oblique, or curvilinear asymptote (if any) for the graph of the given rational function. (c) graph the given rational function.

1. $r_1(x)=\dfrac{3x+5}{2x}$, where $x \neq 0$.

2. $r_2(x)=\dfrac{x}{x^2+x-2}$, where $x \neq -2$, or 1.

3. $r_3(x)=\dfrac{3x^2}{x^2+1}$.

4. $r_4(x)=\dfrac{5(x^2-4)}{2(x^2-1)}$, where $x \neq \pm 1$.

5. $r_5(x)=\dfrac{2x^3}{x^3-4x}$, where $x \neq 0$, or ± 2.

6. $r_6(x)=\dfrac{-x^2+2x-1}{x}$, where $x \neq 0$.

7. $r_7(x) = \dfrac{x^3 - 3x^2 + 2}{x^2 - 4}$, where $x \neq \pm 2$.

8. $r_8(x) = \dfrac{x^3 - 2x^2 - x + 5}{x - 2}$, where $x \neq 2$.

9. Suppose that the number of eggs per day produced by a chicken farm can be modeled by a rational function as

$$N(t) = \frac{10(10+t)}{1+0.01t}, \text{ where } t \geq 0,$$

where t is the time in days.

(a) Find the number of eggs produced when $t = 5$, 15, and 30.

(b) What is the limiting size of the number of the eggs to be produced by this chicken farm as time increases?

10. Suppose that the average cost $\overline{C}$ per unit in producing x units per month of a product is modeled by a rational function as $\overline{C} = \dfrac{0.001 \cdot x^2 + 10x + 250}{x}$, where $x > 0$. By graphing $\overline{C}$, estimate the product level that minimizes the average cost per unit.

15.3 A Partial Fraction Decomposition Technique

Ordinarily, the principle of simplification is applied everywhere in mathematics. For example, when we add (or subtract or multiply or divide or compose) two rational functions, we eventually simplify it to become a single new rational function as shown in Example 15.1.4. However, we are going to learn a new technique, called the method of a partial fraction decomposition whose goal is to decompose a simplified rational function back into the sum of several simpler proper rational functions. The reason why such a technique is needed is that there is no division rule for integration in calculus. Consequently, whenever we are asked to find an anti-derivative of a rational function, we have to use the method of a partial fraction decomposition to break the given rational function into the sum of several proper rational functions and then find an anti-derivative for each of these proper rational functions.

A Procedure for a Partial Fraction Decomposition for Rational Functions

Given a rational function $r(x)$ of Eq. (15.1.1), i.e., $r(x) = \dfrac{p_n(x)}{q_m(x)}$, where $m \geq 1$. Then we proceed as follows:

Step 1. Determine if $r(x)$ is a proper rational function. If yes, go to Step 2. If no, use the long division to obtain for $r(x)$ the quotient polynomial function and the remainder term given by

$$r(x) = \frac{p_n(x)}{q_m(x)} = s_{n-m}(x) + \frac{t(x)}{q_m(x)}, \tag{15.3.1}$$

where $t(x)$ is a polynomial function of degree at most $m - 1$. Then, apply Steps 2 and 3 (below) to the remainder of $\dfrac{t(x)}{q_m(x)}$, which is a proper rational function.

Step 2. Factor completely $q_m(x)$ into the product of real linear and (irreducible) quadratic factors.

Step 3. For any linear factor with a multiplicity no less than one, say, it is of the form $(a_1x + a_0)^j$, $j \geq 1$, the partial fraction decomposition has to include the following sum of j fractions:

$$\frac{A_1}{a_1x + a_0} + \frac{A_2}{(a_1x + a_0)^2} + \ldots + \frac{A_j}{(a_1x + a_0)^j}, \tag{15.3.2}$$

where A_i's, $i = 1, 2, \ldots, j$, are the unknown real constants to be determined. Similarly, for any irreducible quadratic factor with a multiplicity no less than one, say, it is of the form $(b_2x^2 + b_1x + b_0)^k$, $k \geq 1$, the partial fraction decomposition has to include the following sum of k fractions

$$\frac{B_1x + C_1}{b_2x^2 + b_1x + b_0} + \ldots + \frac{B_kx + C_k}{(b_2x^2 + b_{1x} + b_0)^k}, \tag{15.3.3}$$

where B_i's and C_i's, $i = 1, 2, \ldots, k$, are the unknown real constants to be determined.

Remarks

1. The result of Step 2 is assured by Theorem 14.6.3 (Factors of a real polynomial function).

2. Algebraic techniques for determining the unknown real constants A_i's, B_i's and C_i's are demonstrated in the following examples.

Example 15.3.1

Given: $r(x) = \dfrac{2x+1}{x^2-5x+6}$, where $x \neq 2$, or 3. Find a partial fraction decomposition for $r(x)$.

Solution

Since $r(x)$ is a proper rational function, we proceed to find a partial fraction decomposition for $r(x)$ as follows:

Step 1. By factoring the denominator of $r(x)$ completely, we have

$$x^2 - 5x + 6 = (x - 2)\cdot(x - 3). \quad (1)$$

Step 2. Because both of the two linear factors of Eq. (1) have merely a multiplicity of 1, we then set up the following identity:

$$\frac{2x+1}{x^2-5x+6} = \frac{A}{x-2} + \frac{B}{x-3} = \frac{A(x-3)+B(x-2)}{(x-2)(x-3)}, \quad (2)$$

where A and B are the unknown real constants to be determined. Since Eq. (2) is an identity and both sides of Eq. (2) have the same denominator, their numerators have to equal one another, namely,

$$2x + 1 = A\cdot(x - 3) + B\cdot(x - 2). \quad (3)$$

Note that Eq. (3) is again an identify, namely, Eq. (3) holds for any real number x. We want to take an advantage of this feature when solving for the unknown A and B, namely, we should evaluate an identity of Eq. (3) for some special x-values so that only one unknown remains after such an evaluation. Obviously, to solve for A, we have to evaluate both sides of Eq. (3) at $x = 2$:

$$2\cdot2 + 1 = A\cdot(2 - 3) + B\cdot(2 - 2) = A\cdot(-1) + B\cdot0 = -A, \Rightarrow$$
$$5 = -A, \Rightarrow A = -5.$$

To solve for B, we have to evaluate both sides of Eq. (3) at $x = 3$:

$$2\cdot3 + 1 = A\cdot(3 - 3) + B\cdot(3 - 2) = A\cdot0 + B\cdot1 = B, \Rightarrow B = 7.$$

Therefore, the decomposition is obtained as

$$\frac{2x+1}{x^2-5x+6} = \frac{-5}{x-2} + \frac{7}{x-3} = \frac{7}{x-3} - \frac{5}{x-2}.$$

Remarks

1. To check for the validity of the decomposition, we can simplify the right side by following the appropriate operational rules for

rational functions to see if it reduces to the given rational function as follows:

Right side of the decomposition

$$=\frac{7}{x-3}-\frac{5}{x-2}=\frac{7(x-2)-5(x-3)}{(x-3)(x-2)}=\frac{7x-14-(5x-15)}{(x-3)(x-2)}$$

$$=\frac{7x-14-5x+15}{(x-3)(x-2)}=\frac{2x+1}{x^2-5x+6}=r(x)$$

2. By the way, we need to discern the difference between an equation and an identity. An equation is satisfied only by a few x-values depending on the degree of the equation, while an identity holds for all real numbers.

Example 15.3.2

Given: $r(x)=\dfrac{2x+3}{x^3-1}$, where $x\neq 1$.

Find a partial function decomposition for $r(x)$.

Solution

Since $r(x)$ is a proper rational function, we proceed to find a partial fraction decomposition for $r(x)$ as follows:

Step 1. By using the factoring formula for the difference between two cubes in Section 3.2 in Part I, the denominator of $r(x)$ can be factored as

$$x^3-1=(x-1)\cdot(x^2+x+1). \tag{1}$$

Step 2. Since both of the linear and quadratic factors of Eq. (1) have only a multiplicity of 1, we then set up the following identity:

$$\frac{2x+3}{x^3-1}=\frac{A}{x-1}+\frac{Bx+C}{x^2+x+1},$$

$$=\frac{A\cdot(x^2+x+1)+(Bx+C)\cdot(x-1)}{(x-1)\cdot(x^2+x+1)}, \tag{2}$$

where A, B, and C are the unknown real constants to be determined. Since Eq. (2) is an identity and both sides of Eq. (2) have the same denominator, their numerators have to equal each other, namely,

$$2x+3=A\cdot(x^2+x+1)+(Bx+C)\cdot(x-1). \tag{3}$$

Note that Eq. (3) is again an identity, namely, Eq. (3) holds for any real number x. To solve for the unknown A, we evaluate both sides of Eq. (3) at $x=1$:

$$2\cdot 1 + 3 = A\cdot(1^2 + 1 + 1) + (B\cdot 1 + C)\cdot(1 - 1)$$
$$= A\cdot 3 + (B + C)\cdot 0, \Rightarrow 5 = 3\cdot A, \Rightarrow \frac{5}{3} = A. \quad (4)$$

Since the quadratic factor associated with A, $x^2 + x + 1$, has no real roots, we substituting A of Eq. (4), $x = 0$ and $x = -1$ into both sides of Eq. (3) in order to solve for the unknown B and C:

$$2\cdot 0 + 3 = \frac{5}{3}\cdot(0^2 + 0 + 1) + (B\cdot 0 + C)\cdot(0 - 1). \quad (5)$$

$$2\cdot(-1) + 3 = \frac{5}{3}\cdot[(-1)^2 + (-1) + 1)] + [(B\cdot(-1) + C)]\cdot(-1 - 1). \quad (6)$$

From Eq. (5), we have

$$3 = \frac{5}{3} + C\cdot(-1), \Rightarrow 3 - \frac{5}{3} = -C, \Rightarrow C = -\frac{4}{3}. \quad (7)$$

By substituting C of Eq. (7) into Eq. (6), we obtain

$$1 = \frac{5}{3} + (-B - \frac{4}{3})\cdot(-2) = 2B + \frac{13}{3}$$
$$-\frac{10}{3} = 1 - \frac{13}{3} = 2\cdot B, \Rightarrow B = -\frac{10}{2\cdot 3} = -\frac{5}{3}. \quad (8)$$

By substituting Eqs. (4), (7), and (8) into Eq. (2), the decomposition is then given by

$$\frac{2x+3}{x^3-1} = \frac{\frac{5}{3}}{x-1} + \frac{-\frac{5}{3}x - \frac{4}{3}}{x^2+x+1} = \frac{5}{3}\cdot\frac{1}{x-1} - \frac{1}{3}\cdot\frac{5x+4}{x^2+x+1}.$$

Remarks

When solving for the unknown B and C, we again will end with two equations in two variables B and C if we evaluate Eq. (3) at other x-values, say $x = 2$ and $x = 3$ as follows:

$$2\cdot 2 + 3 = \frac{5}{3}\cdot(2^2 + 2 + 1) + (B\cdot 2 + C)\cdot(2 - 1),$$

$$2\cdot 3 + 3 = \frac{5}{3}\cdot(3^2 + 3 + 1) + (B\cdot 3 + C)\cdot(3 - 1),$$

$$7 = \frac{35}{3} + 2B + C,$$

$$9 = \frac{65}{3} + 2\cdot(3B + C),$$

$$2B+C=7-\frac{35}{3}=-\frac{14}{3}, \quad (9)$$

$$3B+C=\frac{1}{2}\cdot(9-\frac{65}{3})=-\frac{19}{3}. \quad (10)$$

By subtracting Eq. (9) from Eq. (10), we have

$$3B+C-(2B+C)=-\frac{19}{3}-(-\frac{14}{3}),\Rightarrow\ B=-\frac{5}{3}. \quad (11)$$

By substituting Eq. (11) into Eq. (9), we have

$$2\cdot(-\frac{5}{3})+C=-\frac{14}{3},\ \Rightarrow\ C=-\frac{14}{3}+\frac{10}{3}=-\frac{4}{3}.$$

The values of B and C obtained here are exactly the same as those obtained earlier. Hence, it does not make matter as far as the choice of x-values for evaluation is concerned. However, the choice of $x = 0$ and $x = -1$ did make the process of solving B and C easier.

Exercises

Using the partial fraction decomposition technique to decompose the following rational functions:

1. $r_1(x)=\dfrac{5}{x^2-1}$.

2. $r_2(x)=\dfrac{3x+1}{x^2-x-6}$.

3. $r_3(x)=\dfrac{1}{x^4-16}$.

4. $r_4(x)=\dfrac{2x+3}{x^3-4x}$.

5. $r_5(x)=\dfrac{5x-2}{x^3+x^2+x+1}$.

15.4 References

1. Boyer, C. B. (1991). A History of Mathematics, 2nd edition, John Wiley and Sons, Inc., New York.
2. Flanigan, F. J. (1972). Complex Variables: Harmonic and Analytic Functions, Dover Publications, Inc., New York.
3. http://en.wikipedia.org/wiki/Category:Rational_functions.
4. http://en.wikipedia.org/wiki/Pad%C3%A9_approximant.
5. http://en.wikipedia.org/wiki/Ideal_gas_law.

College Algebra: Historical Notes

Chapter 16
Exponential and Logarithmic Functions

Among functions there are at least five big families. We already encountered two of them, polynomial and rational functions, in Chapters 14 and 15. The families of polynomial and rational functions belong to the class of algebraic functions. They are obtainable through algebraic operational rules (addition, subtraction, scalar multiplication, multiplication, division, composition, and power) applied to polynomial functions. Functions that are not algebraic are called transcendental, namely, they transcend, or go beyond algebraic functions. Three other families, exponential, logarithmic, and trigonometric function, belong to the class of transcendental functions.

In this chapter we're going to learn the important pair of transcendental functions, that is, exponential and logarithmic functions. Further, it's interesting to note that they are the inverse function of one another. Although the topic of exponential functions is presented first, properties of logarithmic functions are needed in the proof of Theorem 16.1.1 and also in the solution of Examples 16.1.5 and 16.1.7. Note that we cannot treat this pair of functions as two separated topics. Instead, they are entangled with one another as can be seen when we're facing to solve the exponential/logarithmic equations in Section16.3.

16.1 Exponential Functions

Definition 16.1.1

A (real) function of the form

$$f(x) = a^x \qquad (16.1.1)$$

is called an exponential function with base a, where $a > 0$, $a \neq 1$, and x is any real number.

Remarks

1. Note that both $a = 1$ and $a < 0$ are excluded from serving as a base for exponential functions. When $a = 1$, $f(x) = 1^x = 1$ is simply a constant function, Hence, an algebraic function. When $a < 0$, some of the function values are not ral numbers, e.g., if $f(x) = (-2)^x$, then $f(0.5) = (-2)^{0.5} = \sqrt{-2} = \sqrt{-1}\cdot\sqrt{2} = i\cdot\sqrt{2}$ is a complex number.
2. There is a technical difficulty associated with evaluating $f(x) = a^x$ when x is an irrational number. A formal treatment to remove this obstacle is beyond the scope of this book because it requires knowledge from calculus. Here we adopt a more pragmatic approach, namely, just find an approximate function value. For example, if $f(x) = 3^x$, then $f(\pi) = 3^\pi \approx 3^{3.14159} = 31.54418874$. Consequently, we assume in this book that $f(x) = a^x$ exists for all real numbers x and the operational rules for rational exponents described in Section 1.3 in Part I can be extended to cover exponential functions with irrational exponents.
3. Exponential function are widely used in describing economic and physical phenomena such as compound interest, population growth, and decay of radioactive material.
4. We need to distinguish between the exponential function $f(x) = a^x$ and the power function $g(x) = x^n$. An exponential function $f(x) = a^x$ is a function that has a constant base with an independent variable x serving as an exponent. However, a power function $g(x) = x^n$ is a function that has an independent variable x serving as a base with a fixed constant n serving as its exponent. They are completely different. We must not be confused by mistaking them as the same.
5. For a history of exponential function, please see Remark 3 of Definition 16.2.1.

Theorem 16.1.1. (Operational Rules for Exponential Functions) Let $f(x)$ be an exponential function given by Eq. (16.1.1). Then for any real numbers x and y, we have

1. Multiplication Rule

$$a^x \cdot a^y = a^{x+y}. \qquad (16.1.2)$$

2. Division Rule

$$\frac{a^x}{a^y} = a^{x-y}. \tag{16.1.3}$$

3. Power of the Power Rule

$$(a^x)^y = a^{x \cdot y}. \tag{16.1.4}$$

4. Power Rule

$$a^{-x} = \frac{1}{a^x}. \tag{16.1.5}$$

Remarks

1. The proof of this theorem for real exponents has to invoke the technique of powers via logarithms, which is beyond the scope of this book (Reference 10). Nevertheless, when x and y are rational numbers, all the rules here are exactly the same as those operational rules for rational exponents described in Section 1.3 in Part I.
2. Among functions, only exponential functions possess such properties:

$$f(x) \cdot f(y) = f(x+y),$$
$$f(x)/f(y) = f(x-y),$$

and

$$f(-x) = \frac{1}{f(x)}.$$

Example 16.1.1

Use a scientific calculator to evaluate the following function values:

(a) $2^{1.5} = ?$ (b) $(\sqrt{2})^{3.1} = ?$ (c) $4^{\frac{6}{7}} = ?$

Solution

A step-by-step procedure is demonstrated in the following display:

Number	Keystrokes	Display
$2^{1.5}$	2 [y^x] 1.5 [=]	2.828427125
$(\sqrt{2})^{3.1}$	2 [2nd] [x^2] [y^x] 3.1 [=]	2.928171392
$4^{\frac{6}{7}}$	4 [y^x] [( ] 6 [÷] 7 [ )] [=]	3.281341424

Therefore, we have:

(a) $2^{1.5} = 2.828427125$,

(b) $(\sqrt{2})^{3.1} = 2.928171392$,

(c) $4^{\frac{6}{7}} = 3.281341424$.

Remark

The above results were obtained by using Texas Instruments TI35 Plus.

Example 16.1.2

Solve each of the following equations:

(a) $2^{5-2x} = 8$. (b) $(1/27)^{2y-1} = 3^{-y}$.

Solution

(a) Since the given equation can be written as

$2^{5-2x} = 2^3$, we have

$$5 - 2x = 3, \Rightarrow \; -2x = 3 - 5 = -2, \Rightarrow$$
$$x = -2/(-2) = 1.$$

(b) Since the given equation can be written as

$$(3^{-3})^{2y-1} = 3^{-y}, \Rightarrow 3^{-3(2y-1)} = 3^{-y}, \Rightarrow$$
$$3^{-6y+3} = 3^{-y},$$

we have

$$-6y + 3 = -y, \Rightarrow -6y + y = -3, \Rightarrow -5y = -3, \Rightarrow$$
$$y = -3/(-5) = 3/5 = 0.6.$$

Example 16.1.3

Given: $f(x) = 2^x$, where $x \in R$.

(a) Find the horizontal asymptote for $f(x)$ if any exists.

(b) Graph $f(x)$.

Solution

(a) To find a horizontal asymptote, we need to see if $f(x)$ approaches any real number as $x \to \pm\infty$. We use a tabulation approach to calculate its limit as follows:

x	10	100	200	…	$x \to +\infty$
$f(x) = 2^x$	$2^{10} = 1024$	$2^{100} = 1.26768 \cdot 10^{30}$	$2^{200} = 1.607 \cdot 10^{60}$	…	$f(x) \to +\infty$

x	-10	-100	…	$x \to -\infty$

$f(x) = 2^x$	$2^{-10} = \frac{1}{2^{10}} =$ 0.001	$2^{-100} = \frac{1}{-2^{100}} =$ $0.78 \cdot 10^{-32}$	...	$f(x) \to 0^+$

Since $f(x) \to +\infty$ as $x \to +\infty$, there is no horizontal asymptote as $x \to +\infty$. However, since $f(x) \to 0^+$ as $x \to -\infty$, $y = 0$ is a horizontal asymptote as $x \to -\infty$.

(b) To graph $f(x)$, we need to find the y-intercept, the x-intercepts and a few other points. Since $f(0) = 2^0 = 1$, the y-intercepts is $Q(0,1)$. Since $f(x) > 0$, for any real number x, there are no x-intercepts at all. To get a few extra points, we evaluate $f(x)$ at $x = -2, -1, 1$ and 2:

x	-2	-1	1	2
$f(x) = 2^x$	$2^{-2} = \frac{1}{2^2} = \frac{1}{4}$	$2^{-1} = \frac{1}{2^1} = \frac{1}{2}$	$2^1 = 2$	$2^2 = 4$

The four extra points are $P_1(-2, \frac{1}{4})$, $P_2(-1, \frac{1}{2})$, $P_3(1,2)$, and $P_4(2,4)$.

By connecting P_1, P_2, Q, P_3, and P_4 smoothly, plus using the limiting behavior of $f(x)$ as $x \to \pm\infty$, the graph of $f(x)$ is shown in Figure 16.1.1.

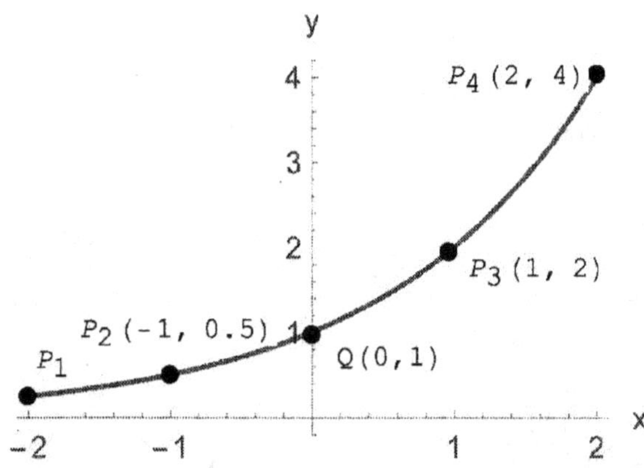

Figure 16.1.1

Remark

$f(x) = 2^x$ is an one-to-one (continuous) increasing function on its domain, Domain(f) = $(-\infty,+\infty)$, with Range(f) = $\{y \in R| \, y > 0\} = (0, +\infty)$. This property is shared by all exponential functions of Eq. (16.1.1) with $a > 1$.

Example 16.1.4

Given: $f(x) = (\frac{1}{2})^x = 0.5^x$, where $x \in R$.

(a) Find the horizontal asymptote for $f(x)$ if any exists.
(b) Graph $f(x)$.

Solution

(a) To find a horizontal asymptote, we need to see if $f(x)$ approaches any real number as $x \to \pm\infty$. We use a tabulation approach to calculate its limit as follows:

x	10	100	...	$x \to +\infty$
$f(x) = 0.5^x$	$0.5^{10} = 0.001$	$0.5^{100} = 0.789 \cdot 10^{-32}$	...	$f(x) \to 0^+$

x	-10	-100	...	$x \to -\infty$
$f(x) = 0.5^x$	$0.5^{-10} = \frac{1}{0.5^{10}} = 1024$	$0.5^{-100} = \frac{1}{0.5^{100}} = 1.268 \cdot 10^{30}$	...	$f(x) \to +\infty$

Since $f(x) \to 0^+$ as $x \to +\infty$, $y = 0$ is a horizontal asymptote as $x \to +\infty$. However, since $f(x) \to +\infty$ as $x \to -\infty$, there is no horizontal asymptote as $x \to -\infty$.

(b) To graph $f(x)$, we need to find the y-intercept, the x-intercepts, and a few other points. Since $f(0) = 0.5^0 = 1$, the y-intercept is $Q(0,1)$. Since $f(x) > 0$ for any real number x, there are no x-intercepts at all. To get a few extra points, we evaluate $f(x)$ at $x = -2, -1, 1$, and 2:

x	-2	-1	1	2
$f(x) = 0.5^x$	$0.5^{-2} = \frac{1}{0.5^2} = 4$	$0.5^{-1} = \frac{1}{0.5} = 2$	$0.5^1 = 0.5$	$0.5^2 = 0.25$

The four extra points are $P_1(-2, 4)$, $P_2(-1, 2)$, $P_3(1, 0.5)$, and $P_4(2, 0.25)$. By connecting P_1, P_2, Q, P_3, and P_4 smoothly, plus using the limiting behavior of $f(x)$ as $x \to \pm\infty$, the graph of $f(x)$ is shown in Figure 16.1.2.

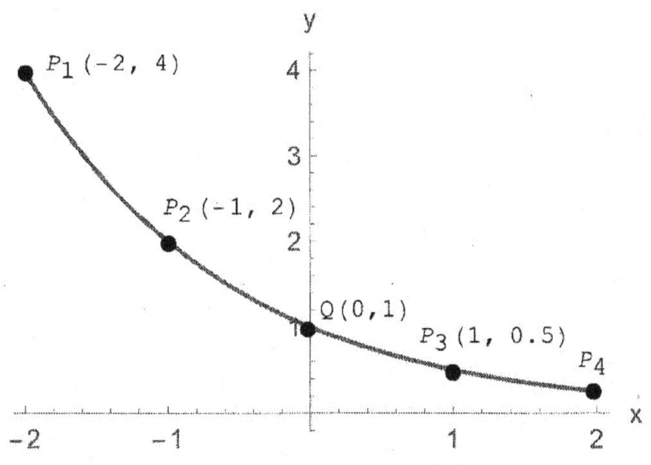

Figure 16.1.2

Remark

$f(x) = (\frac{1}{2})^x = 0.5^x$ is an one-to-one [continuous] decreasing function on its domain $(-\infty,+\infty)$ with Range $(f) = \{y \in R|\ y > 0\} = (0, +\infty)$. This property holds for any exponential function of Eq. (16.1.1) with $0 < a < 1$.

Definition 16.1.2

An irrational number $e \approx 2.718281828\ldots$ is defined as the limit of a function $f(x) = (1+\frac{1}{x})^x$ as $x \to +\infty$, namely, as $x \to +\infty$,

$$(1+\frac{1}{x})^x \to e \qquad (16.1.6)$$

Remark

The exact origin of the irrational number e was shrouded in mystery. One thing for sure was that the number e was known to mathematicians at least half a century before the invention of the calculus. There were two sources that led independently to the same number e. One source was that people was concerned with how money grows with time, i.e., the compound interest problem. The other source was how to find the area under the hyperbola $y = \frac{1}{x}$. The much more familiar role of e as the "natural" base of logarithms was done by L. Euler (1707-1783, Swiss) in an unpublished paper he wrote in 1728. For a more complete story about the irrational number e, please refer to Reference 7, pp. 28-36.

Definition 16.1.3
An exponential function with base e, $f(x) = e^x$, is called a natural exponential function, where e is called the natural base.

Remarks

1. Among all members in the family of exponential functions, $f(x) = e^x$ is the most prominent member because many natural phenomena could be specifically modeled in terms of natural exponential functions like bacteria growth, radioactive decay, and interest compounded continuously. Also, the standard formula for the derivative of an exponential function in calculus is given only for a natural exponential function. For all exponential function with other bases, we have to use the inverse function relationship to re-express it as an exponential function with the natural base e.
2. A family of hyperbolic functions are defined in terms of natural exponential functions (Problem 2 in Exercises for Section 16.1).
3. In de Moivre's theorem, a calculation of the nth root of a complex number involves the natural exponential functions (Reference 12).
4. In statistics, the probability density function of the family of normal distributions is a composition of a natural exponential function and a polynomial function (Reference 9, pp. 73-76).

Theorem 16.1.1 (Interest Compounded Continuously)

Suppose P_0 dollars is invested at an annual interest rate $(100r)\%$. If interest is compounded n times a year at equally spaced time intervals, it can be shown that the value $P_n(t)$ of the investment after t years is given by

$$P_n(t) = P_0(1+\frac{r}{n})^{nt} \tag{16.1.7}$$

If the interest is compounded continuously, that is, compounded at every instant of time, the value $P(t)$ of the initial P_0 dollars after t years is given by

$$P(t) = \lim_{n\to\infty} P_0(1 + r/n)^{nt} = P_0e^{rt}. \tag{16.1.8}$$

Remark
Eq. (16.1.8) follows directly from using Eq. (16.1.6) (Reference 13).

Example 16.1.5
How many years will it take for an initial investment of \$10,000 to grow to \$30,000? Assume that an annual interest rate of 8% compounded quarterly.
Solution
Let t_0 be the required years. Then by using Eq. (16.1.7), we have

$$\$30{,}000 = P_4(t_0) = \$10{,}000\cdot(1+\frac{0.08}{4})^{4\cdot t_0} = \$10{,}000\cdot 1.02^{4t_0}, \Rightarrow$$

$$3 = \frac{30{,}000}{10{,}000} = 1.02^{4\cdot t_0}, \Rightarrow$$

$$\log_{10} 3 = \log_{10} 1.02^{4t_0} = 4t_0\cdot\log_{10} 1.02, \Rightarrow$$

$$t_0 = \frac{\log_{10} 3}{4\cdot\log_{10} 1.02} = \frac{0.477121254}{0.034400687} \approx 13.8695 \approx 13.87 \text{ (years)}.$$

Hence, it will take 13.87 years for an initial investment of \$10,000 to grow to \$30,000.

Remark
In solving for t_0 we need to use the properties of logarithmic function in Section 16.2.

Example 16.1.6

Suppose that the amount of \$10,000 is invested at an interest rate of 7% per year compounded continuously. What will the investment be worth after 5 years?

Solution

By using Eq. (16.1.7) with $P_0 = 10{,}000$, $r = 0.07$ and $t = 5$, we have

$$P(5) = 10{,}000 \times e^{0.07\times 5} = 10{,}000 \times e^{0.35}$$
$$= 10{,}000 \times 1.41907 = 14{,}190.7.$$

Hence, the value of investment after 5 years is \$14,190.7.

Example 16.1.7

An investment is made in a mutual fund at an annual interest rate of 10% compounded continuously. How long will it take for the investment to double in value?

Solution

Let t_0 and P_0 be the required years and an initial investment, respectively. Then, by using Eq. (16.1.8),

$$2P_0 = P(t_0) = P_0 \cdot e^{0.1\cdot t_0}, \Rightarrow\ 2 = e^{0.1t_0}, \Rightarrow$$
$$\ln 2 = \ln e^{0.1\cdot t_0} = 0.1\cdot t_0, \Rightarrow\ t_0 = \frac{\ln 2}{0.1} = 6.9314718 \approx 6.93.$$

Hence, it will take 6.93 years for the investment to double in value.

Remark

In solving for t_0 we need to use the properties of logarithmic function in Section 16.2.

Exercises

1. Solve for each of the following equations:
(a) $625^{2x} = 5$. (b) $(2/3)^{y-1} = 9/4$. (c) $216^{4-3z} = 36^z$. (d) $w^{7/2} = 16$.

Remark

In solving part (d), you need to invoke the properties of the logarithmic functions in Section 16.2 (see Examples 16.1.5-7).

2. Use a calculator to graph each of the following functions:
(a) $f_1(x) = (e^x - e^{-x})/2$. (b) $f_2(x) = (e^x + e^{-x})/2$.

Remark

Actually $f_1(x)$ and $f_2(x)$ are the first two members of a family of hyperbolic functions. $f_1(x) = \sinh(x)$ and $f_2(x) = \cosh(x)$ are called hyperbolic sine and hyperbolic cosine function, respectively. V. Riccati (1707-1775, Italian) had suggested in 1757 a development of hyperbolic function. However, it was J. H. Lambert (1728-1777, Swiss-Germany) who introduced the notation sinh(x) and cosh(x) for the hyperbolic equivalent of the circular (trigonometric) functions. Hyperbolic functions are useful in modeling the curve of an electrical transmission line suspended between two poles (Reference 6).

3. Graph each of the following functions and compare with Example 16.1.3:

(a) $g_1(x) = 2^x + 1$. (b) $g_2(x) = 2^x - 4$.

4. Graph each of the following functions and compare with Example 16.1.4:

(a) $h_1(x) = 0.5^{x+1}$. (b) $h_2(x) = 0.5^{x-2}$.

5. How much would the initial 401k retirement money of $300,000 that is invested at the mutual fund at 6% a year compounded continuously become after (a) 1 year, (b) 5 years, or (c) 10 years?

6. How long does it take for an investment to double in value if it is invested at an annual interest rate of 9% compounded daily?

7. How many years will it take for an initial investment of $100,000 to grow to $250,000 at an annual interest rate of 10% compounded monthly?

8. An investment of $5,000 is compounded continuously. What annual interest rate will produce a balance of $12,000 in 5 years?

9. Suppose that you will buy a new car for $12,000 in 3 years. How much money should you ask your parents for now so that if you invest it at an annual interest rate of 6% compounded continuously, you will have enough money to buy the car?

16.2 Logarithmic Functions

Definition 16.2.1

The inverse function of an exponential function with base a, $f(x) = a^x$, is called a logarithmic function, with base a denoted by

$$g(x) = f^{-1}(x) = \log_a x, \qquad (16.2.1)$$

where $x > 0$, $a > 0$, and $a \neq 1$.

Remarks

1. Rarely in the history of science has an abstract mathematical idea been received more enthusiastically by the entire scientific community than the invention of logarithms. The reason why is that attributed to an enormous expansion of scientific knowledge in astronomy, geography, marine exploration, and physics, these development required scientists to spend much of their time doing tedious numerical calculations. The times called for an invention that would free scientists once and for all from this burden. J. Napier (1550 – 1617, Scottish) took up the challenge. It took him twenty years to complete the task and published it 1614. His idea was to observe that if one uses the exponents of a geometric square to form an arithmetic sequence, then a multiplication/division between terms of the given geometric sequence corresponds an addition/subtraction between the corresponding terms in its associated arithmetic sequence. Hence, if we could write any positive number as a power of some given fixed number, called a base, then multiplication/division of numbers are equivalent to addition/subtraction of their exponents (Reference 7, pp. 5 - 8).
2. Having completed his monumental task, Napier at first called the exponent of each number in the geometric sequence its "artificial number", but later decided on term "logarithm" that means "ratio number" (Reference 7, p. 8).
3. Although we define a logarithmic function as the inverse function of an exponential function, the fact is that logarithm was invented before our modern exponential notation, a^n, was introduced into algebra. For a history on the exponential and logarithmic concepts, see References 2 – 6.

4. Among members in the family of logarithmic functions, the one with base $e \approx 2.71828...$ is the most prominent one, called the natural logarithmic function, denoted as $\log_e(x) = \ln(x)$.
5. Note that $x = 0$ is a vertical asymptote for the graph of $g(x) = \log_a(x)$. See Example 16.2.4 for an illustration.
6. Among applications of the logarithmic function, the logarithmic spiral deserves a special attention because it serves as the preferred growth pattern for numerous natural forms including shells, horns, tusks, sunflowers, and even the galaxy (Reference 7, pp. 134-139). In statistics, the logarithmic transformation is used to transform approximately a positive random variable into a normal random variable.

Definition 16.2.2
The logarithmic function using an irrational number e as its base and denoted by $g(x) = \ln x = \log_e x$ is called the natural logarithmic function.

Remark
Just like the natural exponential function $f(x) = e^x$, $g(x) = \ln x$ plays a dominating role in calculus. The formula for the derivative of a logarithmic function is set up in terms of the natural logarithmic function, for all other members in the family of logarithmic function, their derivatives are obtained through a change-of-base formula to be described in Section 16.3.

Definition 16.2.3
The logarithmic function using a number 10 as its base, denotes by $g(x) = \log_{10} x$, is called a common logarithmic function.

Remark
On any scientific calculator, the keystroke for $\log_{10} x$ is abbreviated as $\log x$. See Example 16.2.2.

Theorem 16.2.1 (Inverse function identities between logarithmic and exponential functions)
Given: $f(x) = a^x$ and $g(x) = \log_a x$, where $a > 0$, $a \neq 1$, Domain(f) = $(-\infty,\infty)$, and Domain $(g) = (0, +\infty)$, where

$$\log_a a^x = x, \tag{16.2.2}$$

and

$$a^{\log_a x} = x. \tag{16.2.3}$$

Proof

Since f and g are inverse functions of one another, we have by using Eqs. (13.4.3) and (13.4.4) in Section 13.4

$$x = (g \circ f)(x) = g(f(x)) = \log_a(f(x)) = \log_a a^x,$$

which leads to Eq. (16.2.2), and

$$x = (f \circ g)(x) = f(g(x)) = a^{g(x)} = a^{\log_a x},$$

which leads to Eq. (16.2.3). Hence, the proof is complete.

Remarks

1. Eq. (16.2.2) states that if a number is of the form a^x, then the logarithm of that number is simply its exponent. Indeed, this is the original definition of the logarithm of a positive number. However, Eq. (16.2.3) does not have such an intuitive interpretation. It simply follows from the fact that $f(x) = a^x$ and $g(x) = \log_a x$ are inverse functions of one another.
2. For $x = 0$ and $x = 1$, we have from Eq. (16.2.2)

   $$\log_a 1 = 0, \text{ (Since } a^0 = 1) \tag{16.2.4}$$

   and

   $$\log_a a = 1. \text{ (Since } a^1 = a) \tag{16.2.5}$$
3. The identities of Eqs. (16.2.2-3) would become useful in solving an equation involving the exponential and logarithmic function that will covered in Section 16.3.

Theorem 16.2.2 (Operational Rules for Logarithmic Functions)

Let $g(x)$ be defined by Eq. (16.2.1). Then, for any $x > 0$ and $y > 0$, we have

1. Addition rule

$$\log_a x + \log_a y = \log_a(xy), \tag{16.2.6}$$

2. Subtraction rule

$$\log_a x - \log_a y = \log_a\left(\frac{x}{y}\right). \tag{16.2.7}$$

3. Power rule

$$\log_a x^r = r \cdot \log_a x, \tag{16.2.8}$$

where r is real number.

4. Change-base rule

$$\log_a x = \frac{\log_b x}{\log_b a}, \qquad (16.2.9)$$

where b is a positive real number with $b \neq 1$.

Proof

Rule 1.

Let $X = \log_a x$ and $Y = \log_a y$. Then, by using Eq. (16.2.3), we have

$a^X = a^{\log_a x} = x,$

and

$a^Y = a^{\log_a y} = y.$

Thus,

$a^X \cdot a^Y = x \cdot y,$

$a^{X+Y} = x \cdot y,$ (By using Eq. (16.1.2))

$\log_a a^{X+Y} = \log_a(x \cdot y),$ (By taking the logarithm of both sides of the above equation)

$X + Y = \log_a(x \cdot y),$

$\log_a x + \log_a y = \log_a(x \cdot y),$

(Since $X = \log_a x$ and $Y = \log_a y$)

which is Eq. (16.2.6). Hence, the proof is complete.

Rule 2. The proof is similar to the one given above; Hence, it is left as a practice in problem 1 of Exercises for Section 16.2.

Rule 3. Let $X = \log_a x$. Then, we have

$a^X = a^{\log_a x} = x,$

$(a^X)^r = x^r,$

$a^{r \cdot X} = x^r,$ (By using Eq. (16.1.4))

$\log_a a^{r \cdot X} = \log_a x^r,$ (By taking the logarithm of both sides of the above equation)

$r \cdot X = \log_a x^r,$ (By using Eq. (16.2.2))

$r \cdot \log_a x = \log_a x^r,$ (Since $X = \log_a x$)

which is Eq. (16.2.8). Hence, the proof is complete.

Rule 4. Let $X = \log_a x$. Then, we have

$a^X = a^{\log_a x} = x,$ (By using Eq. (16.2.3))

$\log_b a^X = \log_b x,$ (By taking the logarithm with base b on both sides of the above equation)

$X \cdot \log_b a = \log_b x$, (By using Eq. (16.2.8))

$$X = \frac{\log_b x}{\log_b a}, \text{ (Dividing both sides by } \log_b a)$$

$$\log_a X = \frac{\log_b x}{\log_b a}, \text{ (Since } X = \log_a x)$$

which is Eq. (16.2.9). Hence, the proof is complete.

Remarks

1. Each of the operational rules given by Eqs. (16.2.6-9) is an identity that can be read either from its left side to its right side, or vice versa. For example, on the one hand, if Eq. (16.2.6) is read from its left side to its right side, it states that the addition of a logarithmic function at two different values in its domain can be simplified as the logarithm of the product of these two values x and y. Oftentimes this aspect of Eq. (16.2.6) is very useful when it comes to solve as equation involving the logarithmic equation to be covered in Section 16.3. On the other hand, if Eq. (16.2.6) is read from its right side to its left side, then it states that the logarithm of the product of two real numbers x and y can be decomposed as the sum of two logarithms of x and y, respectively. The aspect of Eq. (16.2.6) forms a special technique in calculus called a technique of the logarithmic differentiation. The same comments are also applicable to all four identities of Eqs. (16.2.6-9).
2. A warning concerning common mistakes about the addition and the subtraction rules as follows:

   $$\log_a(x \cdot y) \neq \log_a x \cdot \log_a y,$$

   and

   $$\log_a(\frac{x}{y}) \neq \frac{\log_a x}{\log_a y}.$$

Example 16.2.1

Given: $g(x) = \log_9 x$. Evaluate: (a) $g(1) = ?$ (b) $g(3) = ?$ (c) $g(9) = ?$ (d) $g(27) = ?$ (e) $g(\frac{1}{3}) = ?$ (f) $g(\frac{1}{27}) = ?$

Solution

(a) $g(1) = \log_9 1 = 0$ (By using Eq. (16.2.4))

(b) $g(3) = \log_9 3 = \log_9 \sqrt{9} = \log_9 9^{\frac{1}{2}} = \frac{1}{2} \cdot \log_9 9 = \frac{1}{2} \cdot 1 = \frac{1}{2}$.

(c) $g(9) = \log_9 9 = 1$ (By using Eq. (16.2.5)).

(d) $g(27) = \log_9 27 = \log_9 (3 \cdot 9) = \log_9 3 + \log_9 9 = \frac{1}{2} + 1 = \frac{3}{2}$

(By using Eq. (16.2.6) and the results of part (b) and (c)).

(e) $g(\frac{1}{3}) = \log_9 (\frac{1}{3}) = \log_9 1 - \log_9 3 = 0 - \frac{1}{2} = -\frac{1}{2}$

(By using Eq. (16.2.7) and the results of part (a) and (b)).

(f) $g(\frac{1}{27}) = \log_9 (\frac{1}{27}) = \log_9 1 - \log_9 27 = 0 - \frac{3}{2} = -\frac{3}{2}$

(By using Eq. (16.2.7) and the results of part (a) and (d)).

Remark

The reason why we did not use a calculator to figure out the logarithmic function values is that all x-values to be evaluated have a hidden power relationship with the base 9. In general, we do need to use a scientific calculator in coupling with the use of Eqs. (16.2.6-9) to figure out the logarithmic function value as shown in the following example.

Example 16.2.2

Evaluate: (a) $\log_{10} 2 = ?$ (b) $\log_{10} 3 = ?$ (c) $\log_2 3 = ?$

(d) $\log_4 \sqrt{5} = ?$ (e) $\log_5 (6.2) = ?$ (f) $\ln(3.45) = ?$

Solution

Ordinarily, a scientific calculator has two keystrokes "log" and "ln" corresponding to the common and the natural logarithmic function, respectively. To evaluate a logarithmic function with base other than 10 and e, we need to use the change-base identify (Eq. (16.2.9)) to convert into either the common or the natural logarithmic function and then use the calculator to figure out the function value.

Number	Keystrokes	Display
$\log_{10}2$	2 [log] [=]	0.301029995
$\log_{10}3$	3 [log] [=]	0.477121254
$\log_{10}4$	4 [log] [=]	0.602059991
$\log_{10}5$	5 [log] [=]	0.698970004

ln6.2	6.2 [ln] [=]	1.824549292
ln5	5 [ln] [=]	1.609437912
ln3.45	3.45 [ln] [=]	1.238374231

(a) $\log_{10} 2 = 0.301029995$.

(b) $\log_{10} 3 = 0.477121254$.

(c) $\log_2 3 = \dfrac{\log_{10} 3}{\log_{10} 2} = \dfrac{0.477121254}{0.301029995} = 1.584962504$.

(d)$\log_4 \sqrt{5} =$

$$\log_4 5^{\frac{1}{2}} = \frac{1}{2} \cdot \log_4 5 = \frac{1}{2} \cdot \frac{\log_{10} 5}{\log_{10} 4} = \frac{1}{2} \cdot \frac{0.698970004}{0.602059991}$$

$= 0.580482023$.

(e) $\log_5(6.2) = \dfrac{\ln 6.2}{\ln 5} = \dfrac{1.8244929}{1.609437912} = 1.133656215$.

(f) $\ln 3.45 = 1.238374231$.

Remark

In the solution to part (e), it does not make matter if we decide to covert it into the base of 10 as follows:

$$\log_5(6.2) = \frac{\log_{10}(6.2)}{\log_{10}(5)} = \frac{0.792391689}{0.698970004} = 1.133656215 \quad ,$$

which is exactly the same as the value obtained by using the base *e*.

Example 16.2.3

Find the domain of the following logarithmic functions:

(a) $g_1(x) = \log_{10}(3 - x)$. (b) $g_2(x) = \ln(x^2 - 5x + 6)$.

Solution

(a) Since the *x*-value of a common logarithmic function is required to be positive, we have to solve the following inequality:

$$3 - x > 0, \Rightarrow x < 3.$$

Hence, Domain(g_1) = $\{x \in R | x < 3\} = (-\infty, 3)$.

(b) Since the *x*-value of a natural logarithmic function is required to be positive, we have to solve the following inequality:

$$x^2 - 5x + 6 > 0,$$

$$(x - 2)\cdot(x - 3) > 0, \Rightarrow x > 3 \text{ or } x < 2.$$

Hence, Domain(g_2) = $\{ x \in R | x < 2 \text{ or } x > 3\} = (-\infty, 2) \cup (3, +\infty)$.

Remark

For those readers who are not familiar with solving an inequality, please read Chapters 7-8 in Part I.

Example 16.2.4

Given: $g(x) = \log_2 x$, where $x > 0$.

(a) Show that $x = 0$ is a vertical asymptote and there is no horizontal asymptote for the graph of $g(x)$.

(b) Graph $g(x)$.

Solution

(a) To show that $x = 0$ is a vertical asymptote, it suffices to show that $g(x)$ approaches either $+\infty$ or $-\infty$ as $x \to 0^+$. This can be accomplished through a tabulation as follows:

x	0.01	0.000001	…	$x \to 0^+$
$g(x) = \log_2 x$	$\log_2 0.01 = \frac{\log_{10} 0.01}{\log_{10} 2} = -6.644$	-19.932	…	$g(x) \to -\infty$

As it is seen from the above table, $g(x) \to -\infty$ as $x \to 0^+$. Hence, $x = 0$ is a vertical asymptote for the graph of $g(x)$. To show that $g(x)$ has no horizontal asymptote, we will show that $g(x)$ does not approach any finite limit as $x \to +\infty$ through a tabulation as follows:

x	100	10^6	…	$x \to +\infty$
$g(x) = \log_2 x$	$\log_2 100 = \frac{\log_{10} 100}{\log_{10} 2} = 6.644$	19.932	…	$g(x) \to +\infty$

It is seen from the above table that $g(x) \to +\infty$ as $x \to +\infty$. Hence, $g(x)$ has no horizontal asymptote.

(b) To graph $g(x)$, let us first find the y-intercept and the x-intercepts. Since $x = 0$ is not in the domain of $g(x)$, we cannot evaluate $g(x)$ at $x = 0$. Consequently, there is no y-intercept. Since $g(1) = \log_2 1 = 0$, the x-intercept is $Q(1, 0)$. Next, in order to get a few extra point, we evaluate $g(x)$ at $x = 0.5, 2, 4$, and 8:

x	0.5	2	4	8
$g(x) =$ $\log_2 x$	$\log_2 0.5$ $= \log_2 2^{-1}$ $= -1$	1	$\log_2 4$ $= \log_2 2^2$ $= 2$	$\log_2 8 = 3$

The six extra points are $P_2(0.5, -1)$, $P_3(2, 1)$, $P_4(4, 2)$, and $P_5(8, 3)$. By connecting the seven points P_1, P_2, Q, P_3, P_4, and P_5 smoothly and using the knowledge of a vertical asymptote and no horizontal asymptote, the graph of $g(x) = \log_2 x$ is shown in Figure 16.2.1.

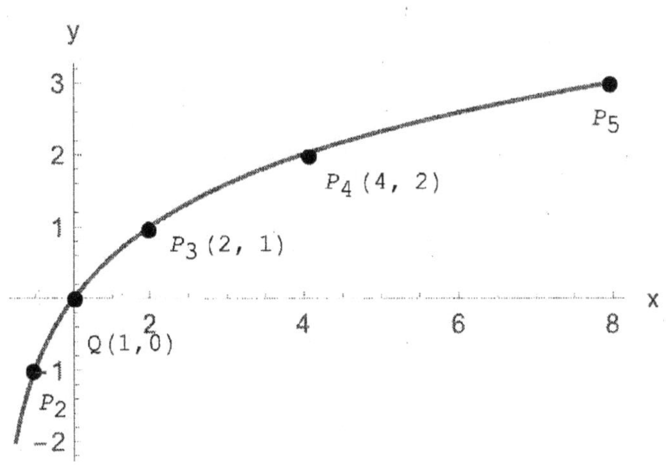

Figure 16.2.1

Remarks

1. $g(x) = \log_2 x$ is an one-to-one (continuous) increasing function on its domain, where Domain(g) = $(0, +\infty)$ and Range(g) = $(-\infty, +\infty)$. This property is shared by all logarithmic functions of Eq. (16.2.1) with $a > 1$. Also note that Figures 16.1.1 and 16.2.1 are symmetric with respect to the 45 degree line of $y = x$.
2. Note that the y-axis in Figure 16.2.1 is, in fact, $x = 1$ rather than $x = 0$. It seems that the "Mathematica" software is designed not to plot $x = 0$ directly as a vertical asymptote.

Example 16.2.5

Given: $g(x) = \log_{0.5} x$, where $x > 0$.
(a) Show that $x = 0$ is a vertical asymptote and there is no horizontal asymptote for the graph of $g(x)$.
(b) Graph $g(x)$.

Solution

(a) The proof will be similar to that of Example 16.2.4 by constructing a table to show that $g(x) \to +\infty$ as $x \to 0^+$ as follows:

x	0.01	0.000001	...	$x \to 0^+$
$g(x) = \log_{0.5} x$	$\log_{0.5} 0.01 = \frac{\log_{10} 0.01}{\log_{10} 0.5} = 6.644$	19.932	...	$g(x) \to +\infty$

As it can be seen from the above table $g(x) \to +\infty$ as $x \to 0^+$. Hence, $x = 0$ is a vertical asymptote for the graph of $g(x)$.
To show that $g(x)$ has no horizontal asymptote, we will show that $g(x)$ does not approach any finite limit as $x \to +\infty$ through a tabulation as follows:

x	100	10^6	...	$x \to +\infty$
$g(x) = \log_{0.5} x$	$\log_{0.5} 100 = \frac{\log_{10} 100}{\log_{10} 0.5} = -6.644$	-19.932	...	$g(x) \to -\infty$

It can be seen from the above table that $g(x) \to -\infty$ as $x \to +\infty$. Hence, $g(x)$ has no horizontal asymptote.
(b) To graph $g(x)$, let us first find the y-intercept and the x-intercepts. Since $x = 0$ is not in the domain of $g(x)$, we cannot evaluate $g(x)$ at $x = 0$. Consequently, there is no y-intercept. Since $g(1) = \log_{0.5} 1 = 0$, the x-intercept is $Q(1, 0)$. Next, in order to get a few extra points, we evaluate $g(x)$ at $x = 0.5$, 2, 4, and 8:

x	0.5	2	4	8
$g(x) = \log_{0.5} x$	$\log_{0.5} 0.5 = 1$	-1	$\log_{0.5} 4 = \log_{0.5} 0.5^{-2} = -2$	$\log_{0.5} 8 = -3$

The six extra points are $P_2(0.5, 1)$, $P_3(2, -1)$, $P_4(4, -2)$, and $P_5(8, -3)$. By connecting P_2, Q, P_3, P_4, and P_5, smoothly and using the knowledge of a vertical asymptote and no horizontal asymptote, the graph of $g(x) = \log_{0.5} x$ is shown in Figure 16.2.2

Remarks

1. $g(x) = \log_{0.5} x$ is an one-to-one (continuous) decreasing function on its domain, where Domain(g) = $(0, +\infty)$ and Range(g) = $(-\infty, +\infty)$. This property is shared by all logarithmic functions of Eq. (16.2.1) with $0 < a < 1$. Also, note that Figures 16.1.2 and 16.2.2 are symmetric with respect to the 45 degree line of $y = x$.
2. The same remark as that of remark 2 for Example 16.2.4 is applicable to Figure 16.2.2.

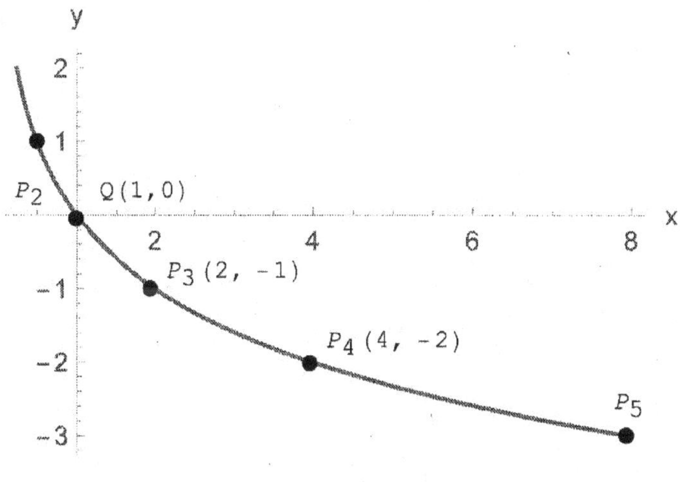

Figure 16.2.2

Example 16.2.6

The human ear is an extremely versatile and sensitive detector of sound. We are able to detect sounds of widely varying intensities and to make rough judgments of comparative loudness. Because of this wide range of varying intensities that can be substantiated by a human ear, a unit, called the decibel (dB), which is used to measure the sound-intensity levels is defined on a logarithmic scale. In physics, the intensity level in decibel for measuring a sound having an intensity of I in watts/m^2 is given by

$$\beta = g(I) = 10 \cdot \log_{10}(\frac{I}{I_0}), \tag{16.2.10a}$$

Where I_0 is the intensity of the "faintest audible sound" and given by

$$I_0 = 10^{-12} \text{ watts/m}^2. \tag{16.2.10b}$$

The intensity of an average whisper approximately equals 10^{-10} watts/m^2, what is its magnitude measured in dB?

Solution

By using Eqs. (16.2.10a-b) with $I = 10^{-10}$ watts/m^2, we have

$$\beta = g(10^{-10}) = 10 \cdot \log_{10}(\frac{10^{-10}}{10^{-12}}) = 20 \text{ (dB)}.$$

Hence, the intensity level of an average whisper is 20 dB.

Remarks

1. For a more detailed discussion on measuring the loudness of sounds, see Reference 1, pp. 308-310.
2. To put a measure on the logarithmic scale is a common practice in empirical sciences, provided that the measure has a very wide range. Many environmental data occurring in statistics are usually put on a logarithmic scale because the data have a wide range of magnitude.

Exercises

1. Show that Eq. (16.2.7) is an identity.

In problems 2–38, evaluate without using a calculator:

2. $\log_2 1024 = ?$

3. $\log_2(\frac{1}{32}) = ?$

4. $\log_{0.5} 128 = ?$

5. $\log_{0.5}(\frac{1}{16}) = ?$

6. $\log_3 243 = ?$

7. $\log_3(\frac{1}{81}) = ?$

8. $\log_{\frac{1}{3}} 27 = ?$

9. $\log_{\frac{1}{3}} (\frac{1}{81}) = ?$

10. $\log_4 2 = ?$

11. $\log_4 8 = ?$

12. $\log_4 (\frac{1}{2}) = ?$

13. $\log_4 (\frac{1}{32}) = ?$

14. $\log_{\frac{1}{4}} 16 = ?$

15. $\log_{\frac{1}{4}} (\frac{1}{32}) = ?$

16. $\log_8 2 = ?$

17. $\log_8 4 = ?$

18. $\log_8 16 = ?$

19. $\log_8 32 = ?$

20. $\log_{\frac{1}{8}} 2 = ?$

21. $\log_{\frac{1}{8}} 16 = ?$

22. $\log_{\frac{1}{8}} 64 = ?$

23. $\log_{\frac{1}{8}} \frac{1}{4} = ?$

24. $\log_{\frac{1}{8}} \frac{1}{32} = ?$

25. $\log_{10} 10{,}000 = ?$

26. $\log_{10} 0.00001 = ?$

27. $\log_{0.1} 1{,}000 = ?$

28. $\log_{0.1} 0.0001 = ?$

29. $\log_{100} 10 = ?$

30. $\log_{100} 1{,}000 = ?$

31. $\log_{0.01} 10 = ?$

32. $\log_{0.01} 100{,}000 = ?$

33. $\ln 1 = ?$

34. $\ln(\frac{1}{e}) = ?$

35. $\log_7 1 = ?$

36. $\log_7 \frac{1}{343} = ?$

37. $\log_\pi 1 = ?$

38. $\log_\pi \frac{1}{\pi^5} = ?$

In problems 39–42, evaluate without a calculator the given expression by using $\log_{10} 2 = 0.3010$ and $\log_{10} 3 = 0.4771$.

39. $\log_{10} 72 = ?$

40. $\log_{10}(9\sqrt{2}) = ?$

41. $\log_{10} \frac{8}{9} = ?$

42. $\log_{10} \frac{3\sqrt{3}}{16} = ?$

In problems 43–46, use a calculator to evaluate the following expression. Round to four decimal places.

43. $\log_{10} 16.45 = ?$

44. $\log_{10} \sqrt[4]{5.2} = ?$

45. $\log_2 27 = ?$

46. $\log_5 15 = ?$

In problems 47–51, find: (a) the domain, (b) the vertical asymptote, (c) the *x*-intercept, and (d) sketch the graph of the given logarithmic function.

47. $g_1(x) = \log_2(-x)$

48. $g_2(x) = \log_3(x-1)$.

49. $g_3(x) = \log_4(2-x)$.

50. $g_4(x) = \ln(\frac{x-1}{2})$.

51. $g_5(x) = \ln(x) - 1$.

In problems 52–55, use Eqs.(16.2.10a-b) to find the intensity level in decibels.

52. The sound of rustle of leaves has an intensity of 10^{-11} watts/m^2.

53. The sound of ordinary conversation has an intensity of 10^{-6} watts/m^2.

54. The sound of busy street traffic has an intensity of 10^{-5} watts/m^2.

55. The sound of pain-producing noise has an intensity of 1 watts/m^2.

16.3 Solving Exponential and Logarithmic Equations

In this section we are going to learn how to solve equations involving exponential and logarithmic functions. The solution procedures are based on a combination of the inverse identities between exponential and logarithmic functions (Theorem 16.2.1) plus the operational properties for exponential functions (Theorem 16.1.1) and logarithmic functions (Theorem 16.2.2), respectively. Also, please note that those identities still hold even when x and y are replaced by a function in x. The detailed guidelines can best be illustrated through concrete examples.

Example 16.3.1

Solve each of the following exponential equations for the unknown root of x. Round your solution to four decimal places.

(a) $4^{3x+2} = \sqrt{2}$.

(b) $2^{3x+4} = 3^{2x-1}$.

(c) $2 = e^{0.05x}$.

(d) $e^{2x} - e^{x} - 6 = 0$.

(e) $(1 + \frac{0.1}{4})^{4x} = 5$.

Solution

(a) Since $4 = 2^2$ and $\sqrt{2} = 2^{\frac{1}{2}}$, we have

$$(2^2)^{3x+2} = 2^{\frac{1}{2}}.$$

By using Eq. (16.1.4),

$$2^{2(3x+2)} = 2^{\frac{1}{2}}.$$

By taking $\log_2$ on both sides,

$$\log_2 2^{2(2x+3)} = \log_2 2^{\frac{1}{2}}.$$

By using Eq. (16.2.2),

$$2(3x+2) = \frac{1}{2}, \Rightarrow \ 6x + 4 = \frac{1}{2}, \Rightarrow$$

$$6x = \frac{1}{2} - 4 = -\frac{7}{2}, \Rightarrow \ x = -\frac{7}{6 \cdot 2} = -\frac{7}{12}.$$

(b) By taking the natural logarithm on both sides,

$$\ln 2^{3x+4} = \ln 3^{2x-1}.$$

By using Eq. (16.2.8),

$$(3x+4) \cdot \ln 2 = (2x-1) \cdot \ln 3$$

$$3 \ln 2 \cdot x + 4 \ln 2 = 2 \ln 3 \cdot x - \ln 3,$$

$$(3 \ln 2 - 2 \ln 3)x = -\ln 3 - 4 \ln 2,$$

$$x = \frac{-\ln 3 - 4 \ln 2}{3 \ln 2 - 2 \ln 3} = -\frac{\ln 3 + 4 \ln 2}{3 \ln 2 - 2 \ln 3} \approx -\frac{1.0986 + 4 \cdot 0.6931}{3 \cdot 0.6931 - 2 \cdot 1.0986},$$

$$= -\frac{3.871}{-0.1179} = 32.832909 .$$

(c) By taking ln on both sides,

$$\ln 2 = \ln e^{0.05x}, \Rightarrow \ \ln 2 = 0.05x, \Rightarrow \ x = \frac{\ln 2}{0.05} \approx 13.86 .$$

(d) Since $(e^x)^2 = e^{2x}$, we have

$$(e^x)^2 - e^x - 6 = 0.$$

By letting $y = e^x$,

$$y^2 - y - 6 = 0, \Rightarrow (y-3)(y+2) = 0.$$

Since $y > 0$, this implies that $y + 2 > 0$. Therefore,

$$y - 3 = 0, \Rightarrow y = 3, \Rightarrow e^x = 3, \Rightarrow \ln e^x = \ln 3, \Rightarrow$$

$$x = \ln 3 = 1.098612289 \approx 1.0986.$$

(e) From the given equation, we have

$$(1+0.025)^{4x} = 5, \Rightarrow 1.025^{4x} = 5, \Rightarrow$$

$$\ln 1.025^{4x} = \ln 5, \Rightarrow 4x \cdot \ln 1.025 = \ln 5, \Rightarrow$$

$$x = \frac{\ln 5}{4 \cdot \ln 1.025} \approx \frac{1.609437912}{0.09877045} = 16.29473093 \approx 16.2947.$$

Remarks

1. For part (a), an alternative way is by taking either the common or the natural logarithm on both sides of the given equation as shown below:

$$\log_{10} 4^{3x+2} = \log_{10} \sqrt{2}$$

$$(3x+2) \cdot \log_{10} 4 = \log_{10} \sqrt{2},$$

$$3 \log_{10} 4 \cdot x + 2 \log_{10} 4 = \log_{10} \sqrt{2},$$

$$3 \log_{10} 4 \cdot x = \log_{10} \sqrt{2} - 2 \log_{10} 4,$$

$$x = \frac{\log_{10} \sqrt{2} - 2 \log_{10} 4}{3 \log_{10} 4},$$

$$\approx \frac{-1.053604985}{1.806179974} = -0.5833333 \ldots = -\frac{7}{12},$$

which is exactly the same as the one obtained earlier.

2. For part (b), we are allowed to take the logarithm with other base on both sides of the equation. We will end up with the same answer. However, we usually take the logarithm with base 10 or e because only the logarithms with these two bases are available on the calculator.

Example 16.3.2

Solve each of the following logarithmic equations for the unknown root of x. Round your solution to four decimal places.

(a) $\log_2(x-5) = 3$, where $x > 5$.

(b) $\log_3 \sqrt{2x+1} = 1$, where $x > -\frac{1}{2}$.

(c) $\log_4 x - \log_4 (x+2) = \frac{1}{2}$, where $x > 0$.

(d) $\ln x + \ln(2x+1) = 0$, where $x > 0$.

(e) $\log_5 (x-1) + \log_5 (x+3) = 1 + \log_5 x$, where $x > 1$.

Solution

(a) By taking the exponential with base 2 on both sides,

$$2^{\log_2 (x-5)} = 2^3.$$

By using Eq. (16.2.3),

$$x - 5 = 8, \Rightarrow x = 8 + 5 = 13.$$

(b) Since $\sqrt{2x+1} = (2x+1)^{\frac{1}{2}}$, we have

$$\log_3 (2x+1)^{\frac{1}{2}} = 1.$$

By using Eq. (16.2.8),

$$\frac{1}{2} \log_3 (2x+1) = 1.$$

By multiplying both sides by a number 2,

$$\log_3 (2x+1) = 2.$$

By taking the exponential with base 3 on both sides,

$$3^{\log_3 (2x+1)} = 3^2.$$

By using Eq. (16.2.3),

$$2x + 1 = 9, \Rightarrow 2x = 9 - 1 = 8, \Rightarrow x = \frac{8}{2} = 4.$$

(c) By using Eq. (16.2.7),

$$\log_4 (\frac{x}{x+2}) = \frac{1}{2}.$$

By taking the exponential on both sides)

$$4^{\log_4 (\frac{x}{x+2})} = 4^{\frac{1}{2}} = \sqrt{4} = 2.$$

By using Eq. (16.2.3),

$$\frac{x}{x+2} = 2,$$

By multiplying both sides by $x + 2$,

$$x = 2 \cdot (x + 2) = 2x + 4,$$

$$x - 2x = 4, \Rightarrow -x = 4, \Rightarrow x = \frac{4}{-1} = -4.$$

Since $x = -4$ is not in the domain, there is no real solution to the given equation.

(d) By using Eq. (16.2.6),

$$\ln[x(2x+1)] = 0, \Rightarrow$$
$$e^{\ln[x(2x+1)]} = e^0 = 1, \Rightarrow$$
$$x\cdot(2x+1) = 1, \Rightarrow 2x^2 + x - 1 = 0, \Rightarrow$$
$$(2x-1)(x+1) = 0, \Rightarrow 2x - 1 = 0, \text{ or } x + 1 = 0, \Rightarrow$$
$$x = \frac{1}{2}, \text{ or } x = -1.$$

Since $x = -1$ is not in the domain, the only solution is $\{\frac{1}{2}\}$.

(e) By using Eqs. (16.2.5) and (16.2.6),

$$\log_5[(x-1)(x+3)] = \log_5 5 + \log_5 x = \log_5(5x).$$

By taking the exponential function with base 5 on both sides,

$$5^{\log_5[(x-1)(x+3)]} = 5^{\log_5(5x)}.$$

By using Eq. (16.2.3),

$$(x-1)\cdot(x+3) = 5\cdot x, \Rightarrow$$
$$x^2 + 2x - 3 = 5x, \Rightarrow$$
$$x^2 - 3x - 3 = 0.$$

By using a quadratic formula (Section 6.3 in Part I),

$$x = \frac{-(-3) \pm \sqrt{(-3)^2 - 4\cdot 1\cdot(-3)}}{2\cdot 1} = \frac{3 \pm \sqrt{21}}{2}.$$

Since $x = \frac{3-\sqrt{21}}{2} = -0.7912879 < 1$ is not in the domain of the given equation, the only solution to the given equation is $\{\frac{3+\sqrt{21}}{2}\}$.

Remark

Unlike the exponential equation, the solution to the logarithmic equation is required to satisfy the domain restriction.

Exercises

In problems 1–5, solve the given exponential equation for the unknown root of x. Round your solution to four decimal places.

1. $2^{3x-4} = \frac{1}{4}$.

2. $3^{2x-1} = 20$.

3. $4^{x+2} = 5^{3x-1}$.

4. $\frac{14}{4+e^{-x}} = 3$.

5. $\sinh(x) = 2$.

In problems 6–10, solve the given logarithmic equation for the unknown root of x. Round your solution to four decimal places.

6. $\log_2(3x-8) = 0$, where $x > \frac{8}{3}$.

7. $\log_3 \sqrt[4]{2x+7} = \frac{1}{2}$, where $x > -\frac{7}{2}$.

8. $\log_4(x+1) + \log_4(x-1) = \frac{1}{2}$, where $x > 1$.

9. $\log_8(x+5) - \log_8(3x-7) = \frac{2}{3}$, where $x > \frac{7}{3}$.

10. $2(\ln x)^2 - 3\ln x - 5 = 0$, where $x > 0$.

16.4 Mathematical Models

In this section we present four basic types of mathematical models based on using the exponential functions: (1) "Law of Growth and Decay", (2) "Method of Carbon Dating", (3) "Newton's Law of Cooling", and (4) "Isothermal Expansion of an Ideal Gas".

1. Law of Exponential Growth and Decay

Let $A(t)$ be the amount of a certain material at any time t that varies with time, where $A_0 = A(0)$ is the initial amount at $t = 0$. If the rate of change of $A(t)$ is directly proportional to the amount present at any time t, then $A(t)$ is given by

$$A(t) = A_0 \cdot e^{k \cdot t}, \qquad (16.4.1)$$

where k, the proportionality constant, is to be determined. If $k > 0$, Eq. (16.4.1) is called a model of exponential growth. If $k < 0$, then Eq. (16.4.1) is called a model of exponential decay.

Remarks

1. In calculus, the rate of change of $A(t)$ is $\frac{dA}{dt}$ that is the first derivative of $A(t)$. Since $\frac{dA}{dt}$ is directly proportional to $A(t)$, there exists a proportional constant k such that
$$\frac{dA}{dt} = k{\cdot}A(t). \qquad (16.4.2)$$
It can be shown that Eq. (16.4.1) is a solution to the first order differential equation Eq. (16.4.2) with the initial condition $A(0) = A_0$.
2. A population growth is a prototypical example for the exponential growth model, while a radioactive decay is a prototypical example for the exponential decay model. When $k > 0$, Eq. (16.4.1) represents an uninhibited growth. However, this model may not be realistic, because growth at an exponential rate is unsustainable due to the influence of factors such as lacking of living space, diminishing food supply, and etc. Usually, Eq. (16.4.1) only reflects the early stages of the population growth.

Example 16.4.1

Suppose that a culture of bacteria grows according to the law of exponential growth. If 200 bacteria are present initially, and there are 300 after 2 hours, how many will be present in the culture after 5 hours?

Solution

By using Eq. (16.4.1) with $A_0 = 200$, $A(2) = 300$, we have
$$300 = A(2) = 200e^{k\cdot 2}, \Rightarrow$$
$$1.5 = \frac{300}{200} = e^{2k}, \Rightarrow$$
$$0.405465108 = \ln 1.5 = \ln e^{2k} = 2k, \Rightarrow$$
$$k = \frac{0.405465108}{2} = 0.202732554 \approx 0.2027.$$
Hence, the model of exponential growth for this bacteria is
$$A(t) = 200e^{0.2027t}. \qquad (1)$$
Now, evaluating Eq. (1) at $t = 5$, we have
$$A(5) = 200e^{0.2027\cdot 5} = 200e^{1.0135} = 551.045491 \approx 552.$$

After 5 hours, there are 552 bacteria present in the culture.

Example 16.4.2
In physics, the half-life is a measure of the stability of a radioactive material. The half-life is simply the time it takes for one-half of the atoms in an initial amount to disintegrate. The longer the half-life of a radioactive material, the more stable it is. The half-life of highly radioactive radium ($R_a^{22\ell}$) is approximately 1700 years. Suppose that a sample contains 5 grams of $R_a^{22\ell}$. How much radium remains after 2,000 years?
Solution
First, let us find a specific model of exponential decay for $R_a^{22\ell}$. By substituting $A(1700) = \frac{5}{2} = 2.5$ into Eq. (16.4.1), we have

$$2.5 = 5 \cdot e^{k \cdot 1700}, \Rightarrow$$
$$0.5 = \frac{2.5}{5} = e^{1700k}, \Rightarrow$$
$$-0.69314718 = \ln 0.5 = \ln e^{1700k} = 1700 \cdot k, \Rightarrow$$
$$k = \frac{-0.69314718}{1,700} = -0.000407733 .$$

Hence, the model of exponential decay for $R_a^{22\ell}$ is given by

$$A(t) = 5 \cdot e^{-0.000407733 \cdot t} . \qquad (1)$$

By evaluating Eq. (1) at $t = 2000$, we have

$$A(2000) = 5 \cdot e^{-0.000407733 \cdot 2000} = 5 \cdot e^{-0.815467271}$$
$$= 2.212162715 = 2.2122 \text{ (grams)}.$$

Therefore, after 2000 years, 2.2122 grams of $R_a^{22\ell}$ remains.

2. Method of Carbon Dating
In 1949 the chemist Dr. W. Libby (1908-1980, American) devised a method of using radioactive carbon-14 (C^{14}) as a means of determining the approximate ages of fossils. This method is called the carbon dating. The idea behind the method of carbon dating is that the isotope C^{14} is produced in the atmosphere by an action of cosmic radiation on nitrogen. The ratio of the amount of C^{14} to ordinary carbon (C^{12}) in the atmosphere appears to be a constant, and as a consequence the proportionate amount of C^{14} present in all living

organisms is the same as that in the atmosphere. When an organism dies, the original amount of C^{12} present remains unchanged, whereas the C^{14} begins to decrease. Therefore, by comparing the proportionate amount of C^{14} present in a fossil with the constant ratio found in the atmosphere, it is possible to obtain a reasonable estimation of its age. For his work, Dr. Libby won a Nobel Prize for Chemistry in 1960 (Reference 15).

Example16.4.3

Suppose that a fossilized bone was found to contain 2% of the original amount of C^{14}. The half-life of C^{14} is 5730 years. Determine the age of the fossil.

Solution

First, let us find a specific model of exponential decay for C^{14}. By substituting $A(5730) = \frac{1}{2}A_0$ into Eq. (16.4.1), we have

$$\frac{1}{2}A_0 = A(5730) = A_0 \cdot e^{k\cdot 5730} = A_0 \cdot e^{5730k}, \Rightarrow$$

$$\frac{1}{2} = \frac{\frac{1}{2}A_0}{A_0} = e^{5730k}, \Rightarrow$$

$$-0.69314718 = \ln\frac{1}{2} = \ln e^{5730\cdot k} = 5730\cdot k, \Rightarrow$$

$$k = \frac{-0.69314718}{5{,}730} = -0.000121.$$

Hence, a model of exponential decay for C^{14} is given by

$$A(t) = A_0\cdot e^{-0.000121\cdot t}. \qquad (1)$$

Let t_0 be the age of the fossil. Then, from Eq. (1), we have

$$0.02A_0 = A(t_0) = A_0 e^{-0.00012t_0}, \Rightarrow$$

$$0.02 = e^{-0.00012 1t_0}, \Rightarrow \ln 0.02 = \ln e^{-0.00012 1t_0} = -0.000121t_0, \Rightarrow$$

$$t_0 = \frac{\ln 0.02}{-0.000121} = 32330.769 \approx 32{,}331.$$

Therefore, the age of this fossil is 32,331 years old.

3. Newton's Law of Cooling

The law says that the rate at which a body cools is proportional to the difference in temperatures between the body and its environment into which it is placed. The temperature T of the body at time t after being placed into an environment having constant temperature C is given by

$$T(t) = C + (T_0 - C)e^{-kt}, \qquad (16.4.3)$$

where the value of k is the rate of cooling to be determined.

Remarks

1. In calculus, the rate of change of the temperature of the cooling body is given by its first derivative $\frac{dT}{dt}$. Thus, Newton's law of cooling can be modeled by a linear first order ordinary differential equation with an initial value as follows:

   $$\frac{dT}{dt} = k(T - C), \qquad (16.4.4a)$$

   $$T(0) = T_0. \qquad (16.4.4b)$$

   It can be shown that Eq. (16.4.3) is a solution of the initial value problem of Eqs. (16.4.4a-b).
2. This law was named after English physicist I. Newton (1642-1727, English). However, the inaccuracies of Newton's law became apparent at high temperature. The corrected version was formulated in 1817 by French physical chemist P. Dulong (1785-1838, French) and physicist A. Petit (1791-1820, French) who, experimenting through temperature ranges as high 243° C, found that the quickness of cooling for a constant excess of temperature, decreases in geometric progression, when the surrounding spaces increases in arithmetic progression (Reference 14).

Example 16.4.4

Suppose that a piece of metal is heated to 200°F and then placed in a cooling liquid at 50°F. After 10 minutes, the metal cooled to 100°F. Find the temperature after 20 minutes.

Solution

By using Eq. (16.4.3) with $T(0) = 200$, $T(10) = 100$, $C = 50$, and $t = 10$, we have

$$100 = 50 + (200 - 50)e^{-k\times 10}, \Rightarrow$$
$$50 = 150e^{-10k}, \Rightarrow 1/3 = e^{-10k}, \Rightarrow$$
$$\ln(1/3) = -10k, \Rightarrow$$
$$k = 1.0986/10 = 0.10986.$$

By substituting $k = 0.10986$ into Eq. (16.1.8) with $t = 20$, we have

$$T(20) = 50 + (200 - 50)e^{-0.10986\times 20}$$
$$= 50 + 150 \times e^{-2.1972}$$
$$= 50 + 16.7 = 66.7.$$

Hence, the temperature of the metal after 20 minutes is 66.7°F.

Example 16.4.5

Suppose that when a cake is removed from a basking oven, its temperature is measured at 250°F and is then allowed to cool in a room whose air temperature is 70°F. Five minutes later its temperature is 150°F. How long will it take to cool off to 80°F?

Solution

Let t_0 be the required time in minutes. By substituting $C = 70$°F, $T_0 = T(0) = 250$°F, and $T(5) = 150$°F into Eq. (16.4.3), we have

$$150 = T(5) = 70 + (250 - 70)\cdot e^{-k\cdot 5}, \Rightarrow$$
$$80 = 180\cdot e^{-5k}, \Rightarrow$$
$$0.444444444 = \frac{80}{180} = e^{-5k}, \Rightarrow$$
$$-0.810930216 = \ln 0.444444444 = \ln e^{-5k} = -5k, \Rightarrow$$
$$k = \frac{-0.810930216}{-5} = 0.162186043 \approx 0.1622.$$

Hence, a specific cooling model for this cake is given by

$$T(t) = 70 + (250 - 70)\cdot e^{-0.1622\cdot t}$$
$$= 70 + 180\cdot e^{-0.1622\cdot t}. \qquad (1)$$

By substituting $T(t_0) = 80$°F into Eq. (1), we have

$$80 = T(t_0) = 70 + 180\cdot e^{-0.1622t_0}, \Rightarrow$$
$$10 = 80 - 70 = 180\cdot e^{-0.1622t_0}, \Rightarrow$$
$$0.0555555 = \frac{10}{180} = e^{-0.1622t_0}, \Rightarrow$$
$$\ln(0.0555555) = \ln(e^{-0.1622t_0}) = -0.1622t_0$$
$$t_0 = \frac{-2.8903728}{-0.1622} = 17.819807 \approx 18.$$

Therefore, it will take approximately 18 minutes for this cake to cool off to 80° F.

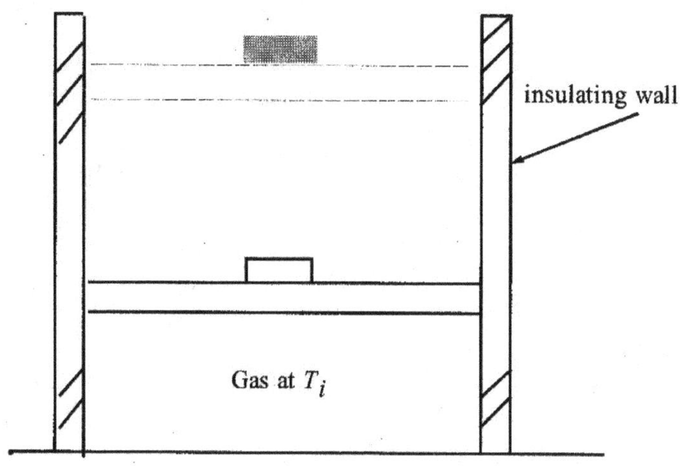

Figure 16.4.1

4. Isothermal Expansion of an Ideal Gas

In physics an expansion or compression of a substance at constant temperature is referred as an isothermal process. The isothermal expansion of a gas can be achieved by placing the gas in thermal contact with a heat reservoir at the same temperature (Figure 16.4.1). Let V_f and V_i denote the final and the initial volume of the gas, respectively. Then, the work done during the isothermal expansion is given by

$$W = g(V_f, V_i, T) = nRT\ln(\frac{V_f}{V_i}), \qquad (16.4.5a)$$

where n is the number of moles of gas, R is the ideal gas constant, given by

$$R = 8.31 J/mol \cdot K, \qquad (16.4.5b)$$

T is the temperature in kelvin degree, and ln is the natural logarithmic function.

Remarks

1. For a discussion on Eqs. (16.4.5a-b), see Reference 8.
2. If the gas is compressed isothermally, namely, $V_f < V_i$, then the work done of Eq. (16.4.5a) would be negative.

Example 16.4.6

What is the work done by 5 mole of an ideal gas that is kept at 27°C in an expansion from 5 liters to 10 liters?

Solution

By substituting $n = 5$, $T = 273 + 27 = 300°K$, $V_i = 5$ liters, and $V_f =$ 10 liters into Eq. (16.4.5a-b), we have

$$W = (5 \text{ mol})\cdot(8.31\ J/\text{mol}\cdot\text{K})\cdot(300°\text{ K})\cdot\ln(\frac{10}{5})$$

$$= 8640.079606\ J \approx 8.64\cdot10^3\ J.$$

Therefore, the heat that must be supplied to the gas from the reservoir to keep T constant is $8.64\cdot10^3\ J$.

Exercises

1. A colony of bacteria grows according to the model of exponential growth. If the number of bacteria doubles in 4 hours, how long will it take for the size of the colony to quadruple?

2. The population P of a small college town is given by $P = 20{,}000\cdot e^{0.005\cdot t}$, where t is the time in years, with $t = 0$ corresponding to 1980. According to this model, in what year will the city have a population of 30,000?

3. A certain car that costs \$19,000 new has a depreciated value of \$16,000 after one year. Assume that the depreciation of a car follows the law of exponential decay. Find the value of the car when it is five years old.

4. Initially, there were 10 grams of a radioactive material present. After 100 days 7.736 grams remain. If the rate of decay follows the law of exponential decay, find the half-life of this radioactive material.

5. The half-life of plutonium (P_u^{239}) is 24,360 years. A sample contains 10 grams of P_u^{239}. How much will P_u^{239} remain after 5,000 years?

6. In a piece of burned wood, it was found that 80% of the C^{14} has decayed. What is the approximate age of the wood?

7. Traces of burned wood found with ancient stone tools in an archaeological dig in Peru were found to contain approximately 1.5% of the original amount of carbon-14. What was the approximate age of the tree cut and burned?

8. A thermometer is removed from a room where the air temperature is 75°F. After $\frac{1}{2}$ minute the thermometer reads 55°F. How long will it take for the thermometer to read 25°F?

9. A pizza baked at 400°F is removed from the oven at 6:00 p.m. into a room with a constant temperature of 70°F. After 10 minutes, the pizza is at 250°F. At what time can you begin eating the pizza if you want its temperature to be 125°F?

10. Boiling water, at 100°C, is placed in a refrigerator at 5°C. The temperature is 50°C after 30 minutes. What is the temperature of the water after one hour?

11. Three months after discontinuing advertising on national television, a manufacturer notices that sales have dropped from 10,000 units per month to 8,500 units. If sales follow the model of exponential decay, what sales will they be after a half year, i.e., another three months?

12. What is the work done if 10 moles of an ideal gas that is kept at 27°C and compressed from 10 liters to 5 liters?

16.5 References

1. Bueche, F. (1972), Principles of Physics, 2nd edition, McGraw–Hill Book Company, New York.

2. Cajori, F. (1913). History of the Exponential and Logarithmic Concepts: I. From Napier to Leibniz and John Bernoulli I, The American Mathematical Monthly, 20, pp. 5–14.
3. _______. History of the Exponential and Logarithmic Concepts: II. The Modern Exponential Notation, The American Mathematical Monthly, 20, pp. 35–47.
4. _______. History of the Exponential and Logarithmic Concepts: III. The Creation of a Theory of Logarithms of Complex Numbers by Euler, The American Mathematical Monthly, 20, pp. 75–84.
5. _______. History of the Exponential and Logarithmic Concepts: IV. From Euler to Wessel and Argand 1749–about 1800, The American Mathematical Monthly, 20, pp. 107–117.
6. _______. History of the Exponential and Logarithmic Concepts: V. Generalizations and Refinements Effected during the Nineteenth Century, The American Mathematical Monthly, 20, pp. 173–182, 205–210.
7. Maor, E. (1994). e: The Story of a Number, Princeton University Press, Princeton, New Jersey, 1994.
8. Serway, R. A. and Faughn, J. S. (1989). College Physics, 2nd edition, Saunders College Publishing Co., pp. 315–316.
9. Stigler, S. M. (1986). The History of Statistics, The Belknap Press of Harvard University Press, Cambridge, MA.
10. http://en.wikipedia.org/wiki/Exponential_function.
11. http://en.wikipedia.org/wiki/Hyperbolic_function.
12. http://en.wikipedia.org/wiki/Abraham_de_Moivre.
13. http://en.wikipedia.org/wiki/Compound_interest.
14. https://en.wikipedia.org/wiki/Newton%27s_law_of_cooling.
15. https://en.wikipedia.org/wiki/Radiocarbon_dating.

Chapter 17
Systems of Linear Equations in Several Variables

In this chapter we're going to learn how to solve n linear equations in n variables. Mainly, the method of Gaussian elimination will be introduced. In particular, we'll use elementary theory of matrix to answer whether a system of linear equations has a unique solution, no solution, or infinitely many solutions. In addition, the Cramer's rule and the Gauss-Jordan method are described and show how to use it to solve a system of linear equations.

17.1 Solving a System of Linear Equations

In Section 9.4 of Part I we presented two methods, the method of elimination, and the method of substitution, for solving a system of two linear equations in two variables. However, those methods are limited to solve a system of two equations in two variables only. In this chapter we are going to learn a general method, called the Gaussian elimination method, for solving a system of n linear equations in n variables, where n ≥ 3.

Definition 17.1.1
The following system of equations is called a system of m linear equations in n variables x_1, x_2 ,…., and x_n:

$$\begin{cases} a_{11}x_1 + a_{12}x_2 + \ldots + a_{1n}x_n = b_1, \\ a_{21}x_1 + a_{22}x_2 + \ldots + a_{2n}x_n = b_2, \\ \ldots\ldots\ldots\ldots\ldots\ldots\ldots\ldots\ldots\ldots, \\ a_{m1}x_1 + a_{m2}x_2 + \ldots + a_{mn}x_n = b_m, \end{cases}, \qquad (17.1.1)$$

where a_{ij}'s, $i = 1, 2, \ldots, m, j = 1, 2, \ldots, n$, coefficients of the unknown x_1, x_2,…., and x_n, and b_i's, $i = 1, 2,\ldots m$, the right side of the equations, are all real numbers, *m* and *n* are positive integers.

Remarks

1. The need for solving a system of linear equations arises out of many practical applications of mathematics. Historically, about 250 B.C., a Chinese mathematical book, "Jiu-Zhang Suan-Shu (九章算术)" ("The Nine Chapters on the Mathematical Art"), already contained a chapter on solving problems formulated in simultaneous linear equations (Reference 4, pp. 196–197; Reference 8, pp. 249-258).
2. For solving a linear programming problem (Section 10.3 in Part I) the simplex method also involves a system of linear equations.
3. To build a statistical prediction model through the data fitting again involves in solving a system of linear equations (Reference 5).
4. In modern days, problems in physics are often modeled by partial differential equations that are difficult to solve analytically. Therefore, a numerical method has to be applied to solve them numerically. This approach ends up with solving a system of linear equations in hundreds of variables (Reference 9, Chapter 17).
5. In fact, the entire subject of "Linear Algebra," a branch of applied mathematics, was evolved from solving a system of linear equations.

Definition 17.1.2

The solution set of Eq. (17.1.1) is the set of all *n*-tuples $(w_1, w_2, \ldots, w_n)$ or $\{ x_1 = w_1, x_2 = w_2, \ldots, x_n = w_n \}$ is a solution, which satisfy Eq. (17.1.1), namely,

$$\begin{cases} a_{11}w_1 + a_{12}w_2 + \ldots + a_{1n}w_n = b_1, \\ a_{21}w_1 + a_{22}w_2 + \ldots + a_{2n}w_n = b_2, \\ \ldots\ldots\ldots\ldots\ldots\ldots\ldots\ldots\ldots\ldots, \\ a_{m1}w_1 + a_{m2}w_2 + \ldots + a_{mn}w_n = b_m. \end{cases}$$

Remarks

1. The solution set of Eq. (17.1.1) has, in general, three possible cases:

   Case 1. When $m < n$, Eq. (17.1.1) has infinitely many solutions. An example is to convert the system of linear inequalities, which serve as the constraints of a linear programming problem, into a system of linear equations by adding the slack variables.

   Case 2. When $m > n$, Eq. (17.1.1) has, in general, no solution To bypass this difficulty, we, instead of finding an exact solution, seek for the best possible approximate solution through the principle of least squares. An example is to build a linear regression model through the data fitting technique in statistics (Reference 5).

   Case 3. When $m = n$, Eq. (17.1.1) has, in general, a unique solution. However, it is also possible to have either no solution or infinitely many solutions. This is the case we will consider in this book. An example of this case is seen in solving a partial differential equation by a numerical method in advanced engineering applications.
2. A solution $(w_1, w_2, \ldots, w_n)$ to Eq. (17.1.1) can be viewed geometrically as a point in R^n, the n-dimensional Euclidean space.

Definition 17.1.3

Two systems of m linear equations in n variables are said to be equivalent if they have the same solution set.

Remarks

1. Based upon this definition, the idea of the Gaussian elimination method for solving a system of m linear equations in n variables is to reduce or transform the original system of equations into a simpler (or specifically, an upper triangular) system of equations, and then solve the obtained simpler system for a solution.
2. To reduce the given system of equation to a simpler equivalent system of equations, we have to apply appropriately any of the three types of elementary row operations as shown in the next definition.

Definition 17.1.4

For any given system of m linear equations in n variables, the following operations between equations are called an elementary row operation (ERO) of

Type 1. Re-arrange the order of equations in the system by interchanging any two equations.

Type 2. Multiply both sides of any equation by a nonzero real number.

Type 3. Replace an equation by adding to it a multiple of another equation.

Remark

Obviously, when an elementary row operation of Type 1 is applied to any system of equations, the new system of equations is equivalent to the old system. As far as an application of elementary row operations of Type 2 or 3 to a given system of equations is concerned, the fact that the new system is still equivalent to the old system can be rationalized or justified by operational rules for equations in Section 5.1 in Part I.

The Gaussian Elimination Method

Given a system of linear equations of Eq. (17.1.1). Then, we proceed to solve Eq. (17.1.1) for the unknown variables as follows:

Step 1. Apply the appropriate elementary row operations on the given system of equations in order to reduce it to an equivalent upper triangular system of equations.

Step 2. Solve the reduced upper triangular system of equation by the method of substitution starting from the last equation upward one-by-one to the first equation.

Remarks

1. This method is named after C. F. Gauss (1777-1855, Germany) developed the Gaussian elimination around 1800 to solve least squares problems in celestial computations. However, the method of Gaussian elimination appears earlier in the Chinese mathematical text "Jiu-Zhang Suan-Shu (九章算术)" ("The Nine Chapters on the Mathematical Art"). Its use is illustrated by solving eighteen problems with two to five equations in Chapter 8 Rectangular Arrays in that book (Reference 11).

2. The idea of reducing the given system of equations to an equivalent upper triangular system of equations systematically is given as follows:

   Step 1. Examine the coefficient of the first variable in the first equation. If it is non-zero, then use the first equation as a pivotal equation in the sense that eliminating the first variable from the second equation down to the mth equation by adding to them an appropriate multiple of the first equation so that the first variable disappears in the resulting equation.

   Step 2. By working with the new system of $m-1$ equations, (from the 2nd equation to the mth equation by ignoring the first equation), repeating Step 1 $m-1$ times to this new system of equations gives us the desired upper triangular system of equations.

Example 17.1.1

Solve the following system of three equations in three variables x, y, and z for the unknown (x, y, z) by the Gaussian elimination method:

$$\begin{cases} x - 2y + z = 0, \\ 3x + y - 2z = 1, \\ 4x - 3y + z = 3. \end{cases}$$

Solution

First, label the equations with (1a)-(1c):

$$x - 2y + z = 0, \tag{1a}$$
$$3x + y - 2z = 1, \tag{1b}$$
$$4x - 3y + z = 3. \tag{1c}$$

Since the coefficient of x in Eq. (1a) is not zero, we proceed to solve the given system of equations by using Eq. (1a) as a pivotal equation as follows:

Step 1. To transform Eqs. (1a-1c) to an upper triangular system of equations, we have by adding Eq. (1a) multiplied by -3 to Eq. (1b),

$$7y - 5z = 1. \tag{2b}$$

By adding Eq. (1a) multiplied by -4 to Eq. (1c),

$$5y - 3z = 3. \tag{2c}$$

Since the coefficient of y in Eq. (2b) is not zero, Eq. (2b) is used as a pivotal equation in the second round of elimination. By adding Eq. (2b) multiplied by $-\frac{5}{7}$ to Eq. (2c),

$$\frac{4}{7}z = \frac{16}{7}. \tag{3c}$$

Step 2. By assembling Eqs. (1a), (2b), and (3c) together, we have an upper triangular system of equations:

$$x - 2y + z = 0, \tag{1a}$$
$$7y - 5z = 1, \tag{2b}$$
$$\frac{4}{7}z = \frac{16}{7}. \tag{3c}$$

From Eq. (3c), we obtain

$$\frac{4}{7}z = \frac{16}{7}, \Rightarrow z = \frac{\frac{16}{7}}{\frac{4}{7}} = 4. \tag{4}$$

Then, we apply the method of back substitution to substitute Eq. (4) into Eq. (2b)

$$7y - 5{\cdot}4 = 1, \Rightarrow 7y - 20 = 1, \Rightarrow$$
$$7y = 1 + 20 = 21, \Rightarrow y = \frac{21}{7} = 3. \tag{5}$$

By substituting both Eqs. (4) and (5) into Eq. (1a),

$$x - 2{\cdot}3 + 4 = 0, \Rightarrow x - 6 + 4 = 0, \Rightarrow$$
$$x - 2 = 0, \Rightarrow x = 0 + 2 = 2.$$

Hence, the given system of equations has a unique solution $\{(x, y, z) = (2, 3, 4)\}$.

Remarks

1. Note that $\{(x, y, z) = (2, 3, 4)\}$ is a solution obtained from the upper triangular system of Eqs. (1a, 2b, 3c). Since Eqs. (1a, 2b, 3c) was resulted from the applications of elementary row operations on the original system of Eqs. (1a – 1c), Eqs.(1a, 2b, 3c) is equivalent to Eqs. (1a–1c). Therefore, Eqs. (1a–1c) have the same solution as that of Eqs. (1a, 2b, 3c). This fact can be confirmed by directly substituting the solution of $\{x = 2, y = 3, z = 4\}$ into the left sides of Eqs.(1a–1c) to see if it satisfy Eqs. (1a–1c):

   Left side of Eq. (1a) $= 2 - 2{\cdot}3 + 4 = 6 - 6 = 0$
   $=$ Right side of Eq. (1a),

Left side of Eq. (1b) $= 3\cdot 2 + 3 - 2\cdot 4 = 9 - 8 = 1$
= Right side of Eq. (1b),
Left side of Eq. (1c) $= 4\cdot 2 - 3\cdot 3 + 4 = 12 - 9 = 3$
= Right side of Eq. (1c).

2. Geometrically, the solution set of each of Eqs. (1a–1c) represents a plane in three-dimensional Euclidean space R^3. The solution of $\{(2, 3, 4)\}$ to a system of Eqs. (1a–1c) simply represents the unique intersection point of three non-parallel planes. In general, there is no guarantee of a unique solution even when the number of equations equals the number of variables. The given system has no solution if two of the three planes are parallel or has infinitely many solutions if two of three planes collapse into one plane. See problems 1–2 in Exercises for Section 17.1 at the end of this section for illustrations.
3. In the next section all the elementary row operations will be performed in the context of an augmented matrix.

Example 17.1.2
Solve the following system of two equations in four variables s, t, u, and v for the unknown (s, t, u, v) by the Gaussian elimination method:

$$\begin{cases} 2s + t - 3u + 5v = 4, \\ 3s - 4t + u - 2v = 1. \end{cases}$$

Solution
First, let us label the two equations in the given system:

$$2s + t - 3u + 5v = 4, \qquad (1a)$$
$$3s - 4t + u - 2v = 1. \qquad (1b)$$

Since the given system of Eqs. (1a–1b) has more variables than equations, it has infinitely many solutions. To express the solution set of these infinitely many solutions, we shall classify the four variables, s, t, u, and v into two types of variables, basic and non-basic (or free), and then express the basic variables in terms of non-basic variables.
Since the coefficient of s is non-zero, Eq. (1a) is used as a pivotal equation to eliminate the term of s in Eq. (1b). By adding Eq. (1a) multiplied by $-\frac{3}{2}$ to Eq. (1b),

$$-\frac{11}{2}t + \frac{11}{2}u - \frac{19}{2}v = -5\,. \qquad (2b)$$

Since the coefficients of s and t in the reduced upper triangular system of Eqs. (1a) and (2b) are non-zero, s and t are chosen as basic variables, while u and v are non-basic variables. From Eq. (2b), we solve for t in terms of u and v:

$$-\frac{11}{2}t = -5 - \frac{11}{2}\cdot u + \frac{19}{2}\cdot v, \Rightarrow$$

$$\frac{-\frac{11}{2}t}{-\frac{11}{2}} = \frac{-5-\frac{11}{2}\cdot u + \frac{19}{2}\cdot v}{-\frac{11}{2}}, \Rightarrow$$

$$t = \frac{10}{11} + u - \frac{19}{11}v. \tag{3}$$

By substituting Eq. (3) into Eq. (1a) and solve for s in terms of u and v, we have

$$2s + \frac{10}{11} + u - \frac{19}{11}\cdot v - 3u + 5v = 4, \Rightarrow$$

$$2s + (u - 3u) + (5v - \frac{19}{11}\cdot v) = 4 - \frac{10}{11}, \Rightarrow$$

$$2s - 2u + \frac{36}{11}\cdot v = \frac{34}{11}, \Rightarrow$$

$$2s = \frac{34}{11} + 2u - \frac{36}{11}\cdot v, \Rightarrow$$

$$s = \frac{1}{2}\cdot(\frac{34}{11} + 2u - \frac{36}{11}\cdot v)$$

$$= \frac{1}{2}\cdot\frac{34}{11} + \frac{1}{2}\cdot 2u - \frac{1}{2}\cdot\frac{36}{11}\cdot v$$

$$= \frac{17}{11} + u - \frac{18}{11}\cdot v. \tag{4}$$

From Eqs. (3) and (4), the solution set is

$\{(\frac{17}{11} + u - \frac{18}{11}\cdot v, \frac{10}{11} + u - \frac{19}{11}\cdot v, u, v)|$ u and v are real numbers$\}$

or is expressed as

$$\begin{cases} s = \frac{17}{11} + u - \frac{18}{11}\cdot v, \\ t = \frac{10}{11} + u - \frac{19}{11}\cdot v, \\ u = \textit{any real number}, \\ v = \textit{any real number}. \end{cases}$$

Remark

There is no regulation on the choice of basic variables except that their coefficients must be nonzero. For this example, it appears that any two out of the four variables can be used as basic variables.

Example 17.1.3

Solve the following system of three equations in two variables x and y for the unknown (x, y) by the Gaussian elimination method:

$$\begin{cases} x + 2y = 1, \\ 3x - y = 2, \\ -2x + y = 3. \end{cases}$$

Solution

First, let us label the three equations as

$$x + 2y = 1, \qquad (1a)$$

$$3x - y = 2, \qquad (1b)$$

$$-2x + y = 3. \qquad (1c)$$

Since the coefficient of x in Eq. (1a) is not zero, Eq. (1a) is used as a pivotal equation to eliminate the variable of x in Eqs. (1b) and (1c). By adding Eq. (1a) multiplied by -3 to Eq. (1b),

$$-7y = -1. \qquad (2b)$$

By adding Eq. (1a) multiplied by 2 to Eq. (1c),

$$5y = 5. \qquad (2c)$$

Since the coefficient of y in Eq. (2b) is not zero, Eq. (2b) is chosen as a pivotal equation to eliminate the variable of y in Eq. (2c).

By adding Eq. (2b) multiplied by $\frac{5}{7}$ to Eq. (2c),

$$0 = \frac{30}{7}. \qquad (3c)$$

However, Eq. (3c) is a contradiction. Hence, the given system has no solution.

Remark

Geometrically, the graph of each of the equations of Eqs. (1a–1c) represents a straight line. No solution implies that three lines do not intersect at a single point.

Example 17.1.4

A small corporation borrowed \$230,000 – some of it at an annual interest rate of 8%, some at 9%, and some at 10%. How much was borrowed at each rate if an annual interest was \$19,500 and the amount borrowed at 8% was 3 times the amount borrowed at 9%?

Solution

Let x, y, and z be the unknown amount borrowed at 8%, 9%, and 10%, respectively. From the given information, we set up a system of linear equations as follows:

$$x + y + z = 230{,}000, \quad (1a)$$
$$0.08 \cdot x + 0.09 \cdot y + 0.1 \cdot z = 19{,}500, \quad (1b)$$
$$x = 3 \cdot y, \Rightarrow x - 3y = 0. \quad (1c)$$

Since the coefficient of x in Eq. (1a) is not zero, Eq. (1a) is used as pivotal equation to eliminate the variables of x in Eqs. (1b–1c).

By adding Eq. (1a) multiplied by – 0.08 to Eq. (1b),

$$0.01 \cdot y + 0.02 \cdot z = 1{,}100. \quad (2b)$$

By adding Eq. (1a) multiplied by –1 to Eq. (1c),

$$-4 \cdot y - z = -230{,}000. \quad (2c)$$

Since the coefficient of y in Eq. (2b) is not zero, Eq. (2b) is used as a pivotal equation to eliminate the variable of y in Eq. (2c).

By adding Eq. (2b) multiplied by 400 to Eq. (2c),

$$7z = 210{,}000, \Rightarrow z = 30{,}000. \quad (3)$$

By substituting Eq. (3) into Eq. (2b),

$$0.01 \cdot y + 0.02 \cdot 30{,}000 = 1{,}100, \Rightarrow$$
$$0.01 \cdot y + 600 = 1{,}100, \Rightarrow$$
$$0.01 \cdot y = 1{,}100 - 600 = 500, \Rightarrow$$
$$y = \frac{500}{0.01} = 50{,}000. \quad (4)$$

By substituting Eqs. (3) and (4) into Eq. (1a),

$$x + 50{,}000 + 30{,}000 = 230{,}000, \Rightarrow$$
$$x + 80{,}000 = 230{,}000, \Rightarrow$$
$$x = 230{,}000 - 80{,}000 = 150{,}000.$$

Hence, this small corporation borrowed \$150,000, \$50,000, and \$30,000 at an annual interest rate of 8%, 9%, and 10%, respectively.

Exercises

In problems 1–9, solve the given system of linear equations for the unknown solution by the Gaussian elimination method.

1. $\begin{cases} 2x+3y=2, \\ 4x+y=-6. \end{cases}$

2. $\begin{cases} x+y+z=3, \\ 2x+3y+z=5, \\ x-y-z=-5. \end{cases}$

3. $\begin{cases} x-y+2z=1, \\ -3x+4y-6z=-5, \\ 2x-3y+4z=6. \end{cases}$

4. $\begin{cases} x-2y+z=3, \\ 5x-3y-z=10, \\ 2x+3y-4z=1. \end{cases}$

5. $\begin{cases} x-2y=1, \\ -\frac{1}{2}x+y=-\frac{1}{2}, \\ 2x+y=-2. \end{cases}$

6. $\begin{cases} 2x+y=1, \\ 3x-2y=2, \\ x+3y=4. \end{cases}$

7. $\begin{cases} x+y+z=-2, \\ 2x+3y+4z=5. \end{cases}$

8. $\begin{cases} y+2z-w=1, \\ x-2y+z+3w=-2. \\ -x+3y+z-4w=3. \end{cases}$

9. $\begin{cases} x-y+z-w=2, \\ 2x+y-3z+w=4, \\ -x+2y-2z+w=3, \\ 3x-y-z+2w=5. \end{cases}$

10. Suppose that you receive a total of \$630 a year in interest from three investments. The interest rates for the three investments are 5%, 6%, and 7%. It is known that the 5% investment is \$1,000 less than the 6% investment, and the sum of 5% and 6% investments equals the 7% investment. What is the amount of each investment?

17.2 (Arithmetic) Operational Rules for Matrices

In this section we are going to implement the Gaussian elimination method in terms of a matrix.

Definition 17.2.1

A matrix of size $m \times n$ (read as m-by-n), denoted by $A_{m\times n}$, if it is a rectangular array of real numbers having m rows and n columns given by

$$A_{m\times n}=[a_{ij}]_{\substack{i=1,\dots,n,\\ j=1,\dots,m.}}=\begin{bmatrix} a_{11} & a_{12} & \cdots & a_{1j} & \cdots & a_{1n} \\ a_{21} & a_{22} & \cdots & a_{2j} & \cdots & a_{2n} \\ \vdots & \vdots & \vdots & \vdots & & \vdots \\ a_{i1} & a_{i2} & \cdots & a_{ij} & \cdots & a_{in} \\ \vdots & \vdots & \vdots & \vdots & & \vdots \\ a_{m1} & a_{m2} & \cdots & a_{mj} & \cdots & a_{mn} \end{bmatrix}\begin{matrix} 1 \\ 2 \\ \vdots \\ i \\ \vdots \\ m \end{matrix}$$

(17.2.1)

where each entry a_{ij} in the matrix $A_{m\times n}$ is identified by two indices, the row index i and the column index j.

Remarks

1. Matrix has a long history of application in solving linear equations, but they were unknown as arrays until 1800s. Although A. Cayley (1821–1895, English) was credited as one of the first men to study matrices, the term "matrix", a Latin word for womb, was introduced by J. J. Sylvester (1814-1897, English) in 1848. Cayley's work on matrix algebra grew out of a memoir of 1858 on the theory of transformations between two planes. Indeed, a matrix $A_{m\times n}$ can be viewed as a linear transformation that maps as vector in R^n (n-dimensional Euclidean space) to a vector in R^m (m-dimensional Euclidean space) (Reference 2, pp. 203-205). For more details, please also see Reference 12.
2. A matrix of the size 1 × 1 is treated as a scalar, i.e., $[a]_{1\times 1}=a$.

Definition 17.2.2

A matrix of size 1 × n is called an n-dimensional row vector denoted by $\vec{r}$. A matrix of size m × 1 is called a m-dimensional column vector, denoted by $\vec{c}$.

Remarks

1. Geometrically, a vector is a directed line segment by connecting the origin to a point with the entries of the vector as its coordinates.
2. The length (or norm) of a vector $\vec{c}=(c_1,c_2,\dots,c_n)^T$ is defined as $\|\vec{c}\|=\sqrt{c_1^2+c_2^2+\dots+c_n^2}$ (Reference 1), the sum of two vectors satisfies the triangle inequality similar to Eq. (1.2.1), that is,

$$\|\vec{x}+\vec{y}\| \leq \|\vec{x}\|+\|\vec{y}\|.$$

Definition 17.2.3
A matrix A of the size $m \times n$ is called a square matrix if $m = n$.

Remark
A square matrix plays an important role in defining the matrix division.

Definition 17.2.4
Two matrices A and B are said to equal to one another, denoted by $A = B$, if the following two conditions are satisfied:
(i) A and B are of the same size.
(ii) All corresponding entries between A and B equal to one another.

Example 17.2.1
Solve for the unknown x_{11}, x_{12}, x_{21}, and x_{22} in the following matrix equation:

$$\begin{bmatrix} x_{11} & x_{12} \\ x_{21} & x_{22} \end{bmatrix} = \begin{bmatrix} -3 & 4 \\ 2 & 5 \end{bmatrix}$$

Solution
By the definition of equality between two matrices, we immediately have

$$x_{11} = -3,\ x_{12} = 4,\ x_{21} = 2, \text{ and } x_{22} = 5.$$

The solution is $\{(x_{11}, x_{12}, x_{21}, x_{22}) = (-3, 4, 2, 5)\}$.

Definition 17.2.5
An entry of a matrix is called a diagonal entry if its row index equals its column index.

Remark
The diagonal entries play important roles in the Gaussian elimination method and in finding the inverse of a square matrix. Only the row containing a nonzero diagonal entry can be used as a pivotal row to eliminate entries in all other rows below the pivotal row.

Definition 17.2.6

A square matrix is called a diagonal matrix if at least one of the diagonal entries is not zero and zero elsewhere.

Definition 17.2.7
A matrix is called an upper triangular matrix if all entries below the diagonal ones are zero and nonzero elsewhere.

Definition 17.2.8
A matrix called a lower triangular matrix if all entries below the diagonal ones are nonzero and zero elsewhere.

Example 17.2.2

(a) $O_{2\times 2} = \begin{bmatrix} 0 & 0 \\ 0 & 0 \end{bmatrix}$ is a 2×2 zero matrix.

(b) $I_{2\times 2} = \begin{bmatrix} 1 & 0 \\ 0 & 1 \end{bmatrix}$ is a 2×2 identity matrix.

(c) $I_{3\times 3} = \begin{bmatrix} 1 & 0 & 0 \\ 0 & 1 & 0 \\ 0 & 0 & 1 \end{bmatrix}$ is a 3×3 identity matrix.

(d) $D_{3\times 3} = \begin{bmatrix} \frac{1}{2} & 0 & 0 \\ 0 & 0 & 0 \\ 0 & 0 & \frac{1}{3} \end{bmatrix}$ is a 3×3 diagonal matrix.

(e) $U_{4\times 4} = \begin{bmatrix} 2 & -1 & 3 & -4 \\ 0 & 5 & 1 & 0 \\ 0 & 0 & 4 & -3 \\ 0 & 0 & 0 & -1 \end{bmatrix}$ is a 4×4 upper triangular matrix.

(f) $L_{4\times 4} = \begin{bmatrix} 0 & 0 & 0 & 0 \\ -1 & 0 & 0 & 0 \\ 2 & 3 & 0 & 0 \\ 4 & -1 & 5 & 0 \end{bmatrix}$ is a 4×4 lower triangular matrix.

Definition 17.2.9
The transpose of the $m \times n$ matrix A denoted by A^T, is the matrix obtained from A by interchanging rows and columns.

Example 17.2.3

Find the transpose of each of the following matrices:

(a) $A_1 = \begin{bmatrix} -1 & 3 & -4 \end{bmatrix}$.

(b) $A_2 = \begin{bmatrix} 2 \\ 5 \\ -6 \end{bmatrix}$.

(c) $A_3 = \begin{bmatrix} 2 & -1 & 3 \\ -4 & 7 & 5 \end{bmatrix}$.

Solution

(a) $A_1^T = \begin{bmatrix} -1 \\ 3 \\ -4 \end{bmatrix}$.

(b) $A_2^T = \begin{bmatrix} 2 & 5 & -6 \end{bmatrix}$.

(c) $A_3^T = \begin{bmatrix} 2 & -4 \\ -1 & 7 \\ 3 & 5 \end{bmatrix}$.

Remark

Conventionally, a vector is always referred to a column vector. Therefore, a row vector is oftentimes written in the form of the transpose of a column vector.

(Arithmetic) Operational Rules for Matrices

Given three matrices $A_{m\times n} = [a_{ij}]$, $B_{m\times n} = [b_{ij}]$, and $C_{n\times e} = [c_{jk}]$, where $i = 1, \ldots m, j = 1,\ldots, n$, and $k = 1, \ldots, \ell$.

1. Addition Rule

$$A \pm B = [a_{ij} \pm b_{ij}]_{\substack{i=1,\ldots,n \\ j=1,\ldots,n}}. \qquad (17.2.2)$$

2. Scalar Multiplication Rule

$$k \cdot A = [k \cdot a_{ij}]_{\substack{i=1,\ldots,n \\ j=1,\ldots,n}}, \text{ where } k \text{ is a real number.} \qquad (17.2.3)$$

3. Matrix Multiplication Rule

$$A_{m\times n} \times C_{n\times e} = [d_{ik}]_{\substack{i=1,\ldots,n \\ k=1,\ldots,l}}, \qquad (17.2.4)$$

where d_{ik} is given by

$$d_{ik} = [\text{the } i^{\text{th}} \text{ row vector of } A]_{1\times n} \times \begin{bmatrix} \textit{the} \\ k^{th} \\ \textit{column} \\ \textit{vector} \\ \textit{of} \\ C \end{bmatrix}_{n\times 1} .$$

$$= a_{i1}\cdot c_{1k} + a_{i2}\cdot c_{2k} + \ldots + a_{in}\cdot c_{nk},\ i = 1, \ldots, m,\ k = 1, \ldots, \ell.$$

Remarks

1. For matrix multiplication, the number of columns of the first matrix must equal the number of rows of the second matrix. Otherwise, the matrix multiplication is undefined.
2. Matrix multiplication is defined as the rows of the first matrix times the columns of the second matrix.
3. Matrix multiplication does not in general obey the commutative law, i.e.,
   $$A \times C \neq C \times A.$$
   Nevertheless, the distributive law holds for matrix, namely,
   $$(A + B)\cdot C = A\cdot C + B\cdot C. \qquad (17.2.5)$$
4. There is no matrix division rule. In fact, it is replaced by the inverse of a matrix to be covered in Section 17.5.

Example 17.2.4

Compute each of the following:

(a) $3\cdot\begin{bmatrix} 1 & -1 & 3 \\ 2 & 4 & 5 \end{bmatrix} - 2\cdot\begin{bmatrix} 1 & 1 & 3 \\ 2 & 0 & 4 \end{bmatrix} = ?$

(b) $\begin{bmatrix} 1 & 2 & -1 \\ 3 & 0 & 4 \end{bmatrix}\cdot\begin{bmatrix} 2 & 0 & 4 \\ 3 & 3 & 1 \\ -1 & 2 & -1 \end{bmatrix} = ?$

Solution

(a) $3\cdot\begin{bmatrix} 1 & -1 & 3 \\ 2 & 4 & 5 \end{bmatrix} - 2\cdot\begin{bmatrix} 1 & 1 & 3 \\ 2 & 0 & 4 \end{bmatrix}$

$= \begin{bmatrix} 3\cdot 1 & 3\cdot(-1) & 3\cdot 3 \\ 3\cdot 2 & 3\cdot 4 & 3\cdot 5 \end{bmatrix} - \begin{bmatrix} 2\cdot 1 & 2\cdot 1 & 2\cdot 3 \\ 2\cdot 2 & 2\cdot 0 & 2\cdot 4 \end{bmatrix}$

$$= \begin{bmatrix} 3 & -3 & 9 \\ 6 & 12 & 15 \end{bmatrix} - \begin{bmatrix} 2 & 2 & 6 \\ 4 & 0 & 8 \end{bmatrix}$$

$$= \begin{bmatrix} 3-2 & -3-2 & 9-6 \\ 6-4 & 12-0 & 15-8 \end{bmatrix}$$

$$= \begin{bmatrix} 1 & -5 & 3 \\ 2 & 12 & 7 \end{bmatrix}.$$

(b) To perform the matrix multiplication, we first partition the first matrix into the form of row vectors and the second matrix into the form of column vectors, and then multiply the row vector of the first matrix with the column vector of the second matrix as follows:

$$\begin{bmatrix} 1 & 2 & -1 \\ 3 & 0 & 4 \end{bmatrix} \cdot \begin{bmatrix} 2 & 0 & 4 \\ 3 & 3 & 1 \\ -1 & 2 & -1 \end{bmatrix}$$

$$= \begin{bmatrix} 1\cdot 1+2\cdot 3+(-1)\cdot(-1) & 1\cdot 0+2\cdot 3+(-1)\cdot 2 & 1\cdot 4+2\cdot 1+(-1)\cdot(-1) \\ 3\cdot 2+0\cdot 3+4\cdot(-1) & 3\cdot 0+0\cdot 3+4\cdot 2 & 3\cdot 4+0\cdot 1+4\cdot(-1) \end{bmatrix}$$

$$= \begin{bmatrix} 1+6+1 & 0+6-2 & 4+2+1 \\ 6+0-4 & 0+0+8 & 12+0-4 \end{bmatrix} = \begin{bmatrix} 8 & 4 & 7 \\ 2 & 8 & 8 \end{bmatrix}.$$

Definition 17.2.10

The nth power of a square matrix A, denoted by A^n, is defined as A multiplies itself n times, namely,

$$A^n = \underbrace{A \cdot A \cdot A \cdot \ldots \cdot A}_{n \text{ times}}. \tag{17.2.6}$$

Definition 17.2.11

A matrix is called a zero matrix, denoted by $O_{m\times n}$, if all of its entries are zero, i.e.,

$$O_{m\times n} = \begin{bmatrix} 0 & \cdots & 0 \\ \vdots & \cdots & \vdots \\ 0 & \cdots & 0 \end{bmatrix}_{m\times n}.$$

Remark

A zero matrix serves as an additive identity element for the matrix addition, namely,

$$A_{m\times n} + O_{m\times n} = A_{m\times n} = O_{m\times n} + A_{m\times n}. \tag{17.2.7}$$

Definition 17.2.12

A square matrix of size $n \times n$ is called an identity matrix, denoted by $I_{n\times n}$, if all of its diagonal entries are one and zero elsewhere.

Remark

An identity matrix $I_{n\times n}$ serves as a multiplicative identity element for the matrix multiplication, namely,

$$A_{n\times n} \cdot I_{n\times n} = A_{n\times n} = I_{n\times n} \cdot A_{n\times n}. \tag{17.2.8}$$

Example 17.2.5

Given a 2×2 matrix A defined by

$$A = \begin{bmatrix} 2 & 4 \\ 0 & 3 \end{bmatrix}.$$

Then for any vector $\vec{v} = [x, y]^T$, a function given by

$$f(\vec{v}) = A \cdot \vec{v}$$

is a linear transformation that maps a vector is R^2 into another vector in R^2. For example,

$$f(\begin{bmatrix} 1 \\ -1 \end{bmatrix}) = \begin{bmatrix} 2 & 4 \\ 0 & 3 \end{bmatrix} \cdot \begin{bmatrix} 1 \\ -1 \end{bmatrix} = \begin{bmatrix} 2\cdot 1 + 4\cdot(-1) \\ 0\cdot 1 + 3\cdot(-1) \end{bmatrix} = \begin{bmatrix} 2-4 \\ 0-3 \end{bmatrix} = \begin{bmatrix} -2 \\ -3 \end{bmatrix}.$$

Geometrically, it is shown in Figure 17.2.1. Moreover,

$$f(\begin{bmatrix} 1 \\ 0 \end{bmatrix}) = \begin{bmatrix} 2 & 4 \\ 0 & 3 \end{bmatrix} \cdot \begin{bmatrix} 1 \\ 0 \end{bmatrix} = \begin{bmatrix} 2 \\ 0 \end{bmatrix} = 2 \cdot \begin{bmatrix} 1 \\ 0 \end{bmatrix}, \tag{1}$$

$$f(\begin{bmatrix} 4 \\ 1 \end{bmatrix}) = \begin{bmatrix} 2 & 4 \\ 0 & 3 \end{bmatrix} \cdot \begin{bmatrix} 4 \\ 1 \end{bmatrix} = \begin{bmatrix} 12 \\ 3 \end{bmatrix} = 3 \cdot \begin{bmatrix} 4 \\ 1 \end{bmatrix}. \tag{2}$$

Generally, a vector will change both its direction and its magnitude when it was mapped by a linear transformation to another vector. However, if a vector changes only its magnitude, but not its direction, this vector is called an eigenvector of the given matrix and its associated multiplier that reflects a change in its magnitude is called an eigenvalue. For Eqs. (1) and (2), the vector $\vec{v}_1 = [1 \quad 0]^T$ is an eigenvector of the matrix A and 2 is the corresponding eigenvalue, while the vector $\vec{v}_2 = [4 \quad 1]^T$ is another eigenvector with the associated eigenvalue of 3. For a detailed treatment on the theory of eigenvalues and eigenvectors, the interested reader should refer to Reference 10.

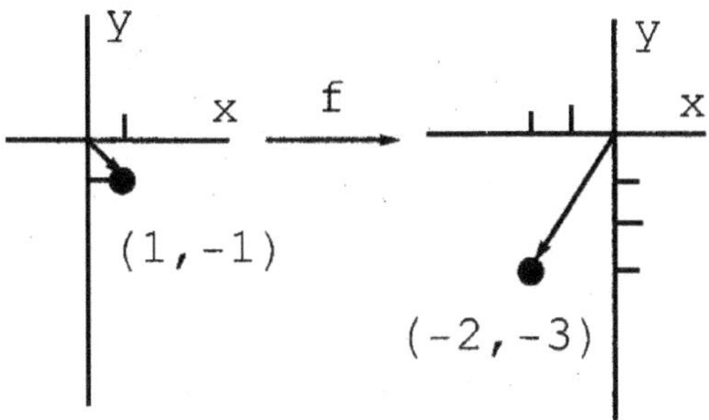

Figure 17.2.1

Example 17.2.6
Given a system of three linear equations in three variables *x*, *y*, and *z* as follows:

$$\begin{cases} x - 2y + z = 0, \\ 3x + y - 2z = 1, \\ 4x - 3y + z = 3. \end{cases}$$

(a) Write the given system into the form of $A \cdot \vec{x} = \vec{b}$, where A is a matrix, $\vec{b}$ is a column vector and $\vec{x}$ is a column vector of the unknown *x*, *y*, and *z*.
(b) Solve for the unknown *x*, *y*, and *z* by using the Gaussian elimination method implemented in the matrix form.
Solution
(a) By using the matrix multiplication and the definition of the equality between two matrices, it can easily be shown that the given system can be written as

$$\begin{bmatrix} 1 & -2 & 1 \\ 3 & 1 & -2 \\ 4 & -3 & 1 \end{bmatrix} \cdot \begin{bmatrix} x \\ y \\ z \end{bmatrix} = \begin{bmatrix} 0 \\ 1 \\ 3 \end{bmatrix}$$

or

$$A \cdot \vec{x} = \vec{b},$$

where A, called the coefficient matrix, $\vec{b}$ and $\vec{x}$ are given, respectively, by

$$A = \begin{bmatrix} 1 & -2 & 1 \\ 3 & 1 & -2 \\ 4 & -3 & 1 \end{bmatrix}, \ \vec{b} = \begin{bmatrix} 0 \\ 1 \\ 3 \end{bmatrix}, \text{ and } \vec{x} = \begin{bmatrix} x \\ y \\ z \end{bmatrix}.$$

(b) To implement the Gaussian elimination method, we first form the augmented matrix as follows:

$$[A \mid \vec{b}] = \left[\begin{array}{ccc|c} 1 & -2 & 1 & 0 \\ 3 & 1 & -2 & 1 \\ 4 & -3 & 1 & 3 \end{array}\right].$$

Next, applying the appropriate elementary row operations on the augmented matrix to transform the part of the coefficient matrix into the upper triangular matrix. Then, use the method of back substitution to solve the equivalent triangular system of equations for the solution.

$$[A \mid \vec{b}] = \left[\begin{array}{ccc|c} 1 & -2 & 1 & 0 \\ 3 & 1 & -2 & 1 \\ 4 & -3 & 1 & 3 \end{array}\right]$$

$$\begin{array}{c} (-3)\cdot[1]+[2] \to [2] \\ \sim \\ (-4)\cdot[1]+[3] \to [3] \end{array} \left[\begin{array}{ccc|c} 1 & -2 & 1 & 0 \\ 0 & 7 & -5 & 1 \\ 0 & 5 & -3 & 3 \end{array}\right]$$

$$\begin{array}{c} (-\frac{5}{7})\cdot[2]+[3] \to [3] \\ \sim \\ \ \end{array} \left[\begin{array}{ccc|c} 1 & -2 & 1 & 0 \\ 0 & 7 & -5 & 1 \\ 0 & 0 & \frac{4}{7} & \frac{16}{7} \end{array}\right].$$

Rewrite the above reduced matrix back as a system of equations.

$$x - 2y + z = 0, \qquad (1)$$

$$7y - 5z = 1, \qquad (2)$$

$$\frac{4}{7}z = \frac{16}{7}. \qquad (3)$$

From Eq. (3), we can solve for z:

$$z = \frac{\frac{16}{7}}{\frac{4}{7}} = 4. \qquad (4)$$

By substituting Eq. (4) into Eq. (2), we can solve for y:

$$7y - 5 \cdot 4 = 1, \Rightarrow$$
$$7y - 20 = 1, \Rightarrow$$
$$7y = 1 + 20 = 21, \Rightarrow y = \frac{21}{7} = 3. \qquad (5)$$

By substituting both Eqs. (4) and (5) into Eq. (1), we can solve for x:

$$x - 2{\cdot}3 + 4 = 0, \Rightarrow x - 6 + 4 = 0, \Rightarrow$$
$$x - 2 = 0, \Rightarrow x = 0 + 2 = 2.$$

Hence, the solution to the given system is $\{(x, y, z) = (2, 3, 4)\}$.

Remarks

1. The meaning of symbols used in part (b) are given as follows: "[3]" means "the 3rd row", "→" means 'assign to", and "~" means "equivalent to". For example, "(–3)·[1]+[2]→[2]" means that row 1 multiplied by (–3) adding to row 2 and the result is assigned to (or replacing) row 2".
2. This example is exactly the same as Example 17.1.1. We simply redo it by using the matrix notation. All Examples in Section 17.1 can be handily done by using the matrix notation.
3. At any stage of elimination, only the row containing the nonzero diagonal entry can serve as a pivotal row to eliminate entries in all rows below the pivotal row.

Example 17.2.7

Find a quadratic function that passes through the three points: $P_1(1, 5)$, $P_2(-1, -3)$, and $P_3(2, 15)$.

Solution

Let the desired quadratic function be given by

$$f(x) = ax^2 + bx + c, \qquad (1)$$

where a, b, and c are to be determined. Since the quadratic function of Eq. (1) is known to pass through $P_1(1, 5)$, and $P_2(-1, -3)$, and $P_3(2, 15)$, we obtain

$$f(1) = a{\cdot}1^2 + b{\cdot}1 + c = 5, \text{ or } a + b + c = 5, \qquad (2)$$
$$f(-1) = a{\cdot}(-1)^2 + b{\cdot}(-1) + c = -3, \Rightarrow$$
$$a - b + c = -3, \qquad (3)$$

$f(2) = a \cdot 2^2 + b \cdot 2 + c = 15, \Rightarrow$
$4a + 2b + c = 15.$ (4)

By applying the method of Gaussian elimination to solve three equations of (2)–(4) in three unknowns a, b, and c, we have

$$\left[\begin{array}{ccc|c} 1 & 1 & 1 & 5 \\ 1 & -1 & 1 & -3 \\ 4 & 2 & 1 & 15 \end{array}\right] \underset{(-4)\cdot[1]+[3]\to[3]}{\overset{(-1)\cdot[1]+[2]\to[2]}{\sim}} \left[\begin{array}{ccc|c} 1 & 1 & 1 & 5 \\ 0 & -2 & 0 & -8 \\ 0 & -2 & -3 & -5 \end{array}\right]$$

$$\overset{(-1)\cdot[2]+[3]\to[3]}{\sim} \left[\begin{array}{ccc|c} 1 & 1 & 1 & 5 \\ 0 & -2 & 0 & -8 \\ 0 & 0 & -3 & 3 \end{array}\right]$$

$$\Rightarrow -3c = 3. \Rightarrow c = \frac{3}{-3} = -1. \tag{5}$$

$$-2b = -8, \Rightarrow b = \frac{-8}{-2} = 4. \tag{6}$$

$a + b + c = 5, \Rightarrow a + 4 + (-1) = 5, \Rightarrow$
$a + 3 = 5, \Rightarrow a = 5 - 3 = 2.$

Hence, the desired quadratic function is

$f(x) = 2x^2 + 4x - 1.$

Exercises

In problems 1–5, find the size of the given matrix:

1. $A_1 = \begin{bmatrix} \frac{1}{3} \end{bmatrix}$. **2.** $A_2 = \begin{bmatrix} 2 & 3 \\ 4 & 5 \end{bmatrix}$.

3. $A_3 = \begin{bmatrix} 2 & 3 \\ 0.1 & -1 \\ 0.4 & 0.6 \end{bmatrix}$. **4.** $A_4 = [-1 \ 2 \ 0 \ 5]$. **5.** $A_5 = \begin{bmatrix} \frac{1}{2} \\ \frac{1}{3} \\ \frac{1}{4} \end{bmatrix}$.

In problems 6–14, perform the indicated matrix operations.

6. Given: $A = \begin{bmatrix} 2 \\ 3 \end{bmatrix}$ and $B = \begin{bmatrix} -1 \\ 4 \end{bmatrix}$. Find (a) $A + B = ?$, (b) $A - B = ?$, (c) $\frac{1}{2} \cdot A = ?$, (d) $\frac{1}{4} \cdot B = ?$, (e) $\frac{1}{2} \cdot A - \frac{1}{4} \cdot B = ?$.

7. Given: $A = \begin{bmatrix} -1 & 2 & 4 \\ 3 & 1 & -5 \end{bmatrix}$ and $B = \begin{bmatrix} 2 & -1 & 3 \\ 1 & 5 & -4 \end{bmatrix}$. Find: (a) $A + B = ?$, (b) $A - B = ?$ (c) $2 \cdot A = ?$ (d) $3 \cdot B = ?$ (e) $2 \cdot A - 3 \cdot B = ?$.

8. Given: $A = [-1 \quad 3]$ and $B = \begin{bmatrix} 2 \\ -5 \end{bmatrix}$. Find (a) $A \cdot B = ?$ (b) $B \cdot A = ?$

9. Given: $A = \begin{bmatrix} 2 & -1 & 3 \\ -3 & 1 & -2 \end{bmatrix}$ and $B = [4 \quad 5]$. Find $B \cdot A = ?$

10. Given: $A = \begin{bmatrix} 2 & -1 & 3 \\ -3 & 1 & -2 \end{bmatrix}$ and $B = \begin{bmatrix} 5 \\ 6 \\ 7 \end{bmatrix}$. Find $A \cdot B = ?$

11. Given: $A = \begin{bmatrix} 2 & -1 & 3 \\ -3 & 1 & -2 \end{bmatrix}$ and $B = \begin{bmatrix} 4 & 5 \\ -1 & 6 \end{bmatrix}$. Find $B \cdot A = ?$

12. Given: $A = \begin{bmatrix} 2 & -1 & 3 \\ -3 & 1 & -2 \end{bmatrix}$ and $B = \begin{bmatrix} 5 & 2 \\ 6 & -1 \\ 7 & 3 \end{bmatrix}$. Find $A \cdot B = ?$

13. Given: $A = \begin{bmatrix} 1 & 3 \\ 2 & 5 \end{bmatrix}$ and $B = \begin{bmatrix} -2 & -1 \\ 1 & -4 \end{bmatrix}$. Find: (a) $A \cdot B = ?$ (b) $B \cdot A = ?$ (c) $A^2 = ?$ (d) $B^3 = ?$

14. Given: $A = \begin{bmatrix} 3 & -1 & 4 \\ 1 & -1 & 0 \\ 2 & 3 & 1 \end{bmatrix}$ and $B = \begin{bmatrix} -5 & 3 & -1 \\ -2 & 4 & 2 \\ 0 & -1 & 1 \end{bmatrix}$. Find: (a) $A \cdot B = ?$ (b) $B \cdot A = ?$ (c) $A^3 = ?$ (d) $B^2 = ?$

15. Given: $A = \begin{bmatrix} 1 & 2 \\ 3 & 2 \end{bmatrix}$. Define $f(\vec{v}) = A \cdot \vec{v}$, where $\vec{v} = \begin{bmatrix} x \\ y \end{bmatrix}$.

Find: (a) $f(\vec{v}_1) = ?$ (b) $f(\vec{v}_2) = ?$ (c) $f(\vec{v}_3) = ?$, where $\vec{v}_1 = [2 \quad 3]^T$, $\vec{v}_2 = [-1 \quad 2]^T$, and $\vec{v}_3 = [1 \quad -1]^T$.

16. Find a quadratic function that passes through the three points $P_1(0, 1)$, $P_2(-1, 6)$, and $P_3(2, 3)$.

17. Find a cubic function that passes through the four points $P_1(-1, -1)$, $P_2(0, 3)$, $P_3(1, 3)$, and $P_4(2, 5)$.

18. Re-do all exercises in Exercise 5.1 by using the Gaussian elimination method implemented in the matrix form.

19. Show that $(A^T)^T = A$.

20. Show that $(A \cdot B)^T = B^T \cdot A^T$.

17.3 The Determinant of a Square Matrix and Cramer's Rule

In this section we will learn another method for solving a system of n linear equations in n variables that involves the determinant of a square matrix.

Definition 17.3.1
The determinant of a 2 × 2 matrix A, (denoted by det(A), or $|A|$) , defined by

$$A = \begin{bmatrix} a_{11} & a_{12} \\ a_{21} & a_{22} \end{bmatrix} \tag{17.3.1}$$

is given by

$$|A| \equiv \det(A) = a_{11} \cdot a_{22} - a_{12} \cdot a_{21}. \tag{17.3.2}$$

Remarks
1. Eq. (17.3.2) states that the determinant of any 2 × 2 matrix is to subtract the product of the off-diagonal entries from the product of the diagonal entries.
2. The determinant of a square matrix can be used as a judging criterion for determining the invertibility of a square matrix to be covered in Section 17.5.

3. To define the determinant of a square matrix of size 3 × 3 and higher, the concepts of minors and cofactors are required as shown in Eqs. (17.3.3-4).
4. The determinant of a 1 × 1 matrix $A = [a]$ is defined as $\det(A) = a$.

Definition 17.3.2

Given a square matrix $A_{n\times n} = [a_{ij}]$., where n ≥ 3 The (i, j)-minor of $A_{n\times n}$ is the $(n-1)\times(n-1)$ matrix obtained by deleting the ith row and j^{th} column of A. The (i, j)-cofactor of the entry a_{ij} is given by

$$C_{ij} = (-1)^{i+j}\det(A_{ij}) \qquad (17.3.3)$$

The determinant of the matrix A is then defined as

$$|A| \equiv \det(A) = a_{i1}C_{i1} + a_{i2}C_{i2} + \ldots + a_{in}C_{in}, \qquad (17.3.4)$$

for some i, $1 \le i \le n$, or

$$|A| \equiv \det(A) = a_{1j}C_{1j} + a_{2j}C_{2j} + \ldots + a_{nj}C_{nj},$$

for some j, $1 \le j \le n$.

Remarks

1. From Eq. (17.3.4), the determinant of an n × n matrix A can be computed by expanding along either a fixed row or a fixed column of A. It does not matter which row or column we choose to expand. Ordinarily, we will choose the row or the column that contains most zero entries to expand. Since a zero entry multiplies its cofactor is always zero, this will reduce somewhat the computational burden.
2. Historically, the concept of the determinant was used in 1693 by G. W. von Leibniz (1646–1716, Germany), co-inventor of calculus that was 150 years before a study of matrices in its own right (Reference 2).

Theorem 17.3.1 (Determinants of Special Matrices)

(a) The determinant of a matrix having a row or column comprised entirely of zero entries is zero.

(b) The determinant of a lower or upper triangular matrix is the product of all of its diagonal entries.

(c) The determinant of a diagonal matrix is the product of its diagonal entries.

Proof

(a) We calculate the determinant by simply expanding the row or the column comprised entirely of zero entries. From Eq. (17.3.4), it is easily seen that the determinant is zero, because all of a_{ij} , $j = 1, \ldots, n$, or a_{ij}, $i = 1, \ldots, n$, are zeroes.

(b) The result follows immediately from Eq. (17.3.4). If the matrix is a lower triangular matrix, we calculate the determinant by expanding its first row. If the matrix is an upper triangular matrix, the determinant is computed by expanding its first column.

(c) A diagonal matrix can be viewed as a special case of either a lower or upper triangular matrix.

Remark

For concrete examples, see problems 7–9, 13–15 in Exercises for Section 17.3.

Example 17.3.1

Find the determinant of each of the following matrices:

(a) $A_1 = \begin{bmatrix} 2 & 5 \\ 4 & 3 \end{bmatrix}$.

(b) $A_2 = \begin{bmatrix} -1 & 0 & 3 \\ 3 & 1 & -2 \\ 4 & 2 & 5 \end{bmatrix}$.

(c) $A_3 = \begin{bmatrix} -1 & 3 & 1 & -2 \\ 4 & 2 & 5 & -1 \\ 1 & 0 & -3 & 2 \\ 0 & -2 & 1 & 5 \end{bmatrix}$.

Solution

(a) By using Eq. (17.3.2), we have

$$\det(A_1) = 2{\cdot}3 - 4{\cdot}5 = 6 - 20 = -14.$$

(b) By expanding the second column of A_2 with Eqs. (17.3.3-4), we have

$$\det(A_2) = 0\cdot(-1)^{2+1}\cdot\det\left(\begin{bmatrix} 3 & -2 \\ 4 & 5 \end{bmatrix}\right) + 1\cdot(-1)^{2+2}\cdot\det\left(\begin{bmatrix} -1 & 3 \\ 4 & 5 \end{bmatrix}\right)+$$

$$2\cdot(-1)^{3+2}\cdot\det\left(\begin{bmatrix} -1 & 3 \\ 3 & -2 \end{bmatrix}\right)$$

$$= 0 + (-1)^4\cdot[(-1)\cdot 5 - 3\cdot 4] + 2\cdot(-1)^5\cdot[(-1)\cdot(-2) - 3\cdot 3]$$

$$= -17 + 2\cdot(-1)\cdot(-7) = -17 + 14 = -3.$$

(c) By expanding the last row of A_3 with Eqs. (17.3.3-4), we have

$$\det(A_3) = 0\cdot(-1)^{4+1}\cdot\det\left(\begin{bmatrix} 3 & 1 & -2 \\ 2 & 5 & -1 \\ 0 & -3 & 2 \end{bmatrix}\right) +$$

$$(-2)\cdot(-1)^{4+2}\cdot\det\left(\begin{bmatrix} -1 & 1 & -2 \\ 4 & 5 & -1 \\ 1 & -3 & 2 \end{bmatrix}\right) + 1\cdot(-1)^{4+3}\cdot\det\left(\begin{bmatrix} -1 & 3 & -2 \\ 4 & 2 & -1 \\ 1 & 0 & 2 \end{bmatrix}\right) +$$

$$5\cdot(-1)^{4+4}\cdot\det\left(\begin{bmatrix} -1 & 3 & 1 \\ 4 & 2 & 5 \\ 1 & 0 & -3 \end{bmatrix}\right). \qquad (1)$$

Let us calculate the determinants of the four 3 × 3 matrices in Eq. (1) separately.

By expanding the first column of the first 3 × 3 matrix in Eq. (1),

$$\det\left(\begin{bmatrix} 3 & 1 & -2 \\ 2 & 5 & -1 \\ 0 & -3 & 2 \end{bmatrix}\right) = 3\cdot(-1)^{1+1}\cdot\det\left(\begin{bmatrix} 5 & -1 \\ -3 & 2 \end{bmatrix}\right) +$$

$$2\cdot(-1)^{2+1}\cdot\det\left(\begin{bmatrix} 1 & -2 \\ -3 & 2 \end{bmatrix}\right) + 0\cdot(-1)^{3+1}\cdot\det\left(\begin{bmatrix} 1 & -2 \\ 5 & -1 \end{bmatrix}\right)$$

$$= 3\cdot 1\cdot(10 - 3) + 2\cdot(-1)\cdot(2 - 6) + 0\cdot 1\cdot(-1 + 10)$$

$$= 21 + 8 + 0 = 29. \qquad (2)$$

By expanding the first row of the second 3 × 3 matrix in Eq. (1),

$$\det\left(\begin{bmatrix} -1 & 1 & -2 \\ 4 & 5 & -1 \\ 1 & -3 & 2 \end{bmatrix}\right) = (-1)\cdot(-1)^{1+1}\cdot\det\left(\begin{bmatrix} 5 & -1 \\ -3 & 2 \end{bmatrix}\right)$$

$$+ 1\cdot(-1)^{1+2}\cdot\det\left(\begin{bmatrix} 4 & -1 \\ 1 & 2 \end{bmatrix}\right) + (-2)\cdot(-1)^{1+3}\cdot\det\left(\begin{bmatrix} 4 & 5 \\ 1 & -3 \end{bmatrix}\right)$$

$= (-1)\cdot 1\cdot(10-3) + 1\cdot(-1)\cdot(8+1) + (-2)\cdot 1\cdot(-12-5)$
$= -7 - 9 + 34 = 18.$ (3)

By expanding the second column of the third 3 × 3 matrix in Eq. (1),

$$\det\left(\begin{bmatrix} -1 & 3 & -2 \\ 4 & 2 & -1 \\ 1 & 0 & 2 \end{bmatrix}\right) = 3\cdot(-1)^{1+2}\cdot\det\left(\begin{bmatrix} 4 & -1 \\ 1 & 2 \end{bmatrix}\right) +$$

$$2\cdot(-1)^{2+2}\cdot\det\left(\begin{bmatrix} -1 & -2 \\ 1 & 2 \end{bmatrix}\right) + 0\cdot(-1)^{3+2}\cdot\det\left(\begin{bmatrix} -1 & -2 \\ 4 & -1 \end{bmatrix}\right)$$

$= 3\cdot(-1)\cdot(8+1) + 2\cdot 1\cdot(-2+2) + 0\cdot(-1)\cdot(1+8)$
$= -27 + 0 + 0 = -27.$ (4)

By expanding the second column of the fourth 3 × 3 matrix in Eq. (1),

$$\det\left(\begin{bmatrix} -1 & 3 & 1 \\ 4 & 2 & 5 \\ 1 & 0 & -3 \end{bmatrix}\right) = 3\cdot(-1)^{1+2}\cdot\det\left(\begin{bmatrix} 4 & 5 \\ 1 & -3 \end{bmatrix}\right) +$$

$$2\cdot(-1)^{2+2}\cdot\det\left(\begin{bmatrix} -1 & 1 \\ 1 & -3 \end{bmatrix}\right) + 0\cdot(-1)^{3+2}\cdot\det\left(\begin{bmatrix} -1 & 1 \\ 4 & 5 \end{bmatrix}\right)$$

$= 3\cdot(-1)\cdot(-12-5) + 2\cdot 1\cdot(3-1) + 0\cdot(-1)\cdot(-5-4)$
$= 51 + 4 + 0 = 55.$ (5)

By substituting Eqs. (2–5) into Eq. (1), we have
$\det(A_3) = 0\cdot(-1)\cdot 29 + (-2)\cdot 1\cdot 18 + 1\cdot(-1)\cdot(-27) + 5\cdot 1\cdot 55$
$= 0 - 36 + 27 + 275 = 266.$

Remark

The determinant possesses such a property that the determinants of two matrices are the same if one matrix is obtained from applying an elementary row operation of Type 3 to another matrix. This property will be covered in the next section. We will show then how to take advantage of this property in calculating the determinant of a square matrix.

Theorem 17.3.2 (Cramer's Rule)

Given a system of n linear equations in n variables as follows:

$$\begin{cases} a_{11}x_1 + a_{12}x_2 + \ldots + a_{1n}x_n = b_1, \\ a_{21}x_1 + a_{22}x_2 + \ldots + a_{2n}x_n = b_2, \\ \ldots\ldots\ldots\ldots\ldots\ldots\ldots\ldots\ldots\ldots, \\ a_{m1}x_1 + a_{m2}x_2 + \ldots + a_{mn}x_n = b_m, \end{cases}$$

or

$$A \cdot \vec{x} = \vec{b}, \qquad (17.3.5)$$

where $A = [a_{ij}]_{\substack{i=1,\ldots,n \\ j=1,\ldots,n}}$, $\vec{x} = (x_1, x_2, \ldots, x_n)^T$, and $\vec{b} = (b_1, b_2, \ldots, b_n)^T$. If det($A$) $\neq 0$, then Eq. (17.3.4) has a unique solution and is given by

$$x_1 = \frac{\det(A_1)}{\det(A)},\ x_2 = \frac{\det(A_2)}{\det(A)},\ \ldots,\ x_n = \frac{\det(A_n)}{\det(A)}, \qquad (17.3.6)$$

where A_i, $i = 1, \ldots, n$ is a matrix obtained by replacing the ith column of the coefficient matrix A with the right side vector $\vec{b}$.

Remarks

1. The proof of this theorem is beyond the scope of this book; hence, it is omitted. However, the interested reader can consult Reference 10.
2. The weak point of Cramer's rule is that if det(A) = 0, Cramer's rule cannot tell whether Eq. (17.3.5) has no solution or infinitely many solutions. But, the Gaussian elimination method can differentiate between the two possible scenarios, that is, it has no solution or infinitely many solutions.

Example 17.3.2

Solve the following system of three equations in three variables for the unknown (x_1, x_2, x_3) by the Cramer's rule:

$$\begin{cases} x_1 - 2x_2 + x_3 = 0, \\ 3x_1 + x_2 - 2x_3 = 1, \\ 4x_1 - 3x_2 + x_3 = 3. \end{cases}$$

Solution

The given system of equations is first written in the form of a matrix equation as follows:

$$\begin{bmatrix} 1 & -2 & 1 \\ 3 & 1 & -2 \\ 4 & -3 & 1 \end{bmatrix} \cdot \begin{bmatrix} x_1 \\ x_2 \\ x_3 \end{bmatrix} = \begin{bmatrix} 0 \\ 1 \\ 3 \end{bmatrix}$$

or

$$\boldsymbol{A} \cdot \vec{\boldsymbol{x}} = \vec{\boldsymbol{b}} .$$

Then, A_1, A_2, and A_3 in Eq. (17.3.6) are given, respectively, by

$$A_1 = \begin{bmatrix} 0 & -2 & 1 \\ 1 & 1 & -2 \\ 3 & -3 & 1 \end{bmatrix},$$

$$A_2 = \begin{bmatrix} 1 & 0 & 1 \\ 3 & 1 & -2 \\ 4 & 3 & 1 \end{bmatrix},$$

and

$$A_3 = \begin{bmatrix} 1 & -2 & 0 \\ 3 & 1 & 1 \\ 4 & -3 & 3 \end{bmatrix}.$$

The determinant of A, A_1, A_2, and A_3 are obtained as follows:

$$\det(A) = \det(\begin{bmatrix} 1 & -2 & 1 \\ 3 & 1 & -2 \\ 4 & -3 & 1 \end{bmatrix})$$

$$= 1 \cdot (-1)^{1+1} \cdot \det(\begin{bmatrix} 1 & -2 \\ -3 & 1 \end{bmatrix}) + 3 \cdot (-1)^{2+1} \cdot \det(\begin{bmatrix} -2 & 1 \\ -3 & 1 \end{bmatrix})$$

$$+ 4 \cdot (-1)^{3+1} \cdot \det(\begin{bmatrix} -2 & 1 \\ 1 & -2 \end{bmatrix})$$

$$= 1 \cdot 1 \cdot (1 - 6) + 3 \cdot (-1) \cdot (-2 + 3) + 4 \cdot 1 \cdot (4 - 1)$$

$$= -5 - 3 + 12 = 4.$$

$$\det(A_1) = \det(\begin{bmatrix} 0 & -2 & 1 \\ 1 & 1 & -2 \\ 3 & -3 & 1 \end{bmatrix})$$

$$= 0 \cdot (-1)^{1+1} \cdot \det(\begin{bmatrix} 1 & -2 \\ -3 & 1 \end{bmatrix}) + 1 \cdot (-1)^{2+1} \cdot \det(\begin{bmatrix} -2 & 1 \\ -3 & 1 \end{bmatrix}$$

$$+ 3\cdot(-1)^{3+1}\cdot\det\left(\begin{bmatrix} -2 & 1 \\ 1 & -2 \end{bmatrix}\right)$$

$$= 0\cdot1\cdot(1-6) + 1\cdot(-1)\cdot(-2+3) + 3\cdot1\cdot(4-1) = 0 - 1 + 9 = 8.$$

$$\det(A_2) = \det\left(\begin{bmatrix} 1 & 0 & 1 \\ 3 & 1 & -2 \\ 4 & 3 & 1 \end{bmatrix}\right)$$

$$= 1\cdot(-1)^{1+1}\cdot\det\left(\begin{bmatrix} 1 & -2 \\ 3 & 1 \end{bmatrix}\right) + 0\cdot(-1)^{1+2}\cdot\det\left(\begin{bmatrix} 3 & -2 \\ 4 & 1 \end{bmatrix}\right)$$

$$+ 1\cdot(-1)^{1+3}\cdot\det\left(\begin{bmatrix} 3 & 1 \\ 4 & 3 \end{bmatrix}\right)$$

$$= 1\cdot1\cdot(1+6) + 0\cdot(-1)\cdot(3+8) + 1\cdot1\cdot(9-4) = 7 + 0 + 5 = 12.$$

$$\det(A_3) = \det\left(\begin{bmatrix} 1 & -2 & 0 \\ 3 & 1 & 1 \\ 4 & -3 & 3 \end{bmatrix}\right)$$

$$= 1\cdot(-1)^{1+1}\cdot\det\left(\begin{bmatrix} 1 & 1 \\ -3 & 3 \end{bmatrix}\right) + (-2)\cdot(-1)^{1+2}\cdot\det\left(\begin{bmatrix} 3 & 1 \\ 4 & 3 \end{bmatrix}\right)$$

$$+ 0\cdot(-1)^{1+3}\cdot\det\left(\begin{bmatrix} 3 & 1 \\ 4 & -3 \end{bmatrix}\right)$$

$$= 1\cdot1\cdot(3+3) + (-2)\cdot(-1)\cdot(9-4) + 0\cdot1\cdot(-9-4)$$

$$= 6 + 10 + 0 = 16.$$

By using Eq. (17.3.6), we have

$$x_1 = \frac{\det(A_1)}{\det(A)} = \frac{8}{4} = 2,$$

$$x_2 = \frac{\det(A_2)}{\det(A)} = \frac{12}{4} = 3,$$

and

$$x_3 = \frac{\det(A_3)}{\det(A)} = \frac{16}{4} = 4.$$

Hence, the unique solution is $\{(x_1, x_2, x_3) = (2,3,4)\}$.

Remark

This example is the same as Example 17.1.1. The solution here is also exactly the same as that of Example 17.1.1.

Exercises

In problems 1–16, find the determinant of the matrix.

1. $A_1 = [-2]$.

2. $A_2 = \begin{bmatrix} 2 & 4 \\ 0 & -3 \end{bmatrix}$.

3. $A_3 = \begin{bmatrix} -3 & 6 \\ -2 & 4 \end{bmatrix}$.

4. $A_4 = \begin{bmatrix} 2 & 6 \\ 1 & -4 \end{bmatrix}$.

5. $A_5 = \begin{bmatrix} 0 & 7 \\ 0 & 9 \end{bmatrix}$.

6. $A_6 = \begin{bmatrix} 4 & 0 & -0.5 \\ -2 & 0 & 0.3 \\ 3 & 0 & 0.2 \end{bmatrix}$.

7. $A_7 = \begin{bmatrix} 0.2 & 1 & -0.4 \\ 0 & 3 & 0.7 \\ 0 & 0 & 5 \end{bmatrix}$.

8. $A_8 = \begin{bmatrix} 2 & 0 & 0 \\ 7 & -3 & 0 \\ 5 & 4 & 1 \end{bmatrix}$.

9. $A_9 = \begin{bmatrix} 7 & 0 & 0 \\ 0 & 9 & 0 \\ 0 & 0 & -5 \end{bmatrix}$.

10. $A_{10} = \begin{bmatrix} 0 & 2 & 5 \\ 4 & -3 & -1 \\ 1 & 0 & -6 \end{bmatrix}$.

11. $A_{11} = \begin{bmatrix} -1 & 2 & 6 \\ -4 & 3 & 7 \\ 8 & -1 & 5 \end{bmatrix}$.

12. $A_{12} = \begin{bmatrix} 2 & -7 & 0 & -8 \\ 5 & 3 & 0 & 2 \\ -3 & 4 & 0 & 1 \\ 1 & 2 & 0 & 3 \end{bmatrix}$.

13. $A_{13} = \begin{bmatrix} -2 & 0 & 0 & 0 \\ 1 & 4 & 0 & 0 \\ -1 & 2 & 5 & 0 \\ 4 & 7 & -3 & -8 \end{bmatrix}$.

14. $A_{14} = \begin{bmatrix} -3 & 0 & 0 & 0 \\ 0 & 0.5 & 0 & 0 \\ 0 & 0 & -0.4 & 0 \\ 0 & 0 & 0 & 3 \end{bmatrix}$.

15. $A_{15} = \begin{bmatrix} 3 & -2 & 0.5 & 1 \\ 0 & -1 & 4 & 2 \\ 0 & 0 & 7 & -3 \\ 0 & 0 & 0 & 5 \end{bmatrix}$.

16. $A_{16} = \begin{bmatrix} 1 & 2 & -3 & -1 \\ 3 & 7 & 2 & 4 \\ 2 & 1 & 6 & 2 \\ 2 & 6 & -1 & 5 \end{bmatrix}$.

In problems 17–18, solve for the unknown x.

17. $\det\left(\begin{bmatrix} 1-x & 2 \\ 2 & 3-x \end{bmatrix}\right) = 0.$ **18.** $\det\left(\begin{bmatrix} 3-x & 0.1 & 0.2 \\ 0 & 1-x & 0.4 \\ 0 & 0 & 5-x \end{bmatrix}\right) = 0.$

In problems 19–23, solve the given system of equations by using the Cramer's rule.

19. $\begin{cases} 2x_1 + 3x_2 = 2, \\ 4x_1 + x_2 = -6. \end{cases}$ **20.** $\begin{cases} x_1 + x_2 + x_3 = 3, \\ 2x_1 + 3x_2 + x_3 = 5, \\ x_1 - x_2 - x_3 = -5. \end{cases}$

21. $\begin{cases} x_1 - x_2 + 2x_3 = 1, \\ -3x_1 + 4x_2 - 6x_3 = -5, \\ 2x_1 - 3x_2 + 4x_3 = 6. \end{cases}$ **22.** $\begin{cases} x_1 - 2x_2 + x_3 = 3, \\ 5x_1 - 3x_2 - x_3 = 10, \\ 2x_1 + 3x_2 - 4x_3 = 1. \end{cases}$

23. $\begin{cases} x_1 - x_2 + x_3 - x_4 = 2, \\ 2x_1 + x_2 - 3x_3 + x_4 = 4, \\ -x_1 + 2x_2 - 2x_3 + x_4 = 3, \\ 3x_1 - x_2 - x_3 + 2x_4 = 5. \end{cases}$

17.4 Operational Rules for Determinants

The determinant of a square matrix possesses some special operational rules that make the calculation of the determinant easier.

Theorem 17.4.1 (Multiplication Rule for Determinants)
Given two square matrices A and B of the size n. Then,

$$\det(A \cdot B) = \det(A) \cdot \det(B) = \det(B \cdot A). \qquad (17.4.1)$$

Remarks

1. The proof of Eq. (17.4.1) is beyond the scope of this book, but it can be found in Reference 7.
2. It can be easily shown that the scalar multiplication rule is given by

$$\det(k \cdot A) = k^n \cdot \det(A). \qquad (17.4.2)$$

3. No addition rule holds for determinants, namely,

$$\det(A \pm B) \neq \det(A) \pm \det(B).$$

4. A verification of Eq, (17.4.1) is given in the following example.

Example 17.4.1

Given: $A = \begin{bmatrix} 1 & 3 \\ -1 & 2 \end{bmatrix}$ and $B = \begin{bmatrix} -2 & 5 \\ 3 & 4 \end{bmatrix}$. Then,

$$\det(A) = 1\cdot 2 - 3\cdot(-1) = 2 + 3 = 5,$$

and

$$\det(B) = (-2)\cdot 4 - 5\cdot 3 = -8 - 15 = -23.$$

Consequently,

$$\det(A)\cdot\det(B) = 5\cdot(-23) = -115. \qquad (1)$$

On the other hand,

$$A\cdot B = \begin{bmatrix} 1 & 3 \\ -1 & 2 \end{bmatrix}\cdot\begin{bmatrix} -2 & 5 \\ 3 & 4 \end{bmatrix}$$

$$= \begin{bmatrix} 1\cdot(-2)+3\cdot 3 & 1\cdot 5+3\cdot 4 \\ (-1)\cdot(-2)+2\cdot 3 & (-1)\cdot 5+2\cdot 4 \end{bmatrix} = \begin{bmatrix} 7 & 17 \\ 8 & 3 \end{bmatrix}.$$

Thus,

$$\det(A\cdot B) = 7\cdot 3 - 17\cdot 8 = 21 - 136 = -115. \qquad (2)$$

From Eqs. (1) and (2), Eq. (17.4.1) is true.
Similarly,

$$B\cdot A = \begin{bmatrix} -2 & 5 \\ 3 & 4 \end{bmatrix}\cdot\begin{bmatrix} 1 & 3 \\ -1 & 2 \end{bmatrix}$$

$$= \begin{bmatrix} (-2)\cdot 1+5\cdot(-1) & (-2)\cdot 3+5\cdot 2 \\ 3\cdot 1+4\cdot(-1) & 3\cdot 3+4\cdot 2 \end{bmatrix} = \begin{bmatrix} -7 & 4 \\ -1 & 17 \end{bmatrix}.$$

Thus,

$$\det(B\cdot A) = (-7)\cdot 17 - 4\cdot(-1) = -119 + 4 = -115. \qquad (3)$$

From Eqs. (1) and (3), Eq. (17.4.1) is true.

Theorem 17.4.2 (Rules for Elementary Row Operations)

Given that A and B are two square matrices. Then the following rules hold:

(i) If B is obtained from A by applying elementary row operation (ERO) of type 1 to A, that is, by interchanging two rows of A, then

$$\det(B) = -\det(A). \qquad (17.4.3)$$

(ii) If B is obtained from A by applying ERO of type 2 to A, that is, by multiplying a row of A with a nonzero constant k, then

$$\det(B) = k\cdot\det(A). \qquad (17.4.4)$$

(iii) If B is obtained from A by applying ERO of type 3 to A, that is by adding a row of A multiplied by a nonzero constant to another row of A, then

$$\det(B) = \det(A). \qquad (17.4.5)$$

Proof

All three types of ERO applied to a matrix A is equivalent to pre-multiply (or multiply on the left of) A by a matrix obtained from applying the same ERO to an identity matrix I of the same size. Let E_I, E_{II}, and E_{III} be the matrices obtained, respectively, by applying ERO of types 1-3 to the identity matrix I. Then we have

(i) $B = E_I A$. By using Eq. (17.4.1),
$\det(B) = \det(E_I)\cdot\det(A) = (-1)\cdot\det(A) = -\det(A)$.

(ii) $B = E_{II} A$. By using Eq. (17.4.1),
$\det(B) = \det(E_{II})\cdot\det(A) = k\cdot\det(A)$,

since E_{II} is a diagonal matrix with one diagonal entry being k and the rest being one; Hence, $\det(E_{II}) = k$ by Theorem 17.3.1.

(iii) $B = E_{III} A$. By using Eq. (17.4.1),
$\det(B) = \det(E_{III})\cdot\det(A) = 1\cdot\det(A) = \det(A)$,

since E_{III} is a lower triangular matrix with all diagonal entry being one and the rest being nonzero; Hence, $\det(E_{II}) = 1$ by Theorem 17.3.1.

Remarks

1. Eq. (17.4.5) provides a practical way to evaluate determinants of matrices of the size being large. The idea is to use the Gaussian elimination method with the ERO of type 3 to reduce the original matrix into an equivalent upper triangular matrix. Then, Eq. (17.4.5) asserts that the determinant of the original matrix equals to that of the reduced upper triangular matrix that is the product of its diagonal entries (Theorem 17.3.1). Example 17.4.3 will illustrate the use of Eq. (17.4.5).
2. Example 17.4.2 illustrates the determinants of three types of elementary row matrices.

Example 17.4.2

Given that $I_{3\times 3}$ be the identity matrix, that is,

$$I_{3\times3} = \begin{bmatrix} 1 & 0 & 0 \\ 0 & 1 & 0 \\ 0 & 0 & 1 \end{bmatrix}.$$

(i) Let F_I be an elementary row matrix of type 1 obtained by interchanging row 2 and row 3 of $I_{3\times3}$. Thus,

$$F_I = \begin{bmatrix} 1 & 0 & 0 \\ 0 & 0 & 1 \\ 0 & 1 & 0 \end{bmatrix}.$$

It is easily shown that $\det(F_I) = -1$ by using Eqs. (17.3.3-4).

(ii) Let F_{II} be an elementary row matrix of type 2 obtained by multiplying row 2 by a nonzero constant k. Thus,

$$F_{II} = \begin{bmatrix} 1 & 0 & 0 \\ 0 & k & 0 \\ 0 & 0 & 1 \end{bmatrix}.$$

It is easily shown that $\det(F_{II}) = k$ by using Theorem 17.3.1(c).

(iii) Let F_{III} be an elementary row matrix of type 3 obtained by adding row 2 multiplied by a nonzero constant k to row 1. Thus,

$$F_{III} = \begin{bmatrix} 1 & k & 0 \\ 0 & 1 & 0 \\ 0 & 0 & 1 \end{bmatrix}.$$

It is easily shown that $\det(F_{III}) = 1$ by using Eqs. (17.3.3-4).

Example 17.4.3

Given that

$$A = \begin{bmatrix} 1 & -2 & 1 \\ 3 & 1 & -2 \\ 4 & -3 & 1 \end{bmatrix}.$$

By using the Gaussian elimination method, find $\det(A) = ?$

Solution

We're going to apply ERO of type 3 repeatedly to reduce A to an upper triangular matrix as follows:

$$A = \begin{bmatrix} 1 & -2 & 1 \\ 3 & 1 & -2 \\ 4 & -3 & 1 \end{bmatrix} \underset{(-4)\cdot[1]+[3]\to[3]}{\overset{(-3)\cdot[1]+[2]\to[2]}{\sim}} \begin{bmatrix} 1 & -2 & 1 \\ 0 & 7 & -5 \\ 0 & 5 & -3 \end{bmatrix}$$

$$\overset{(-5/7)\cdot[2]+[3]\to[3]}{\sim} \begin{bmatrix} 1 & -2 & 1 \\ 0 & 7 & -5 \\ 0 & 0 & \frac{4}{7} \end{bmatrix} = \mathrm{U}. \qquad (1)$$

Thus, we have from using Eq. (1)

$$\det(A) = \det(U) = 1\cdot 7\cdot \frac{4}{7} = 4.$$

Remark

The matrix A here is the coefficient matrix of Example 17.3.2. The determinant of A obtained here is exactly the same as that obtained in Example 17.3.2.

Theorem 17.4.3 (Conditions for Yielding a Zero Determinant)

Given that A be a square matrix. Then $\det(A) = 0$ if either one of the following two conditions is satisfied:

(i) The entire row (or column) of A is zero;

(ii) Two rows (or columns) of A are either equal or proportional to one another.

Proof

If the entire row (or column) is zero, the determinant of A is zero by applying Theorem 17.3.1(a). If two rows (or columns) are either equal or proportional to one another, we can obtain a new matrix by applying a ERO of type 3 to make one entire row to be zero. Since the determinant of this new matrix is zero, that of the matrix A is also zero by Theorem 17.4.2(c). The proof is, therefore, complete.

Remark

The conditions stated in this theorem are only sufficient for the matrix's determinant to be zero, but not necessary. In other words, it does not imply that the matrix A must be of the form that satisfies conditions (i) and (ii) given in the theorem. A counter-example is given below.

Example 17.4.4
Given:

$A = \begin{bmatrix} 1 & -1 & 2 \\ -3 & 4 & -6 \\ 2 & -3 & 4 \end{bmatrix}$. The matrix A does not satisfy conditions (i)-(ii) of Theorem 17.4.3.Yet, the determinant of A can be shown to be zero by using the Gaussian elimination method as follows:

$$A = \begin{bmatrix} 1 & -1 & 2 \\ -3 & 4 & -6 \\ 2 & -3 & 4 \end{bmatrix} \begin{matrix} 3\cdot[1]+[2]\to[2] \\ \sim \\ (-2)\cdot[1]+[3]\to[3] \end{matrix} \begin{bmatrix} 1 & -1 & 2 \\ 0 & 1 & 0 \\ 0 & -1 & 0 \end{bmatrix}$$

$$\begin{matrix} [2]+[3]\to[3] \\ \sim \end{matrix} \begin{bmatrix} 1 & -1 & 2 \\ 0 & 1 & 0 \\ 0 & 0 & 0 \end{bmatrix} = \mathrm{U}. \tag{1}$$

From Eq. (1), we have by using Eq. (17.4.5)

$$\det(A) = \det(U) = 0.$$

Remarks

1. Note that row 2 of the given matrix A is the negative of the sum of rows 1 and 3. Consequently, we found another sufficient condition for zero determinant is that any row is a linear combination of other rows of A.
2. By applying the concept of "linear dependence" in linear algebra, the determinant of a matrix A is zero if the rows of A are linear dependent (Reference 1).

Theorem 17.4.4 (Determinant as the Area of a Triangle)
Given a triangle ΔPQR with vertices $P(x_1, y_1)$, $Q(x_2, y_2)$, and $R(x_3, y_3)$. Then the area of ΔPQR is given by

$$\text{area}(\Delta PQR) = \frac{1}{2}\left|\det\left(\begin{bmatrix} x_1 & y_1 & 1 \\ x_2 & y_2 & 1 \\ x_3 & y_3 & 1 \end{bmatrix}\right)\right|. \tag{17.4.6}$$

Proof
First, we embed the given triangle into the plane $z = 1$ in R^3, a 3-dimensional Euclidean space. Therefore, the three vertices of ΔPQR

become $P'(x_1, y_1, 1)$, $Q'(x_2, y_2, 1)$, and $R'(x_3, y_3, 1)$. By connecting P', Q', and R' with the origin, we get three vectors $\overrightarrow{OP'} = [x_1, y_1, 1]^T$, $\overrightarrow{OQ'} = [x_2, y_2, 1]^T$, and $\overrightarrow{OR'} = [x_3, y_3, 1]^T$. It is well-known in linear algebra that the absolute value of the determinant of a 3×3 matrix whose rows are constituted by three non-coplanar vectors is the volume of a parallelepiped formed by the rows of A. Since the vectors $\overrightarrow{OP'}$, $\overrightarrow{OQ'}$ and $\overrightarrow{OR'}$ lie on the same plane ($z = 1$), the parallelepiped formed by three vectors collapses into a parallelogram; Hence, the absolute value of the determinant becomes the area of this parallelogram. Since the area of the triangle is one-half of the parallelogram, the proof is, therefore, complete.

Remarks

1. The proof that the volume of a parallelepiped formed by the rows of a 3×3 matrix is the absolute value of determinant of A can be found in Reference 10, p. 147.
2. If it happens that the determinant of the triangle is zero, it means that the three vertices of the triangle collapse into lying on the same line. This observation can be used to test whether the given three points lie on the same line as shown in the next theorem.

Theorem 17.4.5 (A Test for Collinear Points)
Three points $P(x_1, y_1)$, $Q(x_2, y_2)$, and $R(x_3, y_3)$ are collinear if and only if the determinant of the matrix formed by the three points is zero, that is,

$$\det\left(\begin{bmatrix} x_1 & y_1 & 1 \\ x_2 & y_2 & 1 \\ x_3 & y_3 & 1 \end{bmatrix}\right) = 0. \qquad (17.4.7)$$

Proof
The "if" part: Suppose that Eq. (17.4.7) holds. Then, from Eq. (17.4.6), the area of the triangle ΔPQR is zero. This implies that ΔPQR collapses into a line. Hence, P, Q, and R lie on the same line.
The "only if" part: Suppose that P, Q, and R lie on the same line. Let m denote the slope of the line. Thus,

$$m = \frac{y_2 - y_1}{x_2 - x_1} = \frac{y_3 - y_1}{x_3 - x_1},$$

or, equivalently,

$$y_2 - y_1 = m(x_2 - x_1), \tag{1}$$

$$y_3 - y_1 = m(x_3 - x_1). \tag{2}$$

By using the Gaussian elimination method, we have by using Eqs. (1) and (2)

$$A = \begin{bmatrix} x_1 & y_1 & 1 \\ x_2 & y_2 & 1 \\ x_3 & y_3 & 1 \end{bmatrix}$$

$$\begin{matrix} (-1)\cdot[1]+[2]\to[2] \\ \sim \\ (-1)\cdot[1]+[3]\to[3] \end{matrix} \begin{bmatrix} x_1 & y_1 & 1 \\ x_2 - x_1 & y_2 - y_1 & 0 \\ x_3 - x_1 & y_3 - y_1 & 0 \end{bmatrix}$$

$$= \begin{bmatrix} x_1 & y_1 & 1 \\ x_2 - x_1 & m(x_2 - x_1) & 0 \\ x_3 - x_1 & m(x_3 - x_1) & 0 \end{bmatrix} = B. \tag{3}$$

By using Eq. (17.3.4) to evaluate the determinant with respect to the last column of Eq. (3), we have det(A) = det(B) = 0. The proof is, therefore, complete.

Remark

Since Eq. (17.4.7) is a necessary and sufficient condition, the three points do not lie in the same line if their associated determinant is not zero.

Example 17.4.5

Given a triangle ΔPQR with three vertices $P(-1, 2)$, $Q(4, -1)$, and $R(4, 6)$. Find the area of ΔPQR.

Solution

By using Eq. (17.4.6),

$$\text{area}(\Delta PQR) = \frac{1}{2}\left|\det\left(\begin{bmatrix} -1 & 2 & 1 \\ 4 & -1 & 1 \\ 4 & 6 & 1 \end{bmatrix}\right)\right|.$$

By using the Gaussian elimination method,

$$A = \begin{bmatrix} -1 & 2 & 1 \\ 4 & -1 & 1 \\ 4 & 6 & 1 \end{bmatrix} \underset{4\cdot[1]+[3]\to[3]}{\overset{4\cdot[1]+[2]\to[2]}{\sim}} \begin{bmatrix} -1 & 2 & 1 \\ 0 & 7 & 5 \\ 0 & 14 & 5 \end{bmatrix}$$

$$\overset{(-2)\cdot[2]+[3]\to[3]}{\sim} \begin{bmatrix} -1 & 2 & 1 \\ 0 & 7 & 5 \\ 0 & 0 & -5 \end{bmatrix} = U.$$

Thus,

$$\text{area}(\Delta PQR) = \frac{1}{2}|\det(A)| = \frac{1}{2}|\det(U)| = \frac{1}{2}\cdot|(-1)\cdot 5\cdot(-5)| = \frac{25}{2}.$$

Example 17.4.6
Determine whether the points $P(-1, 2)$, $Q(1, -2)$, and $R(2, -2)$ lie on the same line.
Solution
By using Eq. (17.4.7),

$$\det(\begin{bmatrix} -1 & 2 & 1 \\ 1 & -2 & 1 \\ 2 & -2 & 1 \end{bmatrix}) = \det(\begin{bmatrix} -1 & 2 & 1 \\ 0 & 0 & 2 \\ 2 & -2 & 1 \end{bmatrix})$$

$$= 2\cdot(-1)^{3+2}\cdot\det(\begin{bmatrix} -1 & 2 \\ 2 & -2 \end{bmatrix}) = 2\cdot(-1)\cdot(2-4) = 4 \neq 0\cdot$$

Since the determinant is not zero, P, Q, and R do not lie on the same line.

Remark
By using Eq. (17.4.6), the area of ΔPQR equals 2.

Exercises

1. Show that $\det(A) = \det(A^T)$.

In problems 2–8, explain why the equation of the determinant holds.

2. $\det(\begin{bmatrix} -1 & 2 & 4 \\ 3 & 5 & 9 \\ 7 & 11 & 13 \end{bmatrix}) = -\det(\begin{bmatrix} 7 & 11 & 13 \\ 3 & 5 & 9 \\ -1 & 2 & 4 \end{bmatrix})$.

3. $\det\left(\begin{bmatrix} -1 & 2 & 4 \\ 3 & 5 & 9 \\ 7 & 11 & 13 \end{bmatrix}\right) = -\det\left(\begin{bmatrix} 3 & 5 & 9 \\ 7 & 11 & 13 \\ -1 & 2 & 4 \end{bmatrix}\right).$

4. $\det\left(\begin{bmatrix} 2 & 8 & 12 \\ 1 & -3 & 5 \\ 9 & 11 & 13 \end{bmatrix}\right) = 2\cdot\det\left(\begin{bmatrix} 1 & 4 & 6 \\ 1 & -3 & 5 \\ 9 & 11 & 13 \end{bmatrix}\right).$

5. $\det\left(\begin{bmatrix} 1 & -3 & 5 \\ 3 & 9 & -18 \\ 12 & -4 & 16 \end{bmatrix}\right) = 12\cdot\det\left(\begin{bmatrix} 1 & -3 & 5 \\ 1 & 3 & -6 \\ 3 & -1 & 4 \end{bmatrix}\right).$

6. $\det\left(4\cdot\begin{bmatrix} 1 & -3 & 5 \\ 2 & 3 & -4 \\ -3 & 7 & 9 \end{bmatrix}\right) = 64\cdot\det\left(\begin{bmatrix} 1 & -3 & 5 \\ 2 & 3 & -4 \\ -3 & 7 & 9 \end{bmatrix}\right).$

7. $\det\left(\begin{bmatrix} 2 & 3 & -4 \\ 1 & -3 & 5 \\ -3 & 7 & 9 \end{bmatrix}\right) = \det\left(\begin{bmatrix} 2 & 3 & -4 \\ 1 & -3 & 5 \\ 0 & -2 & 24 \end{bmatrix}\right).$

8. $\det\left(\begin{bmatrix} 2 & 3 & -4 \\ 1 & -3 & 5 \\ -3 & 7 & 9 \end{bmatrix}\right) = \det\left(\begin{bmatrix} 0 & 9 & -14 \\ 1 & -3 & 5 \\ 0 & -2 & 24 \end{bmatrix}\right).$

In problems 9–15, using EROs as aid for calculating the determinant.

9. $\det\left(\begin{bmatrix} 1 & -3 & 5 \\ -3 & 7 & 9 \\ 2 & 3 & -4 \end{bmatrix}\right) = ?$

10. $\det\left(\begin{bmatrix} 1 & -3 & 5 \\ 2 & 8 & 12 \\ 9 & 11 & 13 \end{bmatrix}\right) = ?$

11. $\det\left(\begin{bmatrix} -1 & 2 & 4 \\ 3 & 5 & 9 \\ 7 & 11 & 13 \end{bmatrix}\right) = ?$

12. $\det\left(\begin{bmatrix} 3 & 9 & -18 \\ 1 & -3 & 5 \\ 12 & -4 & 16 \end{bmatrix}\right) = ?$

13. $\det\left(\begin{bmatrix} 3 & 6 & 2 & 1 \\ 4 & 5 & 7 & -1 \\ 2 & 9 & 11 & 2 \\ -5 & 3 & 4 & -3 \end{bmatrix}\right) = ?$

14. $\det\left(\begin{bmatrix} -1 & 2 & 4 & -3 \\ 3 & 5 & 9 & 7 \\ -2 & 7 & 11 & 13 \\ 4 & 5 & 15 & 17 \end{bmatrix}\right) = ?$

15. $\det\left(\begin{bmatrix} 1 & 2 & -1 & 7 & 4 \\ 3 & 5 & 3 & -5 & -9 \\ -2 & 9 & 2 & 3 & 15 \\ 4 & 11 & -4 & 1 & 2 \\ -7 & 13 & 5 & -2 & 11 \end{bmatrix}\right) = ?$

In problems 16–17, find the area of a triangle with given vertices.

16. $P(2, 2)$, $Q(0, 2)$, and $R(2, 5)$.
17. $P(-1, -1)$, $Q(1, 1)$, and $R(2, 0)$.

In problems 18–19, determine if the given three points lie in the same line.

18. $P(0, 3)$, $Q(2, -1)$, and $R(-1, 5)$.
19. $P(0, -1)$, $Q(2, 0)$, and $R(0, 3)$.

20. Show that $\det\left(\begin{bmatrix} 1 & x & x^2 \\ 1 & y & y^2 \\ 1 & z & z^2 \end{bmatrix}\right) = (x-y)(y-z)(z-x)$.

17.5 The Inverse of a Square Matrix

In this section we're going to study another method for solving a system of n linear equations in n variables. The tool is the (multiplicative) inverse of a square matrix.

Definition 17.5.1
Given a square matrix $A_{n\times n}$. Another matrix $B_{n\times n}$ is called the (multiplicative) inverse of A if A and B satisfy the following equation:

$$A\cdot B = I_{n\times n} = B\cdot A, \qquad (17.5.1)$$

where B is usually denoted by A^{-1} (read as "A inverse"), or vice versa, $A = B^{-1}$ is the inverse of B.

Remarks
1. Not every square matrix has an inverse.

2. If A has an inverse, the inverse is unique (Theorem 17.5.1).
3. If A is not a square matrix or a square matrix with a zero determinant, we can still define its inverse, called the generalized (or pseudo) inverse (Reference 3).

Example 17.5.1

Given that $A = \begin{bmatrix} 1 & 1 \\ 2 & 3 \end{bmatrix}$ and $B = \begin{bmatrix} 3 & -1 \\ -2 & 1 \end{bmatrix}$. Determine if A and B are the inverse of one another.

Solution

To determine if the two matrices are the inverse of one another, all we need to do is to check whether they satisfy Eq. (17.5.1). Let's compute $A \cdot B$ and $B \cdot A$ as follows:

$$A \cdot B = \begin{bmatrix} 1 & 1 \\ 2 & 3 \end{bmatrix} \cdot \begin{bmatrix} 3 & -1 \\ -2 & 1 \end{bmatrix} = \begin{bmatrix} 1\cdot 3 + 1\cdot(-2) & 1\cdot(-1) + 1\cdot 1 \\ 2\cdot 3 + 3\cdot(-2) & 2\cdot(-1) + 3\cdot 1 \end{bmatrix}$$

$$= \begin{bmatrix} 3-2 & -1+1 \\ 6-6 & -2+3 \end{bmatrix} = \begin{bmatrix} 1 & 0 \\ 0 & 1 \end{bmatrix} = I_{2\times 2}. \quad (1)$$

$$B \cdot A = \begin{bmatrix} 3 & -1 \\ -2 & 1 \end{bmatrix} \cdot \begin{bmatrix} 1 & 1 \\ 2 & 3 \end{bmatrix} = \begin{bmatrix} 3\cdot 1 + (-1)\cdot 2 & 3\cdot 1 + (-1)\cdot 3 \\ (-2)\cdot 1 + 1\cdot 2 & (-2)\cdot 1 + 1\cdot 3 \end{bmatrix}$$

$$= \begin{bmatrix} 3-2 & 3-3 \\ -2+2 & -2+3 \end{bmatrix} = \begin{bmatrix} 1 & 0 \\ 0 & 1 \end{bmatrix} = I_{2\times 2}. \quad (2)$$

From Eqs. (1) and (2), A and B are inverse of one another because Eq. (17.5.1) holds.

Definition 17.5.2

A matrix A is said to be invertible (or nonsingular) if its inverse exists. Otherwise, a matrix is non-invertible (or singular).

Remarks

1. Not every square matrix is invertible. A necessary and sufficient condition for A to be invertible is addressed in Theorem 17.5.2.
2. If a matrix is invertible, its inverse must be unique as shown in the following theorem.

Theorem 17.5.1 (Uniqueness of the Matrix Inverse)

If a matrix A is invertible, its inverse is unique.

Proof

Suppose that C is another matrix that is also an inverse of A, that is, C satisfies

$$C \cdot A = I = A \cdot C.$$

Since the associative law holds for the matrix multiplication,

$$C \cdot A \cdot A^{-1} = C \cdot (A \cdot A^{-1}) = C \cdot I = C, \tag{1}$$

and

$$C \cdot A \cdot A^{-1} = (C \cdot A) \cdot A^{-1} = I \cdot A^{-1} = A^{-1}. \tag{2}$$

From Eqs. (1) and (2), we obtain

$$C = A^{-1}. \tag{3}$$

By Eq. (3), it implies that the inverse of A is unique. The proof is, therefore, complete.

Remarks

1. This theorem does not address the question concerning the existence of the matrix inverse. The existence of the matrix inverse is addressed in Theorem 17.5.3.
2. Before addressing the existence of the matrix inverse, we need the following theorem concerning the existence of a solution to the matrix equation, a system of n linear equations in n variables.

Theorem 17.5.2 (The Existence of a Unique Solution to the Matrix Equation)

Given the matrix equation defined by

$$A\vec{x} = \vec{b}, \tag{17.5.2}$$

where the coefficient matrix A is given by Eq. (17.2.1) with $m = n$, $\vec{x}$ and $\vec{b}$ are given, respectively, by

$$\vec{x} = (x_1, x_2, \ldots, x_n)^T, \tag{17.5.3}$$

and

$$\vec{b} = (b_1, b_2, \ldots, b_n)^T. \tag{17.5.4}$$

If the coefficient matrix A in Eq. (17.5.2) is invertible, a unique solution of Eq. (17.5.2) is then given by

$$\vec{x} = A^{-1}\vec{b}. \tag{17.5.5}$$

Proof

Since A is invertible, let A^{-1} denote its inverse. Thus, we have by multiplying both sides of Eq. (17.5.2) on the left

$$A^{-1} \cdot (A\vec{x}) = A^{-1}\vec{b},$$

$$(A^{-1} \cdot A)\vec{x} = A^{-1}\vec{b},$$

$$I \cdot \vec{x} = A^{-1}\vec{b}\,,$$
$$\vec{x} = A^{-1}\vec{b}\,,$$

which is Eq. (17.5.5). The proof is, therefore, complete.

Remarks

1. This theorem could not answer the question: whether Eq. (17.5.2) has a solution if A is not invertible. If A is singular, there are two possible cases for Eq. (17.5.2): (i) it has infinitely many solutions or (ii) it has no solution. Only the Gaussian elimination method can unveil which of the two cases is true.
2. Even if A is invertible, Eq. (17.5.5) is merely a neat theoretical representation of the solution. In fact, Eq. (17.5.5) is not a practical tool for solving Eq. (17.5.2), because it takes more EROs to just find A^{-1} than solving Eq. (17.5.2) directly by using the method of Gaussian elimination. Yet, Eq. (17.5.5) is not without any merit. In case we solve many matrix equations with the same coefficient matrix, but with different right-side vectors. Once we found the inverse matrix A^{-1}, the solutions to those matrix equations are nothing but multiplying A^{-1} by different right-side vectors.
3. By using the result of this theorem, we're ready to provide a necessary and sufficient condition for the existence of the matrix inverse.

Theorem 17.5.3 (The Existence of the Matrix Inverse)

Given a matrix $A_{n\times n}$. A is invertible if and only if $\det(A) \neq 0$.

Proof

The "if" part: Suppose that $\det(A) \neq 0$. Let $X = [\vec{x}_1, \vec{x}_2, ..., \vec{x}_n]$ be the (assumed) inverse of A and the identity matrix $I = [\vec{i}_1, \vec{i}_2, ..., \vec{i}_n]$, where $\vec{i}_j$ is the jth column of I. Then, by Eq. (17.5.1),

$$A \cdot X = I,$$
$$A \cdot [\vec{x}_1, \vec{x}_2, ..., \vec{x}_n] = [\vec{i}_1, \vec{i}_2, ..., \vec{i}_n],$$
$$[A \cdot \vec{x}_1, A \cdot \vec{x}_2, ..., A \cdot \vec{x}_n] = [\vec{i}_1, \vec{i}_2, ..., \vec{i}_n].$$

By using the equality between two matrices,

$$A \cdot \vec{x}_1 = \vec{i}_1,$$
$$A \cdot \vec{x}_2 = \vec{i}_2,$$

$$\ldots\ \ldots\ \ldots\ \ldots, \qquad (17.5.6)$$
$$A\cdot\vec{x}_n = \vec{i}_n.$$

By Theorem 17.3.1, there is a unique solution to each of Eq. (17.5.6) since $\det(A) \neq 0$. Since $\vec{x}_1$, $\vec{x}_2$, …, and $\vec{x}_n$ have solutions, the (assumed) inverse $X = [\vec{x}_1, \vec{x}_2, \ldots, \vec{x}_n]$ becomes the (real) inverse of A. Hence, A is invertible.

The "only if" part: Suppose that A is invertible. Let A^{-1} be its inverse. Thus, by Eq. (17.5.1),

$$A\cdot A^{-1} = I,$$
$$\det(A)\cdot\det(A^{-1}) = \det(I) = 1, \qquad (17.5.7)$$

From Eq. (17.5.7), we have

$$\det(A^{-1}) = 1/\det(A), \qquad (17.5.8)$$

since neither $\det(A)$ nor $\det(A^{-1})$ can be zero; otherwise, it contradicts Eq. (17.5.7). Therefore, the proof is complete.

Remarks

1. If A is invertible, A^{-1} can be expressed in terms of $\det(A)$ and the adjoint matrix of A (Reference 1). However, as the size of A becomes large, the formula for the inverse matrix becomes computationally impractical. Hence, it is instead replaced by the Gauss-Jordan method (Theorem 17.5.5).
2. For a 2×2 matrix, the inverse matrix has a very neat formula as shown in the next theorem.

Theorem 17.5.4 (The Formula for the Inverse of a 2 × 2 Matrix)
Given a 2×2 matrix A defined by

$$A = \begin{bmatrix} a & b \\ c & d \end{bmatrix}. \qquad (17.5.9)$$

If $\det(A) = ad - bc \neq 0$, then A^{-1} is given by

$$A^{-1} = \frac{1}{\det(A)}\cdot\begin{bmatrix} d & -b \\ -c & a \end{bmatrix}. \qquad (17.5.10)$$

Proof

It suffices to show that $A\cdot A^{-1} = I = A^{-1}\cdot A$. Thus,

$$A\cdot A^{-1} = \begin{bmatrix} a & b \\ c & d \end{bmatrix}\cdot\frac{1}{\det(A)}\cdot\begin{bmatrix} d & -b \\ -c & a \end{bmatrix}$$
$$= \frac{1}{\det(A)}\cdot(\begin{bmatrix} a & b \\ c & d \end{bmatrix}\cdot\begin{bmatrix} d & -b \\ -c & a \end{bmatrix})$$

$$= \frac{1}{\det(A)} \cdot \begin{bmatrix} a \cdot d + b \cdot (-c) & a \cdot (-b) + b \cdot a \\ c \cdot d + d \cdot (-c) & c \cdot (-b) + d \cdot a \end{bmatrix}$$
$$= \frac{1}{\det(A)} \cdot \begin{bmatrix} \det(A) & 0 \\ 0 & \det(A) \end{bmatrix}$$
$$= \begin{bmatrix} \frac{1}{\det(A)} \cdot \det(A) & \frac{1}{\det(A)} \cdot 0 \\ \frac{1}{\det(A)} \cdot 0 & \frac{1}{\det(A)} \cdot \det(A) \end{bmatrix}$$
$$= \begin{bmatrix} 1 & 0 \\ 0 & 1 \end{bmatrix} = I_{2\times 2}.$$

Similarly, $A^{-1} \cdot A = I$. The proof is, therefore, complete.

Remarks

1. Eq. (17.5.9) is very easy to memorize and should be memorized, because it oftentimes is encountered in many practical applications. Basically, the inverse matrix of A is merely comprised of three steps: (i) interchanging the diagonal entries, (ii) adding the minus sign to the off-diagonal entries, and (iii) finally divided by the value of det(A).
2. In a case-control study, its fourfold table simply takes the form of a 2×2 matrix (Reference 6, pp. 50-63).

Example 17.5.2

Given that $A = \begin{bmatrix} 1 & 1 \\ 2 & 3 \end{bmatrix}$. Find A^{-1} = ? if it exists.

Solution

First, calculate det(A) as follows:

$$\det(A) = 1 \cdot 3 - 2 \cdot 1 = 1 \neq 0. \tag{1}$$

By Eq. (1), A^{-1} exists and can be obtained by using Eq. (17.5.10)

$$A^{-1} = \frac{1}{1} \cdot \begin{bmatrix} 3 & -1 \\ -2 & 1 \end{bmatrix} = \begin{bmatrix} 3 & -1 \\ -2 & 1 \end{bmatrix}. \tag{2}$$

Remark

The matrix A in this example is the same as that of Example 17.5.1. Therefore, A^{-1} obtained here (Eq. (2)) is exactly the same as B in Example 17.5.1.

Example 17.5.3

Given that $A = \begin{bmatrix} 1 & 1 \\ 2 & 3 \end{bmatrix}$. Find A^{-1} = ? if it exists.

Solution

Since det(A) = $2 \cdot 3 - 1 \cdot 6 = 6 - 6 = 0$, A^{-1} does not exist.

Theorem 17.5.5 (The Gauss-Jordan Method)

Given a matrix $A_{n \times n}$. If A is invertible, its inverse can be computed as follows:

Step 1. Form the augmented matrix $[A|I]$ by adjoining the identity matrix to A.

Step 2. By applying EROs to transform $[A|I]$ into the form of $[I|B]$. Then B is the desirable inverse matrix of A, that is, $B = A^{-1}$.

Proof

Since A is invertible, let $X = [\vec{x}_1, \vec{x}_2, ..., \vec{x}_n]$ be its inverse matrix. Thus, we have by using Eq. (17.5.6)

$$
\begin{aligned}
A\vec{x}_1 &= \vec{i}_1, \\
A\vec{x}_2 &= \vec{i}_2, \\
\vdots &= \vdots, \\
A\vec{x}_n &= \vec{i}_n.
\end{aligned}
$$

Since all of the above n matrix equations have the same coefficient matrix A, we can use the Gaussian elimination method to solve them simultaneously for $\vec{x}_1$, $\vec{x}_2$, ..., and $\vec{x}_n$, that is, form the augmented matrix as

$$[A | \vec{i}_1 \vdots \vec{i}_2 \vdots \cdots \vdots \vec{i}_n] = [A | I]. \quad (1)$$

Thus, not only A is reduced into an upper triangular matrix, but also A can be reduced to become an identity matrix I, that is,

$$[A | I] \sim [I | A^{-1}\vec{i}_1 \vdots A^{-1}\vec{i}_2 \vdots ... \vdots A^{-1}\vec{i}_n] = [I | A^{-1} \cdot I] = [I | A^{-1}] \quad (2)$$

From Eq. (2), the right-side matrix in Eq. (2) is evidently the desired A^{-1}. Therefore, the proof is complete.

Remarks

1. The method described in this theorem for finding the inverse of a general n × n matrix is called the Gauss-Jordan method. Even if

A is not invertible, the Gauss-Jordan method is still applicable, because it will stop at some step to unveil the signal that at least one of the pivotal diagonal entries is zero; Hence, the left-side of the augmented matrix in Eq. (1) cannot be reduced to an identity matrix.

2. The Gauss-Jordan method can also be used to solve a system of linear equations. The only disadvantage is that it takes more EROs than that of the Gaussian elimination method. As a result, it is less efficient than the Gaussian elimination method.
3. The difference between the method of Gaussian elimination and the Gauss-Jordan method is that the method of Gaussian elimination only reduces the left side matrix of $[A|I]$ to the upper triangular matrix, whereas the Gauss-Jordan method reduces the left-side matrix completely into the identity matrix.
4. The Gauss-Jordan method was described by W. Jordan (1842-1899, Germany) in 1887. However the same method also appears in an article by B. I. Clasen published in the same year. Jordan and Clasen probably discovered this method independently (Reference 4).

Example 17.5.4

Given that $A=\begin{bmatrix}1 & -2 & 1\\ 3 & 1 & -2\\ 4 & -3 & 1\end{bmatrix}$. By using the Gauss-Jordan method find
A^{-1} if it exists.

Solution

By Theorem 17.5.5, we have

$$[A|I]=\left[\begin{array}{ccc|ccc}1 & -2 & 1 & 1 & 0 & 0\\ 3 & 1 & -2 & 0 & 1 & 0\\ 4 & -3 & 1 & 0 & 0 & 1\end{array}\right]$$

$$\begin{array}{c}(-3)\cdot[1]+[2]\to[2]\\ \sim\\ (-4)\cdot[1]+[3]\to[3]\end{array}\left[\begin{array}{ccc|ccc}1 & -2 & 1 & 1 & 0 & 0\\ 0 & 7 & -5 & -3 & 1 & 0\\ 0 & 5 & -3 & -4 & 0 & 1\end{array}\right]$$

$$\begin{array}{c} \frac{2}{7}\cdot[2]+[1]\to[1] \\ \sim \\ (-\frac{5}{7})\cdot[2]+[3]\to[3] \end{array} \left[\begin{array}{ccc|ccc} 1 & 0 & -\frac{3}{7} & \frac{1}{7} & \frac{2}{7} & 0 \\ 0 & 7 & -5 & -3 & 1 & 0 \\ 0 & 0 & \frac{4}{7} & -\frac{13}{7} & -\frac{5}{7} & 1 \end{array}\right]$$

$$\begin{array}{c} \frac{1}{7}\cdot[2]\to[2] \\ \sim \\ \frac{7}{4}\cdot[3]\to[3] \end{array} \left[\begin{array}{ccc|ccc} 1 & 0 & -\frac{3}{7} & \frac{1}{7} & \frac{2}{7} & 0 \\ 0 & 1 & -\frac{5}{7} & -\frac{3}{7} & \frac{1}{7} & 0 \\ 0 & 0 & 1 & -\frac{13}{4} & -\frac{5}{4} & \frac{7}{4} \end{array}\right]$$

$$\begin{array}{c} \frac{3}{7}\cdot[3]+[1]\to[1] \\ \sim \\ \frac{5}{7}\cdot[3]+[2]\to[2] \end{array} \left[\begin{array}{ccc|ccc} 1 & 0 & 0 & -\frac{5}{4} & -\frac{1}{4} & \frac{3}{4} \\ 0 & 1 & 0 & -\frac{11}{4} & -\frac{3}{4} & \frac{5}{4} \\ 0 & 0 & 1 & -\frac{13}{4} & -\frac{5}{4} & \frac{7}{4} \end{array}\right] = [I \mid A^{-1}]. \quad (1)$$

By Eq. (1), the inverse of A is given by

$$A^{-1} = \begin{bmatrix} -\frac{5}{4} & -\frac{1}{4} & \frac{3}{4} \\ -\frac{11}{4} & -\frac{3}{4} & \frac{5}{4} \\ -\frac{13}{4} & -\frac{5}{4} & \frac{7}{4} \end{bmatrix}.$$

Remark

The Gauss-Jordan method is similar to that of the Gaussian elimination method except that it requires that the nonzero pivotal diagonal entry being transformed to be one and then eliminate the entries both below and above the diagonal entries.

Example 17.5.5

Given that $A=\begin{bmatrix} 1 & -1 & 2 \\ -3 & 4 & -6 \\ 2 & -3 & 4 \end{bmatrix}$. By using the Gauss-Jordan method find A^{-1} if it exists.

Solution

By Theorem 17.5.5, we thus have

$$[A|I] = \left[\begin{array}{ccc|ccc} 1 & -1 & 2 & 1 & 0 & 0 \\ -3 & 4 & -6 & 0 & 1 & 0 \\ 2 & -3 & 4 & 0 & 0 & 1 \end{array}\right]$$

$$\begin{array}{c} 3\cdot[1]+[2]\to[2] \\ \sim \\ (-2)\cdot[1]+[3]\to[3] \end{array} \left[\begin{array}{ccc|ccc} 1 & -1 & 2 & 1 & 0 & 0 \\ 0 & 1 & 0 & 3 & 1 & 0 \\ 0 & -1 & 0 & -2 & 0 & 1 \end{array}\right]$$

$$\begin{array}{c} [2]+[1]\to[1] \\ \sim \\ [2]+[3]\to[3] \end{array} \left[\begin{array}{ccc|ccc} 1 & 0 & 2 & 4 & 1 & 0 \\ 0 & 1 & 0 & 3 & 1 & 0 \\ 0 & 0 & 0 & 1 & 1 & 1 \end{array}\right]. \qquad (1)$$

Since the 3rd diagonal entry of the left-side matrix in Eq. (1) is zero, it cannot be used as the pivotal entry to eliminate the 3rd entry of the 1st row. Hence, the inverse of A does not exist.

Remark

This example illustrates that we do not need to know if A is invertible. The Gauss-Jordan method in its computational process can reveal whether the inverse of A exists.

Example 17.5.6

Given a system of three linear equations in three variables as follows:

$$\begin{cases} x_1 - 2x_2 + x_3 = 0, \\ 3x_1 + x_2 - 2x_3 = 1, \\ 4x_1 - 3x_2 + x_3 = 3. \end{cases}$$

(a) Write the given system as the matrix equation of $A\cdot\vec{x}=\vec{b}$.

(b) Find the inverse of the coefficient matrix A if it exists.

(c) By using Eq. (17.5.5) solve for the unknown $\vec{x}=[x_1, x_2, x_3]^T$.

Solution

The matrix equation corresponding to the above system of equations is given by

$$\begin{bmatrix} 1 & -2 & 1 \\ 3 & 1 & -2 \\ 4 & -3 & 1 \end{bmatrix} \cdot \begin{bmatrix} x_1 \\ x_2 \\ x_3 \end{bmatrix} = \begin{bmatrix} 0 \\ 1 \\ 3 \end{bmatrix}, \qquad (1)$$

or

$$A \cdot \vec{x} = \vec{b}.$$

From Eq. (1), this leads to

$$A = \begin{bmatrix} 1 & -2 & 1 \\ 3 & 1 & -2 \\ 4 & -3 & 1 \end{bmatrix}, \qquad (2)$$

and

$$\vec{b} = [0 \quad 1 \quad 3]^T. \qquad (3)$$

(b) Since Eq. (2) is exactly the same as that of Example 17.5.5, the inverse of A exists and is given by

$$A^{-1} = \begin{bmatrix} -\frac{5}{4} & -\frac{1}{4} & \frac{3}{4} \\ -\frac{11}{4} & -\frac{3}{4} & \frac{5}{4} \\ -\frac{13}{4} & -\frac{5}{4} & \frac{7}{4} \end{bmatrix}. \qquad (4)$$

(c) By using Eqs. (2-3) and (17.5.5),

$$\vec{x} = \begin{bmatrix} x_1 \\ x_2 \\ x_3 \end{bmatrix} = A^{-1}\vec{b} = \begin{bmatrix} -\frac{5}{4} & -\frac{1}{4} & \frac{3}{4} \\ -\frac{11}{4} & -\frac{3}{4} & \frac{5}{4} \\ -\frac{13}{4} & -\frac{5}{4} & \frac{7}{4} \end{bmatrix} \cdot \begin{bmatrix} 0 \\ 1 \\ 3 \end{bmatrix}$$

$$= \begin{bmatrix} (-\frac{5}{4}) \cdot 0 + (-\frac{1}{4}) \cdot 1 + \frac{3}{4} \cdot 3 \\ (-\frac{11}{4}) \cdot 0 + (-\frac{3}{4}) \cdot 1 + \frac{5}{4} \cdot 3 \\ (-\frac{13}{4}) \cdot 0 + (-\frac{5}{4}) \cdot 1 + \frac{7}{4} \cdot 3 \end{bmatrix} = \begin{bmatrix} 0 - \frac{1}{4} + \frac{9}{4} \\ 0 - \frac{3}{4} + \frac{15}{4} \\ 0 - \frac{5}{4} + \frac{21}{4} \end{bmatrix} = \begin{bmatrix} 2 \\ 3 \\ 4 \end{bmatrix}. \qquad (5)$$

From Eq. (5), the solution is $\{(x_1, x_2, x_3) = (2, 3, 4)\}$.

Remark
The solution is exactly the same as that of Example 17.1.1.

Example 17.5.7
By using the Gauss-Jordan method to solve the system of equations given in Example 17.5.6.
Solution
By using Eqs. (2) and (3) in Example 17.5.6,

$$\left[A\middle|\vec{b}\right]=\left[\begin{array}{ccc|c}1 & -2 & 1 & 0\\ 3 & 1 & -2 & 1\\ 4 & -3 & 1 & 3\end{array}\right]$$

$$\begin{array}{c}(-3)\cdot[1]+[2]\to[2]\\ \sim\\ (-4)\cdot[1]+[3]\to[3]\end{array}\left[\begin{array}{ccc|c}1 & -2 & 1 & 0\\ 0 & 7 & -5 & 1\\ 0 & 5 & -3 & 3\end{array}\right]$$

$$\begin{array}{c}\frac{1}{7}\cdot[2]\to[2]\\ \sim\end{array}\left[\begin{array}{ccc|c}1 & -2 & 1 & 0\\ 0 & 1 & -\frac{5}{7} & \frac{1}{7}\\ 0 & 5 & -3 & 3\end{array}\right]$$

$$\begin{array}{c}2\cdot[2]+[1]\to[1]\\ \sim\\ (-5)\cdot[2]+[3]\to[3]\end{array}\left[\begin{array}{ccc|c}1 & 0 & -\frac{3}{7} & \frac{2}{7}\\ 0 & 1 & -\frac{5}{7} & \frac{1}{7}\\ 0 & 0 & \frac{4}{7} & \frac{16}{7}\end{array}\right]$$

$$\begin{array}{c}\frac{7}{4}\cdot[3]\to[3]\\ \sim\end{array}\left[\begin{array}{ccc|c}1 & 0 & -\frac{3}{7} & \frac{2}{7}\\ 0 & 1 & -\frac{5}{7} & \frac{1}{7}\\ 0 & 0 & 1 & 4\end{array}\right]$$

$$\begin{matrix} \frac{3}{7}\cdot[3]+[1]\to[1] \\ \sim \\ \frac{5}{7}\cdot[3]+[2]\to[2] \end{matrix} \left[\begin{array}{ccc|c} 1 & 0 & 0 & 2 \\ 0 & 1 & 0 & 3 \\ 0 & 0 & 1 & 4 \end{array}\right]. \qquad (1)$$

From Eq. (1), the unique solution is $\{(x_1, x_2, x_3) = (2, 3, 4)\}$.

Exercises

1. Show that if A in invertible, $(A^{-1})^{-1} = A$.

2. Show that if A is invertible, A^T is also invertible. Moreover, $(A^T)^{-1} = (A^{-1})^T$.

3. Show that if both A and B are invertible, their product $A\cdot B$ is also invertible. Moreover, $(A\cdot B)^{-1} = B^{-1}\cdot A^{-1}$.

In problems 4–9, find the inverse of the given matrix if it exists.

4. $A_4 = \begin{bmatrix} -2 & 4 \\ 1 & 3 \end{bmatrix}$.

5. $A_5 = \begin{bmatrix} 3 & 2 \\ 9 & 6 \end{bmatrix}$.

6. $A_6 = \begin{bmatrix} 1 & 1 & 1 \\ 2 & 3 & 1 \\ 1 & -1 & -1 \end{bmatrix}$.

7. $A_7 = \begin{bmatrix} 1 & -1 & 2 \\ -3 & 4 & -6 \\ 2 & -3 & 4 \end{bmatrix}$.

8. $A_8 = \begin{bmatrix} 1 & -1 & 1 & -1 \\ 2 & 1 & -3 & 1 \\ -1 & 2 & -2 & 1 \\ 3 & -1 & -1 & 2 \end{bmatrix}$.

9. $A_9 = \begin{bmatrix} 0 & 1 & 2 & -1 \\ 1 & -2 & 1 & 3 \\ -1 & 3 & 1 & -4 \\ 1 & 0 & 5 & 1 \end{bmatrix}$.

In problems 10–25, solve the given system for the unknown $\vec{x}$ by using either (i) $\vec{x} = A^{-1}\vec{b}$ if A^{-1} exists, or (ii) the Gauss-Jordan method if A^{-1} does not exist.

10. $A_4\vec{x} = \vec{b}_4$, where A_4 is given in problem 4, $\vec{b}_4 = [2 \ \ -6]^T$.

11. $A_5\vec{x} = \vec{b}_5$, where A_5 is given in problem 5, $\vec{b}_5 = [1 \ \ 3]^T$.

12. $A_5\vec{x} = \vec{b}_6$, where A_5 is given in problem 5, $\vec{b}_6 = [2 \ \ 3]^T$.

13. $A_6\vec{x} = \vec{b}_7$, where A_6 is given in problem 6, $\vec{b}_7 = [3 \ \ 5 \ \ -5]^T$.

14. $A_6\vec{x} = \vec{b}_8$, where A_6 is given in problem 6, $\vec{b}_8 = [-1 \ \ 0 \ \ 2]^T$.
15. $A_6\vec{x} = \vec{b}_9$, where A_6 is given in problem 6, $\vec{b}_9 = [2 \ \ -1 \ \ 3]^T$.
16. $A_7\vec{x} = \vec{b}_{10}$, where A_7 is given in problem 7, $\vec{b}_{10} = [1 \ \ -2 \ \ 1]^T$.
17. $A_7\vec{x} = \vec{b}_{11}$, where A_7 is given in problem 7, $\vec{b}_{11} = [2 \ \ -8 \ \ 6]^T$.
18. $A_7\vec{x} = \vec{b}_{12}$, where A_7 is given in problem 7, $\vec{b}_{12} = [1 \ \ -1 \ \ -1]^T$.
19. $A_8\vec{x} = \vec{b}_{13}$, where A_8 is given in problem 8, $\vec{b}_{13} = [2 \ \ 4 \ \ 3 \ \ 5]^T$.
20. $A_8\vec{x} = \vec{b}_{14}$, where A_8 is given in problem 8, $\vec{b}_{14} = [-1 \ \ 0 \ \ 2 \ \ 0]^T$.
21. $A_8\vec{x} = \vec{b}_{15}$, where A_8 is given in problem 8, $\vec{b}_{15} = [0 \ \ -2 \ \ 3 \ \ 0]^T$.
22. $A_9\vec{x} = \vec{b}_{16}$, where A_9 is given in problem 9, $\vec{b}_{16} = [-2 \ \ 4 \ \ 6 \ \ 6]^T$.
23. $A_9\vec{x} = \vec{b}_{17}$, where A_9 is given in problem 9, $\vec{b}_{17} = [0 \ \ 3 \ \ 6 \ \ -3]^T$.
24. $A_9\vec{x} = \vec{b}_{18}$, where A_9 is given in problem 9, $\vec{b}_{18} = [0 \ \ 4 \ \ -4 \ \ 4]^T$.
25. $A_9\vec{x} = \vec{b}_{19}$, where A_9 is given in problem 9, $\vec{b}_{19} = [2 \ \ 2 \ \ 0 \ \ 2]^T$.

17.6 References

1. Anton, H. (1991). Elementary Linear Algebra, 6th edition, Wiley, New York.
2. Bell, E. T. (1940). The Development of Mathematics, Dover Publications, Inc., New York.
3. Boullion, T. L. and Odell, P. L. (1971). Generalized Inverse Matrices, Wiley-Interscience, New York.
4. Boyer, C. R. (1991). A History of Mathematics, John Wiley & Sons, Inc., New York.
5. Draper, N. R. and Smith, H. (1998). Applied Regression Analysis, 3rd edition, Wiley Interscience, New York.
6. Fleiss, J. L., Levin, B., and Paik, M. C. (2003). Statistical Methods for Rates and Proportions, 3rd edition, Wiley, New York.
7. Lang, S. (1971). Linear Algebra, 2nd edition, Addison-Wesley Publishing Co., Reading, MA.
8. Martzloff, J-C. (1997). A History of Chinese Mathematics, Springer-Verlag, New York.
9. Press, W. H., Flannery, B. P., Teukolsky, S. A., and Vetterling, W. T. (1986). Numerical Recipies: The Art of Scientific Computing, Cambridge University Press, New York.

10. Strang, G. (1976). Linear Algebra and Its Applications, Academic Press, New York.
11. http://en.wikipedia.org/wiki//Gaussian_elimination.
12. http://en.wikipedia.org/wiki//Matrix_(mathematics).

Chapter 18
Sequences, Series, Mathematical Induction, and Binomial Theorem

In this chapter we're going to learn the concept of sequences and series as one group, mathematical induction and binomial theorem as two separated topics.

18.1 Sequences and Series

Definition 18.1.1
A sequence is a function with a domain D being a subset of integers, usually denoted by $\{a(n)\}_{n\in D}$ or $\{a_n\}_{n\in D}$, where $D \equiv$ Domain($a(n)$). If D consists of infinitely many integers, then $a(n)$ is called an infinite sequence. Otherwise, $a(n)$ is called a finite sequence.

Remarks

1. Historically, the notion of sequence was as old as the mathematics herself. However, in ancient times, the term "progression" was used for "sequence." The Rhind papyrus, one of the oldest mathematical book, which was published in 1650 B.C., contained an exercise, problem 64, to find an arithmetic progression with 10 terms, with sum 10, and with a common difference $\frac{1}{8}$ (Reference 1, pp. 2-3). Incidentally, problems on arithmetic progression (or sequence) were also given in "Jiu-Zhang Suan-Shu (九章算术)" ("The Nine Chapters on the Mathematical Art") (Reference 9).
2. A sequence is usually represented by listing its value in order: $\{a_1, a_2, a_3, a_4, \ldots\}$. The numbers in this ordered list are called the

"terms" of the sequence. The domain of a sequence is not necessarily restricted to the set of positive integers, provided that the function values of the sequence are well defined. The independent variable n here is used merely as an index of ordering. Because of this attribute, a sequence is a very useful notation in numerical analysis for denoting the fixed-point iterations, numerical solution of differential equations and in statistics for denoting the time series data (Reference 11).

3. Many important constants in mathematics such as π and e are defined as the limit of a sequence. In fact, irrational numbers can be defined as the limit of a sequence of rational numbers. Note that the irrational number $e \approx 2.718281828$ ... is the limit of a sequence defined by $a_n = (1+\frac{1}{n})^n$ of Eq. (16.1.6) in Section 16.1, i.e., $e = \lim_{n\to\infty}(1+\frac{1}{n})^n$. However, the speed of convergence is very slow (see Example 18.1.1(f)). Another common example is that the area of a circle is obtained as the limit of a sequence of the area of the inscribed n-side regular polygon.
4. A discrete version of any continuous function, by restricting the domain of this continuous function to the set of integers, can always be viewed as a sequence. Since the domain of a sequence consists only of the set of integers, a sequence is always a discontinuous function because its range is comprised of only the isolated discrete points.

Example 18.1.1

The following are examples of sequences:

(a) $\{a_n\}_{n=0}^{+\infty} = \{4, 4, 4, 4, 4, 4, \ldots\}$, where $a_n = 4$.

(b) $\{b_n\}_{n=-1}^{+\infty} = \{-1, 1, 3, 5, 7, 9, \ldots\}$, where $b_n = 2n+1$.

(c) $\{c_n\}_{n=1}^{+\infty} = \{1^2, 2^2, 3^2, 4^2, 5^2, 6^2, \ldots\} = \{1, 4, 9, 16, 25, 36, \ldots\}$, where $c_n = n^2$.

(d) $\{d_n\}_{n=0}^{+\infty} = \{1, \frac{1}{2}, \frac{1}{3}, \frac{1}{4}, \frac{1}{5}, \frac{1}{6}, \ldots\}$, where $d_n = \frac{1}{n+1}$.

(e) $\{e_n\}_{n=-2}^{+\infty} = \{\frac{1}{4}, -\frac{1}{2}, 1, -2, 4, -8, \ldots\}$, where $e_n = (-2)^n$.

(f) $\{f_n\}_{n=1}^{+\infty} = \{ f_1 = 2,\ f_2 = 2.25,\ f_3 = 2.3704,\ f_4 = 2.4414,\ f_5 = 2.4883,\ f_6 = 2.5216,\ f_7 = 2.5465,\ f_8 = 2.5658,\ f_9 = 2.5812,\ f_{10} = 2.5937,\ \ldots,\ f_{20} = 2.6533,\ \ldots,\ f_{30} = 2.6743,\ \ldots,\ f_{40} = 2.6851,\ \ldots,\ f_{50} = 2.6916,\ \ldots,\ f_{60} = 2.6960,\ \ldots,\ f_{70} = 2.6991,\ \ldots,\ f_{80} = 2.7015,\ \ldots,\ f_{90} = 2.7033,\ \ldots,\ f_{100} = 2.7048,\ \ldots,\ f_{200} = 2.7115,\ \ldots,\ f_{300} = 2.7138,\ \ldots,\ f_{400} = 2.7149,\ \ldots,\ f_{500} = 2.7156,\ \ldots,\ f_{1000} = 2.7169,\ \ldots,\ f_{5000} = 2.7180,\ \ldots\},\ \ldots,$

where $f_n = (1+\frac{1}{n})^n$.

Remarks

1. These sequences $\{a_n\}_{n=0}^{+\infty}$, $\{b_n\}_{n=-1}^{+\infty}$, $\{c_n\}_{n=1}^{+\infty}$, $\{d_n\}_{n=0}^{+\infty}$, and $\{e_n\}_{n=-2}^{+\infty}$ can be viewed as the discrete version of the (continuous) constant, linear, quadratic, rational, and exponential functions studied in Chapters 14-16. Also, note that the sequence $\{e_n\}$ has a negative base of – 2 that is not allowed for the (continuous) exponential function defined in Section 16.1. This is because for a sequence its function values are well-defined even for a negative base.
2. From Eq. (16.1.6), it is well known that $f_n \to e$ as $n \to +\infty$. Here it is seen that the speed of convergence is very slow. Even for the 5000th term, $f_{5000} = 2.7180$ has only an accuracy of three significant digits. For a comparison with a series approximation, see Example 18.1.5(e).
3. Another irrational number π can be defined as the limit of the sequence $\{g_n\}_{n=1}^{+\infty}$, where g_n is given by

$$g_n = 2\cdot\frac{2\cdot 2\cdot 4\cdot 4\cdot 6\cdot 6\cdot 8\cdot 8\cdot 10\cdot 10\cdot \ldots}{3\cdot 3\cdot 5\cdot 5\cdot 7\cdot 7\cdot 9\cdot 9\cdot 11\cdot 11\cdot \ldots}$$

$$= 2\cdot\frac{2^2\cdot 4^2\cdot 6^2\cdot \ldots\cdot (2n)^2}{3^2\cdot 5^2\cdot 7^2\cdot \ldots\cdot (2n+1)^2}, \qquad (18.1.1)$$

that is,

$\lim_{n\to\infty} g_n = \pi$ (Reference 1, p. 176).

Definition 18.1.2

Let n be a positive integer. Then n-factorial, denoted by $n!$, is defined by

$$n! = 1\cdot 2\cdot 3\cdot 4\cdot\ldots\cdot(n-1)\cdot n. \tag{18.1.2}$$

Specifically, zero factorial is defined as 1, i.e., $0! \equiv 1$.

Remarks

1. n-factorial is merely the product of the consecutive integers from 1 to n.
2. From Eq. (18.1.2) we obtain the following recursive relationship between $n!$ and $(n-1)!$ if we use the associative law by putting a parenthesis before 1 and after $n-1$:
$$n! = (1\cdot 2\cdot 3\cdot 4\cdot\ldots\cdot(n-1))\cdot n$$
$$= (n-1)!\cdot n = n\cdot(n-1)!. \tag{18.1.3}$$
Eq. (18.1.3) is what we usually use to calculate the value of $n!$ recursively as shown below:
$1! = 1$,
$2! = 1\cdot 2 = 2$,
$3! = 3\cdot 2! = 3\cdot 2 = 6$,
$4! = 4\cdot 3! = 4\cdot 6 = 24$,
$5! = 5\cdot 4! = 5\cdot 24 = 120$,
$6! = 6\cdot 5! = 6\cdot 120 = 720$,
$7! = 7\cdot 6! = 7\cdot 720 = 5{,}040$,
$8! = 8\cdot 7! = 8\cdot 5{,}040 = 40{,}320$,
$9! = 9\cdot 8! = 9\cdot 40{,}320 = 362{,}880$,
$10! = 10\cdot 9! = 10\cdot 362{,}880 = 3{,}628{,}800$.
From the above calculations, it can be seen that the sequence of n-factorial grows very fast. For $n = 10$, the value of $10!$ is already approximately 3.63 million. For large n, the Stirling's approximation, given by
$$n! \approx \sqrt{2\pi n}\cdot\left(\frac{n}{e}\right)^n, \tag{18.1.4}$$
is often used to calculate the value of n-factorial (Reference 1, p.185).
3. n-factorial is used in several areas of mathematics, for example, the binomial theorem (Section 18.5), the Taylor's series expansion of an analytic function, and the tools of counting: permutations and combinations (Section 19.2).

4. A generalization of n-factorial to the case that n is a real number becomes the gamma function $\Gamma(x)$ (Reference 10, Chapter 1).

Example 18.1.2
List the first six terms of each of the following sequences:

(a) $a_n = \dfrac{1}{n!}$, $n = 0, 1, 2, \ldots.$

(b) $b_n = \dfrac{(-1)^{n+1}}{n}$, $n = 1, 2, \ldots.$

Solution

(a)
$$a_0 = \frac{1}{0!} = \frac{1}{1} = 1,$$
$$a_1 = \frac{1}{1!} = \frac{1}{1} = 1,$$
$$a_2 = \frac{1}{2!} = \frac{1}{2},$$
$$a_3 = \frac{1}{3!} = \frac{1}{6},$$
$$a_4 = \frac{1}{4!} = \frac{1}{24},$$
$$a_5 = \frac{1}{5!} = \frac{1}{120}.$$

Therefore, the sequence of $\{a_n\}_{n=0}^{+\infty}$ is given as follows:
$$\{1, 1, \frac{1}{2}, \frac{1}{6}, \frac{1}{24}, \frac{1}{120}, \ldots\}.$$

(b)
$$b_1 = \frac{(-1)^{1+1}}{1} = 1,$$
$$b_2 = \frac{(-1)^{2+1}}{2} = -\frac{1}{2},$$
$$b_3 = \frac{(-1)^{3+1}}{3} = \frac{1}{3},$$
$$b_4 = \frac{(-1)^{4+1}}{4} = -\frac{1}{4},$$
$$b_5 = \frac{(-1)^{5+1}}{5} = \frac{1}{5},$$

$$b_6 = \frac{(-1)^{6+1}}{6} = -\frac{1}{6}.$$

Therefore, the sequence of $\{b_n\}_{n=1}^{+\infty}$ is given as follows:

$$\{1, -\frac{1}{2}, \frac{1}{3}, -\frac{1}{4}, \frac{1}{5}, -\frac{1}{6}, \ldots\}.$$

Remark

It can be shown in calculus that the sequences of $\{a_n\}_{n=0}^{+\infty}$ and $\{b_n\}_{n=1}^{+\infty}$ are the coefficients of x^n in the Taylor's series expansion of $f(x) = e^x$ and $g(x) = \ln x$, namely,

$$e^x = \sum_{n=0}^{\infty} \frac{x^n}{n!} = \sum a_n x^n = a_0 + a_1 x + a_2 x^2 + a_3 x^3 + \ldots,$$

and

$$\ln x = \sum_{n=1}^{\infty} (-1)^{n+1} \cdot \frac{x^n}{n} = \sum_{n=1}^{\infty} b_n x^n = b_1 x + b_2 x^2 + b_3 x^3 + \ldots.$$

(Reference 3, pp. 665-666).

Example 18.1.3

Determine the nth term from each of the following sequences beginning with $n = 0$:

(a) $\{2, 5, 8, 11, 14, 17, \ldots\}$.

(b) $\{1, \frac{1}{2}, \frac{1}{4}, \frac{1}{8}, \frac{1}{16}, \frac{1}{32}, \ldots\}$.

(c) $\{1, -\frac{1}{2}, \frac{1}{3}, -\frac{1}{4}, \frac{1}{5}, -\frac{1}{6}, \ldots\}$.

(d) $\{1, -\frac{1}{3!}, \frac{1}{5!}, -\frac{1}{7!}, \frac{1}{9!}, -\frac{1}{11!}, \ldots\}$.

(e) $\{1, -\frac{1}{2!}, \frac{1}{4!}, -\frac{1}{6!}, \frac{1}{8!}, -\frac{1}{10!}, \ldots\}$.

Solution

(a) Let a_n be the nth term of the given sequence. Since the difference between any two terms is always 3, we learned from Example 18.1.1(b) that a_n is likely to be a linear (discrete) function of n, namely,

$$a_n = m \cdot n + b, \tag{1}$$

where m and b are the unknown constants to be determined. By using $a_0 = 2$ and $a_1 = 5$, we have from Eq. (1) that

$$2 = a_0 = m \cdot 0 + b = 0 + b = b, \qquad (2)$$

$$5 = a_1 = m \cdot 1 + b = m + b. \qquad (3)$$

From Eq. (2), we obtain

$$b = 2.$$

By substituting Eq. (2) into Eq. (3),

$$5 = m + 2, \Rightarrow 5 - 2 = m, \Rightarrow m = 3.$$

Hence, the nth term of the given sequence is given by

$$a_n = 3n + 2.$$

(b) Let b_n be the nth term of the given sequence. We notice that the denominator of each term can be written as a power of 2, i.e.,

$$\{\frac{1}{2^0}, \frac{1}{2^1}, \frac{1}{2^2}, \frac{1}{2^3}, \frac{1}{2^4}, \frac{1}{2^5}, \ldots\}.$$

Hence, the nth term is given by

$$b_n = \frac{1}{2^n}.$$

(c) Let c_n be the nth term of the given sequence. We notice that the given sequence can be written as

$$\{(-1)^0 \cdot \frac{1}{0+1}, (-1)^1 \cdot \frac{1}{1+1}, (-1)^2 \cdot \frac{1}{2+1}, (-1)^3 \cdot \frac{1}{3+1}, (-1)^4 \cdot \frac{1}{4+1},$$
$$(-1)^5 \cdot \frac{1}{5+1}, \ldots\}.$$

Hence, the nth term is given by

$$c_n = (-1)^n \cdot \frac{1}{n+1}.$$

(d) Let d_n be the nth term of the given sequence. We notice that the given sequence can be written as

$$\{ (-1)^0 \cdot \frac{1}{(2 \cdot 0 + 1)!}, (-1)^1 \cdot \frac{1}{(2 \cdot 1 + 1)!}, (-1)^2 \cdot \frac{1}{(2 \cdot 2 + 1)!}, (-1)^3 \cdot \frac{1}{(2 \cdot 3 + 1)!},$$
$$(-1)^4 \cdot \frac{1}{(2 \cdot 4 + 1)!}, (-1)^5 \cdot \frac{1}{(2 \cdot 5 + 1)!}, \ldots\}.$$

Hence, the nth term is given by

$$d_n = (-1)^n \cdot \frac{1}{(2n+1)!}.$$

(e) Let e_n be the nth term of the given sequence. We notice that the given sequence can be written as

$$\{(-1)^0\cdot\frac{1}{(2\cdot 0)!},\ (-1)^1\cdot\frac{1}{(2\cdot 1)!},\ (-1)^2\cdot\frac{1}{(2\cdot 2)!},\ (-1)^3\cdot\frac{1}{(2\cdot 3)!},\ (-1)^4\cdot\frac{1}{(2\cdot 4)!},$$

$$(-1)^5\cdot\frac{1}{(2\cdot 5)!},\ \ldots\}.$$

Hence, the nth term is given by

$$e_n = (-1)^n\cdot\frac{1}{(2n)!}.$$

Remarks

1. $\{a_n\}_{n=0}^{+\infty}$ and $\{b_n\}_{n=0}^{+\infty}$ are the arithmetic and the geometric sequence to be covered, respectively, in Sections 18.2 and 18.3.
2. $\{d_n\}_{n=0}^{+\infty}$ and $\{e_n\}_{n=0}^{+\infty}$ are the coefficients of x^{2n+1} and x^{2n} in the Mac-Laurin series for sin x and cos x, respectively, namely,

   $$\sin x = \sum_{n=0}^{\infty} d_n\cdot x^{2n+1},$$

   and

   $$\cos x = \sum_{n=0}^{\infty} e_n\cdot x^{2n}$$

   (Reference 2, pp. 666-668).

Example 18.1.4

The Fibonacci sequence $\{F_n\}_{n=0}^{+\infty}$ is defined by the following (second order) difference equation with two initial conditions:

$$F_{n+2} = F_n + F_{n+1},\ n = 0, 1, 2, \ldots, \qquad (18.1.5)$$

where $F_0 = 1$ and $F_1 = 1$. By using Eq. (18.1.5) with $F_0 = F_1 = 1$, we obtain

$$F_2 = F_0 + F_1 = 1 + 1 = 2,$$
$$F_3 = F_1 + F_2 = 1 + 2 = 3,$$
$$F_4 = F_2 + F_3 = 2 + 3 = 5,$$
$$F_5 = F_3 + F_4 = 3 + 5 = 8,$$
$$F_6 = F_4 + F_5 = 5 + 8 = 13,$$
$$F_7 = F_5 + F_6 = 8 + 13 = 21,$$
$$F_8 = F_6 + F_7 = 13 + 21 = 34,$$

$$F_9 = F_7 + F_8 = 21 + 34 = 55,$$
$$F_{10} = F_8 + F_9 = 34 + 55 = 89.$$

Hence, the Fibonacci sequence is given by

$$\{1, 1, 2, 3, 5, 8, 13, 21, 34, 55, 89, \ldots\}.$$

Remark

The above sequence is an answer to the following problem involving the population growth of the rabbits posed by Fibonacci (1175 – 1250, Italian), Leonardo of Pisa, in his book titled “Liber Abaci” published in 1202:

“How many pairs of rabbits will you have at the beginning of each month if a single pair of newly born rabbits is put into an enclosure at the beginning of January and if each pair bears a new pair at the beginning of the second month following birth and an additional pair at the beginning of each month thereafter?”

The Fibonacci sequence has been found to have applications in number theory, in probability, in electric line theory and atomic hydrocarbons in chemistry (Reference 14).

Definition 18.1.3

The sum of a sequence $\{a_n\}$ is called a series. If $\{a_n\}_{n \in D}$ is infinite, the corresponding series is called an infinite series. Otherwise, it is called a finite series. The sum of the first n terms of $\{a_n\}$ is denoted by

$$\sum_{i=1}^{n} a_i = a_1 + a_2 + \ldots + a_n, \qquad (18.1.6)$$

where *i* is the index of summation, *n* is the upper limit of summation, and 1 is the lower limit of summation.

Remark

A finite series of Eq. (18.1.6) always exists. However, an infinite series may not exist. This involves the problem of convergence of an infinite series. For example, it can be shown that $\sum_{n=1}^{\infty} \frac{1}{n}$ is not a real number; hence, this infinite series, called a harmonic series, does not converge.

Theorem 18.1.1 (Operational Rules for Finite Series)
Given two sequences $\{a_n\}_{n=0}^{+\infty}$ and $\{b_n\}_{n=0}^{+\infty}$. Then

1. Sum Rule (or Addition Rule)

$$\sum_{i=1}^{n}(a_i+b_i)=\sum_{i=1}^{n}a_i+\sum_{i=1}^{n}b_i \ . \tag{18.1.7}$$

2. Scalar Multiplication Rule

$$\sum_{i=1}^{n}(c\cdot a_i)=c\cdot\sum_{i=1}^{n}a_i \ , \tag{18.1.8}$$

where c is a real number.

Proof

Rule 1.

By expanding the left-hand side (LHS) of Eq. (18.1.7),

LHS of Eq. (18.1.7)

$$=(a_1+b_1)+(a_2+b_2)+...+(a_n+b_n)$$
$$=(a_1+a_2+...+a_n)+(b_1+b_2+...+b_n)$$
$$=\sum_{i=1}^{n}a_i+\sum_{i=1}^{n}b_i$$

= RHS of Eq. (18.1.7).

The proof is, therefore, complete.

Rule 2.

By expanding the left-hand side of Eq. (18.1.8),

LHS of Eq. (18.1.8)

$$=c\cdot a_1+c\cdot a_2+...+c\cdot a_n$$
$$=c\cdot(a_1+a_2+...+a_n)=c\cdot\sum_{i=1}^{n}a_i$$

= RHS of Eq. (18.1.8).

The proof is, therefore, complete.

Theorem 18.1.2 (Formula for Finite Series of Discrete Linear, Quadratic, and Cubic Power Functions)

(a) $$\sum_{i=1}^{n} i = 1+2+3+\ldots+n=\frac{n\cdot(n+1)}{2}. \tag{18.1.9}$$

(b) $\sum_{i=1}^{n} i^2 = 1^2 + 2^2 + 3^2 + \ldots + n^2 = \dfrac{n\cdot(n+1)\cdot(2n+1)}{6}$. (18.1.10)

(c) $\sum_{i=1}^{n} i^3 = 1^3 + 2^3 + 3^3 + \ldots + n^3 = (\dfrac{n\cdot(n+1)}{2})^2$. (18.1.11)

Proof

(a) Let $S = \sum_{i=1}^{n} i$. Note that S can be written in two different ways: either as the sum of 1 to n or the sum of n to 1, namely,

$$S = 1 + 2 + 3 + \ldots + (n-1) + n, \tag{1}$$

or

$$S = n + (n-1) + (n-2) + \ldots + 2 + 1. \tag{2}$$

By adding Eqs. (1) and (2),

$$S + S = (1+n) + [2 + (n-1)] + [3 + (n-2)] + \ldots + [(n-1) + 2] \quad + (n+1)$$
$$= (n+1) + (n+1) + (n+1) + \ldots + (n+1) + (n+1), \Rightarrow$$
$$2S = n\cdot(n+1), \Rightarrow$$
$$S = \frac{1}{2}n(n+1) = \frac{n(n+1)}{2},$$

which is Eq. (18.1.9). The proof is, therefore, complete.

(b) To prove Eq. (18.1.10), the following identity is used:

$$(i+1)^3 = i^3 + 3\cdot i^2 + 3\cdot i + 1. \tag{3}$$

By evaluating the above identity of Eq. (3) at $i = 1, 2, \ldots$, n, we have

$$\begin{aligned}
2^3 &= 1^3 + 3\cdot 1^2 + 3\cdot 1 + 1,\\
3^3 &= 2^3 + 3\cdot 2^2 + 3\cdot 2 + 1,\\
4^3 &= 3^3 + 3\cdot 3^2 + 3\cdot 3 + 1,\\
&\vdots \qquad \vdots\\
(n-1)^3 &= (n-2)^3 + 3\cdot(n-2)^2 + 3\cdot(n-2) + 1,\\
n^3 &= (n-1)^3 + 3\cdot(n-1)^2 + 3\cdot(n-1) + 1,\\
(n+1)^3 &= n^3 + 3\cdot n^2 + 3\cdot n + 1.
\end{aligned} \tag{4}$$

By adding the above n equation of Eq. (4), we obtain

$$2^3 + 3^3 + 4^3 + \ldots + (n-1)^3 + n^3 + (n+1)^3$$
$$= 1^3 + 2^3 + 3^3 + \ldots + (n-2)^3 + (n-1)^3 + n^3 + 3\cdot \sum_{i=1}^{n} i^2$$
$$+ 3\cdot\sum_{i=1}^{n} i + \sum_{i=1}^{n} 1.$$

By noting that the terms of $2^3, 3^3, 4^3, \ldots, n^3$ on both sides of the above equation are canceled out, it follows that

$$(n+1)^3 = 1^3 + 3\cdot\sum_{i=1}^{n} i^2 + 3\cdot\sum_{i=1}^{n} i + n,$$

$$(n+1)^3 - 1^3 - 3\cdot\sum_{i=1}^{n} i - n = 3\cdot\sum_{i=1}^{n} i^2 . \qquad (5)$$

By using Eq. (18.1.9), we have from Eq. (5)

$$n^3 + 3n^2 + 3n + 1 - 1 - 3\cdot\frac{n\cdot(n+1)}{2} - n = 3\cdot\sum_{i=1}^{n} i^2 ,$$

$$n^3 + 3n^2 + 2n - \frac{3}{2}n^2 - \frac{3}{2}n = 3\cdot\sum_{i=1}^{n} i^2 , \Rightarrow$$

$$n^3 + \frac{3}{2}n^2 + \frac{1}{2}n = 3\cdot\sum_{i=1}^{n} i^2 , \Rightarrow$$

$$\frac{n}{2}\cdot(2n^2 + 3n + 1) = 3\cdot\sum_{i=1}^{n} i^2 , \Rightarrow$$

$$\frac{n}{2}\cdot(n+1)\cdot(2n+1) = 3\cdot\sum_{i=1}^{n} i^2 , \Rightarrow$$

$$\frac{\frac{n}{2}\cdot(n+1)\cdot(2n+1)}{3} = \sum_{i=1}^{n} i^2 , \Rightarrow$$

$$\frac{n\cdot(n+1)\cdot(2n+1)}{6} = \sum_{i=1}^{n} i^2 ,$$

which is Eq. (18.1.10). The proof is, therefore, complete.
(b) The proof of Eq. (18.1.11) is left as Problem 2 in Exercises for Section 18.1.

Remarks

1. Historically, Eq. (18.1.9) was proved by Pythagoreans, the brotherhood founded by Pythagoras (570-500 B.C., Greek) (Reference 1, pp. 19-20), while Eq. (18.1.10) was long known to Babylonians (2000 B.C.), but first proved by Archimedes (287-212 B.C., Greek) (Reference 1, p. 8). Incidentally, Eq. (18.1.9) was obtained in "Jiu-Zhang Suan-Shu (九章算术)" ("The Nine Chapters on the Mathematical Art") too (Reference 9).
2. Since $i = 0$ does not alter anything, the index i in all of Eqs. (18.1.9-11) can start from i = 0 instead of i =1, namely,

$$\sum_{i=0}^{n} i \equiv \sum_{i=1}^{n} i = \frac{n\cdot(n+1)}{2},$$

$$\sum_{i=0}^{n} i^2 \equiv \sum_{i=1}^{n} i^2 = \frac{n\cdot(n+1)\cdot(2n+1)}{6},$$

and

$$\sum_{i=0}^{n} i^3 \equiv \sum_{i=1}^{n} i^3 = \left(\frac{n\cdot(n+1)}{2}\right)^2.$$

3. If the index i starts from $i = k$ instead of $i = 1$, Eqs. (18.1.9-11) cannot be used immediately. Some algebraic tricks must then be invoked before using Eqs. (18.1.9-11). There are two ways to overcome the difficulty. The first trick is to use the method of subtraction, namely,

$$\sum_{i=k}^{n} a_i = \sum_{i=1}^{n} a_i - \sum_{i=1}^{k-1} a_i. \qquad (18.1.12)$$

The second trick is to use the method of change-of-variable, namely, by letting $j = i - k$, (or $i = j + k$)

$$\sum_{i=k}^{n} a_i = \sum_{j=0}^{n-k} a_{j+k}. \qquad (18.1.13)$$

Example 18.1.5

Find each of the following series:

(a) $\sum\limits_{i=1}^{20}(3i-5) = ?$

(b) $\sum\limits_{i=21}^{50}(3i-5) = ?$

(c) $\sum\limits_{i=1}^{30}(2i^2 - i + 3) = ?$

(d) $\sum\limits_{i=1}^{40}(i^3 - 3i^2 + 4) = ?$

(e) $\sum\limits_{n=0}^{7}\frac{1}{n!} = ?$

Solution

(a) By using Eqs. (18.1.7-9),

$$\sum_{i=1}^{20}(3i-5) = \sum_{i=1}^{20}(3\cdot i) - \sum_{i=1}^{20} 5 = 3\cdot\sum_{i=1}^{20} i - 20\cdot 5$$

$$= 3\cdot\frac{20\cdot(20+1)}{2} - 100 = 630 - 100 = 530\ . \qquad (1)$$

(b) Since the index i begins from $i = 21$ instead of $i = 1$, Eq. (18.1.9) cannot be used immediately. We will demonstrate two methods on how to use Eqs. (18.1.12-13) to evaluate the given series.

"Method 1."

By using Eq. (18.1.12),

$$\sum_{i=21}^{50} (3i-5) = \sum_{i=1}^{50} (3i-5) - \sum_{i=1}^{20} (3i-5). \qquad (2)$$

On the other hand,

$$\sum_{i=1}^{50} (3i-5) = \sum_{i=1}^{50} (3\cdot i) - \sum_{i=1}^{50} 5 = 3\cdot \sum_{i=1}^{50} i - 50\cdot 5$$

$$= 3\cdot \frac{50\cdot(50+1)}{2} - 250 = 3{,}825 - 250 = 3{,}575\ . \qquad (3)$$

By substituting Eq. (1) and Eq. (3) into the right side of Eq. (2), we have

$$\sum_{i=21}^{50} (3i-5) = 3{,}575 - 530 = 3{,}045\ .$$

"Method 2."

By letting $j = i - 20$, or equivalently, $i = j + 20$, we have

$$\sum_{i=21}^{50} (3i-5) = \sum_{j=1}^{30} [3(j+20)-5] = \sum_{j=1}^{30} (3j+60-5)$$

$$= \sum_{j=1}^{30} (3j-55) = 3\cdot \sum_{j=1}^{30} j + \sum_{j=1}^{30} 55 = 3\cdot \frac{30\cdot(30+1)}{2} + 30\cdot 55$$

$$= 1{,}395 + 1{,}650 = 3{,}045.$$

(c) By using Eqs. (18.1.7-8) and (18.1.10),

$$\sum_{i=1}^{30} (2i^2 - i + 3) = 2\cdot \sum_{i=1}^{30} i^2 - \sum_{i=1}^{30} i + \sum_{i=1}^{30} 3$$

$$= 2\cdot \frac{30(30+1)(2\cdot 30+1)}{6} - \frac{30(30+1)}{2} + 30\cdot 3$$

$$= 18{,}910 - 465 + 90 = 18{,}535.$$

(d) By using Eqs. (18.1.7-8) and (18.1.11),

$$\sum_{i=1}^{40} (i^3 - 3i^2 + 4) = \sum_{i=1}^{40} i^3 - 3\cdot \sum_{i=1}^{40} i^2 + \sum_{i=1}^{40} 4$$

$$= [\frac{40(40+1)}{2}]^2 - 3\cdot \frac{40(40+1)(2\cdot 40+1)}{6} + 40\cdot 4$$

$$= (\frac{40\cdot(40+1)}{2})^2 - 3\cdot \frac{40\cdot(40+1)\cdot(2\cdot 40+1)}{2} + 40\cdot 4$$

$$= (20\cdot 41)^2 - 3\cdot 20\cdot 41\cdot 81 + 160 = 672{,}400 - 199{,}260 + 160 = 473{,}300\ .$$

(e) Since none of Eqs. (18.1.9-11) are applicable to the given series, we obtain the given series by a direct expansion as follows:

$$\sum_{n=0}^{7}\frac{1}{n!}=\frac{1}{0!}+\frac{1}{1!}+\frac{1}{2!}+\frac{1}{3!}+\frac{1}{4!}+\frac{1}{5!}+\frac{1}{6!}+\frac{1}{7!}$$
$$=\frac{1}{1}+\frac{1}{1}+\frac{1}{2}+\frac{1}{6}+\frac{1}{24}+\frac{1}{120}+\frac{1}{720}+\frac{1}{5{,}040}$$
$$=1+1+0.16666667+0.04166667+0.00833333$$
$$+0.00138889+0.00019841$$
$$=2.71825397.$$

Remark

The series of (e) is the first eight terms of MacLaurin series of e^x in the remarks of Example 18.1.2 evaluated at $x = 1$, i.e., $e^1=\sum_{n=0}^{\infty}\frac{1^n}{n!}=\sum_{n=0}^{\infty}\frac{1}{n!}$. By comparing the approximate value of the irrational number e obtained here with the result of Example 18.1.1(f), the series converges much faster than that of the sequence given in example 18.1.1(f).

Exercises

1. By using the identity of $(i+1)^2\equiv i^2+2i+1$, prove that Eq. (18.1.9) holds.

2. By using the identity of $(i+1)^4\equiv i^4+4i^3+6i^2+4i+1$, prove that Eq. (18.1.11) holds.

3. Write down the first six terms of the sequence $\{g_n\}_{n=1}^{+\infty}$ defined by Eq. (18.1.1). Note that $\pi = 3.141592654\ldots$.

In problems 4−11, write down the first six terms for each of the sequences, where n is assumed to begin with 0.

4. $a_n=4n-5$.

5. $b_n=n^2-n+1$.

6. $c_n=n^3-3n+1$.

7. $d_n=3n^4-2n^2+n-4$.

8. $e_n=\frac{2n+1}{n^2-3}$.

9. $f_n=(-3)^n$.

10. $g_n = (-\frac{1}{3})^{n-1}$. **11.** $h_n = \frac{1}{(2n+1)!}$.

In problems 12−16, suppose that the pattern of the given sequence continues. Find the nth term a_n of the given sequence, where n is assumed to begin with 0.

12. $\{1, -1, 1, -1, 1, -1, \ldots\}$.
13. $\{3, 1, -1, -3, -5, -7, \ldots\}$.
14. $\{1, \frac{1}{2}, \frac{1}{4}, \frac{1}{8}, \frac{1}{16}, \frac{1}{32}, \ldots\}$.
15. $\{1, -\frac{1}{3}, \frac{1}{9}, -\frac{1}{27}, \frac{1}{81}, -\frac{1}{243}, \ldots\}$.
16. $\{1, \frac{1}{2}, \frac{1}{240}, \frac{1}{720}, \frac{1}{40{,}320}, \frac{1}{3{,}628{,}800}, \ldots\}$.

In problems 17−20, simplify the ratio of factorials.

17. $\frac{8!}{6!} = ?$ **18.** $\frac{17!}{20!} = ?$

19. $\frac{20!}{17! \cdot 3!} = ?$ **20.** $\frac{(n+1)!}{(n+3)!} = ?$

In problems 21−27, find the given series.

21. $\sum_{k=11}^{20} (4k+1) = ?$ **22.** $\sum_{n=6}^{25} (3n^2) = ?$

23. $\sum_{i=1}^{30} (i-1)(i+2) = ?$ **24.** $\sum_{k=21}^{30} (k^2+1)(k-3) = ?$

25. $\sum_{n=1}^{6} \frac{1}{n} = ?$ **26.** $\sum_{i=0}^{5} (-3)^i = ?$

27. $\sum_{k=0}^{4} \frac{1}{(2k+1)!} = ?$

28. By using Eq. (18.1.4), calculate the following factorials and compare them with the exact values:
(a) $10! \approx ?$ (b) $11! \approx ?$ (c) $12! \approx ?$

29. Let $\{a_n\}_{n=0}^{\infty}$ be a sequence defined by

$$a_n = \frac{\sqrt{5}}{5}\left[\left(\frac{1+\sqrt{5}}{2}\right)^{n+1} - \left(\frac{1-\sqrt{5}}{2}\right)^{n+1}\right]. \qquad (18.1.15)$$

(a) Show that (i) $a_0 = a_1 = 1$, and (ii) a_n satisfies Eq. (18.1.5). Hence, a_n of Eq. (18.1.15) is a Fibonacci sequence.

(b) Show that as $n \to +\infty$, $\frac{a_{n+1}}{a_n} \to \phi$, where $\phi = \frac{1+\sqrt{5}}{2}$ is the golden ratio (Reference 15).

30. The Babylonian method of extracting the square root of a positive real number c goes as follows:
Let a_1 be the greatest integer less than $\sqrt{c}$. For $n = 1, 2, 3, \ldots$, calculate $a_{n+1} = \frac{1}{2}(a_n + \frac{c}{a_n})$. Then $a_1, a_2, a_3, \ldots$ is a sequence of better and better approximation to $\sqrt{c}$. Find an approximation to $\sqrt{3}$ by using the above described method with $n = 5$, i.e., $\sqrt{3} \approx a_5$.

18.2 Arithmetic Sequences and Series

Definition 18.2.1

A sequence $\{a_n\}_{n=0}^{\infty}$ is called an arithmetic sequence if all of the differences between consecutive terms are the same, i.e.,

$$a_{i+1} - a_i = a_{i+2} - a_{i+1},\ i = 0, 1, 2, \ldots \qquad (18.2.1)$$

The common difference of the arithmetic sequence is denoted as d.

Remarks

1. $\{a_n\}_{n=0}^{\infty}$ is not an arithmetic sequence as soon as Eq. (18.2.1) is violated by any single value of i.
2. Problem 40 of the Rhind papyrus involves the idea of an arithmetic sequence. The problem will be described in Problem 1 of Exercises for Section 18.2 (Reference 5, p. 32).

Example 18.2.1

Determine if each of the following sequences is arithmetic:
(a) $\{1, -1, 1, -1, 1, -1, \ldots\}$.
(b) $\left\{-\frac{3}{2}, -1, -\frac{1}{2}, 0, \frac{1}{2}, 1, \ldots\right\}$.
(c) $\{1^2, 2^2, 3^2, 4^2, 5^2, 6^2, \ldots\}$.
(d) $\{1^3, 2^3, 3^3, 4^3, 5^3, 6^3, \ldots\}$.
(e) $\left\{1, \frac{1}{2}, \frac{1}{4}, \frac{1}{8}, \frac{1}{16}, \frac{1}{32}, \ldots\right\}$.
(f) $\left\{1, \frac{1}{2!}, \frac{1}{3!}, \frac{1}{4!}, \frac{1}{5!}, \frac{1}{6!}, \ldots\right\}$.

Solution

(a) Let $\{a_n\}_{n=0}^{\infty}$ denote the given sequence, i.e.,
$\{a_0 = 1, a_1 = -1, a_2 = 1, a_3 = -1, a_4 = 1, a_5 = -1, \ldots\}$.
Since $a_1 - a_0 = -1 - 1 = -2 \neq a_2 - a_1 = 1 - (-1) = 2$, $\{a_n\}_{n=0}^{\infty}$ is not an arithmetic sequence because Eq. (18.2.1) is violated for $i = 0$.

(b) Let $\{b_n\}_{n=0}^{\infty}$ denote the given sequence, i.e.,
$\{b_0 = -\frac{3}{2}, b_1 = -1, b_2 = -\frac{1}{2}, b_3 = 0, b_4 = \frac{1}{2}, b_5 = 1, \ldots\}$.
Since

$$b_1 - b_0 = -1 - (-\frac{3}{2}) = \frac{1}{2} = b_2 - b_1 = -\frac{1}{2} - (-1) = b_3 - b_2 = 0 - (-\frac{1}{2})$$

$$= b_4 - b_3 = \frac{1}{2} - 0 = b_5 - b_4 = 1 - \frac{1}{2} = \frac{1}{2},$$

$\{b_n\}_{n=0}^{\infty}$ is an arithmetic sequence because it satisfies Eq. (18.2.1).

(c) Let $\{c_n\}_{n=1}^{\infty}$ denote the given sequence, i.e.,
$\{c_1 = 1^2 = 1, c_2 = 2^2 = 4, c_3 = 3^2 = 9, c_4 = 4^2 = 16, \ldots\}$.
Since $c_2 - c_1 = 4 - 1 = 3 \neq c_3 - c_2 = 9 - 4 = 5$, $\{c_n\}_{n=1}^{\infty}$ is not an arithmetic sequence because Eq. (18.2.1) is violated for $i = 1$.

(d) Let $\{d_n\}_{n=1}^{\infty}$ denote the given sequence, i.e.,
$\{d_1 = 1^3 = 1,\ d_2 = 2^3 = 8,\ d_3 = 3^3 = 27,$
$d_4 = 4^3 = 64, \dots\}$.
Since $d_2 - d_1 = 8 - 1 = 7 \neq d_3 - d_2 = 27 - = 19$,
$\{d_n\}_{n=1}^{\infty}$ is not an arithmetic sequence because Eq. (18.2.1) is violated for $i = 1$.
(e) Let $\{e_n\}_{n=0}^{\infty}$ denote the given sequence, i.e.,

$$\{e_0 = 1,\ e_1 = \frac{1}{2},\ e_2 = \frac{1}{4},\ e_3 = \frac{1}{8},\ e_4 = \frac{1}{16},\ e_5 = \frac{1}{32}, \dots\}.$$

Since $e_1 - e_0 = \frac{1}{2} - 1 = -\frac{1}{2} \neq e_2 - e_1 = \frac{1}{4} - \frac{1}{2} = -\frac{1}{4}$,
$\{e_n\}_{n=0}^{\infty}$ is not an arithmetic sequence because Eq. (18.2.1) is violated for $i = 0$.
(f) Let $\{f_n\}_{n=1}^{\infty}$ denote the given sequence, i.e.,

$$\{f_1 = \frac{1}{1!} = 1,\ f_2 = \frac{1}{2!} = \frac{1}{2},\ f_3 = \frac{1}{3!} = \frac{1}{6},\ f_4 = \frac{1}{4!} = \frac{1}{24}, \dots\}.$$

Since $f_2 - f_1 = \frac{1}{2} - 1 = -\frac{1}{2} \neq f_3 - f_2 = \frac{1}{6} - \frac{1}{2} = -\frac{1}{3}$,
$\{f_n\}_{n=1}^{\infty}$ is not arithmetic sequence because Eq. (18.2.1) is violated for $i = 1$.

Remark
Note that the sequences for (a)-(f) are given, respectively, by:

$$\{a_n = (-1)^n\}_{n=0}^{\infty},\ \{b_n = \frac{n-3}{2}\}_{n=0}^{\infty},\ \{c_n = n^2\}_{n=1}^{\infty},$$

$$\{d_n = n^3\}_{n=1}^{\infty},\ \{e_n = \frac{1}{2^n}\}_{n=0}^{\infty}, \text{ and } \{f_n = \frac{1}{n!}\}_{n=1}^{\infty}.$$

Among them only $\{b_n\}_{n=0}^{\infty}$, a linear function of *n*, is an arithmetic sequence. In fact, it can be shown that only two classes of functions of *n*, constant and linear, are arithmetic sequences (see Problem 3 in Exercises for Section 18.2).

Theorem 18.2.1 (Formula for the (n + 1)st Term of an Arithmetic Sequence)

Let d be the common difference of an arithmetic sequence $\{a_n\}_{n=0}^{\infty}$. Then the $(n + 1)$st term is given by

$$a_n = a_0 + nd \qquad (18.2.2)$$

Proof

Since $\{a_n\}_{n=0}^{\infty}$ is an arithmetic sequence, we have by repeatedly applying Eq. (18.2.1)

$$a_n = a_{n-1} + d = (a_{n-2} + d) + d = a_{n-2} + 2d$$
$$= (a_{n-3} + d) + 2d = a_{n-3} + 3d = \ldots = a_{n-n} + nd$$
$$= a_0 + nd,$$

which is Eq. (18.2.2). The proof is, therefore, complete.

Remarks

1. Note that since a_0 is the first term of $\{a_n\}_{n=0}^{\infty}$, a_n is the $(n + 1)$st term of the sequence, or equivalently, the nth term is given by a_{n-1}.
2. Eq. (18.2.2) indicates that a_n is a discrete linear function of n with d as the slope and a_0 as the y-intercept. In fact, it can be shown that for any arithmetic sequence with a nonzero common difference a linear function of the form given by Eq. (18.2.2) is a necessary and sufficient condition.

Example 18.2.2

(a) Given that an arithmetic sequence whose fourth term is – 3 have a common difference of − 2. Find the 20th term of this arithmetic sequence.

(b) Given that the second term of an arithmetic sequence is – 3 and whose tenth term is 37. Find the 51st term of the arithmetic sequence.

Solution

(a) Let $\{a_n\}_{n=0}^{\infty}$ denote the desired sequence whose fourth term is $a_3 = -3$. By substituting $a_3 = -3$ and $d = -2$ into Eq. (18.2.2), we obtain

$$-3 = a_0 + 3\cdot(-2) = a_0 - 6 \Rightarrow a_0 = 6 - 3 = 3.$$

By substituting $a_0 = 3$ into Eq. (18.2.2), the $(n + 1)$st term of the sequence $\{a_n\}_{n=0}^{\infty}$ is then given by

$$a_n = 3 + n\cdot(-2) = 3 - 2n.$$

The 20th term of $\{a_n\}_{n=0}^{\infty}$ is a_{19} to be given as follows:

$$a_{19} = 3 - 2\cdot 19 = 3 - 38 = -35.$$

(b) Let $\{b_n\}_{n=0}^{\infty}$ denote the desired sequence whose second term is $b_1 = -3$ and its tenth term is $b_9 = 37$. By substituting $b_1 = -3$ and $b_9 = 37$ into Eq. (18.2.2), we obtain

$$-3 = b_1 = b_0 + d, \qquad (1)$$

and

$$37 = b_9 = b_0 + 9\cdot d. \qquad (2)$$

By subtracting Eq. (1) form Eq. (2),

$$37 - (-3) = b_0 + 9d - (b_0 + d), \Rightarrow$$

$$40 = 8d, \Rightarrow \frac{40}{8} = d, \Rightarrow d = 5. \qquad (3)$$

By substituting Eq. (3) into Eq. (1),

$$-3 = b_0 + 5, \Rightarrow -5 - 3 = b_0, \Rightarrow b_0 .= -8. \qquad (4)$$

By substituting both of Eqs. (3) and (4) into Eq. (18.2.2), we have

$$b_n = -8 + n\cdot 5 = -8 + 5n. \qquad (5)$$

From Eq. (5), the 51st term of $\{b_n\}_{n=0}^{\infty}$ is b_{50} to be given as follow:

$$b_{50} = -8 + 5\cdot 50 = -8 + 250 = 242.$$

Remark

Please be reminded that a_n given by Eq. (18.2.2) is actually the $(n + 1)$st term rather than the nth term of the sequence.

Theorem 18.2.2 (Formula for a Finite Arithmetic Series)

Let S_n be sum of the first n terms of an arithmetic sequence $\{a_n\}_{n=0}^{\infty}$. Then S_n is given by

$$S_n = n\cdot\frac{a_0 + a_{n-1}}{2}. \qquad (18.2.3)$$

Proof

Since S_n is the sum of the first n terms of $\{a_n\}_{n=0}^{\infty}$, we have by using Eq. (18.2.2)

$$S_n = a_0 + a_1 + a_2 + ... + a_{n-2} + a_{n-1}$$
$$= a_0 + (a_0 + d) + (a_0 + 2d) + ... + [a_0 + (n-2)d] + [a_0 + (n-1)d] \qquad (1)$$

On the other hand, s_n can also be written in a backward way as

$$S_n = a_{n-1} + a_{n-2} + ... + a_2 + a_1 + a_0$$
$$= a_{n-1} + (a_{n-1} - d) + (a_{n-1} - 2d) + ... + [a_{n-1} - (n-2)d]$$
$$+[a_{n-1} - (n-1)d] . \qquad (2)$$

By adding Eqs. (1) and (2) together, we obtain

$$S_n + S_n = (a_0 + a_{n-1}) + [(a_0 + d) + (a_{n-1} - d)] + ... + [a_0 +$$
$$+(n-2)d + a_{n-1} - (n-2)d] + [a_0 + (n-1)d + a_{n-1} - (n-1)d]$$
$$= (a_0 + a_{n-1}) + (a_0 + a_{n-1}) + ... + (a_0 + a_{n-1}) + (a_0 + a_{n-1}) . \Rightarrow$$
$$2S_n = n(a_0 + a_{n-1}) \Rightarrow S_n = \frac{n(a_0 + a_n)}{2} = n \cdot \frac{a_0 + a_{n-1}}{2} ,$$

which is Eq. (18.2.3). The proof is, therefore, complete.

Remarks

1. Eq. (18.1.9) is a special case of Eq. (18.2.3) because Eq. (18.1.9) is the sum of the first *n* terms of an arithmetic sequence $\{a_n = n\}_{n=1}^{\infty}$.
2. Eq. (18.2.3) depends only on the first term a_0 and the last term a_{n-1}. A more broad interpretation of Eq. (18.2.3) holds for the sum of any n consecutive terms of an arithmetic sequence. The sum equals the average of the first term plus the last term multiplied by *n*, the number of the terms in the sum. Because of this observation, Eq. (18.1.9) can be much easily handled when *i* begins with any integer other than one, namely,
$$\sum_{i=k+1}^{n} i = (n-k) \cdot \frac{(k+1)+n}{2} = \frac{1}{2} \cdot (n-k) \cdot (n+k+1) . \qquad (18.2.4)$$
However, the above comment does not apply to Eqs. (18.1.10) and (18.1.11) since they are not arithmetic sequences.
3. $\sum_{n=0}^{\infty} a_n$ does not exist (or does not converge).

Example 18.2.3

Find: $\sum_{i=21}^{50} (3i-5) = ?$.

Solution

By using Eqs. (18.1.7 – 8) and (18.2.4),

$$\sum_{i=21}^{50}(3i-5) = 3\cdot\sum_{i=21}^{50} i - \sum_{i=21}^{50} 5$$

$$= 3\cdot(50-20)\cdot\frac{21+50}{2} - (50-20)\cdot 5$$

$$= 3\cdot 30\cdot\frac{71}{2} - 30\cdot 5$$

$$= 45\cdot 71 - 150 = 3{,}195 - 150 = 3{,}045.$$

Remark
The result obtained here is exactly the same as Example 18.1.5(b).

Example 18.2.4
Bob accepted a position as an IT manager with a dot.com company at a salary of \$50,000 for the first year. He was guaranteed a raise of \$5,000 per year for the first four years. What would be Bob's total compensation from the company through five full years of employment?

Solution
Let $\{a_n\}_{n=0}^{4}$ be a sequence of Bob's salary for the first five years. Then a_n is given by

$$a_n = 50{,}000 + 5{,}000\cdot n,\ n = 0, 1, 2, 3, 4. \quad (1)$$

As a consequence, $a_0 = 50{,}000$ and $a_4 = 50{,}000 + 5{,}000\cdot 4 = 70{,}000$. If Bob's total compensation from the company through five full years of employment is denoted by S_5, then

$$S_5 = \sum_{n=0}^{4} a_n\ . \quad (2)$$

Since a_n of Eq. (1) is an arithmetic sequence, we obtain S_5 of Eq. (2) by using Eq. (18.2.3)

$$S_5 = 5\cdot\frac{a_0 + a_4}{2} = 5\cdot\frac{50{,}000 + 70{,}000}{2} = 300{,}000.$$

Hence, Bob' total compensation is \$300,000.

Definition 18.2.2
If interest is earned only on the principal P, then the interest is called simple interest.

Remark
In contrast to compound interest covered in Section 16.1, this is another type of interest that is used in some financial circumstances.

Theorem 18.2.3 (Formula for Simple Interest)
Let $0 < r < 1$ be the annual interest rate and P be the principal. Then the amount S_n after n years for simple interest is given as

$$S_n = P \cdot (1 + n \cdot r). \tag{18.2.5}$$

Proof
Since interest is earned only on the principal P, the interest will be $r \cdot P$ for each year. Therefore, the interest earned in n years is $n \cdot r \cdot P$. Since the amount S_n after n years is the original principal plus the interest earned, we therefore have

$$S_n = P + n \cdot r \cdot P = P \cdot (1 + n \cdot r),$$

which is Eq. (18.2.5). Hence, the proof is complete.

Example 18.2.5
Suppose that a principal of \$1,000 be deposited in a local bank that pays only simple interest at an annual rate of 5%. What is the amount in the account after three years?
Solution
By using Eq. (18.2.5),

$$S_3 = \$1{,}000 \cdot (1 + 3 \cdot 0.05) = \$1{,}000 \cdot 1.15 = \$1{,}150.$$

Hence, the amount in the account after three years is \$1,150.

Remark
In Example 16.1.4, the amount in the account were \$1,160.75 and \$1,161.83, respectively if the interest at the same interest rate was compounded quarterly and continuously. After a comparison, the simple interest is the least favorable policy to the customers.

Exercises

1. Problem 40 of the Rhind papyrus is described as follows: "Divide 100 loaves among five men in such a way that the shares received shall be in arithmetic progression (or sequence) and that $\frac{1}{7}$ of the sum of the largest three shares shall be equal to the sum of the smallest two. What is the (common) difference of the shares?"

2. Problem 64 of the Rhind papyrus was to find an arithmetic sequence with 10 terms with sum 10, and with common difference $\frac{1}{8}$ (Reference 1, pp. 2-3).

3. Show that a sequence $\{ a_n \}$ is arithmetic with a nonzero common difference if an only if a_n is a linear function of *n*.

In problems 4–12, determine if the given sequence is arithmetic. If yes, find the (*n*+1)st term of the arithmetic sequence.

4. $\{1.1, 0.8, 0.5, 0.2, -0.1, -0.4, \ldots\}$.
5. $\{-\frac{7}{15}, \frac{1}{5}, \frac{13}{15}, \frac{23}{15}, \frac{11}{5}, \frac{43}{15}, \ldots\}$.
6. $\{-\frac{13}{12}, -\frac{11}{6}, -\frac{31}{12}, -\frac{10}{3}, -\frac{49}{12}, -\frac{29}{6}, \ldots\}$.
7. $\{\frac{9}{4}, -\frac{3}{2}, 1, -\frac{2}{3}, \frac{4}{9}, -\frac{8}{27}, \ldots\}$.
8. $\{-\ln 2, 0, \ln 2, \ln 4, \ln 8, \ln 16, \ldots\}$.
9. $\{-\ln 3, -\ln 4, -\ln 5, -\ln 6, -\ln 7, -\ln 8, \ldots\}$.
10. $\{e^0, e^1, e^2, e^3, e^4, e^5, \ldots\}$.
11. $\{e^{\ln 5}, e^{\ln 6}, e^{\ln 7}, e^{\ln 8}, e^{\ln 9}, e^{\ln 10}, \ldots\}$.
12. $\{\log_3 \frac{1}{9}, \log_3 \frac{1}{3}, 0, \log_3 3, \log_3 9, \log_3 27, \ldots\}$.

In problems 13–18, find a formula for a_n of the given arithmetic sequence, where *n* begins with 0 and $s_n = \sum_{i=0}^{n-1} a_i$.

13. $a_0 = -25, d = 5$.

14. $a_0 = 3,\ a_{10} = \frac{19}{3}$.

15. $a_5 = 0,\ a_{20} = -6$.

16. $a_0 = 1,\ s_5 = 25$.

17. $a_5 = 14,\ s_{10} = 125$.

18. $a_1 = -1,\ s_{20} - s_{10} = 530$.

In problems 19–22, find the given series.

19. $\sum_{k=1}^{n} 2k = ?$

20. $\sum_{k=1}^{n} (2k-1) = ?$

21. $\sum_{n=0}^{20} (3 + \frac{n}{3}) = ?$

22. $\sum_{n=11}^{20} \frac{4-n}{5} = ?$

23. What is the sum of the first 50 odd positive integers?

24. What is the sum of the first 50 even positive integers?

25. Find x so that $x + 2$, $2x + 3$, and $4x + 5$ are three consecutive terms of an arithmetic sequence.

26. How many years does it take for the initial principal invested at an annual interest rate of 8% to double according to the policy of simple interest?

27. Beth accepts a position as a statistical analyst with a pharmaceutical company at a salary of $45,000 for the first year. She is guaranteed a raise of 8% for the first three years. What is her total compensation from the company through four full years of employment?

28. Cinema 1 in a shopping mall has 16 seats in the first row and 20 rows in all. Each successive row contains two additional seats. How many seats are there in Cinema 1?

18.3 Geometric Sequences and Series

Definition 18.3.1

A sequence $\{a_n\}_{n=0}^{\infty}$ is called a geometric sequence if all of the ratios between consecutive terms are the same, i.e.,

$$\frac{a_{i+1}}{a_i} = \frac{a_{i+2}}{a_{i+1}},\ i = 0, 1, 2, \ldots. \qquad (18.3.1)$$

The common ratio of the geometric sequence is denoted as r.

Remarks

1. Historically, geometric progression (or sequence) was long known to Babylonians. In one of Babylonian tablets a geometric progression was summed: $1 + 2 + 2^2 + \ldots + 2^9$ (Reference 3, p. 37).
2. $\{a_n\}_{n=0}^{\infty}$ is not a geometric sequence as soon as Eq. (18.3.1) is violated by any single value of i.
3. The common ratio r is never zero, i.e., $r \neq 0$.
4. Some discrete mathematical models are related to geometric sequences. For example, a probability model for learning, geometric growth, national income and investment within economic dynamics, and inventory analysis are all involved with geometric sequences (Reference 7, Chapter 2).
5. The relationship between a geometric sequence and an arithmetic sequence was discovered by J. Napier (1550-1617, Scottish) in his invention of the concept of logarithm (Reference 6). For a detailed description, see Example 18.3.6.

Example 18.3.1

Determine if each of the following sequences is geometric:

(a) $\{1, -1, 1, -1, 1, -1, \ldots\}$.

(b) $\{-\frac{3}{2}, -1, -\frac{1}{2}, 0, \frac{1}{2}, 1, \ldots\}$.

(c) $\{1^2, 2^2, 3^2, 4^2, 5^2, 6^2, \ldots\}$.

(d) $\{1^3, 2^3, 3^3, 4^3, 5^3, 6^3, \ldots\}$.

(e) $\{1, \frac{1}{2}, \frac{1}{4}, \frac{1}{8}, \frac{1}{16}, \frac{1}{32}, \ldots\}$.

(f) $\{1, \frac{1}{2!}, \frac{1}{3!}, \frac{1}{4!}, \frac{1}{5!}, \frac{1}{6!}, \ldots\}$.

Solution

(a) Let $\{a_n\}_{n=0}^{\infty}$ denote the given sequence, i.e.,

$\{a_0 = 1,\ a_1 = -1,\ a_2 = 1,\ a_3 = -1,\ a_4 = 1,\ a_5 = -1. \ldots\}$.

Since $\frac{a_1}{a_0} = \frac{-1}{1} = -1 = \frac{a_2}{a_1} = \frac{a_3}{a_2} = \frac{a_4}{a_3} = \frac{a_5}{a_4} = \ldots$, $\{a_n\}_{n=0}^{\infty}$ is a geometric sequence because it satisfies Eq. (18.3.1).

(b) Let $\{b_n\}_{n=0}^{\infty}$ denote the given sequence, i.e.,

$\{b_0 = -\frac{3}{2},\ b_1 = -1,\ b_2 = -\frac{1}{2},\ b_3 = 0,\ b_4 = \frac{1}{2},\ b_5 = 1, \ldots\}$.

Since $\frac{b_1}{b_0} = \frac{-1}{-\frac{3}{2}} = \frac{2}{3} \neq \frac{b_2}{b_1} = \frac{-\frac{1}{2}}{-1} = \frac{1}{2}$, $\{b_n\}_{n=0}^{\infty}$ is not a geometric sequence because Eq. (18.3.1) is violated for $i = 0$.

(c) Let $\{c_n\}_{n=1}^{\infty}$ denote the given sequence, i.e.,

$\{c_1 = 1^2 = 1,\ c_2 = 2^2 = 4,\ c_3 = 3^2 = 9,\ c_4 = 4^2 = 16,\ c_5 = 5^2 = 25,$
$c_6 = 6^2 = 36, \ldots\}$.

Since $\frac{c_2}{c_1} = \frac{2^2}{1^2} = 4 \neq \frac{c_3}{c_2} = \frac{3^2}{2^2} = \frac{9}{4}$, $\{c_n\}_{n=1}^{\infty}$ is not a geometric sequence because Eq. (18.3.1) is violated for $i = 1$.

(d) Let $\{d_n\}_{n=1}^{\infty}$ denote the given sequence, i.e.,

$\{d_1 = 1^3 = 1,\ d_2 = 2^3 = 8,\ d_3 = 3^3 = 27,\ d_4 = 4^3 = 64,\ d_5 = 5^3 = 125,$
$d_6 = 6^3 = 216, \ldots\}$.

Since $\frac{d_2}{d_1} = \frac{2^3}{1^3} = 8 \neq \frac{d_3}{d_2} = \frac{3^3}{2^3} = \frac{27}{8}$, $\{d_n\}_{n=1}^{\infty}$ is not a geometric sequence because Eq. (18.3.1) is violated for $i = 1$.

(e) Let $\{e_n\}_{n=0}^{\infty}$ denote the given sequence, i.e.,

$\{e_0 = 1,\ e_1 = \frac{1}{2},\ e_2 = \frac{1}{4},\ e_3 = \frac{1}{8},\ e_4 = \frac{1}{16},\ e_5 = \frac{1}{32}, \ldots\}$.

Since $\frac{e_1}{e_0} = \frac{\frac{1}{2}}{1} = \frac{1}{2} = \frac{e_2}{e_1} = \frac{e_3}{e_2} = \frac{e_4}{e_3} = \frac{e_5}{e_4}, \ldots,$ $\{e_n\}_{n=0}^{\infty}$ is a geometric sequence because Eq. (18.3.1) is satisfied.

(f) Let $\{f_n\}_{n=1}^{\infty}$ denote the given sequence, i.e.,

$\{f_1 = 1,\ f_2 = \frac{1}{2!} = \frac{1}{2},\ f_3 = \frac{1}{3!} = \frac{1}{6},\ f_4 = \frac{1}{4!} = \frac{1}{24},\ f_5 =$
$\frac{1}{5!} = \frac{1}{120},\ f_6 = \frac{1}{6!} = \frac{1}{720}, \ldots\}$.

Since $\frac{f_2}{f_1} = \frac{\frac{1}{2!}}{1} = \frac{1}{2} \neq \frac{f_3}{f_2} = \frac{\frac{1}{3!}}{\frac{1}{2!}} = \frac{1}{3}$, $\{f_n\}_{n=1}^{\infty}$ is not a geometric sequence because Eq. (18.3.1) is violated for $i = 1$.

Remark

Note that the sequences of this example are exactly the same as those of Example 18.2.1. Among them only $\{a_n\}_{n=0}^{\infty}$ and $\{e_n\}_{n=0}^{\infty}$ that are discrete exponential functions of *n* are geometric sequences. In fact, it can be shown that only discrete exponential functions of *n* can be geometric sequences.

Theorem 18.3.1 (Formula for the (n + 1)st Term of a Geometric Sequence)

Let $r \neq 0$ be the common ratio of a geometric sequence $\{a_n\}_{n=0}^{\infty}$. Then the $(n + 1)$st term is given by

$$a_n = a_0 r^n. \qquad (18.3.2)$$

Proof

Since $\{a_n\}_{n=0}^{\infty}$ is a geometric sequence, we have by repeatedly applying Eq. (18.3.1)

$$a_n = a_{n-1}r = (a_{n-2}r)r = a_{n-2}r^2 = \ldots = a_{n-n}r^n = a_0 r^n,$$

which is Eq. (18.3.2). The proof is, therefore, complete.

Remarks

1. Since the first term of $\{a_n\}_{n=0}^{\infty}$ is a_0, a_n is actually the $(n + 1)$st term, or equivalently, the nth term is a_{n-1}.
2. Eq. (18.3.2) indicates that when $r > 0$, a_n is a discrete exponential function of *n* with *r* as a base. Also, it can be shown that the logarithm of any geometric sequence Eq. (18.3.2) with a positive common ratio *r* is an arithmetic sequence (Problem 2 in Exercises for Section 18.3).

Example 18.3.2

(a) Given that a geometric sequence whose fifth term is 405 has a common ratio -3. Find the 16th term of this geometric sequence.

(b) Given that a geometric sequence whose third term is $\frac{16}{9}$ and whose sixth term is $\frac{128}{243}$. Find the ninth term of this geometric sequence.

Solution

(a) Let $\{a_n\}$ denote the desired sequence. By substituting a_4 = 405 and $r = -3$ into Eq. (18.3.2), we obtain

$$405 = a_4 = a_0 \cdot (-3)^4 = 81 \cdot a_0, \Rightarrow a_0 = \frac{405}{81} = 5.$$

By substituting $a_0 = 5$ and $r = -3$ into Eq. (18.3.2), the $(n + 1)$st term of the geometric sequence is given by

$$a_n = 5 \cdot (-3)^n.$$

The 16th term of this geometric sequence is a_{15} to be obtained as follows:

$$a_{15} = 5 \cdot (-3)^{15} = -71{,}744{,}535.$$

(b) Let $\{b_n\}$ denote the desired sequence whose third term is $b_2 = \frac{16}{9}$ and whose sixth term is $b_5 = \frac{128}{243}$. By substituting $b_2 = \frac{16}{9}$ and $b_5 = \frac{128}{243}$ into Eq. (18.3.2), we obtain

$$\frac{16}{9} = b_2 = b_0 r^2, \quad (1)$$

and

$$\frac{128}{243} = b_5 = b_0 r^5. \quad (2)$$

By dividing Eq. (1) from Eq. (2),

$$\frac{\frac{128}{243}}{\frac{16}{9}} = \frac{b_0 \cdot r^5}{b_0 \cdot r^2}, \Rightarrow \frac{8}{27} = r^3, \Rightarrow$$

$$r = \sqrt[3]{\frac{8}{27}} = \frac{2}{3}. \quad (3)$$

By substituting Eq. (3) into Eq. (1),

$$\frac{16}{9} = b_0 \cdot \left(\frac{2}{3}\right)^2 = b_0 \cdot \frac{4}{9},$$

$$b_0 = \frac{\frac{16}{9}}{\frac{4}{9}} = 4. \quad (4)$$

By substituting both of Eqs. (3-4) into Eq. (18.3.2), we have

$$b_n = 4 \cdot \left(\frac{2}{3}\right)^n. \quad (5)$$

From Eq. (5), the ninth term of this geometric sequence is b_8 to be obtained as follows:

$$b_8 = 4 \cdot (\frac{2}{3})^8 = \frac{1{,}024}{6{,}561}.$$

Remark

Please keep it in mind that a_n given by Eq. (18.3.2) is actually the $(n+1)$st term rather than the nth term of the sequence.

Theorem 18.3.2 (Formula for the Finite Geometric Series)

Let s_n be the sum of the first n terms of a geometric sequence $\{a_n\}_{n=0}^{\infty}$ whose common ratio has an absolute value not equal to one, i.e., $|r| \neq 1$. Then s_n is given by

$$s_n = \sum_{i=0}^{n-1} a_i = a_0 \cdot \frac{1-r^n}{1-r}. \qquad (18.3.3)$$

Proof

Since s_n is the sum of the first n terms of $\{a_n\}_{n=0}^{\infty}$, we have by using Eq. (18.3.2)

$$s_n = a_0 + a_1 + a_2 + \ldots + a_{n-1}$$
$$= a_0 + a_0 r + a_0 r^2 + \ldots + a_0 r^{n-1}. \qquad (1)$$

By multiplying both sides of Eq. (1) by r, we obtain

$$r \cdot s_n = r \cdot (a_0 + a_0 r + a_0 r^2 + \ldots + a_0 r^{n-1}), \Rightarrow$$
$$r s_n = a_0 r + a_0 r^2 + a_0 r^3 + \ldots + a_0 r^n. \qquad (2)$$

By subtracting Eq. (2) from Eq. (1), we have

$$s_n - r s_n = (a_0 + a_0 r + \ldots + a_0 r^{n-1}) - (a_0 r + a_0 r^2 + \ldots + a_0 r^n),$$
$$\Rightarrow (1-r) s_n = a_0 - a_0 r^n = a_0 (1 - r^n), \Rightarrow$$
$$s_n = \frac{a_0 (1-r^n)}{1-r} = a_0 \cdot \frac{1-r^n}{1-r},$$

which is Eq. (18.3.3). The proof is, therefore, complete.

Remarks

1. To use Eq. (18.3.2) the index i must begin with 0. If the index i begins with any positive integer k, then a trick must be used, i.e.,

$$\sum_{i=k}^{n-1} a_i = \sum_{i=0}^{n-1} a_i - \sum_{i=0}^{k-1} a_i = s_n - s_k$$

$$= a_0 \cdot \frac{1-r^n}{1-r} - a_0 \cdot \frac{1-r^k}{1-r} = a_0 \cdot \frac{r^k - r^n}{1-r}. \qquad (18.3.4)$$

2. Eq. (18.3.3) holds for both $r < 1$ and $r > 1$. However, if $r > 1$, Eq. (18.3.2) is preferably written as

$$s_n = a_0 \cdot \frac{r^n - 1}{r - 1}.$$

3. In contrast to an arithmetic sequence, the infinite geometric series converges, provided that the absolute value of the common ratio is less than one as shown in the following theorem.

Theorem 18.3.3 (Formula for the Infinite Geometric Series)
Let s_∞ be the sum of infinitely many terms of a geometric sequence $\{a_n\}_{n=0}^{\infty}$ with common ratio $r \neq 0$. If $|r| < 1$, then s_∞ exists and is given by

$$s_\infty = \frac{a_0}{1-r}. \qquad (18.3.5)$$

Proof
By using Eq. (18.3.3) and the fact that $\lim_{n\to\infty} r^n = 0$ for $|r| < 1$,

$$s_\infty = \lim_{n\to\infty} s_n = \lim_{n\to\infty} a_0 \cdot \frac{1-r^n}{1-r}$$

$$= \frac{a_0}{1-r} \cdot \lim_{n\to\infty} (1 - r^n) = \frac{a_0}{1-r} \cdot (\lim_{n\to\infty} 1 - \lim_{n\to\infty} r^n)$$

$$= \frac{a_0}{1-r} \cdot (1-0) = \frac{a_0}{1-r},$$

which is Eq. (18.3.5). The proof is, therefore, complete.

Remarks
1. A condition of $|r| < 1$ is not only sufficient, but also necessary for the convergence of an infinite geometric series.
2. If the index n does not begin with 0, say instead with k, then Eq. (18.3.5) becomes

$$s_\infty = \sum_{n=k}^{\infty} a_n = \frac{a_k}{1-r}. \qquad (18.3.6)$$

Example 18.3.3
Find each of the following geometric series:

(a) $\sum_{n=11}^{20} 1000 \cdot 1.05^n = ?$

(b) $\sum_{n=0}^{\infty} 10 \cdot (\frac{1}{2})^n = ?$

(c) $\sum_{n=10}^{\infty} 10 \cdot (\frac{1}{2})^n = ?$

Solution

(a) By using Eq. (18.1.8) and Eq. (18.3.4), we have

$$\sum_{n=11}^{20} 1000 \cdot 1.05^n = 1000 \cdot \sum_{n=11}^{20} 1.05^n$$
$$= 1000 \cdot \frac{1.05^{10} - 1.05^{20}}{1 - 1.05}$$
$$= 1000 \cdot \frac{1.628894627 - 2.653297705}{-0.05}$$
$$= 1000 \cdot \frac{-1.024403078}{-0.05}$$
$$= 1000 \cdot 20.48806156 = 20{,}488.06156 \approx 20{,}488.06.$$

(b) By using Eq. (18.1.8) and Eq. (18.3.5), we have

$$\sum_{n=0}^{\infty} 10 \cdot (\frac{1}{2})^n = 10 \cdot \sum_{n=0}^{\infty} (\frac{1}{2})^n = 10 \cdot \frac{1}{1 - \frac{1}{2}} = 10 \cdot \frac{1}{\frac{1}{2}} = 10 \cdot 2 = 20.$$

(c) By using Eq. (18.1.8) and Eq. (18.3.6), we have

$$\sum_{n=10}^{\infty} 10 \cdot (\frac{1}{2})^n = 10 \cdot \sum_{n=10}^{\infty} (\frac{1}{2})^n = 10 \cdot \frac{(\frac{1}{2})^{10}}{1 - \frac{1}{2}} = 10 \cdot \frac{(\frac{1}{2})^{10}}{\frac{1}{2}}$$
$$= 10 \cdot (\frac{1}{2})^9 = \frac{10}{2^9} = \frac{5}{2^8} = \frac{5}{256}.$$

Example 18.3.4

The United States banking law requires most banks to maintain a reserve equivalent to a certain proportion of their outstanding deposits. This enables such banks, when they wish and when they find borrowers, to loan out a certain proportion of the funds that have been deposited in their banks. Let us assume that this proportion is 80% (or 0.8). Suppose that Mary deposits her savings of \$10,000 in a local bank that, subsequently, is able to loan the maximum legally allowable amount, and this loan is re-deposited in another bank, and this process of deposit-loan is continued indefinitely. What is the total effect of \$10,000 on the economy?

Solution

The total effect of \$10,000 can be modeled as the sum of an infinite geometric series as follows:

$$s_n = \$10{,}000 + \$10{,}000\cdot 0.8 + (\$10{,}000\cdot 0.8)\cdot 0.8 + [(\$10{,}000\cdot 0.8)\cdot 0.8]\cdot 0.8 + \ldots$$

$$= \$10{,}000\cdot(1 + 0.8 + 0.8^2 + 0.8^3 + \ldots)$$

$$= \$10{,}000\cdot\frac{1}{1-0.8} = \$10{,}000\cdot\frac{1}{0.2} = \$10{,}000\cdot 5 = \$50{,}000.$$

The sum of \$50,000 is what is referred to in economics as the multiplier effect.

Definition 18.3.2

Amortization is a method of repaying a debt, including both principal and interest, by a series of periodic payments, usually in equal amount, each of that is part payment of interest and part payment to reduce outstanding principal.

Theorem 18.3.4 (Formula for Amortization of Loans)

Let A be the loan to be repaid in n equal payments, the charged interest at a period rate i compounded per payment period, and the periodic payment (made at the end of each payment period) is R. Then R is given by

$$R = \frac{A}{a_{n\perp i}}, \qquad (18.3.7)$$

where $a_{n\perp i}$ is defined by

$$a_{n\perp i} = \frac{1-(1+i)^{-n}}{i}. \qquad (18.3.8)$$

Proof

By using Eq. (16.1.7), the present loan A will have a future value F at the end of n payments given by

$$F = A\cdot(1 + i)^n. \qquad (1)$$

On the other hand, F is also the sum of the periodic payments defined by a geometric sequence $\{b_k = R(1+i)^{n-k-1}\}_{k=0}^{n-1}$, namely,

$$F = \sum_{k=0}^{n-1} b_k = b_0 + b_1 + b_2 + \ldots + b_{n-2} + b_{n-1}$$

$$= R(1+i)^{n-1} + R(1+i)^{n-2} + R(1+i)^{n-3} + \ldots + R(1+i)^1 + R(1+i)^0$$

$$= R[(1+i)^{n-1} + (1+i)^{n-2} + (1+i)^{n-3} + ... + (1+i)^1 + 1]$$

$$= R \cdot \sum_{k=0}^{n-1} (1+i)^k = R \cdot \frac{(1+i)^n - 1}{(1+i) - 1} = R \cdot \frac{(1+i)^n - 1}{i}. \qquad (2)$$

By substituting Eq. (2) into Eq. (1), we have

$$R \cdot \frac{(1+i)^n - 1}{i} = A(1+i)^n, \Rightarrow \quad R = \frac{A(1+i)^n}{\frac{(1+i)^n - 1}{i}}, \Rightarrow$$

$$R = \frac{A}{\frac{(1+i)^n - 1}{i(1+i)^n}} = \frac{A}{\frac{1}{i} \cdot [\frac{(1+i)^n}{(1+i)^n} - \frac{1}{(1+i)^n}]} = \frac{A}{\frac{1}{i} \cdot [1 - (1+i)^{-n}]}, \Rightarrow$$

$$R = \frac{A}{\frac{1-(1+i)^{-n}}{i}},$$

which is Eq. (18.3.7). The proof is, therefore, complete.

Example 18.3.5

Suppose that you borrowed a car loan of \$19,000 from a bank that charged an annual interest rate of 9% and you chose a 5-year (or 60-month) payment period. What is your monthly payment?

Solution

Since the annual interest rate is 9%, the monthly interest rate is $\frac{9\%}{12}$ $= \frac{3}{4}\% = 0.75\% = 0.0075$. By substituting n = 60 and $i = 0.0075$ into Eq. (18.3.8), we obtain

$$a_{60\perp 0.0075} = \frac{1-(1+0.0075)^{-60}}{0.0075} = \frac{1-0.638699698}{0.0075} = \frac{0.361300301}{0.0075}$$

$= 48.17337352$.

By substituting the above $a_{60\perp 0.0075}$ into Eq. (18.3.7), we have

$$R = \frac{19{,}000}{48.17337352} = 394.4087493 \approx 394.41.$$

Hence, your monthly payment is \$394.41.

Example 18.3.6

After we have learned both the concepts of arithmetic and geometric sequences, we are now ready to study the history about the invention of the logarithmic concept. As was stated in the remarks about the logarithmic function with base a, $\log_a x$, of Eq. (16.2.1), J. Napier observed that if one used the exponents of a geometric sequence to form an arithmetic sequence, then a multiplication/division between terms of the given geometric sequence would correspond an addition/subtraction between the terms of its associated arithmetic sequence. Specifically, the geometric and arithmetic sequence of Napier may be exhibited in modern notation as follows:

$$\{a_k = M(1-M^{-1})^k\}_{k=0}^{\infty} = \{M, M(1-M^{-1}),\ldots, M(1-M^{-1})^n,\ldots\}, \tag{18.3.9}$$

and

$$\{b_k = k\}_{k=0}^{\infty} = \{0, 1, 2, \ldots, n, \ldots\}, \tag{18.3.10}$$

where $M = 10^7$. Note that $\{a_k\}_{k=0}^{\infty}$ of Eq. (18.3.9) is a decreasing geometric sequence with common ratio $r = 1 - M^{-1} = 1 - 10^{-7}$, while $\{b_k\}_{k=0}^{\infty}$ of Eq. (18.3.10) is an increasing arithmetic sequence. The terms in $\{b_k\}_{k=0}^{\infty}$ of Eq. (18.3.10) is called the logarithm (in the sense of Napier and denoted by Log) of the corresponding terms in $\{a_k\}_{k=0}^{\infty}$ of Eq. (18.3.9), namely, $b_k = \text{Log } a_k$. Consequently, we have that Log M = Log $10^7 = 0$. This differs greatly from what we know today, i.e., $\log_a 1 = 0$ for any base $a > 0$. Indeed, Napier's definition of a logarithm is very different from the modern definition of Eq. (16.2.1) because the notion of a "base" is inapplicable to his system. To force the concept of a "base" upon his system, both a_k of Eq. (18.3.9) and b_k of Eq. (18.3.10) must be divided by M so that 0 would be the logarithm of 1, and also 1 is the logarithm of $(1-M^{-1})^M$, that is nearly equal to e^{-1}, where e is the irrational number of Eq. (16.1.6). Hence, the base of Napier's logarithms, when modified as indicated above, is very near e^{-1}. However, if we look at it from another angle, Napier's logarithmic concept was in fact more general than those of the present day because we could choose the value of M freely. Indeed, Jobst Burgi (1552-1632, Germany) invented logarithms independently by choosing $M = 10^8$. The reason why J. Napier chose $M = 10^7$ was given in Reference 3, pp. 312-313. At first J. Napier called the exponent n in the geometric sequence of a_k of Eq. (18.3.9) its "artificial number", but later changed to the term of "logarithm" that means "number of ratios". This is because the $(n+1)^{\text{st}}$ term of $\{a_k\}_{k=0}^{\infty}$, $a_n = M\cdot(1-M^{-1})^n$, is obtained from the first term M by n successive applications of the common ratio r $(= 1 - M^{-1})$. Hence, n that is the logarithm of $M\cdot(1 - M^{-1})^n$, indicates "the number of ratios" (Reference 6, p. 7). For other stories about J. Napier, please read Reference 1, Chapter 26.

Remark

From the idea of J. Napier, it is clear that there is an intimate relationship between a geometric sequence and an arithmetic sequence, namely, the logarithm of any geometric sequence with a positive common ratio is always an arithmetic sequence (See Problem 2 in Exercises for Section 18.3).

Exercises

1. Show that any discrete exponential function, $f(n) = c\cdot a^n$, where both a and c are nonzero real numbers, and n begins with any integer, defines a geometric sequence.

2. Show that the logarithm with base a of a geometric sequence of Eq. (18.3.2) with a common ratio $r > 0$ is an arithmetic sequence with a common difference $d = \log_a r$.

In problems 3 – 13, determine whether the given sequence is geometric. If yes, find the $(n+1)^{st}$ term of the geometric sequence.

3. $\{3, 6, 12, 24, 48, 96, \ldots\}$.
4. $\{5, -10, 20, -40, 80, -160, \ldots\}$.
5. $\{12, -6, 3, -\frac{3}{2}, \frac{3}{4}, -\frac{3}{8}, \ldots\}$.
6. $\left\{\frac{9}{2}, 3, 2, \frac{4}{3}, \frac{8}{9}, \frac{16}{27}, \ldots\right\}$.
7. $\left\{\frac{1}{4}, \frac{1}{9}, \frac{1}{16}, \frac{1}{25}, \frac{1}{36}, \frac{1}{49} \ldots\right\}$.
8. $\left\{\frac{2}{3}, \frac{4}{5}, \frac{6}{7}, \frac{8}{9}, \frac{10}{11}, \frac{12}{13}, \ldots\right\}$.
9. $\{\ln 2, (\ln 2)^2, (\ln 2)^3, (\ln 2)^4, (\ln 2)^5, (\ln 2)^6, \ldots\}$.
10. $\{\ln 2, \ln 4, \ln 8, \ln 16, \ln 32, \ln 64, \ldots\}$.
11. $\{1, e^2, e^4, e^6, e^8, e^{10}, \ldots\}$.

12. $\left\{\frac{1}{1!},\frac{1}{2!},\frac{1}{3!},\frac{1}{4!},\frac{1}{5!},\frac{1}{6!},\ldots\right\}$.

13. $\left\{1,\frac{1}{3!},\frac{1}{(3!)^2},\frac{1}{(3!)^3},\frac{1}{(3!)^4},\frac{1}{(3!)^5},\ldots\right\}$.

In problems 14−19, find a formula for the $(n + 1)^{st}$ term a_n of the given geometric sequence, where n begins with 0 and $s_\infty = \sum_{i=0}^{\infty} a_i$.

14. $a_0 = 5$, $r = 5$.

15. $a_3 = 1$, $r = 3$.

16. $a_0 = \frac{3}{5}$, $a_5 = \frac{32}{405}$.

17. $a_4 = \frac{9}{8}$, $a_{10} = \frac{6{,}561}{512}$.

18. $a_3 = \frac{40}{27}$, $\frac{a_{10}}{a_6} = \frac{16}{81}$.

19. $a_2 = \frac{4}{9}$, $s_\infty = 3$.

In problems 20−30, find the given (finite or infinite) geometric series.

20. $\sum_{n=0}^{10} 2^{n-3} = ?$

21. $\sum_{n=5}^{15} 2^{n-3} = ?$

22. $\sum_{n=0}^{10} 5\cdot(-2)^{n-2} = ?$

23. $\sum_{n=-2}^{8} 5\cdot(-2)^{n-2} = ?$

24. $\sum_{n=0}^{15} \frac{2^n}{3^{n-1}} = ?$

25. $\sum_{n=-3}^{10} \frac{1}{e^n} = ?$

26. $\sum_{n=0}^{20} (\ln 2)^{n+1} = ?$

27. $\sum_{n=0}^{\infty} \frac{2^n}{3^{n-1}} = ?$

28. $\sum_{n=4}^{\infty} \frac{2^n}{3^{n-1}} = ?$

29. $\sum_{n=0}^{\infty} 3\cdot\frac{(-1)^n}{2^n} = ?$

30. $\sum_{n=-3}^{\infty} 3\cdot\frac{(-1)^n}{2^n} = ?$

31. Find x so that $x - 1$, x and $x + 3$ are three consecutive terms of a geometrical sequence.

32. John accepts a job with a salary of \$40,000 for the first year. Suppose that during the next 9 years John would receive a 10% raise each year. What will John's total compensation be over the 10-year period?

33. Larry deposits \$200 into a savings account at the end of each month for 10 years at an annual interest rate of 6% compounded monthly. What is the balance in his account at the end of the 10 years?

34. David purchased a single-family house by taking out a 30-year \$100,000 mortgage at an annual interest rate of 9% compounded monthly. The mortgage payments are made at the end of each month. What was his monthly payment?

35. A city of 50,000 people is growing at the rate of 1.5% per year. What will be the population of the city 20 years from now?

18.4 Mathematical Induction

Theorem 18.4.1 (Method of Mathematical Induction)
Given a mathematical statement involves a natural number n. Suppose that the following two conditions hold:
1. The statement is true when $n = 1$, (18.4.1)
2. The statement is also true when $n = k + 1$, provided that the statement is true when $n = k$. (18.4.2)

Then the statement is true for all natural numbers.
Proof
By letting $k = 1$ in Eq. (18.4.2), we are assured that the statement is true for $k = 2$ since the statement is known to be true for $k = 1$ by Eq. (18.4.1). By letting $k = 2$, we are assured that the statement is true for $k = 3$ since the statement is known to be true for $k = 2$ as just shown above. By repeating indefinitely the process along this line of argument, the statement can then be shown to be indeed true for all natural numbers.

Remark

Historically, this method of mathematical (or complete) induction was probably already known to the Pythagoreans (Reference 18, p. 126). The method was used implicitly in Euclid's proof of the infinitude of the number of primes. F. Maurolycus (1494-1575, Italian) was the first person to make fairly explicit use of it to prove $1 + 3 + 5 + \ldots + (2n - 1) = n^2$ in his "Arithmetic" of 1575. B. Pascal (1623-1662, French) was the next person to use this method in his Traité du triangle arithmetiqué (1665), wherein we now call the Pascal's arithmetic triangle (Example 18.5.3) (Reference 8, p. 272).

Example 18.4.1

The following examples taken from the number theory that shows some mathematical statements are not true for all natural numbers:

(a) P. de Fermat (1601-1665, French) asserted that the integers of the form $2^{2^n}+1$ are always prime numbers for all natural numbers n. However, L. Euler (1707-1783, Swiss) showed in 1732 that $2^{2^5} + 1 = 4,294,967,297 = 6,700,417 \cdot 641$; Hence, Fermat's conjecture is not true for $n = 5$ (Reference 3, p. 456).

(b) A. M. Legendre (1752-1833, French) noted that $n^2 + n + 17$ is prime for $n = 1, 2, \ldots, 16$, while $2n^2 + 29$ is prime for $n = 1, 2, \ldots, 28$ (Reference 3, p. 488).

(c) The famous Fermat's last theorem states that there are no positive integer solution to the equation: $x^n + y^n = z^n$ if n is greater than 2. Fermat wrote a note in the margin of his copy of the Latin translation of Diophantus Arithmetica: "I have assuredly found an admirable proof of this [theorem], but the margin is too narrow to contain it." Fermat gave elsewhere a proof for $n = 4$. It was not until the next century that L. Euler (1707-1783, Swiss) supplied the proof for $n = 3$ (Reference 18, pp. 36-40). A. M. Legendre ((1752-1833, French) proved for the case $n = 5$, and in 1832, P. G. L. Dirichlet (1805-1859, Germany) provided a proof for $n = 7$. In 1849, E. E. Kummer (1810-1893, Germany) verified Fermat's claim for all $n < 100$, except 37, 59, and 67. Finally, in 1993, A. Wiles (1953- , English) presented work that led to a final proof of Fermat's last theorem (Reference 1, p. 165).

Example 18.4.2

By using the method of mathematical induction (Theorem 18.4.1), show that the following formula is true for all natural numbers n:

$$\sum_{i=1}^{n} i^3 = \frac{n^2 \cdot (n+1)^2}{4}. \qquad (1)$$

Proof

By Theorem 18.4.1, it suffices to show that Eqs. (18.4.1-2) are satisfied by Eq. (1). When $n = 1$,

$$\text{Left-hand side (LHS) of Eq. (1)} = \sum_{i=1}^{n} i^3 = 1^3 = 1,$$

$$\text{Right-hand side (RHS) of Eq. (1)} = \frac{1^2 \cdot (1+1)^2}{4} = \frac{1 \cdot 4}{4} = 1,$$

$\Rightarrow$ LHS of Eq. (1) = RHS of Eq. (1).

Therefore, Eq. (1) is true for $n = 1$, namely, Eq. (18.4.1) is satisfied. Suppose that Eq. (1) is true for $n = k$, that is

$$\sum_{i=1}^{k} i^3 = \frac{k^2 \cdot (k+1)^2}{4}. \qquad (2)$$

Now, for $n = k + 1$, we have

$$\text{LHS of Eq. (1)} = \sum_{i=1}^{k+1} i^3 = \sum_{i=1}^{k} i^3 + (k+1)^3. \qquad (3)$$

By substituting Eq. (2) into Eq. (3), we obtain

$$\text{LHS of Eq. (1)} = \frac{k^2 \cdot (k+1)^2}{4} + (k+1)^3$$

$$= (k+1)^2 \cdot [\frac{k^2}{4} + (k+1)]$$

$$= (k+1)^2 \cdot \frac{k^2 + 4 \cdot (k+1)}{4}$$

$$= (k+1)^2 \cdot \frac{k^2 + 4k + 1}{4} = (k+1)^2 \cdot \frac{(k+2)^2}{4}$$

$$= \frac{(k+1)^2 \cdot (k+2)^2}{4} = \text{RHS of Eq. (1)}.$$

Therefore, Eq. (18.4.2) is satisfied.

Since both Eqs. (18.4.1) and (18.4.2) are satisfied, Eq. (1) is proved by Theorem 18.4.1 to be true for all natural numbers n. The proof is, therefore, complete.

Remark

Eq. (1) is exactly the same as Eq. (18.1.11) that was derived by an alternative method as shown in Problem 2 of Exercises for Section 18.1.

Example 18.4.3

By using the method of mathematical induction, prove that the following inequality is true for all natural numbers n:

$$e^n > n, \qquad (1)$$

where e^n is the discrete natural exponential function.

Proof

When $n = 1$,

$$\text{LHS of Eq. (1)} = e^1 = e,$$

and

$$\text{RHS of Eq. (1)} = 1.$$

Since $e > 1$, LHS of Eq. (1) > RHS of Eq. (1). Therefore, Eq. (1) is true for $n = 1$ because Eq. (18.4.1) is satisfied.

Assume that Eq. (1) is true for $n = k$, that is,

$$e^k > k\,. \qquad (2)$$

Now, for $n = k + 1$, we have

$$\text{LHS of Eq. (1)} = e^{k+1} = e^k \cdot e^1 = e^k \cdot e. \qquad (3)$$

By using Eq. (2), we have from Eq. (3)

$$\text{LHS of Eq. (1)} > k \cdot e > k \cdot 2 = k + k > k + 1$$
$$= \text{RHS of Eq. (1). (Since } k > 1)$$

Therefore, Eq. (18.4.2) is satisfied. Since both of Eqs. (18.4.1-2) are satisfied, Eq. (1) is proved by Theorem 18.4.1 to be true for all natural numbers n. The proof is, therefore, complete.

Remark

Note that the continuous version of the inequality of Eq. (1) is given by

$$e^x > x,$$

which is true for all $x > 0$. This can easily be seen by visualizing that the graph of $f(x) = e^x$ lies above the graph of $i(x) = x$ for all $x > 0$.

Example 18.4.4

By using the method of mathematical induction, prove that a discrete function of n defined by

$$f(n) = n \cdot (n + 1) \cdot (2n + 1) \qquad (1)$$

is divisible by 6.

Proof

For $n = 1$, we have from Eq. (1)

$$f(1) = 1\cdot(1 + 1)\cdot(2\cdot1 + 1) = 1\cdot2\cdot3 = 6.$$

Since $f(1) = 6$ is indeed divisible by 6, this implies that the given statement is true for $n = 1$ because Eq.(18.4.1) is satisfied.

Assume that the given statement is true for $n = k$, $k > 1$. Thus, there exists a number $g(k)$ such that

$$f(k) = k\cdot(k + 1)\cdot(2k + 1) = 6\cdot g(k). \qquad (2)$$

Now, for $n = \mathrm{k} + 1$, we have from Eq. (1)

$$\begin{aligned} f(k + 1) &= (k + 1)\cdot[(k + 1) + 1]\cdot[(2\cdot(k + 1) + 1] \\ &= \{(k + 1)\cdot[(k + 1) + 1]\}\cdot[2k + 3] \\ &= [k\cdot(k + 1) + (k + 1) + k + 1] \cdot [(2k + 1) + 2] \\ &= [k\cdot(k + 1) + 2(k + 1)]\cdot[(2k + 1) + 2] \\ &= k\cdot(k + 1)\cdot(2k + 1) + 2(k + 1)\cdot(2k + 1) + \\ &\quad 2k\cdot(k + 1) + 4(k + 1) \\ &= f(k) + 2\cdot(k + 1)\cdot[(2k + 1) + k + 2] \\ &= f(k) + 2\cdot(k + 1)\cdot(3k + 3) \\ &= f(k) + 6\cdot(k + 1)^2. \end{aligned} \qquad (3)$$

By substituting Eq. (2) into Eq. (3), we obtain

$$\begin{aligned} f(k + 1) &= 6\cdot g(k) + 6\cdot(k + 1)^2 \\ &= 6\cdot[g(k) + (k + 1)^2]. \end{aligned} \qquad (4)$$

From Eq. (4), $f(k + 1)$ is divisible by 6 too. Hence, the statement is also true for $n = k + 1$. Since both of Eqs. (18.4.1-2) are satisfied, the given statement is true for all natural numbers n. The proof is, therefore, complete.

Remark

$f(n)$ of Eq. (1) is exactly the same as the numerator of the right-side of Eq. (18.1.10).

Exercises

1. Show that a sequence of $f(n) = n^2 + n + 17$ are prime numbers only for $n = 1, 2, 3, \ldots, 16$.

2. Show that a sequence of $g(n) = 2n^2 + 29$ are prime numbers only for $n = 1, 2, \ldots, 28$.

In problems 3 − 12, use the method of mathematical induction to prove the given formula that is true for all natural numbers n.

3. $\sum_{i=1}^{n} i = \frac{n(n+1)}{2}$.

4. $\sum_{i=1}^{n} (2i-1) = n^2$.

5. $\sum_{i=1}^{n} 2 \cdot i = n^2 + n$.

6. $\sum_{i=0}^{n-1} 2^i = 2^n - 1$.

7. $\sum_{i=1}^{n} i^2 = \frac{n \cdot (n+1) \cdot (2n+1)}{6}$.

8. $\sum_{i=1}^{n} i^3 = \frac{n^2 \cdot (n+1)^2}{4}$.

9. $\sum_{i=1}^{n} i^4 = \frac{n \cdot (n+1) \cdot (2n+1)(3n^2+3n-1)}{30}$.

10. $\sum_{i=1}^{n} i^5 = \frac{n^2 \cdot (n+1)^2 \cdot (2n^2+2n-1)}{12}$.

11. $\sum_{i=1}^{n} i \cdot (i+1) = \frac{n \cdot (n+1) \cdot (n+2)}{3}$.

12. $\sum_{i=1}^{n} \frac{1}{i \cdot (i+1)} = \frac{n}{n+1}$.

In problems 13−18, use the method of mathematical induction to prove the given inequality that is true for the indicated integers of n.

13. $\sum_{k=1}^{n} \frac{k+1}{k+2} < n$, $n = 1, 2, 3, \ldots$.

14. $3^n > n$, $n = 1, 2, 3, \ldots$.

15. $(1 + \frac{1}{n})^n < 3$, $n = 1, 2, 3, \ldots$.

16. $\sum_{k=0}^{n} \frac{1}{k!} < 3$, $n = 1, 2, 3, \ldots$.

17. $\ln n < n$, $n = 1, 2, 3, \ldots$.

18. $n! > \frac{n^n}{e^{n-1}}$, $n = 2, 3, \ldots$.

In problems 19−22, use the method of mathematical induction to prove that the given statement is true for all natural numbers *n*.

19. $a(n) = n \cdot (n+1)$ is divisible by 2.

20. $b(n) = n\cdot(n + 1)\cdot(n + 2)$ is divisible by 3.
21. $c(n) = n\cdot(n + 1)\cdot(n + 2)\cdot(3n^2 + 3n - 1)$ is divisible by 30.
22. $d(n) = n^2\cdot(n + 1)^2\cdot(2n^2 + 2n - 1)$ is divisible by 12.

18.5 The Binomial Theorem

Definition 18.5.1
The symbol $\binom{n}{k}$, read as " out of n objects takes k objects at a time", is defined by

$$\binom{n}{k} = \frac{n!}{k!\cdot(n-k)!}, \tag{18.5.1}$$

where n is a positive integer and k is an integer with $0 \le k \le n$.

Remarks
1. Eq. (18.5.1) represents the total number of different possible "combinations" of selecting k objects out of n distinct objects in which the order of selection is not important. See also Section 19.2 for a further explanation.
2. Eq. (18.5.1) possesses a symmetric property, i.e.,

$$\binom{n}{k} = \binom{n}{n-k}. \tag{18.5.2}$$

In Eq. (18.5.2), $\binom{n}{n-k}$ is called a complementary formula for $\binom{n}{k}$. For practical computations, whenever $k > \frac{n}{2}$, to compute $\binom{n}{k}$ is better to switch to compute $\binom{n}{n-k}$ (See Example 18.5.1. The reader is asked to prove Eq. (18.5.2) in Problem 1 of Exercises for Section 18.5).
3. Eq. (18.5.1) is defined for computing any value of k, in particular, for $k = 0$ and $k = n$,

$$\binom{n}{0} = \binom{n}{n} = 1. \tag{18.5.3}$$

However, for $1 \le k \le n - 1$, an alternative compact form of Eq. (18.5.1) given by

$$\binom{n}{k} = \frac{n \cdot (n-1) \cdot (n-2) \cdot \ldots \cdot (n-k+1)}{1 \cdot 2 \cdot 3 \cdot \ldots \cdot k}, \qquad (18.5.4)$$

is often used. A way to remember Eq. (18.5.4) is to note that the numerator of Eq. (18.5.4) is the product of k consecutive positive integers decreasing from n to $n - (k - 1)$ $(= n - k + 1)$, while the denominator of Eq. (18.5.4) is the product of k consecutive natural numbers increasing from 1 to k. The reader is asked to prove Eqs. (18.5.3) and (18.5.4) in Problems 2 and 3, respectively, in Exercises for Section 18.5.

Example 18.5.2

Find: (a) $\binom{10}{1} = ?$ (b) $\binom{10}{2} = ?$ (c) $\binom{10}{3} = ?$ (d) $\binom{10}{6} = ?$

Solution

(a) By using Eq. (18.5.4), we have

$$\binom{10}{1} = \frac{10}{1} = 10 \cdot$$

(b) By using Eq. (18.5.4), we have

$$\binom{10}{2} = \frac{10 \cdot 9}{1 \cdot 2} = 5 \cdot 9 = 45 \cdot$$

(c) By using Eq. (18.5.4), we have

$$\binom{10}{3} = \frac{10 \cdot 9 \cdot 8}{1 \cdot 2 \cdot 3} = 10 \cdot 3 \cdot 4 = 120 \cdot$$

(d) By using Eq. (18.5.2), we have

$$\binom{10}{6} = \binom{10}{4} = \frac{10 \cdot 9 \cdot 8 \cdot 7}{1 \cdot 2 \cdot 3 \cdot 4} = 10 \cdot 3 \cdot 7 = 210 \cdot$$

Remarks

1. In all computations of (a)-(d), we ordinarily simplify first the given fraction by dividing both the numerator and the denominator by their common factors and then perform the final multiplication. For example, in the fraction of part (d), we divide both the numerator and the denominator by 3 first and then by 8.
2. Part (d) demonstrates how to use Eq. (18.5.2).

Theorem 18.5.1 (Formula for Successive Rows of Pascal's Triangle)

Let both n and k be positive integers with $1 \le k \le n$. Then

$$\binom{n}{k-1} + \binom{n}{k} = \binom{n+1}{k}. \tag{18.5.5}$$

Proof

By using Eq. (18.5.4), we have

$$\binom{n}{k-1} + \binom{n}{k} = \frac{n\cdot(n-1)\cdot\ldots\cdot(n-k+2)}{1\cdot 2\cdot\ldots\cdot(k-1)} + \frac{n\cdot(n-1)\cdot\ldots\cdot(n-k+1)}{1\cdot 2\cdot\ldots\cdot k}$$

$$= \frac{k\cdot n\cdot(n-1)\cdot\ldots\cdot(n-k+2)+n\cdot(n-1)\cdot\ldots\cdot(n-k+1)}{1\cdot 2\cdot\ldots\cdot k}$$

$$= \frac{n\cdot(n-1)\cdot\ldots\cdot(n-k+2)\cdot[k+(n-k+1)]}{1\cdot 2\cdot\ldots\cdot k}$$

$$= \frac{n\cdot(n-1)\cdot\ldots\cdot(n-k+2)\cdot(n+1)}{1\cdot 2\cdot\ldots\cdot k}$$

$$= \frac{(n+1)\cdot n\cdot(n-1)\cdot\ldots\cdot(n-k+2)}{1\cdot 2\cdot 3\cdot\ldots\cdot k} = \binom{n+1}{k},$$

which is Eq. (18.5.5). The proof is, therefore, complete.

Remarks

1. Eq. (18.5.5) demonstrates a relationship between the binomial coefficients in the Pascal's arithmetic triangle as shown in Example 18.5.3.
2. Eq. (18.5.5) is used in the proof of the following binomial Theorem.

Theorem 18.5.2 (The Binomial Theorem for Positive Integral Powers)

Let n be any positive integer. Then, for any real numbers x and y,

$$(x+y)^n = \binom{n}{n}x^n + \binom{n}{n-1}x^{n-1}y + \ldots + \binom{n}{n-k}x^{n-k}y^k + \ldots + \binom{n}{1}xy^{n-1} + \binom{n}{0}y^n \tag{18.5.6}$$

Proof

We'll prove Eq. (18.5.6) by using the method of mathematical induction of Theorem 18.4.1. For $n = 1$,

LHS of Eq. (18.5.6) $= (x+y)^1 = x + y$,

and

RHS of Eq. (18.5.6)

$= \binom{1}{1}\cdot x^1 + \binom{1}{1-1}\cdot x^{1-1}\cdot y = 1\cdot x + \binom{1}{0}\cdot x^0\cdot y$

$= x + 1\cdot 1\cdot y = x + y,$

$\Rightarrow$ LHS of Eq. (18.5.6) = RHS of Eq. (18.5.6).

Hence, Eq. (18.5.6) is satisfied.

Assume that Eq. (18.5.6) is true for $n = m$, i.e.,

$$(x + y)^m = \binom{m}{m}\cdot x^m + \binom{m}{m-1}\cdot x^{m-1}\cdot y + \ldots + \binom{m}{m-k}\cdot x^{m-k}\cdot y^k + \ldots + \binom{m}{1}\cdot x\cdot y^{m-1} + \binom{m}{0}\cdot y^m. \quad (1)$$

Now, for $n = m + 1$, we have

LHS of Eq. (18.5.6) $= (x + y)^{m+1} = (x + y)^1\cdot(x + y)^m$

$= (x + y)\cdot(x + y)^m$

$= x\cdot(x + y)^m + y\cdot(x + y)^m. \quad (2)$

By substituting Eq. (1) into Eq. (2),

$$\text{LHS of Eq. (18.5.6)} = x\cdot[\binom{m}{m}\cdot x^m + \binom{m}{m-1}\cdot x^{m-1}\cdot y + \ldots + \binom{m}{m-k}\cdot x^{m-k}\cdot y^k + \ldots + \binom{m}{1}\cdot x\cdot y^{m-1} + \binom{m}{0}\cdot y^m\ ] + y\cdot[\binom{m}{m}\cdot x^m + \binom{m}{m-1}\cdot x^{m-1}\cdot y + \ldots + \binom{m}{m-k}\cdot x^{m-k}\cdot y^k + \ldots + \binom{m}{1}\cdot x\cdot y^{m-1} + \binom{m}{0}\cdot y^m]$$

$$= \{\binom{m}{m}\cdot x^{m+1} + \binom{m}{m-1}\cdot x^m\cdot y + \ldots + \binom{m}{m-k}\cdot x^{m-k+1}\cdot y^k + \ldots + \binom{m}{1}\cdot x^2\cdot y^{m-1} + \binom{m}{0}\cdot x\cdot y^m\ \} + \{\binom{m}{m}\cdot x^m\cdot y + \binom{m}{m-1}\cdot x^{m-1}\cdot y^2 + \ldots + \binom{m}{m-k}\cdot x^{m-k}\cdot y^{k+1} + \ldots + \binom{m}{1}\cdot x\cdot y^m + \binom{m}{0}\cdot y^{m+1}\}$$

$$= \binom{m}{m}\cdot x^{m+1} + [\binom{m}{m-1} + \binom{m}{m}]\cdot x^m\cdot y + \ldots + [\binom{m}{m-k-1} + \binom{m}{m-k}]\cdot x^{m-k}\cdot y^{k+1} + [\binom{m}{m-k} + \binom{m}{m-k+1}]\cdot x^{m-k+1}\cdot y^k + \ldots + [\binom{m}{0} + \binom{m}{1}]\cdot x\cdot y^m + \binom{m}{0}\cdot y^{m+1}. \quad (3)$$

By using Eq. (18.5.5), we obtain from Eq. (3)

LHS of Eq. (18.5.6) = $\binom{m}{m}\cdot x^{m+1} + \binom{m+1}{m}\cdot x^{m}\cdot y + \ldots + \binom{m+1}{m-k}\cdot x^{m-k}\cdot y^{k+1} + \binom{m+1}{m-k+1}\cdot x^{m-k+1}\cdot y^{k} + \ldots + \binom{m+1}{1}\cdot x\cdot y^{m} + \binom{m}{0}\cdot y^{m+1}$

$= \binom{m+1}{m+1}\cdot x^{m+1} + \binom{m+1}{m}\cdot x^{m}\cdot y + \ldots + \binom{m+1}{m-k}\cdot x^{m-k}\cdot y^{k+1}$

$+ \binom{m+1}{m-k+1}\cdot x^{m-k+1}\cdot y^{k} + \ldots + \binom{m+1}{1}\cdot x\cdot y^{m} + \binom{m+1}{0}\cdot y^{m+1}$

= RHS of Eq. (18.5.6). (Since $\binom{m}{m} = \binom{m+1}{m+1}$, $\binom{m}{0} = \binom{m+1}{0}$)

Hence, Eq. (18.4.2) is satisfied. By Theorem 18.4.1, Eq. (18.5.6) is true for all natural numbers n. The proof is, therefore, complete.

Remarks

1. Because of Eq. (18.5.2), the expansion of Eq. (18.5.6) can be written in either way.
2. The binomial coefficients of Eq. (18.5.6) for different values of n form the Pascal's triangle that is discussed in Example 18.5.3.
3. Note that the term of $\binom{n}{n-k}\cdot x^{n-k}\cdot y^{k}$ in Eq.(18.5.6) is the $(k + 1)$st term rather than the kth term. Also, the largest binomial coefficient occurs at the middle term. When n is even, the middle term is unique. However, when n is odd, there are two middle terms.
4. Historically, although the expansion of Eq. (18.5.6) was known by the Arabs of the 13th century, the binomial coefficients were not expressed in terms of the combinatorical symbol $\binom{n}{k}$ (Reference 1, p. 272). The binomial coefficients to be arranged in the form of Pascal's triangle (Figure 18.5.1) was published in Europe by P. Apianus (1495-1552, Germany) in 1529. In 1570, the Pascal's triangle appeared in both forms and various applications of G. Cardano's (1501-1576, Italian) "Opus novum de proportionibus", where he gave the numbers in the triangle through $n = 17$ and pointed out the relationship between the binomial coefficients given by Eq. (18.5.5). G. Cardano cited M. Stifel (1487-1567, Germany) as the putative discoverer of the Pascal triangle. Nevertheless, B. Pascal (1623-1662, French) was the first person

to study many of the properties of the Pascal triangle and applied it to the questions in the theory of probability (Reference 4, p. 388).

5. However, the first adequate proof of Eq. (18.5.6) in its current form was given in the J. Bernoulli's (1654-1705, Swiss) "Ars Conjectandi" ("Art of Conjecturing") published in 1713. Issac Newton (1642-1727, English) generalized the binomial theorem in 1664 to the case when n is any rational number rather than a positive integer. But Newton did not give a proof. The proof was later supplied by L. Euler (1707-1783, Swiss) (Reference 3, p. 419).
6. As far as applications of Eq. (18.5.6) are concerned, Chinese mathematicians of the 13th and early 14th centuries already used Eq. (18.5.6) through the use of the Pascal triangle in finding an approximation of any desired accuracy to a real root of a polynomial equation of any degree (Reference 9). In a similar way, we will show how to use the binomial theorem in deriving Tartaglia-Cardano's formula for solving cubic equations (Theorem 14.5.4) and Ferrari's formula for solving quartic equations (Theorem 14.5.5), respectively, in Examples 18.5.4 and 18.5.5.
7. When $x = p$, the probability of successes in a Bernoulli trial and $y = q = 1 - p$, the probability of failure in a Bernoulli trial, Eq. (18.5.6) becomes the probability distribution of the binomial random variable in statistics (Reference 13, pp. 169-176).

Example 18.5.2

(a) Use Eq. (18.5.6) to expand $(x + y)^7 = ?$

(b) Find the 7th term in the expansion of $(x + 2)^{10}$.

Solutions

(a) By using Eq. (18.5.6), we have

$$(x + y)^7 = \binom{7}{7}\cdot x^7 + \binom{7}{6}\cdot x^6\cdot y^1 + \binom{7}{5}\cdot x^5\cdot y^2 + \binom{7}{4}\cdot x^4\cdot y^3 + \binom{7}{3}$$
$$\cdot x^3\cdot y^4 + \binom{7}{2}\cdot x^2\cdot y^5 + \binom{7}{1}\cdot x^1\cdot y^6 + \binom{7}{0}\cdot y^7. \quad (1)$$

By using Eqs. (18.5.2-4), we obtain

$$\binom{7}{7} = \binom{7}{0} = 1,$$

$$\binom{7}{6} = \binom{7}{1} = \frac{7}{1} = 7,$$
$$\binom{7}{5} = \binom{7}{2} = \frac{7\cdot 6}{1\cdot 2} = 7{\cdot}3 = 21, \qquad (2)$$
$$\binom{7}{4} = \binom{7}{3} = \frac{7\cdot 6\cdot 5}{1\cdot 2\cdot 3} = 7{\cdot}5 = 35.$$

By substituting Eq. (2) into Eq. (1), we have

$$(x+y)^7 = x^7 + 7{\cdot}x^6{\cdot}y + 21{\cdot}x^5{\cdot}y^2 + 35{\cdot}x^4{\cdot}y^3 + 35{\cdot}x^3{\cdot}y^4 + 21{\cdot}x^2{\cdot}y^5 + 7{\cdot}x{\cdot}y^6 + y^7$$
$$= x^7 + 7x^6y + 21x^5y^2 + 35x^4y^3 + 35x^3y^4 + 21x^2y^5 + 7xy^6 + y^7.$$

(b) The 7th term in the expansion of $(x + 2)^{10}$ is given by

$$\binom{10}{6}{\cdot}x^6{\cdot}2^4 = \binom{10}{4}{\cdot}x^6{\cdot}16 = \frac{10\cdot 9\cdot 8\cdot 7}{1\cdot 2\cdot 3\cdot 4}{\cdot}x^6{\cdot}16 = 210{\cdot}x^6{\cdot}16 = 3{,}360x^6.$$

Remark

Note that the binomial coefficients are symmetric with respect to the middle term. Since $n = 7$ is odd, $35x^4y^3$ and $35x^3y^4$ are the two middle terms with the largest coefficient 35.

```
n=0                  1
  1               1     1
  2            1     2     1
  3         1     3     3     1
  4      1     4     6     4     1
  5   1     5    10    10     5     1
  6 1    6    15    20    15     6     1
  . ...................................
```

Figure 18.5.1

Example 18.5.3

If we list the binomial coefficients of Eq. (18.5.6) for different values of n in the form of a triangle, called the Pascal arithmetic triangle

(Figure 18.5.1). Note that beginning from $n = 1$, the entries in the next row are always the sum of two neighboring entries in the previous row, except the first and the last entries that must be one. This relationship was confirmed by Eq. (18.5.5).

Remarks

1. Not only the Pascal triangle was known by many mathematicians in the west before Pascal, but it was also known as early as 1100 A.D. by Chinese mathematicians. The Pascal triangle up to $n = 8$ was depicted in 1303 A.D. on the title page of "Si-Yuan Yu-Jin (四元玉鉴)" by Zhu Shi-Jie (1280-1303, Chinese) (Reference 12, p. 153). However, Zhu disclaims credit for the triangle. A similar triangle through $n = 6$ had appeared in the work of Yang Hui (1238-1298, Chinese), about whose life almost nothing is known and whose work has survived only in part (Reference 3, pp. 204-206).
2. An interesting fact was pointed out by L. V. Robinson that the diagonal entries along the direction from the northeast to the southwest are coefficients for the (infinite series) binomial expansion of $(x - y)^{-n}$. For example, for $x > y$,
$$(x - y)^{-1} = x^{-1} + x^{-2}\cdot y + x^{-3}\cdot y^2 + \ldots + x^{-n-1}\cdot y^n + \ldots.$$
(Reference 17, p. 540).
3. The connection between golden rectangles and the Fibonacci sequence related to the Pascal triangle was also observed (References 15-16).

Example 18.5.4

After having learned the binomial theorem of Eq. (18.5.6), we would like to show how to use it in deriving Tartaglia-Cardano's formula of Theorem 14.5.4 for solving cubic equations given as follows:

$$x^3 + b_2^* x^2 + b_1^* x + b_0^* = 0, \; b_0^* \neq 0, \qquad (18.5.7)$$

where Eq. (18.5.7) is exactly the same as Eq. (14.5.5). The idea of solving Eq. (18.5.7) for x is to first remove the term of $b_2^* x^2$ in Eq. (18.5.7) through a change-of-variable by setting

$$x = y - \frac{1}{3} b_2^* \qquad (18.5.8)$$

After substituting Eq. (18.5.8) into Eq. (18.5.7) and then using Eq. (18.5.6), we obtain

$$(y-\frac{1}{3}b_2^*)^3+b_2^*(y-\frac{1}{3}b_2^*)^2+b_1^*(y-\frac{1}{3}b_2^*)+b_0^*=0,$$

$$\binom{3}{3}y^3+\binom{3}{2}y^2\cdot(-\frac{b_2^*}{3})+\binom{3}{1}y\cdot(-\frac{b_2^*}{3})^2+\binom{3}{0}(-\frac{b_2^*}{3})^3+b_2^*[\binom{2}{2}y^2$$

$$+\binom{2}{1}y\cdot(-\frac{b_2^*}{3})+\binom{2}{0}(-\frac{b_2^*}{3})^2]+b_1^*y-\frac{b_1^*b_2^*}{3}+b_0^*=0,$$

$$y^3-b_2^*y^2+\frac{b_2^{*2}}{3}y-\frac{b_2^{*3}}{27}+b_2^*y^2-\frac{2b_2^{*2}}{3}y+\frac{b_2^{*3}}{9}+b_1^*y-\frac{b_1^*b_2^*}{3}$$

$$+b_0^*=0,$$

$$y^3+c_1^*y+c_0^*=0, \qquad (18.5.9)$$

where c_0^* and c_1^* are given, respectively, by Eqs. (14.5.6a-b). Now, let us concentrate on solving Eq. (18.5.9) for y. Again, use the trick of change-of-variable by setting

$$y=u+v, \qquad (18.5.10)$$

where u and v are two (unknown) variables to be determined. By substituting Eq. (18.5.10) into Eq. (18.5.9) and then using Eq. (18.5.6), we obtain

$$(u+v)^3+c_1^*(u+v)+c_0^*=0,$$

$$\binom{3}{3}u^3+\binom{3}{2}u^2v+\binom{3}{1}uv^2+c_1^*(u+v)+c_0^*=0,$$

$$u^3+3u^2v+3uv^2+v^3+c_1^*(u+v)+c_0^*=0,$$

$$u^3+v^3+3uv(u+v)+c_1^*(u+v)+c_0^*=0,$$

$$u^3+v^3+(3uv+c_1^*)(u+v)+c_0^*=0 \qquad (18.5.11)$$

Next, let us choose u and v such that

$$3uv+c_1^*=0. \qquad (18.5.12)$$

After substituting Eq. (18.5.12) into Eq. (18.5.11), Eq. (18.5.11) reduces to

$$u^3+v^3+c_0^*=0,$$

$$u^3+v^3=-c_0^*. \qquad (18.5.13)$$

From Eq. (18.5.12), we obtain by taking cubic power after moving c_1^* to the right side of the equation

$$(3uv)^3=(-c_1^*)^3,$$

$$27u^3v^3=-c_1^{*3},$$

$$u^3v^3 = -\frac{c_1^{*3}}{27}. \tag{18.5.14}$$

Then, note that Eq. (18.5.13) and Eq. (18.5.14) are two equations in two variables u^3 and v^3. We can solve Eq. (18.5.13) and Eq. (18.5.14) for u^3 and v^3 by using the method of elimination (or substitution). From Eq. (18.5.13), we solve for u^3:

$$u^3 = -v^3 - c_0^*. \tag{18.5.15}$$

By substituting Eq. (18.5.15) into Eq. (18.5.14), Eq. (18.5.14) becomes

$$(-v^3 - c_0^*)v^3 = -\frac{c_1^{*3}}{27},$$

$$-(v^3)^2 - c_0^* v^3 = -\frac{c_1^{*3}}{27},$$

$$0 = (v^3)^2 + c_0^* v^3 - \frac{c_1^{*3}}{27}. \tag{18.5.16}$$

Since Eq. (18.5.16) is a quadratic equation in v^3, we obtain the solution for v^3 by using the quadratic formula, namely,

$$v^3 = \frac{-c_0^* \pm \sqrt{c_0^{*2} - 4\cdot 1\cdot(-\frac{c_1^{*3}}{27})}}{2} = -\frac{c_0^*}{2} \pm \frac{1}{6}\sqrt{\frac{\Delta^*}{3}}, \tag{18.5.17}$$

when Δ^* is given by Eq. (14.5.6c).

Due to Eq. (18.5.13) and Eq. (18.5.14) are symmetric in u^3 and v^3, the two roots given by Eq. (18.5.17) are the solutions for u^3 and v^3. It does not matter how we designate them. Hence, we let

$$u^3 = -\frac{c_0^*}{2} + \frac{1}{6}\sqrt{\frac{\Delta^*}{3}}, \tag{18.5.18}$$

and

$$v^3 = -\frac{c_0^*}{2} - \frac{1}{6}\sqrt{\frac{\Delta^*}{3}}. \tag{18.5.19}$$

By taking the cubic root of Eqs. (18.5.18-19), we obtain

$$u = \sqrt[3]{-\frac{c_0^*}{2} + \frac{1}{6}\sqrt{\frac{\Delta^*}{3}}} = \sqrt[3]{A}, \tag{18.5.20}$$

and

$$v = \sqrt[3]{-\frac{c_0^*}{2} - \frac{1}{6}\sqrt{\frac{\Delta^*}{3}}} = \sqrt[3]{B}, \tag{18.5.21}$$

where A and B are given, respectively, by Eqs. (14.5.7d) and (14.5.7e). By substituting both of Eqs. (18.5.20-21) into Eq. (18.5.10), we have

$$y = \sqrt[3]{A} + \sqrt[3]{B}. \tag{18.5.22}$$

After substituting Eq. (18.5.22) into Eq. (18.5.8),

$$x = -\frac{1}{3}b_2^* + \sqrt[3]{A} + \sqrt[3]{B}. \tag{18.5.23}$$

Depending on which of the three cases holds: (i) $\Delta^* > 0$, (ii) $\Delta^* = 0$, and (iii) $\Delta^* < 0$, Eqs. (14.5.7a-e), (14.5.8a-b), and (14.5.9a-d) can be obtained from Eq. (18.5.23).

Remark

A change of variable given by Eq. (18.5.8) is the crucial first step not only in deriving Tartaglia-Cardano's formula, but also in deriving Ferrari's formula in the next example.

Example 18.5.5

In this example, we would like to show how to use the binomial theorem of Eq. (18.5.6) in deriving the Ferrari's formula for solving quartic equations given as follows:

$$x^4 + b_3x^3 + b_2x^2 + b_1x + b_0 = 0,\ b_0 \neq 0, \tag{18.5.24}$$

where Eq. (18.5.24) is exactly the same as Eq. (14.5.10). Similar to Example 18.5.4, the first step in solving Eq. (18.5.24) for x is to remove the term b_3x^3 in Eq. (18.5.24) through a change-of-variable by setting

$$x = y - \frac{1}{4}b_3. \tag{18.5.25}$$

After substituting Eq. (18.5.25) into Eq. (18.5.24), and then using Eq. (18.5.6), we obtain

$$(y-\frac{b_3}{4})^4 + b_3(y-\frac{b_3}{4})^3 + b_2(y-\frac{b_3}{4})^2 + b_1(y-\frac{b_3}{4}) + b_0 = 0,$$

$$\binom{4}{4}y^4 + \binom{4}{3}y^3\cdot(-\frac{b_3}{4}) + \binom{4}{2}y^2\cdot(-\frac{b_3}{4})^2 + \binom{4}{1}y\cdot(-\frac{b_3}{4})^3$$

$$+\binom{4}{0}(-\frac{b_3}{4})^4+b_3[\binom{3}{3}y^3+\binom{3}{2}y^2\cdot(-\frac{b_3}{4})+\binom{3}{1}y\cdot(-\frac{b_3}{4})^2$$

$$+\binom{3}{0}(-\frac{b_3}{4})^3]+b_2[\binom{2}{2}y^2+\binom{2}{1}y\cdot(-\frac{b_3}{4})+\binom{2}{0}(-\frac{b_3}{4})^2]$$

$$+b_1 y-\frac{b_1 b_3}{4}+b_0=0,$$

$$y^4-b_3y^3+\frac{3b_3^2}{8}y^2-\frac{b_3^3}{16}y+\frac{b_3^4}{256}+b_3y^3-\frac{3b_3^2}{4}y^2+\frac{3b_3^3}{16}y$$

$$-\frac{b_3^4}{64}+b_2y^2-\frac{b_2b_3}{2}y+\frac{b_2b_3^2}{16}+b_1y-\frac{b_1b_3}{4}+b_0=0,$$

$$y^4+(\frac{3b_3^2}{8}-\frac{3b_3^2}{4}+b_2)y^2+(-\frac{b_3^3}{16}+\frac{3b_3^3}{16}-\frac{b_2b_3}{2}+b_1)y+\frac{b_3^4}{256}$$

$$-\frac{b_3^4}{64}+\frac{b_2b_3^2}{16}-\frac{b_1b_3}{4}+b_0=0,$$

$$y^4+(-\frac{3b_3^2}{8}+b_2)y^2+(\frac{b_3^3}{8}-\frac{b_2b_3}{2}+b_1)y-\frac{3b_3^4}{256}+\frac{b_2b_3^2}{16}$$

$$-\frac{b_1b_3}{4}+b_0=0,$$

$$y^4+c_2y^2+c_1y+c_0=0, \qquad (18.5.26)$$

where c_2, c_1, and c_0 are given, respectively, by Eq. (14.5.11a-c). Now, let us concentrate on solving Eq. (18.5.26) for y. By moving both of the linear and the constant terms of Eq. (18.5.26) to the right-hand side, and completing the square with respect to y^2 on the left-hand side, we obtain

$$(y^2+\frac{c_2}{2})^2=-c_1y-c_0+\frac{1}{4}c_2^2. \qquad (18.5.27)$$

By adding a new variable u to the expression squared in the left-hand side of Eq. (18.5.27), Eq. (18.5.27) becomes

$$(y^2+\frac{c_2}{2}+u)^2=-c_1y-c_0+\frac{1}{4}c_2^2+2uy^2+c_2u+u^2, \qquad (18.5.28)$$

where u is to be determined. Now, we are going to choose u in such a way that the right-hand side of Eq. (18.5.28) also becomes a square. It is easy to see that if the right-hand side of Eq. (18.5.28) is a square, then it must be the square of $\sqrt{2u}\cdot y-\frac{c_1}{2\sqrt{2u}}$, where $u\neq 0$. Therefore, we should have the following identity:

$$-c_1y-c_0+\frac{c_2^2}{4}+2uy^2+c_2u+u^2\equiv(\sqrt{2u}\,y-\frac{c_1}{2\sqrt{2u}})^2. \tag{18.5.29}$$

By expanding the right-hand side of Eq. (18.5.29), we get

$$2uy^2-c_1y-c_0+\frac{c_2^2}{4}+c_2u+u^2\equiv 2uy^2-2\sqrt{2u}\,y\cdot\frac{c_1}{2\sqrt{2u}}$$

$$+(\frac{c_1}{2\sqrt{2u}})^2=2uy^2-c_1y+\frac{c_1^2}{8}. \tag{18.5.30}$$

Since Eq. (18.5.30) is an identity in y, the constant term on both sides of the equation must equal to one another, namely, u must be chosen to satisfy the following cubic equation in u:

$$-c_0+\frac{1}{4}c_2^2+c_2u+u^2=\frac{c_1^2}{8u},$$

$$8u(-c_0+\frac{1}{4}c_2^2+c_2u+u^2)=c_1^2,$$

$$8u^3+8c_2u^2+(2c_2^2-8c_0)u-c_1^2=0,$$

$$u^3+c_2u^2+(\frac{1}{4}c_2^2-c_0)u-\frac{1}{8}c_1^2=0, \tag{18.5.31}$$

where Eq. (18.5.31) that is exactly the same as Eq. (14.5.12) is called the resolvent cubic equation associated with Eq. (18.5.26), because a solution for y to Eq. (18.5.26) depends on a solution for u to Eq. (18.5.31).

Now, by substituting Eq. (18.5.29) into Eq. (18.5.28), we obtain

$$(y^2+\frac{c_2}{2}+u)^2=(\sqrt{2u}\,y-\frac{c_1}{2\sqrt{2u}})^2. \tag{18.5.32}$$

By taking the square root on both sides of Eq. (18.5.32), we obtain

$$y^2+\frac{c_2}{2}+u=\pm(\sqrt{2u}\,y-\frac{c_1}{2\sqrt{2u}}). \tag{18.5.33}$$

Consequently, the four solutions for y to Eq. (18.5.26) can be obtained by solving two quadratic equations given by Eq. (18.5.33) that are given by Eqs. (14.5.13a-f) (Problem 6 in Exercises for Section 18.5).

Remarks

1. From Example 18.5.4 and 18.5.5, it is easy to see that the second leading term in any polynomial equation of the form

$$x^n + b_{n-1}x^{n-1} + b_{n-2}x^{n-2} + ... + b_1 x + b_0 = 0,\ b_0 \neq 0, \tag{18.5.34}$$

can be eliminated by using the binomial theorem of Eq. (18.5.6) through a change of variable defined by

$$x = y - \frac{1}{n} \cdot b_{n-1}. \tag{18.5.35}$$

The proof is left as Problem 5 in Exercises for Section 18.5.

2. However, for over 200 years, people tried to find similar methods for solving a polynomial equation of Eq. (18.5.34) with $n = 5$ so that the roots would have a closed-form solution in terms of the coefficients $\{b_i\}$, i = 0, 1, 2, 3, 4. It was only in 1799 that P. Ruffini (1765-1822, Italian) proved that there is no (closed-form) algebraic solutions for a general polynomial equation of degree five or higher. His proof was not clear and rigorous, but it was soon supplemented by the work of N. H. Abel (1802-1829, Norwegian) and E. Galois (1811-1832, French). In other words, it is impossible to find a closed-form solution expressed, in terms of the coefficients of the terms, for the roots a general polynomial equation of degree five or higher (Reference 1, p. 151).

Exercises

1. Show that Eq. (18.5.2) holds.
2. Show that Eq. (18.5.3) holds.
3. Show that Eq. (18.5.4) holds.
4. Given a quadratic equation of the form $x^2 + b_1 x + b_0 = 0$, where $b_1 = \frac{b}{a}$, $b_0 = \frac{c}{a}$, $a \neq 0$. Derive the quadratic formula (Section 6.3 in Part I) by using the binomial theorem of Eq. (18.5.6) in coupling with a change of variable $x = y - \frac{1}{2}b_1$, and then solve the resulting equation in y.
5. Show that a substitution of Eq. (18.5.35) into Eq. (18.5.34) will make the second leading term of y^{n-1} disappears in the new polynomial equation of y.
6. Show that Eqs. (14.5.13 a-f) can be obtained by solving Eq. (18.5.33).

In problems 7−10, use Eq. (18.5.4) to evaluate each expression.

7. (a) $\binom{13}{2} = ?$ (b) $\binom{13}{3} = ?$ (c) $\binom{13}{4} = ?$ (d) $\binom{13}{5} = ?$

8. (a) $\binom{14}{2} = ?$ (b) $\binom{14}{3} = ?$ (c) $\binom{14}{4} = ?$ (d) $\binom{14}{5} = ?$

9. (a) $\binom{15}{5} = ?$ (b) $\binom{15}{6} = ?$ (c) $\binom{15}{12} = ?$ (d) $\binom{15}{11} = ?$

10. (a) $\binom{20}{3} = ?$ (b) $\binom{20}{4} = ?$ (c) $\binom{20}{15} = ?$ (d) $\binom{20}{14} = ?$

In problems 11−19, use the Binomial Theorem of Eq. (18.5.6) to expand the given expression.

11. $(x + y)^8 = ?$
12. $(x + y)^9 = ?$
13. $(x + y)^{10} = ?$
14. $(2x + 3y)^4 = ?$
15. $(3x - 2y)^4 = ?$
16. $(2x + 3)^5 = ?$
17. $(3x - 2)^5 = ?$
18. $(\sqrt{x} + \sqrt{3})^6 = ?$
19. $(\sqrt{x} - \sqrt{2})^6 = ?$

In problems 20−30, use the Binomial Theorem of Eq. (18.5.6) to find the indicated term or coefficient.

20. The fifth term in the expansion of $(x + y)^8$.
21. The second term in the expansion of $(3x - 2y)^4$.
22. The fourth term in the expansion of $(2x + 3)^5$.
23. The largest coefficient in the expansion of $(x + y)^8$.
24. The largest coefficient in the expression of $(x + y)^9$.
25. The largest coefficient in the expansion of $(x + y)^{10}$.
26. The coefficient of x^2y^2 in the expansion of $(2x + 3y)^4$.
27. The coefficient of xy^3 in the expansion of $(3x - 2y)^4$.
28. The coefficient of $x^{\frac{3}{2}}$ in the expansion of $(\sqrt{x} + \sqrt{3})^6$.
29. The coefficient of $x^{\frac{5}{2}}$ in the expansion of $(\sqrt{3} + \sqrt{x})^6$.
30. The coefficient of x^2 in the expansion of $(\sqrt{x} - \sqrt{2})^6$.

18.6 References

1. Anglin, W. S. (1994). Mathematics: A Concise History and Philosophy, Springer-Verlag, New York.
2. Anton, H. (1988). Calculus with Analytic Geometry, 3rd edition, John Wiley & Sons, New York.
3. Boyer, C. B. (1991). A History of Mathematics, second edition, John Wiley & Sons, Inc., New York.
4. Boyer, C. B. (1950). Cardan and the Pascal triangle, American Mathematical Monthly, 57, pp. 387-340.
5. Bunt, L. N. H., Jones, P. S. and Bedient, J. D. (1976). The Historical Roots of Elementary Mathematics, Prentice-Hall, Inc., Englewood Cliffs, New Jersey.
6. Cajori, F. (1913). History of the exponential and logarithmic concepts: I. From Napier to Leibniz and John Bernoulli, The American Mathematical Monthly, 20, pp. 5-14.
7. Goldberg, S. (1958). Introduction to Difference Equations, John Wiley & Sons, Inc., New York.
8. Kline, M. (1972). Mathematical Thought from Ancient to Modern Times, Oxford University Press, New York.
9. Lam, L-Y. (1980). The Chinese connection between the Pascal triangle and the slution of numerical equations of any degree, Historia Mathematica, 7, pp. 407-424.
10. Lebedev, M. N. (1972). Special Functions and Their Applications, Dover Publications, Inc, New York.
11. Mills, T. C. (1990). Time Series Techniques for Economists, Cambridge University Press, Cambridge, England.
12. Neddham, J. (1968). Science and Civilization in China III: Mathematics and the Sciences of Heavens and the Earth, Cambridge University Press, Cambridge, England.
13. Olkin, I., Gleser, L. J., and Derman, C. (1980). Probability Models and Applications, Macmillan Publishing Co., Inc., New York.
14. Philippou, A. N., Bergum, G. E., and Horadam, A. F. (1986). Fibonacci Numbers and Their Applications, D. Reidel Publishing Company, Dordrecht, Holland.
15. Raab, J. A. (1962). The golden rectangle and Fibonacci sequence, as related to the Pascal triangle, The Mathematics Teacher, 55, pp. 538-543.

16. Raab, J. A. (1963). A generalization of the connection between the Fibonacci sequence and Pascal's triangle, Fibonacci Quarterly, 1, pp. 21-31.
17. Robinson, L.V. (1947). Pascal's triangle and negative exponents, American Mathematical Monthly, 54, pp. 540-541.
18. Struik, D. J. (1969). A Source Book in Mathematics: 1200–1800, Harvard University Press, Cambridge, Massachusetts.
19. van der Waerden, B. L. (1961). Science Awakening, Oxford University Press, New York.

Chapter 19
Probability

In this chapter we are going to study the most fascinating and exciting branch of mathematics, namely (elementary) theory of probability. In particular, through a connection with the law of errors and the data variation in statistics (Reference 11) and the law of chance in physics (Reference 31), we may conclude that in the long run everything depends on the probability because uncertainty rules in life. It is no wonder, P. S. Laplace (1749-1827, French) said: "We see … that the theory of probabilities is at bottom only common sense reduced to calculation; it makes appreciate with exactitude what reasonable minds feel by a sort of instinct, often without being able to account for it. It is remarkable that (this) science, that originated in the consideration of games of chance, should have become the most important object of human knowledge (Reference 38, p. 43)." A question immediately arises: what is the probability anyway?

19.1 Random Experiments, Sample Spaces, and Events

Definition 19.1.1
A probability is a real number p with $0 \le p \le 1$ that is used to quantify the degree of certainty about the occurrence of an uncertain event in the future.

Remarks

1. Probability has a dual nature. On the one side it is statistical, concerning itself with stochastic laws of chance processes. On the other side it is epistemological, dedicated to assessing reasonable degrees of belief in proposition quite devoid of statistical background (Reference 16, p. 12).

2. Both the distinction and the connection must be made between "probability" and "possibility". In the logic of certainty about an uncertain event, three alternative conclusions can be reached: either certain (certainly true), or impossible (certainly false), or else possible. Hence, the concept of possibility is objective because it belongs to the domain of the logic of certainty. In contrast to possibility, probability is an additional notion that one applies within the range of possibility to assess the degree of certainty (more or less probable) and, therefore, a subjective notion (Reference 9, pp. 26-27). By using statistical jargons, we may say that probability is a quantitative concept, while possibility is a qualitative notion. Indeed, the theory of probability was viewed by C. S. Peirce (1839-1914, American) as the science of logic quantitatively treated (Reference 23, p. 100).
3. The probability of zero is used to describe an impossible event that never occurs, while the probability of one is used to describe a certain event that certainly occurs. Any number p with $0 < p < 1$ (read as $100 \cdot p\%$ of the times) is used to quantify the degree of our belief concerning the likelihood of the occurrence of the event that is of interest to us. The larger the probability, the higher the likelihood of the occurrence. Also, note that the probability is used usually for prediction, namely, before the occurrence of the event. Once the event occurs, the probability is one. On the contrary, if the event does not occur, the probability is zero.
4. Historically, the birth of the mathematical theory of probability calculus (or Lady Luck) is commonly credited to B. Pascal (1625-1662, French) when he started to study two probability problems in the gambling: the dice problem and the division problem (Reference 38, pp. 46-52). However, in his book of "Liber de Ludo Aleae" (or Book on "Games of Chance"), G. Cardano (1501-1576, Italian) introduced concepts of combinations into calculations of probability and defined probability as "the number of favorable outcomes divided by the number of possible outcomes". However, the book was not published until 1663, 87 years after Cardano's death. Had the publication not been delayed, G. Cardano would be known as "the father of the theory of probability" (Reference 28).

5. The formulation of basic concepts and initial methods of probability theory also occurs in other sciences and problems. Indeed, in the framework of Jewish law, questions involving probability consideration occur frequently in the ancient rabbinic literature. For example, the acceptance rule of "follow the majority" involves the (relative) frequency interpretation and the inheritance problem has everything to do with the concept of mathematical expectation (Reference 29, pp. 36-47, 161-164).
6. The question of how to calculate the probability is not specified in this definition. Indeed, there are at least three methods for calculating the probability of an uncertain event: (i) Bernoulli's law of large numbers, (ii) the assumption of equiprobability (or equipossibility), and (iii) personal (or subjective) probability.

Definition 19.1.2
A random experiment is an activity with observable, but unpredictable outcomes. Each repetition of a given random experiment is called a trial.

Remarks
1. Here the emphasis is that this activity must have an inherent uncertainty. Such a random experiment is presumably supposed that it can be performed repeatedly and indefinitely under the same circumstance. In each trial we can observe clearly the outcome of the experiment.
2. By the way, the word, "experiment", used here is very different from other laboratory experiments conducted in the physical sciences, e.g., chemistry and/or physics.
3. Flip a coin, throw a die, and playing cards are typical examples of random experiments.

Definition 19.1.3
A random experiment is called a Bernoulli trial if the experiment has only two possible outcomes, called "success" and "failure", respectively.

Remarks

1. The words, "success" and "failure", are merely generic terms for the two possible outcomes of a Bernoulli trial. Whenever the outcome is of interest (or favorable) to us, it will be called a success. Otherwise, it is called a failure.
2. In statistical sampling inspection of manufactured items from the mass production, the defective rate is of concern to the inspector. Thus, spotting the defective item is considered as a success.

Definition 19.1.4
If a Bernoulli trial is repeatedly performed n (≥ 2) times independently, then it is called a binomial trial with a parameter n.

Remark
A Bernoulli trial may be considered as a binomial trial with $n = 1$.

Theorem 19.1.1 (Bernoulli's Law of Large Numbers)
Let p denote the probability of success in a Bernoulli trial and $\hat{p}_n$ be the fraction of successes in n Bernoulli trials. Then, as $n \to +\infty$,

$$\hat{p}_n \to p. \tag{19.1.1}$$

Remarks
1. The proof of this theorem is beyond the scope of this book. But it can be found in Reference 8, pp. 203-204.
2. This theorem provides a rational basis for the development of the frequency theory of probability. Statistical probabilities are basically built upon the frequency theory of probability (Reference 11, Chapter 2).

Definition 19.1.5
The collection of all possible outcomes from the given random experiment is called the sample space, denoted by S.

Remarks
1. The sample space is merely a set whose elements are the outcomes of the given random experiment.
2. For any random experiment, the associated sample space is unique.

Example 19.1.1
Find the sample space of the given random experiment:
(a) Flip a coin once and observe the side that faces upward.
(b) Flip a coin twice independently and observe each time the side that faces upward.
(c) Flip a coin three times independently and observe sequentially the side that faces upward.
Solution
(a) Obviously, flipping a coin is a random experiment. Let H and T denote the head and the tail of the flipped coin, respectively. Then the associated sample space S_1 is given by

$$S_1 = \{H, T\}.$$

(b) Flipping a coin twice is evidently a random experiment. Yet, it is different from flipping a coin once. There are four possible outcomes resulted from performing this random experiment that can be described in terms of the tree diagram. By expressing the outcomes in Figure 19.1.1 as two-letter words, the associated sample space S_2 is given by

$$S_2 = \{HH, HT, TH, TT\}.$$

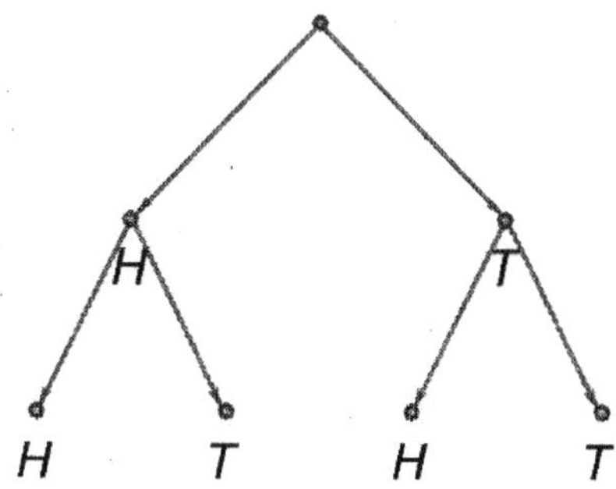

Figure 19.1.1

(c) Flipping a coin three times is a random experiment that is surely different from the two previous random experiments. In terms of the tree diagram, eight possible outcomes resulted from this random experiment are depicted as follows: By expressing the outcomes in

Figures 19.1.2 as three-letter words, the associated sample space is then given by

$$S_3 = \{\text{HHH, HHT, HTH, HTT, THH, THT, TTH, TTT}\}$$
$$= \{\text{HHH, HHT, HTH, THH, HTT, THT, TTH, TTT}\}.$$

Remarks

1. The outcomes in the first collection of S_3 are obtained by reading the tree diagram from top down to bottom and from left to right, while the outcomes in the second collection of S_3 are listed in a group according to the number of heads in the outcomes.
2. The random experiment in (a) is a Bernoulli trial, while random experiments in (b) and (c) are binomial trials with $n = 2$ and $n = 3$, respectively.

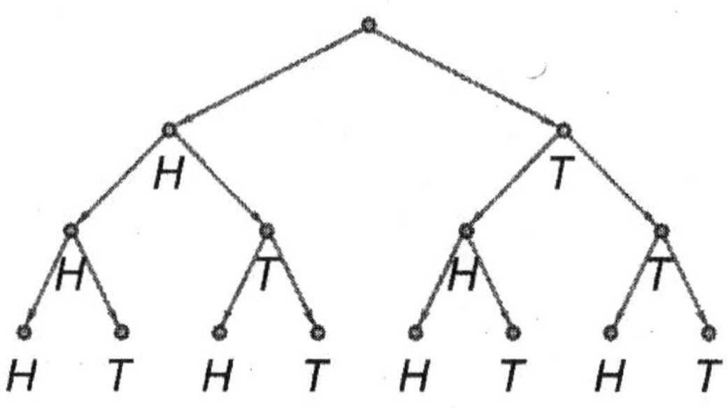

Figure 19.1.2

Definition 19.1.6

The assumption of equiprobability (or equipossibility) is said to hold if all outcomes of a random experiment have the same chance to occur.

Remarks

1. The concept of the equiprobable assumption was commonly supposed to be originated with Laplace around the end of eighteenth century. Laplace did define the probability as the ratio of the number of favorable cases to that of all the cases possible that implicitly assumed various cases to be equally possible, but so did G. W. von Leibniz (1646-1716, Germany) in 1678. In a memorandum headed "De Incerti Aestimatione," dated September

1678, von Leibniz asserts that probability is degree of possibility (Reference 4, p. 125). Nevertheless, the theory of probability developed on the assumption of equiprobability is often called the Laplacian theory of probability.
2. In order for the assumption of equiprobability applicable, a requirement is that the associated sample space of a random experiment must be finite; hence, it can be enumerated.
3. Since the fairness was implicitly implied by the assumption of equiprobability, random mechanisms were employed in the allocation of daily duties of the priests to serve in the temple (Reference 19, pp.318-319).
4. Almost all of the random experiments used in this chapter are chosen in such a way that the assumption of equiprobability holds for the given random experiment.

Definition 19.1.7

A personal probability is a real number p with $0 \leq p \leq 1$ assigned to an uncertain event by an individual based upon his/her personal rationale.

Remarks

1. A personal probability just reflects our personal opinion about an uncertain event. It is quite subjective. Generally, no two persons will have exactly the same personal probability regarding an uncertain event. Since it is hard to find a common ground to define a personal probability, it will not be further used in this chapter to illustrate the calculation of probability.
2. Personal probabilities are nevertheless used all the times by us in dealing with our personal business.

Definition 19.1.8

The collection of favorable outcomes is called an event, usually denoted by E.

Remarks

1. An event is just a subset of the sample space.
2. An empty event, denoted by a Greek letter ϕ, is a set that is comprised of no outcomes from the given random experiment. It is

called an impossible event because its probability is zero. Incidentally, the sample space S itself can also be viewed as an event. Since its probability will be one, S is called a certain event.

Example19.1.2

(a) Throw a six-side die once and observe the uppermost face. Let E be an event of the even number of dots observed. Find the associated sample space S_1 and the event E.

(b) Throw two-colored (red and blue) dice simultaneously and observe the uppermost faces of the two dice. Let F be an event that the sum of the number of dots of the two dice is 10. Find the associated sample space S_2 and the event F.

Solution

(a) Obviously, throwing a die is a random experiment. Let the number of dots on the faces of a die be denoted by the integers from 1 to 6, respectively. In terms of the tree diagram (Figure 19.1.3), the six possible outcomes are given as follows:

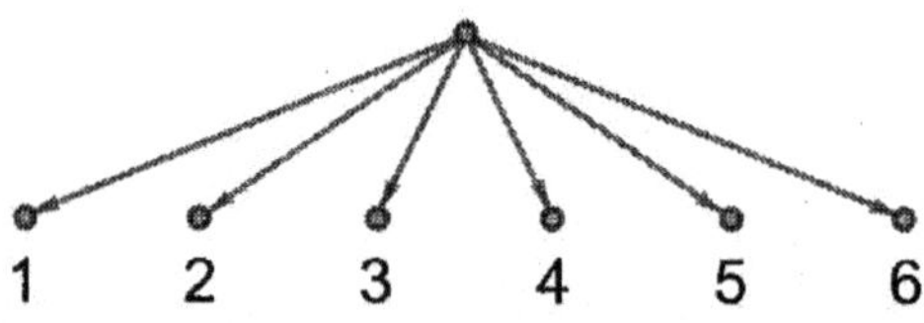

Figure 19.1.3

Thus, the associated sample space S_1 is given by

$$S_1 = \{1, 2, 3, 4, 5, 6\},$$

while the event E given by

$$E = \{2, 4, 6\}.$$

(b) Throwing two dice is a random experiment that is different from throwing just one die. The 36 possible outcomes resulted from this random experiment could be expressed in terms of the tree diagram. However, we are going to use a matrix table of ordered pairs to depict the outcomes as follows (Figure 19.1.4). Thus, the associated sample space S_2 is given by

$S_2 = \{(1,1), (1,2), (1,3), (1,4), (1,5), (1,6), (2,1), (2,2), (2,3),$
$(2,4), (2,5), (2,6), (3,1), (3,2), (3,3), (3,4), (3,5)\ (3,6), (4,1),$

(4,2), (4,3), (4,4), (4,5), (4,6), (5,1), (5,2), (5,3), (5,4), (5,5), (5,6), (6,1), (6,2), (6,3), (6,4), (6,5), (6,6)},

while the event F is given by

$$F = \{(4,6), (5,5), (6,4)\}.$$

Remark

The reader is asked to construct a tree diagram for throwing two dice.

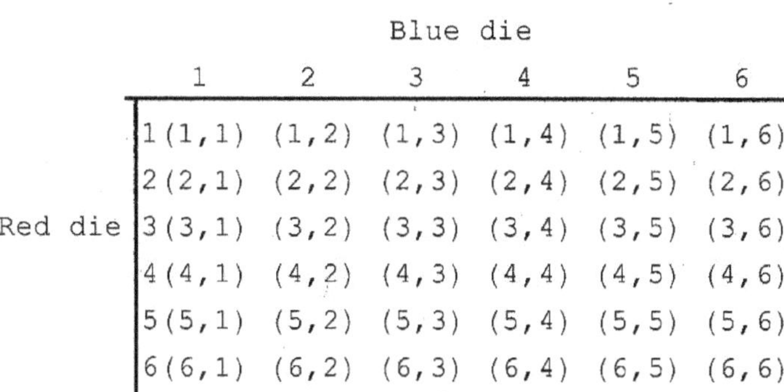

		Blue die					
		1	2	3	4	5	6
	1	(1,1)	(1,2)	(1,3)	(1,4)	(1,5)	(1,6)
	2	(2,1)	(2,2)	(2,3)	(2,4)	(2,5)	(2,6)
Red die	3	(3,1)	(3,2)	(3,3)	(3,4)	(3,5)	(3,6)
	4	(4,1)	(4,2)	(4,3)	(4,4)	(4,5)	(4,6)
	5	(5,1)	(5,2)	(5,3)	(5,4)	(5,5)	(5,6)
	6	(6,1)	(6,2)	(6,3)	(6,4)	(6,5)	(6,6)

Figure 19.1.4

Example 19.1.3

Draw a card at random from a standard deck of 52 playing cards. Let G be an event of all face cards. Find the sample space S and the event G.

Solution

A standard deck of 52 playing cards is comprised of 4 suits, spade (S), heart (H), club (C), and diamond (D). Each suit is comprised of 13 denominations, from A to J (Jack), Q (Queen), and K (King) as shown in Figure 19.1.5. The face cards are Jack, Queen, and King in each suit. Let us express each card by attaching the suit as a subscript of each denomination. For example, a card of heart 5 is denoted by 5_H. Drawing one card at random is certainly a random experiment. Thus, the associated sample space is given by

$$S = \{A_S, 2_S, 3_S, 4_S, 5_S, 6_S, 7_S, 8_S, 9_S, 10_S, J_S, Q_S, K_S, A_H, 2_H, 3_H, 4_H, 5_H, 6_H, 7_H, 8_H, 9_H, 10_H, J_H, Q_H, K_H, A_C, 2_C, 3_C, 4_C, 5_C, 6_C, 7_C, 8_C, 9_C, 10_C, J_C, Q_C, K_C, A_D, 2_D, 3_D, 4_D, 5_D, 6_D, 7_D, 8_D, 9_D, 10_D, J_D, Q_D, K_D\},$$

while the event G is given by

$G = \{J_S, Q_S, K_S, J_H, Q_H, K_H, J_C, Q_C, K_C, J_D, Q_D, K_D\}$.

Club	Diamond	Heart	Spade
A	A	A	A
2	2	2	2
3	3	3	3
4	4	4	4
.	.	.	.
.	.	.	.
.	.	.	.
10	10	10	10
J	J	J	J
Q	Q	Q	Q
K	K	K	K

Figure 19.1.5

Exercises

In problems 1−10, decide whether the given activity is a random experiment.

1. Flip a biased coin and observe the side of the coin that faces upward.

2. Throw a loaded die and observe the uppermost side of the die.

3. Draw two cards one-by-one without replacement from a standard deck of 52 playing cards and observe the two cards.

4. A teacher gives grades to the students in his class based on the result of their tests.

5. A high school graduate applies to several colleges for admission.

6. A couple is expecting from a new birth without taking any memogram test.

7. Lightening never strikes the same place twice.

8. A student receives his grade based on guessing on a multiple-choice test.

9. A sharp shooter fires a shot at the target.

10. A good driver will not have a car accident on a trip from Peoria to Chicago.

In problems 11 −20, decide if the given random experiment is a Bernoulli or binomial trial or neither.

11. Throw a loaded die.

12. Flip a biased coin.

13. Throw two unloaded dice.

14. Flip a biased coin twice independently.

15. Draw a card at random from a standard deck of 52 playing cards.

16. A voter is randomly selected to be questioned about his opinion on the United States President's job performance.

17. A pollster uses a telephone survey about the opinion of 1500 randomly selected citizens on the United States President's job performance.

18. You get a total of three heads when flipping a coin five times independently.

19. You get a full house from playing a poker hand.

20. A baseball player gets two hits in his next three times at bat.

In problems 21 −25, find the associated sample space and the given event.

21. Toss three coins (a penny, a nickel, and a dime) simultaneously. We are interested in having exactly two heads. Hint. The sample spaces for flipping a coin three times independently and tossing three distinct coins simultaneously are the same.

22. Throw two dice simultaneously. We are interested in observing that the sum of two dice is at least 9.

23. Flip a coin four times independently. We are interested in having exactly two heads.

24. An urn contains five colored marbles: Blue (B), Green (G), Red (R), White (W), and Yellow (Y). Select two marbles at random from the urn. We are interested in having exactly one white marble.

25. Select two letters randomly to form a two-letter word form a set of $\{a, b, c, d\}$. We are interested in having a letter b in the formed word. Note that a word here is just a string of letters that needs not have any English meaning.

19.2 Permutations and Combinations

In order to apply the equipossibility assumption in calculating the probability of an event, we need to know how to enumerate all possible outcomes resulting from performing a random experiment. In this section we will learn certain counting techniques to accomplish that purpose. The first tool is the fundamental counting rule shown below.

Theorem 19.2.1 (The Multiplication Principle)
If k different choices can be made in succession, and the first choice can be made in n_1 possibilities, the second choice in n_2 possibilities, the third choice in n_3 possibilities, etc., then the total number of possibilities for the overall k choices equals to

$$n_1 \cdot n_2 \cdot n_3 \cdot \ldots \cdot n_k . \qquad (19.2.1)$$

Remarks

1. Since the multiplication principle is intuitively appealing, a formal proof would amount to repeating what is described in the theorem in more cut-and-dried terms; Hence, it is omitted.
2. This principle is similar to the permutation of taking k objects out of n distinct objects that is covered after Example 19.2.2. The only difference is that the sources of k choices in this principle are different, whereas the sources of choice for the permutation are the same.

Example 19.2.1

A local appliance store carries five different types of washing machines and four different types of dryers. Suppose that a customer arrives at the store and would like to purchase a washing machine and a dryer. How many different choices are possible for this customer to purchase?

Solution

Let n_1 and n_2 be the number of possible choices for selecting a washing machine and a dryer, respectively. Thus, we have

$$n_1 = 5, \text{ and } n_2 = 4.$$

By Eq. (19.2.1), the total number of different possible choices for purchasing a pair of washing machine and dryer is then equal to $5 \cdot 4 = 20$.

Example 19.2.2

The car license plate of a certain state was designed to consist of two letters followed by four digits. How many different possible plates are there?

Solution

Note that the car license plate consists of a sequence of six choices: two for choosing letters and four for choosing digits as shown in Figure 19.2.1. So, the total number of different possible license plates is 5,760,000.

Definition 19.2.1

A selection (or sampling) without replacement occurs when an object is not replaced after it has been selected.

Remark
This kind of selection without replacement is often used in survey sampling in statistics (Reference 33).

W	C	2	4	6	5
A	A	0	0	0	0
B	B	1	1	1	1
C	C	2	2	2	2
.	.	.	.	.	.
.	.	.	.	.	.
.	.	.	.	.	.
Y	Y	8	8	8	8
Z	Z	9	9	9	9

$$26 \times 26 \times 10 \times 10 \times 10 \times 10 = 5{,}760{,}000$$

Figure 19.2.1

Definition 19.2.2
A permutation of taking k objects with $1 \leq k \leq n$ from a collection of n distinct objects is a way of selection without replacement in that the order of selection is important.

Remarks
1. To distinguish a permutation from a combination that is defined later, a concrete way to grasp the idea of a permutation is to imagine that you are trying to construct a k-letter word by selecting k letters one-by-one in succession without replacement from a collection of n distinct letters.
2. Historically, in the West, Jewish mathematicians were especially fond of permutations and combinations because the twenty-two letters of the Hebrew alphabet were viewed to represent the building block of creation. Much of the "Sefer Yetsirah" ("Book of Creation") was devoted to enumerating various permutations of

the letters of the alphabet. Sefer Yetsirah was written no later than the second century, but possible much earlier (Reference 29, p.143). In addition, the influential twelfth century astrologer and mathematician Rabbi Abraham Ben Ezra (1093-1167) developed a theory of permutations because of his interest in studying the influence of the stars on the destinies of the world (Reference 12). In the east, Chinese book "Yijing (易经)" (or "I Ching") that was written around 2200 B.C. used a combination of "yin (阴)" ("female") and "yang (阳)" ("male") as a method of divination (Reference 20, pp. 34-44). Indian mathematicians also found some elementary properties of permutations and combinations during the time period from around the middle of 300 B.C. to 500 A.D. (Reference 14, p. 98).

3. Permutations and combinations are not only intimately connected with probability, but also useful in expressing the genetic code of the living organism (Reference 2, pp. 125-141). In statistics, permutations can be used in testing the hypotheses (Reference 13).
4. A curious question immediately arises: How many different permutations are possible? This is answered in the following Theorem.

Theorem 19.2.2 (Total Number of Permutations by Selecting Without Replacement)

If k objects are randomly selected without replacement with $1 \le k \le n$ from a collection of n distinct objects, the total number of different possible permutations, denoted by ${}_nP_k$, is given by

$$ {}_nP_k = n \cdot (n-1) \cdot (n-2) \cdot \ldots \cdot (n-k+1) . \qquad (19.2.2)$$

In particular, when $k = n$, we have

$$ {}_nP_n = n \cdot (n-1) \cdot (n-2) \cdot \ldots \cdot 1 = n! . \qquad (19.2.3)$$

Proof

As we commented, a permutation of taking k objects without replacement from a collection of n distinct objects is equivalent to constructing a k-letter word from a collection of n distinct letters. There would be n letters available to be selected for the first letter of a k-letter word. Regardless of which letter was selected, it cannot be used again once it was selected. There remains $n - 1$ letters available for selecting the second letter of a k-letter word. Regardless of which letter was selected for the second letter of a k-letter word, it cannot be used for the selection of the third letter of a k-letter word. Thus, there remains only $n-2$ letters available for selecting the third letter of a k-letter word. By repeating this process of selection, there are eventually only $n-(k-1)=n-k+1$ letters available for selecting the kth letter of a k-letter word. By applying the multiplication principle of Theorem 19.2.1, we, therefore, obtain Eq. (19.2.2) by using Eq. (19.2.1). To prove Eq. (19.2.3), we substitute $k = n$ into Eq. (19.2.2) and then use the definition of n-factorial given by Eq. (18.1.2). The proof is, therefore, complete.

Remarks

1. If the sampling of "selecting without replacement" is replaced by "selecting with replacement", then the total number of permutations is not given by Eq. (19.2.2). Instead, it is given by Eq. (19.2.5).
2. If n objects in the given collection are not all distinct, this also makes Eq. (19.2.2) invalid. See Example 19.2.6.
3. Eq. (19.2.2) can also be written alternatively as

$$ {}_nP_k = \frac{n!}{(n-k)!} . \tag{19.2.4}$$

Note that Eq. (19.2.4) holds for $k=0$ since ${}_nP_0 = \frac{n!}{(n-0)!} = 1$.

Example 19.2.3

Compute each of the following:

(a) ${}_8P_4 = ?$ (b) ${}_7P_7 = ?$ (c) ${}_{50}P_3 = ?$

Solution

(a) By using Eq. (19.2.2), we have

$$ {}_8P_4 = 8\cdot 7\cdot 6\cdot 5 = 1{,}680 .$$

(b) By using Eq. (19.2.3), we have

$$ {}_7P_7 = 7! = 5{,}040 .$$

(c) By using Eq. (19.2.2), we have

$$ {}_{50}P_3 = 50\cdot 49\cdot 48 = 117{,}600 .$$

Example 19.2.4

How many different words can be constructed from the letters of the word "QUALITY" by taking without replacement (a) 6 letters at a time? (b) 4 letters at a time? (c) 2 letters at a time?

Solution

Since the order of selection is important in constructing a word plus the word "QUALITY" has seven distinct letters: {A, L, I, Q, T, U, Y}, we have

(a) ${}_7P_6 = 7\cdot 6\cdot 5\cdot 4\cdot 3\cdot 2 = 5{,}040$.

(b) ${}_7P_4 = 7\cdot 6\cdot 5\cdot 4 = 840$.

(c) ${}_7P_2 = 7\cdot 6 = 42$.

Example 19.2.5

A science club in a local high school has 10 members. Suppose that every member is eligible for being selected to be one of four officers: president, vice-president, secretary, and treasurer. How many different ways are possible for them to choose a slate of four officers? Assume that no one can occupy two positions.

Solution

This problem is equivalent to construct a 4-letter word with the first letter being president, the second letter being vice president, the third letter being secretary, and the fourth letter being treasurer. The 10 members are equivalent to 10 distinct letters. Hence, the total number of different possible ways for choosing a slate of four officers is ${}_{10}P_4 = 10\cdot 9\cdot 8\cdot 7 = 5{,}040$.

Definition 19.2.3

A selection (or sampling) with replacement occurs when an object is selected and then replaced before the next object is selected.

Theorem 19.2.3 (Total Number of Permutations by Selection with Replacement)

If k objects are randomly selected with replacement from a collection of n distinct objects, then the total number of different possible permutations equals

$$n^k. \qquad (19.2.5)$$

Proof

Since we are here concerned with permutations, each permutation can be viewed as a k-letter word. The given collection of n distinct objects is viewed as n distinct letters. Since the sampling was done with replacement, there would be n letters available for each letter of the k-letter word. Hence, by the multiplication principle of Theorem 19.2.1, the total number of different possible permutations (or k-letter words) is given by

$$\underbrace{n \cdot n \cdot n \cdot \ldots \cdot n}_{k \ \ times} = n^k,$$

which is Eq. (19.2.5). The proof is, therefore, complete.

Example 19.2.6
Two cards are drawn randomly one-by-one from a standard deck of 52 playing cards. How many different hands are possible if the selection is done (a) without replacement? (b) with replacement?
Solution
(a) There are 52 cards available for the first drawing. Since the sampling is done without replacement, there remains only 51 cards available for the second drawing. Hence, the total number of different possible hands is obtained by either using the multiplication principle of Eq. (19.2.1) (or Eq. (19.2.2)) with $k = 2$, namely,

$${}_{52}P_2 = 52 \cdot 51 = 2{,}652\,.$$

(b) Since the sampling is done with replacement, there would be 52 cards available for both of the first and the second drawing. Hence, the total number of different possible hands is obtained by using Eq. (19.2.5) with $n = 52$ and $k = 2$, i.e.

$$52 \cdot 52 = 52^2 = 2{,}704\,.$$

Remark
There is a big difference between the sampling without replacement and with replacement in the theory of probability. The former is a dependent concept, whereas the latter an independent concept. See Section 19.4.

Theorem 19.2.4 (Total Number of Circular Permutations)
If n distinct objects are randomly arranged in a circle, then the total number of different possible circular permutations is given by

$$(n-1)!. \qquad (19.2.6)$$

Proof

Since all n objects are randomly chosen, there are $n!$ circular permutations by Eq. (19.2.3). However, not all these $n!$ permutations are different. By taking any particular circular permutation, you can switch their location in succession and will generate n circular permutations. However, all of these n circular permutations are exactly the same; Hence, they should be viewed as one circular permutation. Clearly, $n!$ circular permutations can thus be grouped into $(n-1)$ groups. In each group, there are n indistinguishable circular permutations. Hence, as far as different possible circular permutations are concerned, there are only $(n-1)!$ permutations. The proof is, therefore, complete.

Example 19.2.7
In how many different possible ways can a science club of 10 members be arranged to sit around a circular picnic table?
Solution
Since 10 members of the science club are being arranged to sit around a circular picnic table, this is equivalent to finding the total number of different possible circular permutations. By using Eq. (19.2.6) with $n = 10$, the total number of different possible ways of seating arrangements for the 10 member of this science club is $(10 - 1)! = 9! = 362{,}880$.

Definition 19.2.4
A combination is a way of selection by taking k objects simultaneously $(1 \le k \le n)$ out of n distinct objects.

Remarks

1. Since a combination is carried out by taking k objects out of n distinct objects simultaneously rather than one-by-one in succession, the order of selection clearly becomes not important.
2. A permutation and a combination are like the twin brother. The only difference between them is the order of selection. The order of selection is of concern for a permutation, but not for a combination.
3. In statistics, a construction of the experimental design involves combinatorial problems (Reference 30).

4. The total number of different possible combinations is given in the following theorem.

Theorem 19.2.5 (Total Number of Different Possible Combinations)
If k $(1 \le k \le n)$ objects are randomly selected simultaneously out of n distinct objects, then the total number of different possible combinations, denoted by ${}_nC_k$, is given by

$$ {}_nC_k = \frac{{}_nP_k}{k!} = \frac{n \cdot (n-1) \cdot (n-2) \cdot \ldots \cdot (n-k+1)}{1 \cdot 2 \cdot 3 \cdot \ldots \cdot k}. \qquad (19.2.7) $$

Proof
To prove Eq. (19.2.7) is equivalent to explaining why the first equality of Eq. (19.2.7) is true. First, let us pretend that the order of selection is important for the time being and then consider how to take care of the matter that the order of selection is in fact not important for combinations. If the order of selection is important, we learned from Theorem 19.2.2 that the total number of different possible permutations is ${}_nP_k$. Now, let us take any particular permutation and view it as a k-letter word. Then, let us permute all k letters of this k-letter word. We are going to have $k!$ words generated by this k-letter word. However, when the order of selection is not important, all these $k!$ words has to be regarded as the same combination. Therefore, in order to obtain the total number of different possible combinations, all we have to do is to partition ${}_nP_k$ different permutations into $\frac{{}_nP_k}{k!}$ groups of permutations so that each group of permutations is generated by a specific permutation; Hence, all permutations in the group are regarded as the same combination when the order of selection becomes unimportant. Consequently, the total number of different possible combinations, ${}_nC_k$, is equal to the total number of groups of permutations that are partitioned by us, namely,

$$ {}_nC_k = \frac{{}_nP_k}{k!}, $$

which is the first equality in Eq. (19.2.7). The proof is, therefore, complete.

Remarks

1. It can be shown that Eq. (19.2.7) is exactly the same as Eq. (18.5.1), namely,

$$\binom{n}{k} \equiv {}_nC_k = \frac{n!}{k!\cdot(n-k)!}. \qquad (19.2.8)$$

From Eq. (19.2.8), it follows that ${}_nC_0 = {}_nC_0 = 1$.

2. As given by Eq. (18.5.2), we have

$${}_nC_k = {}_nC_{n-k}. \qquad (19.2.9)$$

Example 19.2.8

Compute each of the following:

(a) ${}_8C_4 = ?$ (b) ${}_6C_4 = ?$ (c) ${}_{50}C_3 = ?$

Solution

(a) By using Eq. (19.2.7), we have

$${}_8C_4 = \frac{8\cdot 7\cdot 6\cdot 5}{1\cdot 2\cdot 3\cdot 4} = 7\cdot 2\cdot 5 = 70.$$

(b) By using Eq. (19.2.9) and (19.2.7), we have

$${}_6C_4 = {}_6C_2 = \frac{6\cdot 5}{1\cdot 2} = 3\cdot 5 = 15.$$

(c) By using Eq. (19.2.7), we have

$${}_{50}C_3 = \frac{50\cdot 49\cdot 48}{1\cdot 2\cdot 3} = 50\cdot 49\cdot 8 = 19{,}600.$$

Example 19.2.9

The mathematics department of a state university has 10 tenured full professors. In how many different possible ways can the department choose 5 tenured full professors to form a personnel committee?

Solution

In forming a departmental personnel committee, the order of selection is not important. So, this is a combination problem. Therefore, the total number of different possible ways in forming a departmental personnel committee is obtained by using Eq. (19.2.7) with $n = 10$ and $k = 5$.

$${}_{10}C_5 = \frac{10\cdot 9\cdot 8\cdot 7\cdot 6}{1\cdot 2\cdot 3\cdot 4\cdot 5} = 3\cdot 2\cdot 7\cdot 6 = 252.$$

Example 19.2.10

(a) How many different possible 5-card poker hands can be dealt at random from a standard deck of 52 playing cards?
(b) A full house is a poker hand consisting of a pair (two of a kind) and three of a kind. How many different full houses are there that consist of 3 aces and 2 kings?

Solution

(a) Since the order of selection is unimportant in such a hand, this is a combination of taking 5 cards randomly out of 52 playing cards. Hence, the total number of different possible 5-card poker hands is obtained by using Eq. (19.2.7) with $n = 52$ and $k = 5$:

$${}_{52}C_5 = \frac{52 \cdot 51 \cdot 50 \cdot 49 \cdot 48}{1 \cdot 2 \cdot 3 \cdot 4 \cdot 5} = 13 \cdot 17 \cdot 5 \cdot 49 \cdot 48 = 2{,}598{,}960.$$

(b) There are 4 aces in the deck and ${}_4C_3$ ways to obtain 3 of them. There are 4 kings in the deck and ${}_4C_2$ ways to obtain 2 of them. By the multiplication principle of Theorem 19.2.1 the total number of different possible full houses of 3 aces and 2 kings is then given by

$${}_4C_3 \cdot {}_4C_2 = {}_4C_1 \cdot {}_4C_2 = \frac{4}{1} \cdot \frac{4 \cdot 3}{1 \cdot 2} = 4 \cdot 2 \cdot 3 = 24.$$

Example 19.2.11

Suppose that a coin is flipped 6 times independently and then the sequence of the upside of the coin (head or tail) is observed.
(a) How many different possible outcomes are there in the sample space?
(b) How many different outcomes have exactly three heads?
(c) How many different outcomes have at most two heads?
(d) How many different outcomes have at least four heads?

Solution

(a) Let H and T denote head and tail, respectively. In each flip of a coin, there are two possible outcomes, i.e., either H or T. By applying the multiplication principle, the total number of different possible outcomes in the sample space from flipping a coin 6 times independently is

$$\underbrace{2 \cdot 2 \cdot 2 \cdot 2 \cdot 2 \cdot 2}_{6 \ times} = 2^6 = 64 \cdot$$

(b) Each outcome can be viewed as a 6-letter word being constructed from a collection of two letters {H, T}. Those outcomes having exactly three heads are those 6-letter words having exactly 3 H's, e.g., {HTTHTH, THHTTH, ...}. Now the question becomes to ask how many different 6-letter words have exactly 3 H's. There are two approaches to answer the question. One approach is to use the concept of permutation. Take a 6-letter word and permute all six letters. We'll get 6! words. But not all of these 6! words are distinct because there are 3 H's and 3 T's. If permute those 3 H's and 3 T's, we'll get 3! ·3! words that are exactly the same word. Hence, out of 6! words only $\frac{6!}{3!\cdot 3!} = 20$ are different 6-letter words. The other approach is to use the technique of combination. Let us form a 6-letter word having exactly 3 H's as follows. Use a string of 6 blank boxes to represent a 6-letter word. Label the first blank box by 1, the second blank box by 2, etc. (Figure 19.2.2).

[1]–[2]–[3]–[4]–[5]–[6]
H T H H T T

Figure 19.2.2

Taking 3 blank boxes randomly out of 6 blank boxes and fill in the selected boxes with the letter H and the remaining boxes with the letter of T. This assures us that the obtained 6-letter words have exactly 3 H's. This is a combination of taking 3 objects (or blank boxes) out of 6 distinct objects (or blank boxes labeled with numbers). Hence, the total number of different possible combinations is given by

$$ {}_6C_3 = \frac{6\cdot 5\cdot 4}{1\cdot 2\cdot 3} = 5\cdot 4 = 20 , $$

which is exactly the same as the number obtained by the first approach of using permutations. Therefore, there are 20 different possible outcomes having exactly three heads.
(c) We note that a set of different possible outcomes having "at most" two heads can be decomposed as the union of three mutually disjoint sets of those outcomes having exactly 0 head, 1 head, or 2 heads.

Hence, the total number of different possible outcomes having at most 2 heads is given by

$$_6C_0 + _6C_1 + _6C_2 = 1 + 6 + 15 = 22.$$

(d) We note that a set of different possible outcomes having "at least" 4 heads can be decomposed as the union of three mutually disjoint sets of those outcomes having exactly 4 heads, or 5 heads, or 6 heads. Hence, the total number of different possible outcomes having at least 4 heads is given by:

$$_6C_4 + _6C_5 + _6C_6 = 15 + 6 + 1 = 22.$$

Remarks

1. As a convention, we just refer outcomes to number of heads. This automatically implies that the remaining are tails. For example, {outcomes have exactly 2 heads} = {outcomes have exactly 2 heads and 4 tails}.
2. You may want to list 20 different possible outcomes having exactly 3 heads by using 6-letter words (Problem 20 in Exercises for Section 19.2).

Example 19.2.12

A bag of ten apples contains three rotten apples and seven good apples. A shopper randomly selects a sample of two apples simultaneously from the bag.

(a) How many different possible samples are there in the sample space?

(b) How many different samples contain all good apples?

(c) How many different samples contain at least one good apple?

Solution

(a) Since two apples are selected simultaneously, the order of selection is not important. This is a combination problem. Hence, the total number of different possible samples of two apples is obtained by using Eq. (19.2.7) with $n = 10$ and $k = 2$:

$$_{10}C_2 = \frac{10 \cdot 9}{1 \cdot 2} = 5 \cdot 9 = 45.$$

(b) Since we would like to have samples containing good apples, we have to select two apples specifically from a collection of seven good apples. Therefore, the total number of different possible samples containing good apples is

$${}_7C_2 = \frac{7 \cdot 6}{1 \cdot 2} = 7 \cdot 3 = 21.$$

(c) Since the set of samples containing at least one good apple can be decomposed as the union of two mutually disjoint sets of samples containing exactly one good or exactly two good apples. Therefore, the total number of different possible samples containing at least one good apple is

$${}_7C_1 \cdot {}_3C_1 + {}_7C_2 = 7 \cdot 3 + 21 = 42.$$

Exercises

1. Suppose that Jill wants to go to a party. She has seven dresses and four pairs of shoes. In how many different ways can she possibly dress herself up to go to the party?

2. Two people get into an AMTRAK train where there are five vacant seats. In how many different ways can they seat themselves?

3. In how many different ways can we make a two-letter word that consisting of one consonant and one vowel?
Hint: There are 20 consonants and six vowels among the 26 alphabetical letters.

4. A company has 680 employees. Show that at least two employees will have the same pair of initials.

5. Twenty cross-country runners run a race for three prizes. In how many different ways may the prizes be given?

6. How many license plates consisting of three letters followed by three digits are there?

7. In how many different ways can eight books be arranged on a bookshelf?

8. How many different words can possibly be obtained from permuting the word "Mississippi"?

9. In how many different ways can four letters be put into three mailboxes?

10. In how many ways can four couples be seated around a circular table so that husband and wife are seated next to each other?

11. Suppose that 5 boys and 4 girls are arranged to sit around a circular table. In how many different ways can they be seated so that no two girls are seated next to one another?

12. How many different outcomes of "winner" and "runner-ups" are possible if there are ten contestants in a beauty-queen contest?

13. A university offers 4 science courses, 5 humanity courses, and 3 literature courses. In how many ways can a student possibly choose 2 science courses, 2 humanity courses, and 2 literature courses?

14. Suppose that the expression $x^3y^5z^2$ is written without exponents as a 10-letter word. How many different 10-letter words are possible?

15. A peculiar professor decided to grade 20 students in his class on a curve without regard to the student's test performance. There will be 2 A's, 2 B's, 10 C's, 2 D's, and 4 F's. In how many ways can this be done?

16. How many committees can be formed from a set of 10 senators and 8 representatives if each committee contains 4 senators and 3 representatives?

17. How many different five-card poker hands containing two cards of one denomination and two cards of another different denomination and one card of a third denomination are possible?

18. An urn contains 12 numbered balls, of that 8 are red and 4 are white. A sample of 3 balls is randomly selected from the urn simultaneously.

(a) How many different samples are possible in the sample space?

(b) How many different samples contain all red balls?

(c) How many different samples contain exactly 2 red balls?

(d) How many different samples contain at most 2 red balls?
(e) How many different samples contain at least 2 white balls?

19. A coin is flipped 8 times independently and observing the sequence of heads and tails.
(a) How many different outcomes are possible in the sample space?
(b) How many different outcomes have exactly four heads?
(c) How many different outcomes have at most three heads?
(d) How many different outcomes have at least six heads?

20. Refer to Example 19.2.11. Use 6-letter words to list the set of different outcomes that have exactly three heads.

19.3 Operational Rules for Probability (I)

In this section, we are going to define the calculation of the probability by invoking the equipossibility assumption. Also, some operational rules for calculating the probability of compounding two events are set up accordingly.

Definition 19.3.1
The symbol $n(A)$ is defined as the number of elements in a set A. In particular, $n(\phi) = 0$, where ϕ is the empty set.

Definition 19.3.2
An event E is said to occur if the outcome of a random experiment belongs to E.

Theorem 19.3.1 (Equipossibility Concept of Probability)
If the equipossibility assumption holds for a random experiment, then the probability that an event E occurs, denoted by $P(E)$, is given by

$$P(E) = \frac{n(E)}{n(S)}, \tag{19.3.1}$$

where S is the finite sample space associated with the given random experiment.
Proof

Since the equipossibility assumption holds, every outcome has the same probability $\frac{1}{n(S)}$ to occur. In addition, outcomes of a random experiment possess the property of mutual exclusiveness, namely, all the other outcomes cannot occur if one outcome occurs. Hence, the probability that an event E occurs is simply the sum of the probability of different outcomes in E or the number of outcomes in E times the probability that each outcome will occur, that is,

$$P(E) = n(E) \cdot P(\{\text{a single outcome occurs}\})$$
$$= n(E) \cdot \frac{1}{n(S)} = \frac{n(E)}{n(S)},$$

which is Eq. (19.3.1). The proof is, therefore, complete.

Remarks

1. Eq. (19.3.1) was used by S. P. Laplace (1749-1827, French) to define the probability of any event as the ratio of the number of cases favorable to the event to the total number of all possible cases (Reference 24, p. 11). In his definition of probability, Laplace implicitly assumed that all cases were equally possible to occur. Critics had long considered that Laplace's definition of probability represents a vicious circle because it defined probability in terms of probability. As explained in equipossibility theories of probability, I. Hacking provided two interpretations of probability, which were associated with two well-established senses of the word "possible", that is, "epistemic" and "physical" possibility. The epistemic possibility is relative to our state of knowledge, while the physical possibility is link to something physically possible. Hence, there are two different concepts of probability, that is, epistemological and physical. Laplace's definition of probability is epistemic, while Bernoulli's is physical (Reference 17).
2. In practice, whether the equipossibility assumption holds is difficult to verify. Even in flipping a coin or throwing a die, there is no assurance that the equipossibility assumption will hold because a coin could possibly be biased or a die could possibly be loaded. In order to use Eq. (19.3.1), we often add additional adjectives such as "a fair coin", "a perfect (or unloaded) die", or

"select at random" to assert the validity of the equipossibility assumption for the given random experiment.

3. In order for Eq. (19.3.1) to be applied, the sample space for the given random experiment has to be finite.
4. For many practical applications, Eq. (19.3.1) cannot be used in calculating the probability. Indeed, the probability in statistical applications cannot be calculated form Eq. (19.3.1). Instead, Bernoulli's definition of probability is oftentimes being used. For example, it is known that the birth ratios of male and female do not equal to another. In fact, the observed proportion of male to female human birth is the ratio 18:17 (Reference 11, p. 275).
5. Eq. (19.3.1) can be viewed as a theoretical approach in calculating the probability. In addition, an experimental approach was also attempted (Reference 22).

Example 19.3.1

(a) Flip a fair coin once. What is the probability that the head occurs?
(b) Throw a perfect die once. What is the probability that the even number of dots on the top face of the die occurs?

Solution

(a) Since the coin is fair, the equipossibility assumption holds for the random experiment of flipping a coin. As shown in Example 19.1.1(a), the sample space is given by

$$S = \{H, T\}.$$

Thus, $n(S) = 2$. Let E be the event that the head occurs. Then, $E = \{H\}$. As a consequence, $n(E) = 1$. By using Eq. (19.3.1), we have

$$P(\{\text{the head occurs}\}) = P(E) = \frac{n(E)}{n(S)} = \frac{1}{2} = 0.5.$$

Hence, the probability that the head occurs is 0.5.

(b) Because the die is perfect, the equipossibility assumption holds for the random experiment of throwing a die. As shown in Example 19.1.2 (a), the sample space is given by

$$S = \{1, 2, 3, 4, 5, 6\}.$$

Thus, $n(S) = 6$. Let E be the event that the even number of dots on the top face of the die occurs. Then, $E = \{2, 4, 6\}$. Consequently, $n(E) = 3$. By using Eq. (19.3.1), we have

$$P(E) = \frac{n(E)}{n(S)} = \frac{3}{6} = \frac{1}{2} = 0.5.$$

Hence, the probability that the even number of dots occurs is 0.5.

Remark

Whether a coin is fair or not is never known to us. In reality, Bernoulli's frequency approach is probably more appropriate to be used in calculating the probability. This was what J. Kerrich, a mathematician did when he was interned in Denmark during World War II. He actually tossed a coin 10,000 times, keeping a tally of the number of heads. After 10,000 tosses he had 5067 heads (Reference 32, pp. 7-8). Hence, the (empirical) probability that the head occurs is 0.5067. You may want to repeat a similar experiment in problems 1 and 2 in Exercises for Section 19.3.

Example 19.3.2

Flip a fair coin 6 times independently and observe the sequence of heads and tails. Find the probability that outcomes have
(a) exactly 3 tails occur.
(b) at most 2 tails occur.
(c) at least 4 tails occur.

Solution

(a) Since the coin is fair, the equipossibility assumption holds for this random experiment. Let S be the sample space for this random experiment. Thus, as shown in Example 19.2.11(a), $n(S) = 2^6 = 64$. Let E_1 be the event that outcomes have exactly 3 tails occur. Note that outcomes having exactly 3 tails are the same as those outcomes having exactly 3 heads. As shown in Example 19.2.11(b), we have $n(E_1) = \binom{6}{3} = 20$. By using Eq. (19.3.1), we obtain

$$P(E_1) = \frac{n(E_1)}{n(S)} = \frac{20}{64} = \frac{5}{16} = 0.3125.$$

Hence, the probability of the event that outcomes having exactly 3 tails occur is 0.3125.

(b) Let E_2 be the event that outcomes having at most 2 tails occur. Note that E_2 is also the same as the event that outcomes having at least 4 heads occur. As shown in Example 19.2.11(d), we have $n(E_2)$ = 22. Thus, by using Eq. (19.3.1), we obtain

$$P(E_2)=\frac{n(E_2)}{n(S)}=\frac{22}{64}=\frac{11}{32}=0.34375\,.$$

Hence, the probability of the event that outcomes having at most 2 tails occur is 0.34375.

(c) Let E_3 be the event that outcomes having at least 4 tails occur. Note that E_3 is also the same as the event that outcomes having at most 2 heads occur. As shown in Example 19.2.11(c), we have $n(E_3) = 22$. Thus, by using Eq. (19.3.1), we obtain

$$P(E_3)=\frac{n(E_3)}{n(S)}=\frac{22}{64}=\frac{11}{32}=0.34375\,.$$

Hence, the probability of the event that outcomes having at least 4 tails occur is 0.34375.

Remarks

1. We don't have to link the events E_i, $i = 1, 2, 3$ in this example, back to those of Example 19.2.11. We can tackle the problem by simply dealing with the number of tails directly. However, we would like to remind the readers of the connection between the number of heads and the number of tails.
2. Two events E_2 and E_3 are totally different, though their probabilities are the same. This is attributed to the symmetry of the binomial distribution (Section 18.5).

Example 19.3.3

A bag of 10 apples contains 3 rotten apples and 7 good apples. A shopper randomly selects a sample of 2 apples simultaneously from the bag. Find the probability that samples contain

(a) all good apples?

(b) at most 1 rotten apple?

Solution

(a) Since this shopper randomly selects 2 apples from the bag, the equipossibility assumption holds for this random experiment. Let S be the sample space for this random experiment. As shown in Example 19.2.12(a), $n(S) = 45$. Let E_1 be the event that samples contain all good apples. As shown in Example 19.2.12(b), $n(E_1) = 21$. Thus, by using Eq. (19.3.1), we obtain

$$P(E_1)=\frac{n(E_1)}{n(S)}=\frac{21}{45}=\frac{7}{15}=0.4666\ldots\approx 0.47.$$

Hence, the probability that samples contain all good apples is 0.47.
(b) Let E_2 be the event that samples contain at most 1 rotten apple. Since E_2 also equals the event that samples contains at least 1 good apple, we have from Example 19.2.12(c) $n(E_2) = 42$. By using Eq. (19.3.1),

$$P(E_2)=\frac{n(E_2)}{n(S)}=\frac{42}{45}=\frac{14}{15}=0.93333\ldots\approx 0.93.$$

Hence, the probability that samples contain at most 1 rotten apple is 0.93.

Example 19.3.4
In playing a poker hand, a 5-card hand with {A, K, Q, J, 10} of the same suit is called a royal flush. What is the probability of getting a royal flush?
Solution
Let S be the sample space of playing 5-card poker hands. As shown in Example 19.2.10(a), $n(S)={}_{52}C_5$ = 2,598,960. Let E be the event that 5-card hands are royal flush. Since there are 4 suits and in each suit there is only one poker hand that is a royal flush,

$$n(E)={}_4C_1\cdot{}_5C_5=4\cdot 1=4.$$

Since there is no sufficient reason that the equipossibility assumption does not hold for playing a game of 5-card poker hands, we have by using Eq. (19.3.1)

$$P(E)=\frac{n(E)}{n(S)}=\frac{4}{2{,}598{,}960}=0.0000015\ .$$

Hence, the probability of getting a royal flush is 0.0000015.

Remark
For other special poker hands, see problems 13-20 in Exercises for Section 19.3.

Definition 19.3.3
Given two events E and F.
(a) The union of the events E and F, denoted by $E \cup F$ (read E union F), is said to occur if either E or F (or both) occur.

(b) The intersection of the events E and F, denoted by $E \cap F$ (read E intersect F), is said to occur if both E and F occur.
(c) The complement of the event E, denoted by $\overline{E}$ (read E bar) is said to occur if E does not occur.

Remarks

1. The symbols "$\cup$" and "$\cap$" are exactly the same as those used for set operations. Consequently, for those outcomes, which are common to both E and F, we only include one of them into the union or the intersection of two events E and F.
2. By using Venn's diagram, the union, the intersection, and the complement can be illustrated by the shaded region in Figures 19.3.1-19.3.3, respectively.
3. Other symbols like E^c or $E^{'}$ are also used by other authors for denoting the complement of an event E.
4. The event E and its complement $\overline{E}$ satisfy

$$E \cup \overline{E} = S, \tag{19.3.2}$$

$$E \cap \overline{E} = \phi. \tag{19.3.3}$$

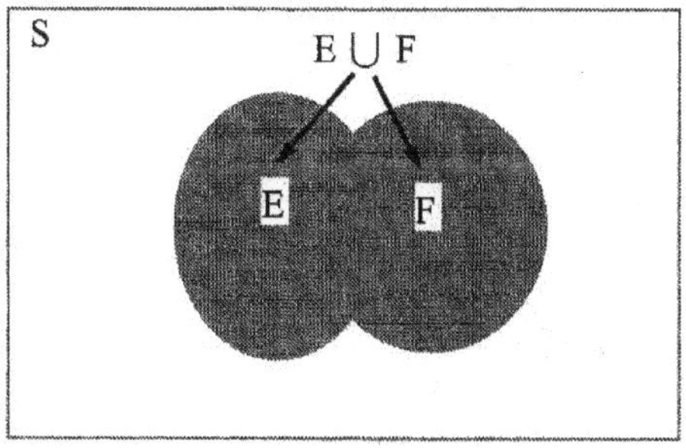

Figure 19.3.1

Example 19.3.5

Consider the random experiment throwing a perfect die in Example 19.3.1(b). Let E and F be, respectively, the events that even numbers and numbers less than 4 occur. Thus, E and F are given by $E = \{2, 4, 6\}$ and $F = \{1, 2, 3\}$. Then, we have $E \cup F = \{1, 2, 3, 4, 5, 6\}$, $E \cap F = \{2\}$, $\overline{E} = \{1, 3, 5\}$, and $\overline{F} = \{4, 5, 6\}$.

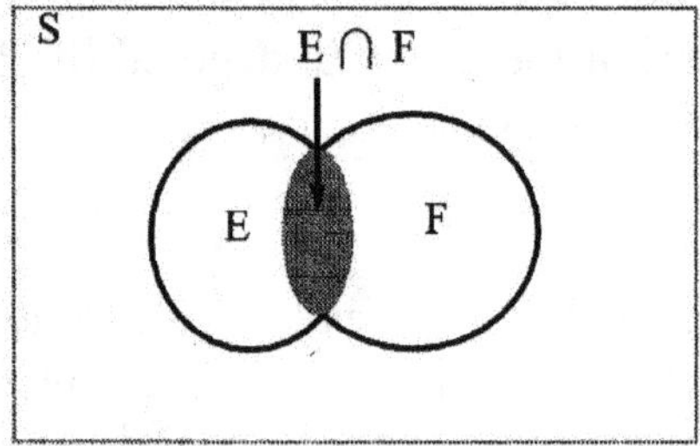

Figure 19.3.2

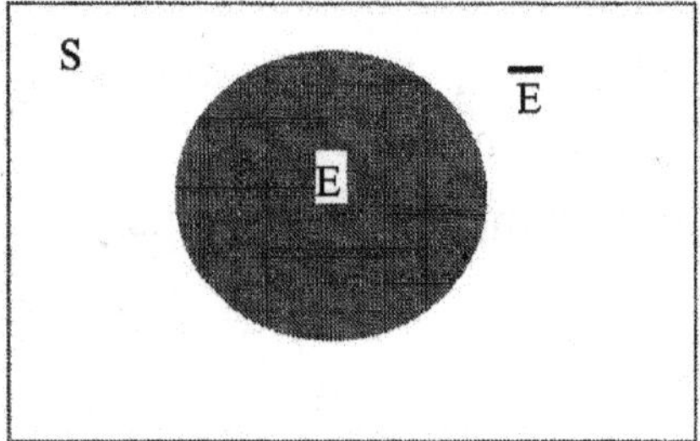

Figure 19.3.3

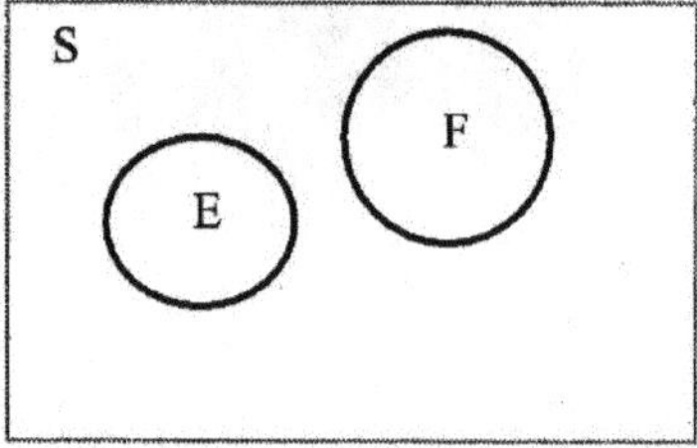

Figure 19.3.4

Definition 19.3.4
Two events E and F are said to be mutually exclusive (or disjoint) if $E \cap F = \phi$.

Remarks

1. By using Venn's diagram, the mutual exclusiveness of two events E and F are illustrated in Figure 19.3.4.
2. By Eq. (19.3.3), E and $\overline{E}$ are always mutually exclusive.
3. Those events consisting of only a single outcome are mutually exclusive.

Theorem 19.3.2 (Operational Rules for Probability)
Given two events E and F. Then

1a. The Addition Rule (General Case)

$$P(E \cup F) = P(E) + P(F) - P(E \cap F). \qquad (19.3.4)$$

1b. The Addition Rule (Special Case)

If E and F are mutually exclusive, then

$$P(E \cup F) = P(E) + P(F). \qquad (19.3.5)$$

2. The Complement Rule

$$P(\overline{E}) = 1 - P(E) \text{ or } P(E) = 1 - P(\overline{E}). \qquad (19.3.6)$$

Proof

1a. Assume that the equipossibility holds for the given random experiment. By the definition of the union of two events E and F as shown by Figure 19.3.1, it is easy to show that

$$n(E \cup F) = n(E) + n(F) - n(E \cap F). \qquad (19.3.7)$$

By dividing both sides of Eq. (19.3.7) by $n(S)$, we have

$$\frac{n(E \cup F)}{n(S)} = \frac{n(E) + n(F) - n(E \cap F)}{n(S)} = \frac{n(E)}{n(S)} + \frac{n(F)}{n(S)} - \frac{n(E \cap F)}{n(S)},$$

or

$$P(E \cup F) = P(E) + P(F) - P(E \cap F),$$

which is exactly Eq. (19.3.4). The proof is, therefore, complete.

1b. Since E and F are mutually exclusive, we have $E \cap F = \phi$. Since $P(\phi) = 0$, Eq. (19.3.4) reduces to Eq. (19.3.5). The proof is, therefore, complete.

2. Since E and $\overline{E}$ are mutually exclusive, we have by using Eq. (19.3.5)

$$P(S) = P(E \cup \overline{E}) = P(E) + P(\overline{E}), \text{ (By Eq. (19.3.2))} \Rightarrow$$

$$1 = P(E) + P(\overline{E}) \text{ (Since } P(S) = 1).$$

From the above equation Eq. (19.3.6) follows. The proof is, therefore, complete.

Remarks

1. Although the proof given here hinges on the validity of the equipossiblity assumption, Eqs. (19.3.4-19.3.6) hold without this assumption. For a more general proof, see Reference 8, pp. 35-36.
2. Eq. (19.3.7) is known as an addition rule for counting (Problem 3 in Exercises for Section 19.3).
3. In Eq. (19.3.4) there are four probabilities. As long as three of them are given, we can solve for the fourth one (See Example 19.3.7).
4. Unless two events E and F satisfy the additional requirement of mutual exclusiveness, we are not allowed to use Eq. (19.3.5).
5. An addition rule for three events is given in Problem 4 in Exercises for Section 19.3.

Example 19.3.6

Flip a fair coin 6 times independently and observe the sequence of heads and tails. Let E be the event that outcomes having at most 4 heads occur and F the event that outcomes having at least 3 heads occur. What is the probability that
(a) both E and F occur?
(b) either E or F occur?

Solution

(a) Obviously, the equipossibility assumption holds for this random experiment. Note that since

$$E = \{\text{at most 4 heads}\} = \{\text{exactly 0 head}\} \cup \{\text{exactly 1 head}\} \cup \{\text{exactly 2 heads}\} \cup \{\text{exactly 3 heads}\} \cup \{\text{exactly 4 heads}\}, \quad (1)$$

and

$$F = \{\text{at least 3 heads}\} = \{\text{exactly 3 heads}\} \cup \{\text{exactly 4 heads}\} \cup \{\text{exactly 5 heads}\} \cup \{\text{exactly 6 heads}\}, \quad (2)$$

we have from Eqs. (1) and (2)

$$E \cap F = \{\text{exactly 3 heads}\} \cup \{\text{exactly 4 heads}\}. \quad (3)$$

Thus, we obtain from Eq. (3) that $n(E \cap F) = \binom{6}{3} + \binom{6}{4} = 20 + 15 = 35$.

As shown in Example 19.2.11(a), $n(S) = 64$. By using Eq. (19.3.1),

$$P(E \cap F) = \frac{n(E \cap F)}{n(S)} = \frac{35}{64} = 0.546875 \approx 0.55 \quad (4)$$

Hence, the probability that both E and F occur is approximately 0.55.
(b) From Eqs. (1) and (2),

$$n(E)=\binom{6}{0}+\binom{6}{1}+\binom{6}{2}+\binom{6}{3}+\binom{6}{4}=1+6+15+20+15=57,$$

and

$$n(F)=\binom{6}{3}+\binom{6}{4}+\binom{6}{5}+\binom{6}{6}=20+15+6+1=42.$$

Therefore, we have

$$P(E)=\frac{n(E)}{n(S)}=\frac{57}{64}, \tag{5}$$

and

$$P(F)=\frac{n(F)}{n(S)}=\frac{42}{64}=\frac{21}{32}. \tag{6}$$

By using the general case of the addition rule given by Eq. (19.3.4) in coupling with Eqs. (4-6),

$$P(E\cup F)=\frac{57}{64}+\frac{21}{32}-\frac{35}{64}=1.$$

Therefore, the probability that either E or F occur is 1.

Remark
Since the probability is 1, $E\cup F$ surely occurs.

Example 19.3.7
A student is taking both a biology examination and a chemistry examination on the same day. The probability of passing both is 0.7, and the probability of passing at least one is 0.9. If the probability of passing the biology exam is 0.85, what is the probability of passing the chemistry exam?
Solution
Let E and F be the events of passing the biology exam and the chemistry exam, respectively. From the given information, we have

$$P(E\cap F)=0.7,\ P(E\cup F)=0.9 \text{ and } P(E)=0.85.$$

By substituting the above probabilities into Eq. (19.3.4), we obtain

$$0.90=0.85+P(F)-0.7,\ \Rightarrow$$
$$P(F)=0.9+0.7-0.85=0.75.$$

Therefore, the probability of passing the chemistry exam is 0.75.

Example 19.3.8

A group of four people is selected at random. What is the probability that at least two of them have the same birthday?

Solution

For simplicity, let us ignore leap years. Thus, a year has 365 days. Assume that the equipossibility assumption holds for this random experiment, namely, every day in a year is equally likely to be a birthday. Let E be the event that at least two of the four selected people have the same birthday. It is difficult to calculate directly the number of different possible outcomes in E. On the contrary, it is easy to calculate the number of outcomes in $\overline{E}$, the complement of E. This is because $\overline{E}$ is the event that all four selected people have different birthdays. To compute $n(\overline{E})$ is simply a permutation of taking 4 days without replacement out of 365 days. Thus,

$$n(\overline{E}) = {}_{365}P_4 = 365 \cdot 364 \cdot 363 \cdot 362 = 1.7459 \cdot 10^{10}.$$

Let S be the sample space for this random experiment. To compute $n(S)$ is a permutation of taking 4 days with replacement out of 365 days. By Eq. (19.2.5) (Theorem 19.2.3), we have that $n(S) = 365^4 = 1.7749 \cdot 10^{10}$.

By using Eq. (19.3.1),

$$P(\overline{E}) = \frac{n(\overline{E})}{n(S)} = \frac{1.7459 \cdot 10^{10}}{1.7749 \cdot 10^{10}} = \frac{1.7459}{1.7749} = 0.9836611 \approx 0.984.$$

By the complement rule of Eq. (19.3.6),

$$P(E) = 1 - P(\overline{E}) = 1 - 0.984 = 0.016.$$

Therefore, the probability that at least two of the four randomly selected people have the same birthday is 0.016.

Remarks

1. Although the probability of at least two people having the same birthday among a group of four people is very small, it can be shown that the probability that at least two people among a group of 23 or more people have the same birthday is greater than 0.5 (See problem 32 in Exercises for Section 19.3).
2. In reality, the equipossibility assumption does not hold for the birthday problem. There are more births in the months of November and December and fewer in April.

Definition 19.3.5
If the odds for an event E are a to b, then the probability that E occurs is $\frac{a}{a+b}$, i.e.,

$$P(E) = \frac{a}{a+b}. \tag{19.3.8}$$

Remark
Since $P(E) = \frac{a}{a+b}$, we have by the complement rule of Eq. (19.3.6) that $P(\overline{E}) = 1 - \frac{a}{a+b} = \frac{b}{a+b}$. Hence, the odds for an event E is equivalent to the ratio of $P(E) : P(\overline{E})$. Similarly, the odds against an event E is equal to $P(\overline{E}) : P(E)$. It thus follows Theorem 19.3.3.

Theorem 19.3.3 (The Odds for (or against) the Occurrence of an Event)
If $P(E) = p$, then
(a) the odds for the event E is given by the ratio
$p{:}(1-p)$;
(b) the odds against the event E is given by the ratio
$(1-p){:}p$.

Remark
The concept of the odds ratio is very important in Epidemiologic studies (Reference 10).

Example 19.3.9
One card is drawn at random from a well-shuffled standard deck of 52 playing cards.
(a) What are the odds for the occurrence of a face card, i.e, {J, Q, K}?
(b) What are the odds against the occurrence of a face card?
Solution
(a) Obviously, the equipossibility assumption holds for this random experiment. Let E be the event that the drawn card is a face card. Since there are altogether 12 face cards, we have

$$n(E) = \binom{12}{1} = 12.$$

Let S be the sample space for the random experiment. Thus, $n(S) = \binom{52}{1} = 52$. By using Eq. (19.3.1),

$$P(E) = \frac{n(E)}{n(S)} = \frac{12}{52} = \frac{3}{13}.$$

By the complement rule of Eq. (19.3.6), $P(\overline{E}) = 1 - \frac{3}{13} = \frac{10}{13}$. By Theorem 19.3.3(a), the odds for the occurrence of a face card is

$$P(E) : P(\overline{E}) = \frac{3}{13} : \frac{10}{13} = 3:10 \text{ (or 3 to 10).}$$

(b) By Theorem 19.3.3(b), the odds against the occurrence of a face card is

$$P(\overline{E}) : P(E) = \frac{10}{13} : \frac{3}{13} = 10:3 \text{ (or 10 to 3).}$$

Exercises

1. Take a coin and flip it 30 times independently. Record the outcomes. Repeat this experiment 5 times. Calculate separately the (empirical) probability that the head occurs for each time and compare them with the (overall) probability that head occurs in 150 flips. What are your observations?

2. Take a die and throw it 60 times independently. Repeat this experiment 5 times. Calculate separately the (empirical) probabilities that each of the six possible outcomes, namely, {1, 2, 3, 4, 5, 6} occurs in 60 throws for each of the 5 experiments. Compare these probabilities with the corresponding probabilities that each of the six outcomes occur in 300 throws. What are your observations?

3. Show that the addition rule for counting given by Eq. (19.3.7) holds.

4. For any given three events E, F, and G, show that

$$P(E\cup F\cup G)=P(E)+P(F)+P(G)-P(E\cap F)-P(F\cap G)-P(E\cap G)$$
$$+P(E\cap F\cap G)$$

5. By using addition rule and the complement rule for counting, prove that
(a) $\overline{E\cup F}=\overline{E}\cap\overline{F}$. (b) $\overline{E\cap F}=\overline{E}\cup\overline{F}$.

Remarks

The results of (a) and (b) are called de Morgan's Law for set operations.

6. Throw two dice (a red die and a blue die) once at random. (a) What is the probability of getting a sum of 10? (b) which sum has the highest probability?

7. Roll a perfect die four times independently. Are you more likely, or less likely, to have at least one six? (Hint: Use the complement rule of Eq. (19.3.6) to show that the odds for at least one six are 671 to 625.)

Remark

Historically, this is the first problem of de Mere (Reference 38, pp. 115 –116).

8. What is the probability of getting at least one double six in twenty-four throws of two dice? Hint: Use the complement rule of Eq. (19.3.6).

9. Roll three perfect dice once. Let X be the sum of three dice. Why is the sum of 10 more likely to occur than the sum of 9? Hint: Show that the probabilities for the sum of 3 to 18 is given by

$X=x$	3	4	5	6	7	8	9
$P(X=x)$	$\frac{1}{216}$	$\frac{3}{216}$	$\frac{6}{216}$	$\frac{10}{216}$	$\frac{15}{216}$	$\frac{21}{216}$	$\frac{25}{216}$

10	11	12	13	14	15	16	17
$\frac{27}{216}$	$\frac{27}{216}$	$\frac{25}{216}$	$\frac{21}{216}$	$\frac{15}{216}$	$\frac{10}{216}$	$\frac{6}{216}$	$\frac{3}{216}$

18
$\frac{1}{216}$

Remark

There are more discussions on different kind of dice games in Reference 7, pp. 152-175.

In problems 10-12, one card is drawn at random from a well-shuffled standard deck of 52 playing cards. What is the probability if drawing:

10. a spade? **11.** a red card? **12.** a 7 or a queen?

In problems 13-20, five cards are dealt at random from a well-shuffled standard deck of 52 playing cards. What is the probability of dealing:

13. A straight flush? (Hint: a straight flush consists of 5 cards in a sequence in the same suit, but does not include royal flush.)

14. A flush? (Hint: A flush is a 5-card hand in that all the cards are of the same suit, but neither a straight flush nor a royal flush.)

15. A straight? (Hint: A straight is a 5-card hand in sequence, but not in the same suit.)

16. A full house? (Hint: A full house consists of a pair (two of a kind) and three of a different kind.)

17. four-of-a-kind? (Hint: Four-of-a-kind is a 5-card hand in that 4 of the cards are of the same denomination, such as {Q, Q, Q, Q, 3}.)

18. three-of-a-kind? (Hint: Three-of-a-kind consists of 5-card hand in that 3 of the cards are of the same denomination and the other 2 cards are not, such as {J, J, J, 4, 7}.)

19. two pairs? (Hint: Two pair is a 5-card hand in that 2 sets of 2 cards of a kind and the fifth card that does not match the others, such as {K, K, 7, 7, 2}.)

20. a pair? (Hint: A pair is a 5-card hand in that just 2 of the cards are of the same denomination, such as {K, K, 10, 5, 2}.)

In problems 21-24, flip a fair coin 8 times independently. What is the probability that:

21. exactly 4 tails occur?

22. at most 3 tails occur?

23. at least 6 tails occur?

24. either exactly 3 tails or exactly 5 tails occur

In problems 25-27, an urn contains six red balls and four white balls. A sample of three balls is selected simultaneously at random from the urn. What is the probability that the selected sample contains:

25. exactly 1 red ball?

26. at least 1 white ball?

27. at least 2 white balls?

28. A factory produces fuses, which are packaged in boxes of 20 fuses. Four fuses are selected at random from each box for inspection. The box is rejected if at least one of these four fuses is found to be defective. What is the probability that a box containing three defective fuses will be rejected?

29. Let E and F be events with $P(E) = 0.5$, $P(F) = 0.7$, and $P(E \cup F) = 0.8$. Find $P(E \cap F) = ?$

30. Suppose that 65% of all customers of a large insurance company have automobile policies with the company, 50% have home-owner's policies, and 30% have both types of policies. If a customer is randomly selected, what is the probability that the selected customer has at least one of these types of policies with the company?

31. Four couples are seated at random around a circular table. What is the probability that husband and wife are seated next to each other?

32. A group of 23 people is to be selected at random. Assume that the equipossibility assumption holds and a year has 365 days. Show that the probability of having at least 2 people to have the same birthday is greater than 0.5. (Hint: Use the complement rule of Eq. (19.3.6).)

Remark
Even the equipossibility assumption does not hold, it can be shown that the non-uniform distribution of the birthdays only increases the probability of having at least 2 people to have the same birthday (See Reference 18, p. 126).

33. An exam contains ten "true or false" questions. What is the probability that a student guessing at the answers will get six or more answers correct?

34. Suppose that the odds for raining tomorrow are 7 to 2. What is the probability that it will rain tomorrow?

35. It is known that the odds for a male birth are 18 to 17. A couple decides to have five children. What is the probability they will have more boys than girls?

19.4 Conditional Probability and Independent Events

In this section we are going to learn the concept of conditional probability, one of the most important concepts in probability theory.

Sometimes, we are given additional information about the given random experiment, namely, we are told that the event F has occurred. By conditioning on that F has occurred, we are asked to find the probability that the event E occurs. The probability of E given F, denoted by $P(E|F)$, is called the conditional probability. Conditional probability plays a key role in many practical applications of probability. Two examples are hereby cited to illustrate its application. The first example is that in medical sciences, physicians to detect the presence or absence of a disease often use screening tests. However, none of the screening tests are absolutely reliable. It is possible that the test indicates a person to have the disease when this person in fact does not have it (false positive), or the test indicates a person does not have the disease when this person in fact has it (false negative). Note that both the probability of false positive and the probability of false negative are the conditional probabilities. A good screening test is then defined as a test that has a very low probability of false negative. A second example is that in the mass production of industrial product manufactured by different plants we would like to know which plant has the higher defective rate. Again, the plant's defective rate is a conditional probability that can be calculated by using Bayes' theorem in Section 19.5. Before giving a formal definition of the conditional probability, the following example provides a concrete illustration.

Example 19.4.1

Roll a perfect die once. Suppose that we are told that it is an odd number. What is the probability that a number 5 will occur?

Solution

Since the die is perfect, the equipossibility assumption holds. Let F be the event that odd numbers occur and E be the event that a number 5 occurs. Thus, $F = \{1, 3, 5\}$ and $E = \{5\}$. The desired probability is equal to finding the conditional probability of E given F. Since we are told that the outcome is an odd number, this indicates that F has occurred. Now, F partitions the original sample space $S = \{1, 2, 3, 4, 5, 6\}$ into the two mutual exclusive events F and $\overline{F}$. Since F has occurred, $\overline{F}$ does not occur. As a consequence, S reduces to F, or F becomes the new sample space upon that you are based to make a guess about the chance that E will occur. Therefore, the conditional probability of E given F is obtained by using Eq. (19.3.1)

$$P(E \mid F) = \frac{n(E \cap F)}{n(F)} = \frac{n(\{5\})}{n(\{1,3,5\})} = \frac{1}{3}.$$

Hence, the conditional probability that a number of 5 occurs, given that an odd number has occurred is $\frac{1}{3}$.

Remarks

1. This example is similar to the one given in Reference 26, p. 44. Von Mises (1883-1953, Austrian) also used another example to illustrate the concept of conditional probability. Suppose that a bus stop is served by six lines (Lines #1-#6) with two types of buses, single-decked (Lines #1, #3, #5) and double-decked (Lines #2, #4, #6). When a bus approaches the bus stop, you recognize from afar that it is a dobule-decked bus, what is the (conditional) probability that the coming bus is a Line #2?
2. In contrast to the conditional probability $P(E \mid F)$, $P(E)$ in Section 19.3 may be called the unconditional probability. In fact, $P(E)$ could be viewed as the unconditional counterpart of the conditional probability $P(E \mid F)$. Note that $P(E)$ and $P(E \mid F)$ are called the initial and the final probability, respectively, in Reference 9.
3. Although $P(E \mid F) = \frac{1}{3} > P(E) = \frac{1}{6}$ in this example, all three cases are possible between $P(E \mid F)$ and $P(E)$, which are: (i) $P(E \mid F) > P(E)$, (ii) $P(E \mid F) = P(E)$, and (iii) $P(E \mid F) < P(E)$. Clearly, Case (iii) holds if $F = \overline{E}$. Case (ii) is used to define the concept of independence between two events E and F.

Theorem 19.4.1 (The Equipossibility Definition of the Conditional Probability)

Given two events E and F with $n(F) > 0$. If the equipossibility assumption holds for the given random experiment, then the conditional probability of E given F is given by

$$P(E \mid F) = \frac{n(E \cap F)}{n(F)}. \tag{19.4.1}$$

Proof

Since F has occurred, F becomes the new sample space for the random experiment. The occurrence of E must be restricted to this conditional sample space F that is expressed as the intersection of E and F. Now, since the equipossibilty holds for the given random experiment, the conditional probability is just the ratio of favorable outcomes in $E \cap F$ to all possible outcomes in F, which gives us Eq. (19.4.1). The proof is, therefore, complete.

Remarks

1. By dividing both the numerator and the denominator of the right side of Eq. (19.4.1) by $n(S)$, we obtain

$$P(E \mid F) = \frac{\dfrac{n(E \cap F)}{n(S)}}{\dfrac{n(F)}{n(S)}} = \frac{P(E \cap F)}{P(F)}. \tag{19.4.2}$$

2. Eq. (19.4.2) provides us a general definition of the conditional probability $P(E \mid F)$ as given below, regardless of whether the equipossibility holds for a random experiment or not.

Definition 19.4.1

For any two events E and F with $P(F) > 0$, the conditional probability of E given that F has occurred is defined by Eq. (19.4.2).

Remarks

1. In terms of causality, there are two types of conditional probabilities. If F denotes the event of cause and E denotes the event of effect, then $P(E \mid F)$ (= P(Effect | Cause)) is called the direct (conditional) probability, whereas $P(F \mid E)$ (= P(Cause | Effect)) is the inverse (conditional) probability (Reference 25). The inverse (conditional) probability is also known as the probability of cause (Reference 5).
2. Historically, A. de Moivre (1667 –1754, French) introduced the concept of conditional probability via the notion of independent and dependent events in "The Doctrine of Chances" (Reference 27, pp. 6-7). However, P. R. de Montmort (1678-1719, French) had already demonstrated how to use the concept of conditional probability in calculating the probability of getting four aces without replacement from a pack of 40 cards in his book of "Essai

d'Analyses" (1708) (Reference 5, pp. 146-147). In addition, the concept of conditional probability also appeared in the Jewish Talmud (Reference 29, pp. 51-54).

3. By multiplying both sides of Eq. (19.4.2) by $P(F)$, we obtain
$$P(E \cap F) = P(E \mid F)P(F). \tag{19.4.3}$$
Eq. (19.4.3) is called the multiplication rule for $P(E \cap F)$ that will be used in Section 19.5.
4. In general, $P(E \mid F) \neq P(F \mid E)$.
5. As far as its application is concerned, conditional probability was used in building probability models, e.g., Markov chain (Reference 8) or in setting up a school of Bayesian statistical inference (Reference 34).

Example 19.4.2

A statistics class has 30 students. Of these 20 are senior, 15 are law enforcement major, and 7 are neither. Suppose that a student is selected at random from the class. We are given the additional information that the selected student is a law enforcement major. What is the probability that he/she is also a senior?

Solution

Since a student is selected at random, the equipossibility assumption obviously holds for this random experiment. Also, since only one student is selected, the sample space consists of all 30 students, i.e., $n(S) = 30$. Let E and F be the events that the selected students is a senior and a law enforcement major, respectively. Thus, $n(E) = \binom{20}{1} = 20$ and $n(F) = \binom{15}{1} = 15$. Also, since 7 students are neither senior nor law enforcement majors, $n(\overline{E \cup F}) = 7$. This implies that
$$n(E \cup F) = n(S) - n(\overline{E \cup F}) = 30 - 7 = 23.$$
By using the addition rule for counting given by Eq. (19.3.7),
$$23 = n(E \cup F) = n(E) + n(F) - n(E \cap F) = 20 + 15 - n(E \cap F) \Rightarrow$$
$$n(E \cap F) = 35 - 23 = 12.$$
Since we have the additional information that the selected student is a law enforcement major, the probability that he/she is also a senior is the conditional probability $P(E \mid F)$. By using Eq. (19.4.1),

$$P(E \mid F)=\frac{n(E\cap F)}{n(F)}=\frac{12}{15}=\frac{4}{5}=0.8\,.$$

Hence, the probability that he/she is also a senior is 0.8.

Remark

In practical applications to the real world problems, the equipossibility assumption oftentimes does not hold. Yet, we still can calculate the conditional probability by using Eq. (19.4.2) as shown in the following example. Also, see problem 5 in Exercises for Section 19.4.

Example 19.4.3

To study the efficiency of inoculation again cholera, the data from the 818 cases studied are given in Table 19.4.1.

Table 19.4.1 Incidence of cholera among inoculated and non-inoculated populations.

	Not attacked	Attacked	Total
Inoculated	276	3	279
Not inoculated	473	66	539
Total	749	69	818

From the data given in the above table, do you think that the inoculation help people to fend off the attack of cholera?

Solution

Let E be the event that people were not attacked by the cholera and F be the event that people were inoculated. By using Bernoulli's frequency approach in calculating the probability,

$$P(E)=\frac{749}{818}=0.916\,,$$

and

$$P(E \mid F)=\frac{P(E\cap F)}{P(F)}=\frac{\frac{276}{818}}{\frac{279}{818}}=\frac{276}{279}=0.989\,.$$

Since $P(E \mid F) > P(E)$, the inoculation is effective in helping people to fend off the attack of cholera.

Remarks

1. The data of Table 19.4.1 is taken from Reference 15.
2. For an alternative statistical analysis of Table 19.4.1, see Reference 21, p. 568.

Example 19.4.4

The composition of the 107th Congress of the United States is given in Table 19.4.2.

Table 19.4.2 The composition of the U. S. 107th Congress.

	Democrats	Republicans	Independent
Senate	50	49	1
House of Representatives	211	221	2
Total	261	270	3

Suppose that a person is randomly selected from the 107th Congress. What is the probability that the selected person is
(a) a Democrat, if he/she is a Senator?
(b) a Senator, if he/she is a Democrat?
(c) a Republican, if he/she is a House Representative?
(d) a House Representative, if he/she is a Republican?

Solution

Let
D = the event that the selected person is a democrat,
R = the event that the selected person is a republican
S = the event that the selected person is a senator,
and
H = the event that the selected person is a House Representative.
Since the person is randomly selected, the equipossibility assumption holds for this random experiment.
(a) By using Eq. (19.4.1), the probability that the selected person is a Democrat, given that he/she is a Senator, is given by

$$P(D \mid S) = \frac{n(D \cap S)}{n(S)} = \frac{50}{100} = 0.5.$$

(b) The probability that the selected person is a Senator, given that he/she is a Democrat, is given by

$$P(S \mid D) = \frac{n(S \cap D)}{n(D)} = \frac{50}{261} = 0.1915709 \approx 0.19.$$

(c) The probability that the selected person is a Republican, given that he/she is a House Representative, is given by

$$P(R \mid H) = \frac{n(R \cap H)}{n(H)} = \frac{221}{434} = 0.5092166 \approx 0.51.$$

(d) The probability that the selected person is a House Representative given that he/she is a Republican, is given by

$$P(H \mid R) = \frac{n(H \cap R)}{n(R)} = \frac{221}{270} = 0.8185185 \approx 0.82.$$

Remarks

1. One seat in the House of Representatives is vacant. Therefore, the total number of the Representatives is 434 rather than 435.
2. Note that $P(D \mid S) \neq P(S \mid D)$ and $P(R \mid H) \neq P(H \mid R)$.

Definition 19.4.2

Two events E and F are said to be independent if either one of the following equations holds:

$$P(E \mid F) = P(E) \text{ or } P(F \mid E) = P(F). \qquad (19.4.4)$$

Otherwise, E and F are said to be dependent.

Remarks

1. By re-phrasimg this definition in the words of de Moivre, two events E and F are independent when these two events have no connection one with another, and the happening of one neither forwards nor obstructs the happening of the other, in other words, neither $P(E \mid F) > P(E)$ nor $P(E \mid F) < P(E)$ (Reference 27, p.163).
2. In general, it is not easy to tell on the surface whether the given two events are independent, in particular, when those two events are defined within the same random experiment (See Example 19.4.5). Only when two events are defined, respectively, within two independent random experiments, it is assured that these two events are independent. For example, flip a coin twice independently. The events that the head occurs in the first flipping and the event that the tail occurs in the second flipping are surely two independent events. Also, see Example 19.4.6.

3. The notion of independence is very important in engineering design. Note that the main brake system in a car is designed by using a hydraulic pressure. However, the auxiliary brake system, namely the hand brake is designed by using a mechanical device.

Example 19.4.5
Roll a perfect dice once. Let E, F, and G be defined by
$E = \{1, 2, 3, 4\}$,
$F = \{1, 2, 3\}$,
and
$G = \{1, 3, 5\}$
(a) Are E and F independent?
(b) Are E and G independent?
(c) Are F and G independent?
Solution
Since the die is perfect, the equipossibility assumption holds for this random experiment. Let S denote the sample space. Then, $n(S) = 6$. Also, $n(E) = 4$, $n(F) = 3$, and $n(G) = 3$. Since

$$E \cap F = \{2, 3\},$$
$$E \cap G = \{3\},$$

and

$$F \cap G = \{3\},$$

we have

$$n(E \cap F) = 2,$$
$$n(E \cap G) = 1,$$

and

$$n(F \cap G) = 1.$$

(a) By using Eq. (19.4.1),

$$P(E \mid F) = \frac{n(E \cap F)}{n(F)} = \frac{2}{3}.$$

By using Eq. (7.3.1),

$$P(E) = \frac{n(E)}{n(S)} = \frac{4}{6} = \frac{2}{3}.$$

Since $P(E \mid F) = P(E)$, Eq. (19.4.4) is satisfied. Hence, E and F are independent.
(b) By using Eq. (19.4.1),

$$P(E \mid G) = \frac{n(E \cap G)}{n(G)} = \frac{1}{3}.$$

Since $P(E \mid G) \neq P(E)$, Eq. (19.4.4) is not satisfied. Hence, E and G are not independent, i.e., E and G are dependent.

(c) By using Eq. (19.4.1),

$$P(F \mid G) = \frac{n(F \cap G)}{n(G)} = \frac{1}{3}.$$

By using Eq. (19.3.1),

$$P(F) = \frac{n(F)}{n(S)} = \frac{3}{6} = \frac{1}{2}.$$

Since $P(F \mid G) \neq P(F)$, Eq. (19.4.4) is not satisfied. Hence, F and G are not independent, i.e., F and G are dependent.

Remark

By just taking a glimpse on E, F, and G, we have no idea whether they are independent or not. The only way to determine is to see if Eq. (19.4.4) is satisfied.

Example 19.4.6

Rolling a perfect dice twice independently. Let E and F be the events defined, respectively, by

E = {The outcome of the first rolling of a die is 4},

and

F = {The outcome of the second rolling of a die is 3}.

Are E and F independent?

Solution

Since the die is perfect, the equipossibility assumption holds for this random experiment. Also, by listing the favorable outcomes in E and F, we have

$$E = \{(4, 1), (4, 2), (4, 3), (4, 4), (4, 5), (4, 6),$$

and

$$F = \{(1, 3), (2, 3), (3, 3), (4, 3), (5, 3), (6, 3)\}.$$

Thus,

$$E \cap F = \{(4, 3)\}.$$

Let S denote the sample space for this random experiment. By noting that the total number of different possible outcomes associated with rolling a die twice is the same as that of rolling two dice once. Thus, $n(S) = 6{\times}6 = 36$.

By using Eq. (19.3.1),

$$P(E)=\frac{n(E)}{n(S)}=\frac{6}{36}=\frac{1}{6},$$

and

$$P(E\mid F)=\frac{n(E\cap F)}{n(F)}=\frac{1}{6}.$$

Since $P(E) = P(E \mid F)$, Eq. (7.4.4) is satisfied. Hence, E and F are independent.

Remark

For this example, since E and F are defined on two independent random experiments, namely, the first rolling and the second rolling of a die, E and F are apparently independent even without a check to see if Eq. (19.4.4.) is satisfied.

Example 19.4.7

Drawing two cards successively at random from a standard deck of 52 playing cards. Let E and F be the events defined by

$$E=\{\text{The first card is a spade}\},$$

and

$$F=\{\text{The second card is a heart}\}.$$

Are E and F independent if the draws are done: (a) with replacement? (b) without replacement?

Solution

(a) Since two cards are drawn randomly, the equipossibility assumption holds for this random experiment. Because the samplings are done with replacement, the sample space for the second drawing remains unchanged. Hence, by using Eq. (19.4.1),

$$P(F\mid E)=\frac{13}{52}=\frac{1}{4}=0.25.$$

To compute $P(F)$, we note that since

$$F=F\cap S=F\cap(E\cup\overline{E})=(F\cap E)\cup(F\cap\overline{E}),$$

$$P(F)=P[(F\cap E)\cup(F\cap\overline{E})]=P(F\cap E)+P(F\cap\overline{E})$$

(By using Eq. (19.3.5))

$$=P(F\mid E)P(E)+P(F\mid\overline{E})P(\overline{E})$$

(By using Eq. (19.4.3))

$$= \frac{13}{52} \cdot \frac{13}{52} + \frac{13}{52} \cdot \frac{39}{52} = \frac{1}{4} \cdot \frac{1}{4} + \frac{1}{4} \cdot \frac{3}{4} = \frac{4}{16} = \frac{1}{4} = 0.25 .$$

Since $P(F \mid E) = P(F)$, Eq. (19.4.4) is satisfied. Hence, E and F are independent.

(b) Since the samplings are done without replacement, the sample space for the second drawing has changed, namely, it has only 51 cards available for the second drawing. Since the first card is a spade, there still remains 13 hearts in the deck. By taking Eq. (19.4.1),

$$P(F \mid E) = \frac{13}{51} = 0.254902 \approx 0.255 .$$

Since $P(F \mid E) \neq P(F)$, Eq. (19.4.4) is not satisfied. Hence, E and F are not independent, namely, E and F are dependent.

Remark

In general, the events from samplings without replacement are dependent. Consequently, the events involving the opinion polls are dependent. However, due the population is so large, they could be considered as approximately independent.

Exercises

1. A local high school choir has 40 students. Of these, 9 are freshmen, 28 are girls, and 5 are neither. A student is randomly selected from this group. If the selected student is known to be a girl, what is the probability that she is also a freshman?

2. A survey of 70 high school students revealed that 10 like mathematics, 30 like chemistry, and 7 like both courses. A student is randomly selected from this group. If the selected student is found to like mathematics, what is the probability that he/she also likes chemistry?

3. One hundred college students were surveyed after voting in a local election involving a Democratic and a Republican candidate. Sixty students are freshmen, 55 voted Republican, and 30 are non-freshmen who voted Democratic. A student is randomly selected from this group. What is the probability that he/she voted Republican, given that the selected student was a freshman?

4. One hundred and seventy business executives were surveyed to determine if they regularly read Fortune, Business Week, or Money magazines. 130 read Fortune, 115 read Business Week, 110 read Money, 95 read Fortune and Business Week, 85 read Fortune and Money, 75 read Business Week and Money, and 60 read all three magazines. A person is randomly selected from this group. What is the probability that the selected business executive reads (a) Business Week, given that he/she read Fortune and Money? (b) Money, given that he/she read Business Week and Fortune?

5. Suppose that at a local self-service gas station 30% of all customers check their oil level, 15% check their tire pressure, and 5% check both oil level and tire pressure. If a random selected customer is known to check his/her tire pressure, what is the probability that he/she did not check his/her oil level?

6. Suppose that 80% of all customers of a large insurance agency have automobile policies with the agency, 50% have homeowner's policies, and 85% have at least one of these two types of policies. If a randomly selected customer is found to have a homeowner's policy, what is the probability that he/she also has an automobile policy with this agency?

7. The following table shows 42 children according to the nature of their teeth and type of feeding:

	Normal teeth (N)	Mal-occcluded teeth ($\overline{N}$)	Total
Breast-fed (B)	4	16	20
Bottle-fed ($\overline{B}$)	1	21	22
Total	5	37	42

Does the breast-fed help to prevent the children from having the mal-occluded teeth?

Remark

This data was taken from Reference 38.

8. The following table contains the relative frequency of a medical survey of 1000 people over the age of 55 that was made to investigate the relationship between smoking and lung cancer.

	Cancer (C)	No cancer ($\overline{C}$)
Smoking (or S)	0.05	0.20
Not smoking (or $\overline{S}$)	0.03	0.72

A person is randomly selected from this group. What is the probability that the selected person has
(a) cancer if he/she is also smoking?
(b) no cancer if he/she is also smoking?
(c) cancer if he/she does not smoke?
(d) no cancer if he/she does not smoke?

Remark
For a discussion on the relationship between smoking and lung cancer, see Reference 3.

9. The following table is the data on the graduate admission for the fall of 1973 of at University of California at Berkeley:

	Admit (A)	Deny ($\overline{A}$)
Men (M)	3738	4704
Women ($\overline{M}$)	1494	2829

An applicant is randomly selected. What is the probability that the selected person is
(a) a male if he is also admitted?
(b) a female if she is also admitted?

Remark
For a discussion on the sex bias in graduate admissions, see Reference 4.

10. What is the probability of drawing three clubs in succession from a standard deck of 52 playing cards if the draws are made at random and without replacement?

11. A bag contains 7 good and 3 rotten apples. If the apples are selected at random from the bag, one at a time without replacement, what is the possibility that the third rotten apple is the eighth apple selected?

12. A grade school boy has three red and four green chips in his right pocket and four red and three green chips in his left pocket. If he transfers one chip at random from his right to his left pocket, what is the probability of his then drawing a green chip from his left pocket?

13. Determine whether each pair of the following events seems to be independent:
(a) The weather today and the weather tomorrow.
(b) The weather today and the weather in six months.
(c) The statistics professor arrives at class on time and the philosophy professor arrives at class on time.
(d) The failure of a washer and the failure of a dryer.
(e) Bob Smith buys a Toyota Camry and Jeff Williamson buys a Toyota Camry, where Bob and Jeff do not know each other.
(f) A couple's first child is a boy and their second child is still a boy.
(g) A person has a high blood pressure and he/she will have a heart attack.
(h) Smoking and diseases.
(i) The successive selections in the sampling inspection of industrial quality control.

14. In Example 19.4.1, are the event that an odd number occurs and the event that a number 3 occurs independent?

15. In Example 19.4.3, are the inoculation and the attack by cholera independent?

16. In Example 19.4.4, (a) Are the events that the selected person is a democrat and he/she is also a senator independent? (b) Are the events that the selected person is a house representative and he/she is also a republican independent?

19.5 Operational Rules for Probability (II)

Basically, there are three operational rules and one formula in probability theory. In Section 19.3, we have learned the addition rule and the complement rule (Theorem 19.3.2). In this section we are going to learn the third operation rule, i.e., the multiplication rule, and one formula, namely, the Bayes' formula.

Theorem 19.5.1 (The Multiplication Rule)
Given any two events E and F with $P(E) > 0$ and $P(F) > 0$. Then
3a. The multiplication rule (general case):

$$P(E \cap F) = P(E \mid F)P(F) = P(F \mid E)P(E). \qquad (19.5.1)$$

3b. The multiplication rule (special case):
If E and F are independent, then

$$P(E \cap F) = P(E)P(F). \qquad (19.5.2)$$

Proof
3a. By using the definition of $P(E \mid F)$ defined by Eq. (19.4.2),

$$P(E \mid F) = \frac{P(E \cap F)}{P(F)}.$$

By multiplying both sides of the above equation with $P(F)$, we have

$$P(E \mid F)P(F) = P(E \cap F). \qquad (1)$$

Similarly, by using the definition of $P(F \mid E)$ defined by Eq. (19.4.2),

$$P(F \mid E) = \frac{P(F \cap E)}{P(E)}.$$

Again, by multiplying both sides of the above equation with $P(E)$, it follows that

$$P(F \mid E)P(E) = P(F \cap E). \qquad (2)$$

Since $P(E \cap F) = P(F \cap E)$, Eq. (19.5.1) follows immediately from Eqs. (1) and (2). The proof is, therefore, complete.
"3b." Since E and F are independent, we have by Eq. (19.4.4)

$$P(E \mid F) = P(E) \qquad (3)$$

and

$$P(F \mid E) = P(F) \tag{4}$$

By substituting Eqs. (3) and (4) into Eq. (19.5.1), Eq. (19.5.2) follows. The proof is, therefore, complete.

Remarks

1. Note that the multiplication rule of Eq. (19.5.1) is nothing but the other face of the conditional probability of Eq. (19.4.2).
2. In Eq. (19.5.1), there are two formulas for calculating $P(E \cap F)$. Which one to be used depends on which of the two conditional probabilities [$P(E \mid F)$ or $P(F \mid E)$] is given. See Example 19.5.1 (a).
3. Unless E and F satisfy the additional qualification, namely, E and F are independent, Eq. (19.5.2) must not be used.
4. From Eq. (19.5.1), we can obtain
$$P(F \mid E) = \frac{P(E \mid F)P(F)}{P(E)}. \tag{19.5.3}$$
5. In fact, Eq. (19.5.2) is often used as a judging criterion to determine if E and F are independent besides using Eq. (19.4.4) (See Example 19.5.3).
6. The multiplication rule involves three events is given in problem 1 in Exercises for Section 19.5.

Example 19.5.1

Suppose that 90% of all families with young kids who took vacation in Florida visited the Disney World at last Christmas. Of those families who visited the Disney World, 70% of them also visited the Universal Studio. If a family is randomly selected from this group, what is the probability that this family visited both Disney World and Universal Studio?

Solution

Let D denote the event that a family visits the Disney World and U the event that a family visits the Universal Studio. From the given information, we have

$$P(D) = 0.90,$$

and

$$P(U \mid D) = 0.70.$$

By using Eq. (19.5.1), the probability that the selected family visited both the Disney World and the Universal Studio is then given by

$$P(D \cap U) = P(U \mid D)P(D) = 0.9 \cdot 0.7 = 0.63.$$

Example 19.5.2

A pharmaceutical company has a new test to detect cervical cancer. Of those women having cervical cancer, this new test will detect it in 98% of the cases. Of those not having cervical cancer, 5% test positive when the test is applied. Suppose that 30% of a certain group of women is known to have cervical cancer. A woman is randomly selected from this group. What is the probability that a woman who tests

(a) positive has cervical cancer?
(b) positive actually has no cervical cancer?
(c) negative has no cervical cancer?
(d) negative actually has cervical cancer?

Solution

Let C and N denote, respectively, the events that a woman has cervical cancer and the test is negative. Thus, the complement events $\overline{C}$ and $\overline{N}$ are the events that a women has no cervical cancer and the test is positive. From the given information, we have

$$P(\overline{N} \mid C) = 0.98,$$

$$P(\overline{N} \mid \overline{C}) = 0.05,$$

and

$$P(C) = 0.3.$$

(a) The probability that a woman has cervical cancer, given that her test was positive is given by

$$P(C \mid \overline{N}) = \frac{P(C \cap \overline{N})}{P(\overline{N})} = \frac{P(\overline{N} \mid C)P(C)}{P(\overline{N} \mid C)P(C) + P(\overline{N} \mid \overline{C})P(\overline{C})}$$

(Since $\overline{N} = \overline{N} \cap S = \overline{N} \cap (C \cup \overline{C}) = (\overline{N} \cap C) \cup (\overline{N} \cap \overline{C})$))

$$= \frac{0.98 \cdot 0.3}{0.98 \cdot 0.3 + 0.05 \cdot (1 - 0.3)} = \frac{0.294}{0.294 + 0.035} = \frac{0.294}{0.329}$$

$$= 0.893617 \approx 0.89.$$

(b) The probability that a woman who tests positive actually has a cervical cancer, i.e., the false positive rate, is the conditional probability is obtained by using the complementary rule for the conditional probability

$$P(\overline{C} \mid \overline{N}) = 1 - P(C \mid \overline{N}) = 1 - 0.89 = 0.11.$$

(c) The probability that a woman has no cervical cancer, given that her test was negative is obtained by using Eq. (19.5.1):

$$P(\overline{C} \mid N) = \frac{P(\overline{C} \cap N)}{P(N)} = \frac{P(N \mid \overline{C})P(\overline{C})}{P(N \mid C)P(C) + P(N \mid \overline{C})P(\overline{C})}$$

$$= \frac{[1 - P(\overline{N} \mid \overline{C})][1 - P(C)]}{[1 - P(\overline{N} \mid C)]P(C) + [1 - P(\overline{N} \mid \overline{C})][1 - P(C)]}$$

$$= \frac{(1 - 0.05)(1 - 0.3)}{(1 - 0.98) \cdot 0.3 + (1 - 0.05)(1 - 0.3)} = \frac{0.95 \cdot 0.7}{0.02 \cdot 0.3 + 0.95 \cdot 0.7}$$

$$= \frac{0.665}{0.671} \approx 0.991 .$$

(d) The probability that a woman actually has cervical cancer, given that her test was negative i.e., the false negative rate, is the conditional probability that is obtained by using the complementary rule for the conditional probability

$$P(C \mid N) = 1 - P(\overline{C} \mid N) = 1 - 0.991 = 0.009 .$$

Remarks

1. To avoid the use of the capital letter P to denote the test is positive, which might confuse with the symbol of probability P, this is why we define first the event N that the test is negative and treat the event that the test is positive as the complement of N.
2. A criterion for a good test is tha the false negative rate is as low as possible. Since the false negative rate of this new test for cervical cancer is only 0.009 (or 0.9%) that is low, we conclude that this new test is somewhat a good test, at least not bad.

Example 19.5.3

An electronic store sells both computers and monitors that are manufactured by different companies. Form the past record, it is known that 6% of the computers sold require service while under warranty, where only 4% of the monitors sold need such service. If a customer purchases randomly a computer and a monitor from this store, what is the probability that both machines need warranty service?

Solution

Let C denote the event that a computer purchased from this store requires a service while under warranty and M the event that a monitor purchased from this store needs a warranty service. From the given information, we have

$$P(C) = 0.06,$$

and

$$P(M) = 0.04.$$

Since C is clearly independent of M, the probability that both machines need warranty service is obtained by using Eq. (19.5.2):

$$P(C \cap M) = P(C)P(M) = 0.06 \cdot 0.04 = 0.0024 .$$

Remark

Since different companies manufacture computers and printers, the failure of computers has no effect on the failure of monitors. Hence, C and M are independent.

Definition 19.5.1

A set of n nonempty events $\{E_1, E_2, \ldots, E_n\}$ is called a partition of the sample space S if they satisfy the following conditions:

(i) $E_i \cap E_j = \phi$, if $i \neq j$.

(ii) $E_1 \cup E_2 \cup \ldots \cup E_n = S$.

Remark

Condition (i) says that $\{E_1, E_2, \ldots, E_n\}$ are mutually exclusive among themselves, while condition (ii) says that they are altogether exhaustive for the sample space S.

Theorem 19.5.2 (Bayes' Formula)

Let $\{E_i\}$, $i = 1, 2, \ldots, n$ be a partition of the sample space S. Then, for any event F,

(a) $P(F) = \sum_{j=1}^{n} P(F \mid E_j)P(E_j)$. (19.5.4)

(b) $P(E_i \mid F) = \dfrac{P(F \mid E_i)P(E_i)}{\sum_{j=1}^{n} P(F \mid E_j)P(E_j)}$, $i = 1, 2, \ldots, n$. (19.5.5)

Proof

Since $\{E_j\}$, $j = 1, 2, \ldots, n$, is a partition of the sample space S, we have

$$F = F \cap S = F \cap (E_1 \cup E_2 \cup \ldots \cup E_n)$$
$$= (F \cap E_1) \cup (F \cap E_2) \cup \ldots \cup (F \cap E_n) .$$

From the equation, it is so easy to see that $\{F \cap E_j\}$,$j = 1, 2, \ldots$, n, forms a partition of F. Thus,

$$P(F) = P(F \cap E_1) + P(F \cap E_2) + \ldots + P(F \cap E_n) .$$

By applying Eq. (19.5.1) to each of the n terms on the right side of the above equation, we have

$$P(F) = P(F \mid E_1)P(E_1) + P(F \mid E_2)P(E_2) + \ldots + P(F \mid E_n)P(E_n) ,$$

which is Eq. (19.5.4). The proof is, therefore, complete.

(b) By using Eq. (19.4.1),

$$P(E_i \mid F) = \frac{P(E_i \cap F)}{P(F)} = \frac{P(F \mid E_i)P(E_i)}{P(F)} .$$

(using Eq. (19.5.1))

Eq. (19.5.5) follows immediately by substituting Eq. (19.5.4) for $P(F)$ in the denominator of the right side of the above equation. The proof is, therefore, complete.

Remarks

1. Theorem 19.5.2 is known as Bayes' theorem, that appeared as Proposition 5 (not exactly the same as Eq. (19.5.5)) in Reference 1, p. 301. The research motivation for T. Bayes (1702 –1761, English) is that he was attempted to answer a question related to Bernoulli's law of large numbers (Theorem 19.1.1), namely, when n does not go to infinite, but rather stops at some finite value. If m successes are observed in n trials, then what is the probability that the unknown probability p lies between two given real numbers a and b, i.e., calculate the following conditional probability:

   $P(a \le p \le b \mid m$ successes are observed in n trials) = ?

   (Reference 37, pp. 294-300). Not much was known about the life of T. Bayes. For a brief biography, see Reference 25, pp. 87-88.
2. Note that $P(E_i \mid F)$ of Eq. (19.5.5) is viewed as the inverse (conditional) probability. If $\{E_i\}$, $i = 1, 2, \ldots, n$, are regarded as different causes, then $P(E_i \mid F)$ is also interpreted as the probability of cause (Reference 5).

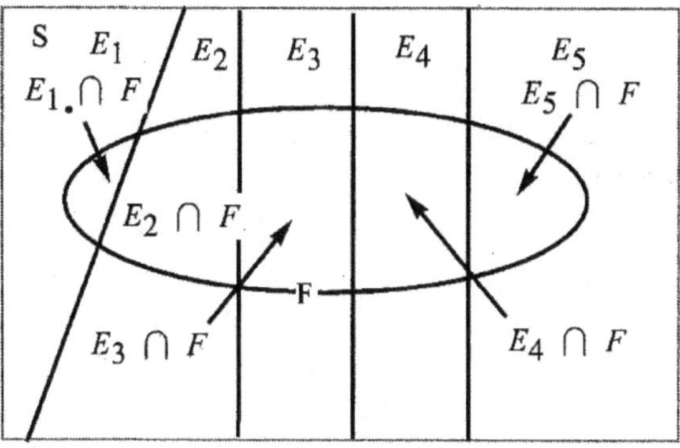

Figure 19.5.1

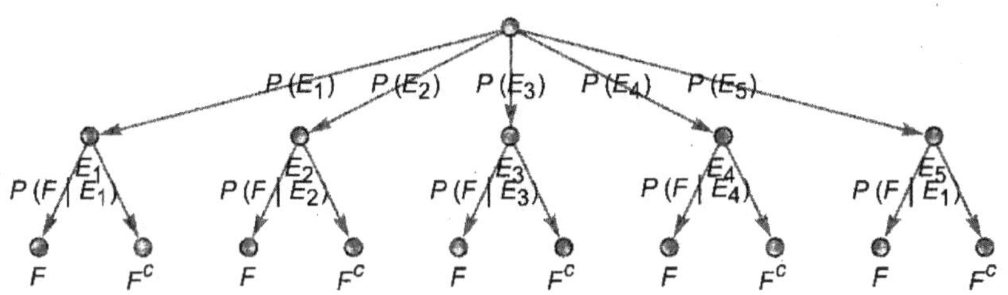

Figure 19.5.2

3. Historically, Bayes' formula of Eq. (19.5.5) had already occurred in a 1749 book by D. Hartley(1705-1757, English), "Observation on Man." This was 24 years earlier than Bayes' paper that was published in 1763 (Reference 1). Hartley confides in his book, "an ingenious friend has communicated to me a solution of the inverse problem …,". It seems that D. Hartley did not know T. Bayes. Who then is Hartley's ingenious friend? Stigler bets that the odds is 3 to 1 in favor of another mathematician, N. Saunderson (1682-1739, English) (Reference 35).
4. In Bayesian statistics, $P(E_i)$ and $P(E_i \mid F)$ are called the prior and the posterior probability, respectively (Reference 34).
5. A Venn diagram to illustrate a partition of F with $n = 5$ is given in Figure 19.5.1. A tree diagram to illustrate Eq. (19.5.4) is given in Figure 19.5.2.

Example 19.5.4
A tire manufacturer has three plants. 50% of its tires are manufactured by Plant I, 30% by Plant II, and 20% by Plant III. It is known that 2% of the tires produced by Plant I is defective, whereas the defective rate at Plant II and Plant III are 4% and 10% respectively. A tire is randomly selected.
(a) What is the probability that the selected tire is found to be defective?
(b) If the selected tire is found to be defective, which of the three plants should be blamed?
Solution
Let E_i, $i = 1, 2, 3$ denote, respectively, the event that the selected tire was manufactured by Plant I, II, and III. Let D denote that the selected tire is defective. From the given information, we have

$$P(E_1) = 0.5,$$
$$P(E_2) = 0.3,$$
$$P(E_3) = 0.2,$$
$$P(D \mid E_1) = 0.02,$$
$$P(D \mid E_2) = 0.04,$$

and

$$P(D \mid E_3) = 0.1.$$

(a) The probability that the selected tire is defective is obtained by using Eq. (19.5.4):

$$P(D) = P(D \mid E_1)P(E_1) + P(D \mid E_2)P(E_2) + P(D \mid E_3)P(E_3)$$
$$= 0.02 \cdot 0.5 + 0.04 \cdot 0.3 + 0.1 \cdot 0.2 = 0.042.$$

(b) By using Eq. (19.5.5), let us compute the probability of causes as follows:

$$P(E_1 \mid D) = \frac{P(D \mid E_1)P(E_1)}{P(D)} = \frac{0.01}{0.042} = 0.2380952 \approx 0.24,$$

$$P(E_2 \mid D) = \frac{P(D \mid E_2)P(E_2)}{P(D} = \frac{0.012}{0.042} = 0.2857143 \approx 0.29,$$

$$P(E_3 \mid D) = \frac{P(D \mid E_3)P(E_3)}{P(D} = \frac{0.02}{0.042} = 0.4761905 \approx 0.48.$$

Since Plant III has the highest probability of cause, $P(E_3 \mid D) = 0.48$, Plant III should probably be blamed for producing this defective tire.

Definition 19.5.2

k events $E_1, E_2, \ldots, E_k$ are said to be independent if for every subset of indices $i_1, i_2, \ldots, i_k$, $E_{i_1}, E_{i_2}, \ldots, E_{i_k}$, $2 \le k \le n$, the following equation holds:

$$P(E_{i_1} \cap E_{i_2} \cap \ldots \cap E_{i_k}) = P(E_{i_1})P(E_{i_2}) \cdot \ldots \cdot P(E_{i_k}). \qquad (19.5.6)$$

Remarks

1. For $n \ge 3$, the use of the (special case) multiplication rule is handier than the use of the conditional probability to define the concept of independence among events.
2. When $n = 3$, in order to show that three events E_1, E_2, and E_3 are independent, the following four equations must all be satisfied:

$$P(E_1 \cap E_2) = P(E_1)P(E_2), \qquad (19.5.7a)$$

$$P(E_1 \cap E_3) = P(E_1)P(E_3), \qquad (19.5.7b)$$

$$P(E_2 \cap E_3) = P(E_2)P(E_3), \qquad (19.5.7c)$$

and

$$P(E_1 \cap E_2 \cap E_3) = P(E_1)P(E_2)P(E_3). \qquad (19.5.7d)$$

Example 19.5.5

Two perfect dice, one red and one blue, are randomly rolled. Let E denote the event that the sum of the dice is 8. Let F denote the event that the red die is 3 and the let G denote the event that the blue die is 5. Are three events E, F, and G independent?

Solution

In order to show that three events E, F, and G are independent, we have to show that all four equations of Eqs. (19.5.7a-d) are satisfied. From the given information, we have

$$E = \{(2, 6), (3, 5), (4, 4), (5, 3), (6, 2)\},$$

$$F = \{(3, 1), (3, 2), (3, 3), (3, 4), (3, 5), (3, 6)\},$$

and

$$G = \{(1, 5), (2, 5), (3, 5), (4, 5), (5, 5), (6, 5)\}.$$

Since the dice are perfect, the equipossibility assumption holds for this random experiment. On the one hand, we have by using Eq. (19.3.1),

$$P(E) = \frac{n(E)}{n(S)} = \frac{5}{36},$$

$$P(F) = \frac{n(F)}{n(S)} = \frac{6}{36} = \frac{1}{6},$$

and

$$P(G) = \frac{n(G)}{n(S)} = \frac{6}{36} = \frac{1}{6}.$$

On the other hand,

$$E \cap F = \{(3, 5)\},$$
$$E \cap G = \{(3, 5)\},$$
$$F \cap G = \{(3, 5)\},$$

and

$$E \cap F \cap G = \{(3, 5)\}.$$

Thus,

$$P(E \cap F) = P(E \cap G) = P(F \cap G) = P(E \cap F \cap G) = \frac{1}{36}.$$

Since

$$P(E)P(F) = \frac{5}{36} \cdot \frac{1}{6} = \frac{5}{216} \neq P(E \cap F) = \frac{1}{36},$$

$$P(E)P(G) = \frac{5}{36} \cdot \frac{1}{6} = \frac{5}{216} \neq P(E \cap G) = \frac{1}{36}$$

$$P(F)P(G) = \frac{1}{6} \cdot \frac{1}{6} = \frac{1}{36} = P(F \cap G) = \frac{1}{36},$$

and

$$P(E)P(F)P(G) = \frac{5}{36} \cdot \frac{1}{6} \cdot \frac{1}{6} = \frac{5}{1296} \neq P(E \cap F \cap G) = \frac{1}{36},$$

three out of four equations of Eq. (19.5.7 a-d) do not hold. Hence, E, F, and G are not independent.

Remark

By following a convention used in Example 19.1.2(b), the ordered pair (x, y) is used to represent the outcome of rolling two dice, where x is the outcome of a red die and y the outcome of a blue die.

Definition 19.5.3

In repeated trials, T_1, T_2, …, T_n, of a random experiment are said to be independent if the knowledge of the outcome on any one trial does not change the probabilities of outcomes on other trials. Moreover, if event E_i is associated with i^{th} trial, and $P(E_i)$ is its probability, $i = 1, 2, \ldots, n$, then

$$P(E_1 \cap E_2 \cap \ldots \cap E_n) = P(E_1)P(E_2) \cdot \ldots \cdot P(E_n).$$

Trials that are not independent are said to be dependent.

Example 19.5.6
Suppose that 60% of all individuals holding VISA cards issued by a bank pay in full each month. Out of those customers who hold VISA card, five customers are randomly selected.
(a) What is the probability that all five customers pay their bills in full each month?
(b) What is the probability that at least one out of five selected customers pay his/her bill in full each month?
Solution
(a) Let E_i, $i = 1, 2, 3, 4, 5$ denote, respectively, the events that customer #1, #2, #3, #4 and #5 pay his/her bills in full each month, From the given information, we have

$$P(E_i) = 0.6\,,\ i = 1, 2, 3, 4, 5.$$

Since these 5 customers are randomly selected, E_1, E_2, E_3, E_4, and E_5 are likely to be independent. By using Eq. (19.5.6), the probability that all five customers pay their bills in full each month is

$$P(E_1 \cap E_2 \cap E_3 \cap E_4 \cap E_5) = P(E_1)P(E_2)P(E_3)P(E_4)P(E_5)$$
$$= 0.6 \cdot 0.6 \cdot 0.6 \cdot 0.6 \cdot 0.6 = 0.6^5 = 0.07776 \approx 0.08\,.$$

(b) The complement $\overline{E}_i$, $i = 1, 2, 3, 4, 5$, are the events that customer #1, #2, #3, #4, and #5 do not pay his/her bill in full each month is given by

$$1 - P(\overline{E}_1 \cap \overline{E}_2 \cap \overline{E}_3 \cap \overline{E}_4 \cap \overline{E}_5)$$
$$= 1 - P(\overline{E}_1)P(\overline{E}_2)P(\overline{E}_3)P(\overline{E}_4)P(\overline{E}_5)$$
$$= 1 - 0.4 \cdot 0.4 \cdot 0.4 \cdot 0.4 \cdot 0.4 = 1 - 0.4^5 = 1 - 0.01024 \approx 0.99.$$

Exercises

1. Given three events E, F, and G with $P(E) > 0$, $P(F) > 0$, and $P(G) > 0$. Show that

$$P(E \cap F \cap G) = P(G \mid E \cap F) \cdot P(F \mid E) \cdot P(E)\,.$$

2. Show that if E and F are independent, so are E and $\overline{F}$.
3. Show that if E and F are independent, so are $\overline{E}$ and $\overline{F}$.
4. Show that if E and F are mutually exclusive, E and F are not independent.
5. Show that if E_1, E_2, …, E_n are independent events, then

$$P(E_1 \cup E_2 \cup \ldots \cup E_n) = 1 - (1 - P(E_1))(1 - P(E_2)) \cdot \ldots \cdot (1 - P(E_n))\,.$$

6. If $P(G) = 0.6$, $P(H) = 0.5$ and $P(G \cap H) = 0.4$, are G and H independent?
7. If two events E and F are dependent and it is known that $P(E) = 0.4$, $P(F) = 0.7$, $P(E \mid F) = 0.8$, find: $P(E \cup F) = ?$
8. If two events F and G are independent and it is known that $P(\bar{F}) = 0.3$ and $P(\bar{G}) = 0.1$, find: $P(E \cup F) = ?$
9. Let E and F be independent events with $P(E) = 0.6$ and $P(F) = 0.7$. Find: (a) $P(E \cap \bar{F}) = ?$, (b) $P(\overline{E \cap F}) = ?$.

10. Suppose that there is a correct diagnostic test that is 98 percent accurate on those who have a certain cancer, but 93 percent accurate on those who do not have the cancer. If 0.3 percent of the populations have this type of cancer, what is the probability that a tested person has the cancer, given that his/her test result is positive?

11. A drug-testing laboratory produces false positive results 6% of the time and false negative 1% of the time. This laboratory was hired by a company in that 8% of the employees use drugs. If an employee was tested positive for drug use, what is the probability that he/she actually used drugs?

12. At a certain stage on a criminal investigation the inspector in charge feels 75% certain that the suspect being held for murder is guilty. Suppose now that the murder be discovered to be left-handed. Twenty percent of the population are left-handed, and the suspect is also left-handed. In light of this new evidence, what probability should the inspector now assign to the guilt of the suspect?

13. Suppose that 4 percent of men and 0.2 percent of women are color blind. A color blind person is chosen at random. What is the probability that the chosen person is a female? Assume that there are equal number of males and females.

14. An automobile insurance company classifies each insured driver as a high risk (E_1), a medium risk (E_2), or a low risk (E_3). Of those currently insured, 25% are high risks, 60 % are medium risks, and 15% are low risks. In any given year the probability that an insured driver has at least one accident is 0.5 for a high risk, 0.3 for a medium risk, and 0.1 for a low risk. If a randomly selected driver insured by this company has an accident during the next year, what is the probability that the driver is actually: (a) a high risk? (b) a medium risk? (c) a low risk?

15. An automobile insurance company has determined the accident rate, i. e., the probability of having at least one accident during a year, for various age groups (See Table 19.5.1). A policyholder calls in to report an accident. What is the probability that he/she is in the age group of 25-50?

Table 19.5.1 The accident rate for various age groups.

Age group	Percent of total insured	Accident rate
Under 25	0.1	0.25
25 – 50	0.55	0.15
51 – 60	0.2	0.1
Over 60	0.15	0.2

16. The distribution of incomes and that of families having at least two-car by income level for a certain suburb of a large city is given in Table 19.5.2. A randomly chosen family has at least two cars. What is the probability that its income is in the range of $50,000 - < $60,000?

Table 19.5.2 The percent of families having two or more cars for various income levels.

Annual family income	Percent of people	Percent having at least two cars
< $30,000	0.1	0.05
$30,000 - < $40,000	0.2	0.25
$40, 000 - < $50,000	0.25	0.5
$50,000 - < $60,000	0.35	0.75
≥ $60,000	0.1	0.9

17. The distribution of voters' racial ethnicity and their turn out rates for a certain city is given in Table 19.5.3. A randomly chosen person is questioned at the polls. What is the probability that the person is a Hispanic?

Table 19.5.3 The voter turnout rates for various ethnic groups.

Ethnic group	Percent of racial groups	Percent turnout
Caucasian	0.49	0.85
Black	0.18	0.8
Hispanic	0.27	0.8
Asian	0.05	0.75
Other	0.01	0.65

18. There are three cards identical in form except that both sides of the first card are colored yellow, both sides of the second card are colored green, and one side of the third card is colored yellow and the other side green. The three cards are mixed up in a hat, and one card is randomly selected and put down on a pool table. If the upper side of the card on the pool table is colored green, what is the probability that the other side is colored yellow?

19. Two perfect dice (one red and one blue) are rolled randomly. Let E denote the event that the sum of the dice is seven and let F denote the event that the red die is four. Are E and F independent?

20. A family has three children. Let E be the event "At most one boy" and F the event "at least one child of each sex". Are E and F independent?

21. The probability that a small town is hit by a tornado is 0.0005 and the probability that this small town will produce a Miss USA is 0.003. What is the probability that this small town (a) is hit by a tornado and produces a Miss USA? (b) hit by a tornado or produces a Miss USA?

22. The probability that a mathematics professor becomes sick on the first day of the class is 0.03 and the probability that one third of his class fail to pass the first quiz is 0.8. What is the probability that (a) this professor becomes sick on the first day of the class and one third of his class fail to pass the first quiz? (b) either this professor becomes sick on the first day of the class or one third of his class fail to pass the first quiz?

23. It is known that 20% of a certain company's washing machines require service under warranty, whereas only 8% of its dryer need such service. If someone purchases both a washer and a dryer made by this company, what is the probability that both machines need such warranty service?

24. The life support system in the space shuttle has a built-in redundant auxiliary system. An engineering design was devised so that the failure of the auxiliary system is independent of that of the main system. The probability of failure of the main system is 0.003, whereas the probability of failure of the auxiliary system is 0.02. What is the probability that at least one of the main and the auxiliary systems do not fail during a particular mission?

25. Each of three person, A, B, and C fires one shot at a target. The probability that A, B, and C hit the target are given, respectively, by 0.9, 0.7, and 0.8. What is the probability that exactly two people hit the target? Assume that their performances in shooting at a target are independent of one another.

19.6 References

1. Bayes, T. (1763). An essay towards solving a problem in the doctrine of chances, Biometrika, 45 (1958), pp. 296-315.
2. Baldi, P. and Brunak, S. (1998). Bioinformatics: The Machine Learning Approach, The MIT Press, Cambridge, Massachusetts.
3. Berkson, J. (1963). Smoking and lung cancer, The American Statistician, 17, pp. 15-22.
4. Bickel, P. J., Hammel, E. A. and O'Connell, H.W. (1975). Sex bias in graduate admissions : data from Berkeley, Science, 187, pp. 398-404.
5. Daston, L. (1988). Classical Probability in the Enlightenment, Princeton University Press, Princeton, New Jersey.

6. David, F.N. (1962). Games, God and Gambling: A History of Probability and Statistical Ideas, Dover Publications, Inc., 1998, New York.
7. Epstein, R. A. (1967). The Theory of Gambling and Statistical Logic, Academic Press, New York.
8. Feller, W. (1968). Introduction to Probability Theory and Its Applications, Third edition, Vol. 1, John Wiley & Sons, Inc., New York.
9. de Finetti, B. (1974). Theory of Probability, Vol. 1, Wiley Interscience Publishers, New York.
10. Fleiss, J. L., Levin, B., and Paik, M. C. (2003). Statistical Methods for Rates and Proportions, 3rd edition, Wiley, New York.
11. Gigerenzer, G., et al (1989). The Empire of Chance, Cambridge University Press, Cambridge, England.
12. Ginsburg, J. (1922). Rabbi Ben Ezra on permutations and combinations, The Mathematics Teacher, 15, pp. 347-356.
13. Good, P. (1994). Permutation Tests: A Practical Guide to Resampling Testing Hypotheses, Springer-Verlag, New York.
14. Grattan-Guinness, I. (1997). The Rainbow of Mathematics: A History of the Mathematical Sciences, W.W. Norton & Company, New York.
15. Greenwood, M., and Yule, G. U. (1915). The statistics of antityphoid and anti-cholera inoculations, and the interpretation of such statistics in general, Proceedings of the Royal Society of Medicine, 8, pp. 113-190.
16. Hacking, I. (1975). The Emergence of Probability: A Philosophical Study of Early Ideas about Probability, Induction, and Statistical Inference, Cambridge University Press, Cambridge, England.
17. Hacking, I. (1971). Equipossibility Theories of Probability, British Journal for the Philosophy of Sciences, 22, pp. 339-355.
18. Haigh, J. (1999). Taking Chances: Winning with Probability, Oxford University Press, Oxford, England.
19. Hasofer, A. M. (1967). Random mechanisms in talmudic literature, Biometrika, 54, pp. 316-321.
20. Ho, P.Y. (1985). Li, Qi and Shu: An Introduction to Science and Civilization in China, Dover Publications, Inc., New York.

21. Kendall, M. G. and Stuart, A. (1961). The Advanced Theory of Statistics, Vol. 2: Inference and Relationship, Charles Griffin & Company Limited, London.
22. Kerrich, J. E. (1950). An Experimental Introduction to the Theory of Probability, Belgisk Import Compagni, Copenhagen, Denmark.
23. King, A. C. and Read, C. B. (1963). Pathways to Probability, Holt, Rinehart, and Winston, Inc., New York.
24. Laplace, P. S. M. (1814). A Philosophical Essay on Probabilities, Dover Publications, Inc., New York, 1951.
25. Maistrov, L.E. (1974). Probability Theory: A Historical Sketch, Academic Press, New York.
26. von Mises, R. (1957). Probability, Statistics, and Truth, Dover Publications, Inc., New York.
27. de Moivre, A. (1738). The Doctrine of Chances, Frank Cass & Co. Ltd, 1967, London, England.
28. Ore, O. (1953). Cardano, the Gambling Scholar, Princeton University Press, Princeton, New Jersey.
29. Rabinovitch, N. L. (1973). Probability and Statistical Inference in Ancient and Medieval Jewish Literature, University of Toronto Press, Toronto, Canada.
30. Raghavarao, D. (1971). Constructions and Combinatorial Problems in Design of Experiments, Dover Publications, Inc., New York.
31. Ruhla, C. (1989). The Physics of Chance: From Blaise Pascal to Niels Bohr, Oxford University Press, Oxford, England.
32. Scheaffer, R.L. (1990). Introduction to Probability and Its Application, PWS-KENT Publishing Company, Boston, Massachusetts.
33. Scheaffer, R.L., Mendenhall (III), W., Ott, L. R. (1996). Elementary Survey Sampling, Duxbury Press, Belmont, California.
34. Sivia, D. S. (1996). Data Analysis: A Bayesian Tutorial, Clarendon Press, Oxford, England.
35. Stigler, S. M. (1983). Who discovered Bayes's Theorem?, The American Statistician, 37, pp. 290-296.
36. Stigler, S. M. (1986). The History of Statistics: The Measurement of Uncertainty before 1900, Harvard University Press, Cambridge, Massachusetts.

37. Todhunter, I. (1949). A History of the Mathematical Theory of Probability: From the time of Pascal to that of Laplace, Chelsea Publishing Company, New York.
38. Weaver, W. (1963). Lady Luck: The Theory of Probability, Dover Publications, Inc., New York.
39. Yates, F. (1934). Contingency tables involving small numbers and the χ^2 test, Supplement of Journal of Royal Statistical Society, 1, p. 217.

Selected Answers

Chapter 1

§1.2 **1.** 11. **2.** 2. **3.** 14. **4.** $\sqrt{5}-2$. **5.** 15. **6.** –7.
7. $\frac{1}{6}$. **8.** $\frac{1}{12}$. **9.** $\frac{28}{15}$. **10.** $\frac{14}{5}$.

§1.3 **1.** 81. **2.** 4. **3.** $\sqrt[12]{128}$. **4.** 5/16. **5.** $-15\sqrt{2}$.
6. $13\sqrt{3}$. **7.** $x^{\frac{3}{2}}=\sqrt{x^3}$. **8.** $1/(x^{10}y^{15})$. **9.** $12x^4y^4$.
10. $\frac{x^{\frac{1}{3}}}{y^{\frac{1}{3}}}=\sqrt[3]{\frac{x}{y}}$.

§1.4 **1.** (a) –10, (b) $243i$, (c) –12, (d) $-\sqrt{ab}$.
2. (a) $-5-7i$, (b) $4i$, (c) $3+8i$, (d) 9.
3. (a) $\sqrt{34}$, (b) 4, (c) $\sqrt{65}$, (d) 6.
4. (a) $5+10i$, (b) $-3-3i$, (c) $-10+2i$,
(d) -41, (e) $15-9i$, (f) $\frac{14}{25}+\frac{23}{25}i$, (g) $\frac{5}{4}-\frac{1}{2}i$.

Chapter 2

§2.1 **1.** 10. **2.** 16. **3.** 4. **4.** 7. **5.** 12.

§2.2 **1.** x^3+4x^2-4x+3. **2.** $-2x^2+15x-42$.
3. x^2y+4xy^2+4. **4.** $3x^3y-6x^2y^2+12xy^3$.
5. $8x^2-14x-15$. **6.** $x^2+2xy+y^2$.
7. $x^2-2xy+y^2$. **8.** $x^3+4x^2+16x+64$.
9. $x^2-10x+25$. **10.** x^3+y^3.
11. x^3-y^3. **12.** $2x-3+\frac{4}{x}-\frac{5}{x^2}$.
13. $2x+1+\frac{7}{x+3}$. **14.** $x^2+5x-10+\frac{20}{2x-1}$.
15. $4x-3-\frac{2x}{x^2+x+1}$.

Chapter 3

§3.1 **1.** $3\cdot(x+2)$. **2.** $3a\cdot(x+2)$. **3.** $3x\cdot(x+5)$.

4. $2x(2x^2+6x+9)$. **5.** $2y(2y^4-5y^2+3)$.
6. $(x+3)\cdot(x+4)$. **7.** $3\cdot(x-1)\cdot(4-3x)$.
8. $x\cdot(x-2)\cdot(x-1)$. **9.** $2xy\cdot(x+4y)$.
10. $2xy^2(4y^3-5x^3y+3x^5)$.

§3.2 **1.** $(x+4)(x-4)$. **2.** $(5+2x)(5-2x)$.
3. $3(x+5y)(x-5y)$. **4.** $(2x+y+3)(2x-y-3)$.
5. $(2y+1)^2$. **6.** $(2x-3y)^2$. **7.** $x(x-6)^2$.
8. $3(2x-1)(4x^2+2x+1)$. **9.** $2(3-y)(9+3y+y^2)$.
10. $(2z+3)(4z^2-6z+9)$.
11. $2(4w+5)(16w^2-20w+25)$. **12.** $(x^2+4)(x-2)(x+2)$.
13. $3(x^2+y^2)^2$. **14.** $3(x+y)^2(x-y)^2$.
15. $6xy^2(x-1)^2$.

§3.3 **1.** $(3x+1)(x+1)$. **2.** $(3x+2)(x+3)$.
3. $(2x+1)(3x+4)$. **4.** $(2x-1)(4x+5)$.
5. $-(3x-2)(2x+5)$. **6.** $-(4x-3)(2x+7)$.
7. $x(3x-2)(4x+3)$. **8.** $x(3x+4)(5x-1)$.
9. $(x-1+\sqrt{5})(x-1-\sqrt{5})$. **10.** $(x+2+\sqrt{3})(x-2-\sqrt{3})$.
11. $2(x-1+\sqrt{\tfrac{3}{2}})(x-1-\sqrt{\tfrac{3}{2}})$. **12.** $3(x+1+\sqrt{\tfrac{5}{3}})(x+1-\sqrt{\tfrac{5}{3}})$.
13. $(x-2y)(2x+y)$. **14.** $(3x+2y)(2x-3y)$.
15. $4(x+\sqrt{\tfrac{1}{2}})(x-\sqrt{\tfrac{1}{2}})(x+\sqrt{\tfrac{3}{2}})(x-\sqrt{\tfrac{3}{2}})$.

Chapter 4

§4.1 **1.** $\dfrac{3a^2b^2}{4c^3}$. **2.** $\dfrac{3x^2}{7y}$. **3.** $x(2xy+3)$. **4.** $\dfrac{4}{9(x-y)}$.

5. $\dfrac{3x^2}{y^2(5x+4y)}$. **6.** $\dfrac{x+3}{x-2}$. **7.** $\dfrac{x+2}{x^2+2x+4}$.

8. $\dfrac{x^2+ax+a^2}{a-x}$. **9.** $\dfrac{x(x-3)}{(x-2)(x^2+4)}$.

10. $\dfrac{x}{(x^2+2x+4)(x^2-2x+4)}$.

§4.2 **1.** (a) LCD = $45x^2$, (b) $\frac{8x+5}{15x}=\frac{24x^2+15x}{45x^2}$; $\frac{2x+3}{9x^2}=\frac{10x+15}{45x^2}$.

2. (a) LCD = $(x-2)(x+3)$,

(b) $\frac{3x}{x-2}=\frac{3x^2+9x}{(x-2)(x+3)}$; $\frac{5}{x+3}=\frac{5x-10}{(x-2)(x+3)}$.

3. (a) LCD = $(x-3)(x-2)(x+2)$,

(b) $\frac{3x+5}{x^2-4}=\frac{3x^2-4x-15}{(x-3)(x-3)(x+2)}$, $\frac{2-5x}{x^2-x-6}=\frac{-5x^2+12x-4}{(x-3)(x-2)(x+2)}$.

4. (a) LCD = $x^2(x-a)$,

(b) $\frac{-1}{x^2-ax}=\frac{-x}{x^2(x-a)}$; $\frac{x+a}{x^2(x-a)}=\frac{x+a}{x^2(x-a)}$.

5. (a) LCD = $30x^2y^2$,

(b) $\frac{x-2}{3x^2}=\frac{10xy^2-20y^2}{30x^2y^2}$; $\frac{4}{5y}=\frac{24x^2y}{30x^2y^2}$; $\frac{y-x}{2xy^2}=\frac{15xy-15x^2}{30x^2y^2}$.

§4.3 **1.** $x-y$. **2.** $\frac{x^2-2x+15}{x^2-9}$. **3.** $\frac{2x^2+10x+12}{x^2-1}$.

4. $\frac{5x-2}{x^3+x^2-4x-4}$. **5.** $\frac{3x^2+21x+6}{x^3+x^2-4x-4}$. **6.** $\frac{x}{2x-3}$.

7. x^2-1. **8.** $\frac{-(x+5)}{x}$, $x\neq 0$.

9. $\frac{x^2+2x}{2x^2+x-1}$, $x\neq -1,\frac{1}{2}$.

10. $\frac{2x^2y(x-y)}{3(x+y)}$, $x\neq -y$.

Chapter 5

§5.2 **1.** $x=5$. **2.** $y=\frac{7}{2}=3.5$. **3.** $z=6$.

4. $x=\frac{32}{5}=6.4$. **5.** $x=-29$. **6.** $x=\frac{5}{4}=1.25$.

7. $x=-8$. **8.** $y=\frac{7}{3}=2\frac{1}{3}$. **9.** $z=1.76$. **10.** $x=6,-1$.

11. $x = 4$. **12.** $x = \frac{23}{2} = 11.5$. **13.** No Solution.

14. $x = 7$. **15.** $x = 4$.

§5.3 **1.** {17}. **2.** {25}. **3.** {12}. **4.** {9, 27}.

5. {60}. **6.** {–3, 11}. **7.** {14}. **8.** {69}.

9. Mary's age = 24, Beth's age = 12.

10. 20 years old.

11. #{nickels} = 50, #{dimes} = 40.

12. Wendy = 10 hours; Jill = 16 hours.

13. $ invested respectively at 8% and 10% = $5000, $15,000.

14. {$120,900}. **15.** {$140,000}.

16. {$200}. **17.** {length = 48 inches, width = 12 inches}.

18. {length = 200 ft, width = 100 ft}.

19. {10 gallons}. **20.** {45 students}.

Chapter 6

§6.1 **1.** {–2, –3}. **2.** $\{-\frac{2}{5}, 3\}$. **3.** $\{-\frac{1}{2}, -\frac{4}{3}\}$. **4.** $\{-\frac{5}{4}, \frac{1}{2}\}$.

5. {3, 3}. **6.** {±3}. **7.** $\{\pm\frac{4\sqrt{5}}{5}\}$. **8.** {–1, 5}.

9.{–9,11}. **10.** $\{\frac{5 \pm 2\sqrt{7}}{3}\}$.

§6.2 **1.** {–2, –3}. **2.** $\{-\frac{1}{2}, 2\}$. **3.** $\{-\frac{2}{5}, 3\}$. **4.** $\{-\frac{1}{2}, -\frac{4}{3}\}$.

5. $\{-\frac{5}{4}, \frac{1}{2}\}$. **6.** $\{-\frac{1}{2} \pm \frac{\sqrt{5}}{2}\}$. **7.** $\{-2 \pm \sqrt{2}\}$. **8.** $\{\frac{5}{2} \pm \frac{\sqrt{17}}{2}\}$.

9.$\{\frac{3}{4} \pm \frac{\sqrt{17}}{4}\}$. **10.** $\{-\frac{5}{6} \pm \frac{\sqrt{37}}{6}\}$.

§6.3 **1.** Omitted. **2.** $\{\frac{1}{3}, 2\}$. **3.** $\{\frac{2 \pm \sqrt{2}}{2}\}$. **4.** $\{\frac{9 \pm \sqrt{101}}{2}\}$.

5. $\{\frac{-1 \pm \sqrt{51}}{5}\}$.

§6.4 **1.** $\{-\frac{1}{4} \pm i\frac{\sqrt{23}}{4}\}$. **2.** $\{\frac{1}{6} \pm i\frac{\sqrt{11}}{6}\}$. **3.** $\{\frac{3}{8} \pm i\frac{\sqrt{71}}{8}\}$.

4. $\{\frac{1}{5} \pm i\frac{2\sqrt{10}}{5}\}$. **5.** $\{\frac{9}{10} \pm i\frac{\sqrt{339}}{10}\}$.

§6.5 **1.** $\{0, \frac{5}{4} \pm \frac{\sqrt{33}}{4}\}$.**2.** $\{\pm 2, \pm 2i\}$. **3.** $\{0, 1, \frac{1 \pm \sqrt{2}}{2}\}$.

4. $\{\frac{-3 \pm \sqrt{17}}{4}\}$. **5.** $\{3 \pm \frac{\sqrt{6}}{3}\}$. **6.** $\{-3 \pm \sqrt{10}\}$.

7. $\{1, \frac{5}{2}\}$. **8.** $\{\frac{7 \pm 4\sqrt{2}}{2}\}$. **9.** $\{\frac{66 \pm 24\sqrt{3}}{9}\}$.

10. $\{1 \pm \sqrt{2}, 1 \pm i\sqrt{3}\}$.

§6.6 **1.** $\{7, 9\}$. **2.** $\{9\}$. **3.** $\{3\}$. **4.** $\{\frac{5+\sqrt{29}}{2}\}$.
5. {length = 8 ft, width = 5 ft}.
6. {length = 17 ft, height = 8 ft}.
7. {father's age = 53, son's age = 16}.
8. {his speed in still water = $2\sqrt{5}$ miles per hour}.
9. {440 miles per hour}. **10.** {16 hours}.

Chapter 7

§7.1 **1.** (a) $[-1, 3]$, (b) $(5, 10)$, (c) $(-\infty,-17]$, (d) $(-\frac{4}{3},+\infty)$. (e) $(-\infty,9]$, (f) $[1, 8)$, (g) $(3, 12]$, (h) $[-6,+\infty)$.
2. (a) a half-open interval, (b) a closed interval, (c) a closed left-half-line, (d) an open interval, (e) an open left-half-line, (f) an open interval, (g) an open right-half-line, (h) a half-closed interval, (i) a closed right-half-line, (j) an open left-half-line.

§7.3 **1.** $(-\infty,\frac{9}{5}]$. **2.** $(\frac{3}{2},+\infty)$. **3.** $(-\infty,\frac{2}{3})$. **4.** $[5,+\infty)$.
5. $(-\infty,-1)$. **6.** $(-\frac{1}{12},\frac{3}{4})$. **7.** $(\frac{1}{2},\frac{7}{2}]$. **8.** $[-1, 1]$.
9. $(-\frac{15}{2},-\frac{9}{2}]$. **10.** $(-\frac{9}{8},\frac{3}{8}]$.

§7.4 **1.** $(-1, 5)$. **2.** $[-\frac{3}{2},\frac{5}{2}]$. **3.** $(-16, 4)$. **4.** $[-\frac{7}{2},\frac{5}{2}]$.
5. $(-\frac{1}{36},\frac{1}{36})$. **6.** $(-\infty,-3)\cup(1,+\infty)$. **7.** $(-\infty,-\frac{7}{2}]\cup[\frac{1}{2},+\infty)$.
8. $(-\infty,\frac{2}{5}]\cup[\frac{14}{15},+\infty)$. **9.** $(-\infty,-\frac{9}{4})\cup(\frac{9}{4},+\infty)$.
10. $(-\infty,-11)\cup(4,+\infty)$.

Chapter 8

§8.1 **1.** $(-\infty,-3]\cup[-2,+\infty)$. **2.** $[-2\sqrt{2},2\sqrt{2}]$.
3. $\{3\}$. **4.** $(-\infty,-3)\cup(3,+\infty)$.
5. $(-\infty,-1]\cup[4,+\infty)$. **6.** $(-\infty,\frac{5-\sqrt{17}}{4})\cup(\frac{5+\sqrt{17}}{4},+\infty)$.
7. $(-2-\sqrt{6},-2+\sqrt{6})$. **8.** No Solution.
9. $(-\infty,-8)\cup(-1,+\infty)$. **10.** No Solution.

§8.2 **1.** $(-\infty,-\sqrt{5})\cup(\sqrt{5},+\infty)$. **2.** $[1-\sqrt{5},1+\sqrt{5}]$.

3. No Solution. **4.** $(-\infty,-\frac{27}{8}]\cup[-\frac{1}{8},+\infty)$.

5. $[-1,\frac{2}{3}]$. **6.** $(-\infty,-1-\frac{\sqrt{6}}{2})\cup(-1+\frac{\sqrt{6}}{2},+\infty)$.

7. $(-\infty,\frac{1-\sqrt{41}}{2}]\cup[\frac{1+\sqrt{41}}{2},+\infty)$. **8.** $(\frac{3-\sqrt{69}}{10},-\frac{2}{5})\cup(1,\frac{3+\sqrt{69}}{10})$.

9. $(-\infty,+\infty)$. **10.** No Solution.

Chapter 9

§9.1 **1.** (a) omitted, (b) $4\sqrt{2}$, (c) $R(2, 4)$.

2. (a) omitted, (b) $\sqrt{29}$, (c) $R(-0.5, 6)$.

3. (a) omitted, (b) $8\sqrt{2}$, (c) $R(3, 6)$.

4. (a) omitted, (b) $\frac{\sqrt{109}}{60}$, (c) $R\left(\frac{5}{12},\frac{9}{40}\right)$.

5. (a) omitted, (b) $4\sqrt{5}$, (c) $R(0.3, 1.4)$.

§9.2 **1.** $y=-3x-1$. **2.** $y=-2x+5$. **3.** $x=3$.

4. $y=-1$. **5.** $y=-\frac{1}{4}x+\frac{13}{4}$. **6.** $x=-1$

7. $y=-\frac{4}{5}x$. **8.** $x=-3$. **9.** $y=\frac{1}{2}x-1$. **10.** $x=-2$.

§9.3 **1.** $\{(x, x) \mid x$: any real number$\}$ or $\{(y, y) \mid y$: any real number$\}$.

2. $\{(x, x-2) \mid x$: any real number$\}$ or $\{(y+2, y) \mid y$: any real number$\}$.

3. $\{(x, -x+3) \mid x$: any real number$\}$ or $\{(-y+3, y) \mid y$: any real number$\}$.

4. $\{(x, \frac{3}{4}x-3) \mid x$: any real number$\}$ or $\{(\frac{4}{3}y+4, y) \mid y$: any real number$\}$.

5. $\{(x, 2x+4) \mid x$: any real number$\}$ or $\{(\frac{1}{2}y-2, y) \mid y$: any real number$\}$.

6. $\{(x, \frac{1}{2}x) \mid x$: any real number$\}$ or $\{(2y, y) \mid y$: any real number$\}$.

7. $\{(\frac{5}{2}, y) \mid y$: any real number$\}$.

8. $\{(x, 2) \mid x$: any real number$\}$.

9. $\{(0, y) \mid y$: any real number$\} = y$-axis.

10. $\{(x, 0) \mid x\text{: any real number}\} = x$-axis.

§9.4 **1.** (a) $\{(2, 1)\}$, (b) consistent. **2.** (a) $\{ (\frac{8}{13}, -\frac{1}{13}) \}$, (b) consistent.

3. (a) $\{(1, -1)\}$, (b) consistent.

4. (a) $\{ (\frac{7}{5}, \frac{6}{5}) \}$, (b) consistent.

5. (a) $\{(x, 2 - x) \mid x\text{: any real number}\}$, (b) consistent.

6. (a) No solution, (b) inconsistent.

7. (a) $\{ (-\frac{5}{7}, \frac{11}{7}) \}$, (b) consistent.

8. (a) $\{ (\frac{5}{3}, \frac{22}{3}) \}$, (b) consistent. **9.** (a) $\{ (\frac{5}{12}, -\frac{3}{4}) \}$, (b) consistent.

10. (a) $\{ (\frac{6}{17}, \frac{60}{17}) \}$, (b) consistent.

§9.5 **1.** $\{43, 59\}$. **2.** $\{29, 37\}$. **3.** $\{31\}$.

4. Amanda's age = 32, Bob's age = 38.

5. {\$1.55-coffee = 70 pounds, \$1.85-coffee = 30 pounds}.

6. {8 quarters and 12 dimes}.

7. {\$10,000 at 7% and \$6,000 at 8%}.

8. {Bill needs 72 hours, while John needs 36 hours}.

9. {wind's speed = 50 mph and plane's speed in still air = 450 mph}.

10. {length = 21 ft, width = 17 ft}.

Chapter 10

§10.1 Omitted.

§10.2 **1.** The feasible set is not empty and has three corner points $O(0, 0)$, $A(0, 2)$, and $B(2, 0)$.

2. The feasible set is empty.

3. The feasible set is not empty and has four corner points $O(0, 0)$, $A(0, 2)$, $B(\frac{6}{5}, \frac{6}{5})$, and $C(2, 0)$.

4. The feasible set is not empty and has two corner points $A(0, \frac{3}{2})$, and $B(3, 0)$.

5. The feasible set is not empty and has four corner points $A(0, 2)$, $B(0, 1)$, $C(1, 0)$, and $D(3, 0)$.

§10.3 **1.** The couple should invest \$10,000 in corporate bonds and \$20,000 in Treasury bills with the maximum yearly return of \$2,010.
2. The car dealer should ship 3 cars from Chicago to Springfield and 2 cars from Indianapolis to Springfield, whereas he ships no cars from Chicago to Louisville and 8 cars from Indianapolis to Louisville with the least shipping cost of \$1,650.
3. She should work 12 hours in the math help center and 8 hours in the library with the maximum weekly income of \$144.
4. The furniture maker should produce 6 chairs and 1 desk with the maximum profit of \$250.
5. $\frac{12}{7}$ pounds of type I food and $\frac{20}{7}$ pounds of type II food should be purchased with the least cost of \$ $33\frac{1}{7}$.

Chapter 11

§11.1 **1.** {13:4}. **2.** $\{b' = 24, c' = 30\}$.
3. $\{4(\sqrt{5}+1)$ inches}. **4.** {18,000 students}.
5. {\$0.99}. **6.** {500 miles}.
7. {20 pairs of twins}. **8.** {34 ft}.
9. (\$150,000}. **10.** (\$73.80}.

§11.2 **1.** (i) $y = 0.8018$ when $x = 2.23$, (ii) $y = 0.9011$ when $x = 2.46$.
2. (i) $p = 0.0394$ when $z = -1.76$, (ii) $p = 0.0258$ when $z = -1.94$.

§11.3 **1.** {27}. **2.** {192}. **3.** $\{\frac{27}{5}\}$. **4.** $\{\frac{27}{32}\}$.
5. $\{10\pi$ cm}. **6.** {4 inches}. **7.** {1,600 ft}.
8. {0.02 ohms}. **9.** {\$9,000}. **10.** {800 units}.

Chapter 12

§12.1 **1.** $\{(x, \frac{1}{2}x^2 - \frac{1}{4}x - \frac{5}{4}) \mid x:$ any real number}.

2. $\{(\frac{3}{5}y^2 - \frac{2}{5}y - \frac{1}{5}, y) \mid y:$ any real number}.

3. $\{\ (x, \frac{3}{2} \pm \sqrt{(x + \frac{\sqrt{15}}{2} - 1)(\frac{\sqrt{15}}{2} + 1 - x)} \mid 1 - \frac{\sqrt{15}}{2} \le x \le 1 + \frac{\sqrt{15}}{2}\ \}$,

or

$\{(1 \pm \sqrt{(y + \frac{\sqrt{15}-3}{2})(\frac{\sqrt{15}+3}{2} - y)}, y) \mid \frac{3-\sqrt{15}}{2} \le y \le \frac{3+\sqrt{15}}{2}\}$.

4. $\{(x, 1\pm\sqrt{(\frac{\sqrt{2}}{2}x+\frac{\sqrt{15}}{2}-\sqrt{2})(\frac{\sqrt{15}}{2}+\sqrt{2}-\frac{\sqrt{2}}{2}x)})\mid 2-\frac{\sqrt{30}}{2}\le x\le 2+\frac{\sqrt{30}}{2}\}$, or

$\{(2\pm\sqrt{(\sqrt{2}y+\frac{\sqrt{30}}{2}-\sqrt{2})(\frac{\sqrt{30}}{2}+\sqrt{2}-\sqrt{2}y)}, y)\mid 1-\frac{\sqrt{15}}{2}\le y\le 1+\frac{\sqrt{15}}{2}\}$.

5. $\{(x, -2\pm\sqrt{(\sqrt{2}x-\frac{3\sqrt{2}-\sqrt{6}}{2})(\sqrt{2}x-\frac{3\sqrt{2}+\sqrt{6}}{2})})\mid x\le\frac{3-\sqrt{3}}{2}$ or $x\ge\frac{3+\sqrt{3}}{2}\}$, or $\{(\frac{3}{2}\pm\sqrt{\frac{1}{2}(y+2)^2+\frac{3}{4}}, y)\mid y:$ any real number$\}$.

§12.2 **1.** (a) vertex: $Q(0,-\frac{5}{4})$, focus: $F(0,-\frac{3}{4})$, directrix: $y=-\frac{7}{4}$, line of symmetry: $x = 0$ (or y-axis), (b) omitted.

2. (a) vertex: $Q(3,0)$, focus: $F(\frac{19}{6},0)$, directrix: $x=\frac{17}{6}$, line of symmetry: $y = 0$ (or x-axis), (b) omitted.

3. (a) vertex: $Q(1,2)$, focus: $F(1,\frac{49}{24})$, directrix: $y=\frac{47}{24}$, line of symmetry: $x = 1$, (b) omitted.

4. (a) vertex: $Q(-\frac{11}{6},\frac{3}{2})$, focus: $F(-\frac{35}{24},\frac{3}{2})$, directrix: $x=-\frac{53}{24}$, line of symmetry: $y=\frac{3}{2}$, (b) omitted.

5. (a) vertex: $Q(-1,-7)$, focus: $F(-1,-\frac{111}{16})$, directrix: $y=-\frac{113}{16}$, line of symmetry: $x = -1$, (b) omitted.

§12.3 **1.** A circle with center at $Q(-2, -3)$ and a radius 1.

2. A circle with center at $Q(2, 3)$ and a radius 4.

3. An ellipse with center at $Q(0, 0)$, the y-axis is the major axis, the semi-major axis $a = 5$, the semi-minor axis $b = 4$, and foci: $F_1(0, -3)$ and $F_2(0, 3)$.

4. An ellipse with center at $Q(-1, 2)$, the major axis is parallel to the x-axis, the semi-major axis $a = 4$, the semi-minor axis $b = 3$, and foci: $F_1(-1-\sqrt{7}, 2)$ and $F_2(-1+\sqrt{7}, 2)$.

5. An ellipse with center at $Q(\frac{1}{2},-\frac{3}{2})$, the major axis is parallel to the y-axis, the semi-major axis $a = 3$, the semi-minor axis $b = 2$, and foci: $F_1(\frac{1}{2},-\frac{3}{2}-\sqrt{5})$ and $F_2(\frac{1}{2},-\frac{3}{2}+\sqrt{5})$.

§12.4 **1.** (a) center: $Q(0, 0)$, vertices: $V_1(0, -3)$, $V_2(0, 3)$, focal axis: $x = 0$ (or y-axis), foci: $F_1(0, -5)$, $F_2(0, 5)$, asymptotes: $y = \frac{3}{4}x$ and $y = -\frac{3}{4}x$, (b) omitted.

2. (a) center: $Q(0, 0)$, vertices: $V_1(-\sqrt{3}, 0)$, $V_2(\sqrt{3}, 0)$, focal axis: $y = 0$ (or x-axis), foci: $F_1(-\sqrt{5}, 0)$, $F_2(\sqrt{5}, 0)$, asymptotes: $y = \frac{\sqrt{6}}{3}x$ and $y = -\frac{\sqrt{6}}{3}x$, (b) omitted.

3. (a) center: $Q(-2, 1)$, vertices: $V_1(-2, -2)$, $V_2(-2, 4)$, focal axis: $x = -2$, foci: $F_1(-2, 1-\sqrt{13})$, $F_2(-2, 1+\sqrt{13})$, asymptotes: $y = \frac{3}{2}x + 4$ and $y = -\frac{3}{2}x - 2$, (b) omitted.

4. (a) center: $Q(1, -1)$, vertices: $V_1(1-\sqrt{5}, -1)$, $V_2(1+\sqrt{5}, -1)$, focal axis: $y = -1$, foci: $F_1(-2, -1)$, $F_2(4, -1)$, asymptotes: $y = \frac{2\sqrt{5}}{5}x - 1 - \frac{2\sqrt{5}}{5}$ and $y = -\frac{2\sqrt{5}}{5}x + \frac{2\sqrt{5}}{5} - 1$, (b) omitted.

5. (a) center: $Q(-1, 1)$, vertices: $V_1(-1, 1-\sqrt{6})$, $V_2(-1, 1+\sqrt{6})$, focal axis: $x = -1$, foci: $F_1(-1, 1-\sqrt{11})$, $F_2(-1, 1+\sqrt{11})$, asymptotes: $y = \frac{\sqrt{30}}{6}x + 1 + \frac{\sqrt{30}}{6}$ and $y = -\frac{\sqrt{30}}{6}x + 1 - \frac{\sqrt{30}}{6}$, (b) omitted.

Chapter 13

§13.1 **1.** Yes, this equation defines y as a function of x: $y = f(x) = \frac{2}{3}(x+2)$.

2. Yes, this equation defines y as a function of x: $y = f(x) = \frac{1}{4}(-x^2 + 3x + 5)$.

3. No, this equation does not define y as a function of x. Instead, it defines y as two distinct functions of x: $y = f_1(x) = 2 + \sqrt{5-2x}$ and $y = f_2(x) = 2 - \sqrt{5-2x}$. However, note that this equation defines x as a function of y: $x = g(y) = \frac{1}{2}(1 + 4y - y^2)$.

4. No. Instead, it defines y as two distinct functions of x: $y = f_1(x) = -3 + \sqrt{-(x+1)(x+3)}$ and $y = f_2(x) = -3 - \sqrt{-(x+1)(x+3)}$.

5. No. Instead, it defines y as two distinct functions of x: $y = f_1(x) = 1 + \sqrt{\frac{2}{3}(-x^2 + 2x - 2)}$ and $y = f_2(x) = 1 - \sqrt{\frac{2}{3}(-x^2 + 2x - 2)}$.

6. No. Instead, it defines y as two distinct functions of x: $y = f_1(x) = -3 + \sqrt{(x-1)(x-3)}$ and $y = f_2(x) = -3 - \sqrt{(x-1)(x-3)}$.

7. Yes, this equation defines y as a function of x: $y = f(x) = \frac{5+x}{4+x^2}$.

8. (a) $f(2) = 1$, (b) $f(2 + \Delta x) = 3\Delta x + 1$, (c) $\dfrac{f(2+\Delta x) - f(2)}{\Delta x} = 3$.

9. (a) $g(3) = 2$, (b) $g(3 + \Delta t) = \sqrt{4 + \Delta t}$, (c) $\dfrac{g(3+\Delta t) - g(3)}{\Delta t} = \dfrac{1}{\sqrt{4+\Delta t} + 2}$.

10. (a) $h(1) = \frac{1}{2}$, (b) $h(1+\Delta s) = \dfrac{1}{(\Delta s)^2 + 2\Delta s + 2}$, (c) $\dfrac{h(1+\Delta s) - h(1)}{\Delta s} = \dfrac{-(\Delta s + 2)}{(\Delta s)^2 + 2\Delta s + 2}$.

11. (a) $k(-3) = 1$, (b) $k(3) = 1$, (c) $k(-5) = 3$, (d) $k(-1) = -1$.

12. (a) $p(0) = 3$, (b) $p(1) = 4$, (c) $p(3) = -5$, (d) $p(-2) = 1$.

13. Domain(f) = $\{x \in R \mid x \geq 3\} = [3, +\infty)$.

14. Domain(g) = $\{t \in R \mid t \neq -1\} = (-\infty, -1) \cup (-1, +\infty)$.

15. Domain(h) = $\{s \in R \mid s \geq 2, s \neq 5\} = [2,5) \cup (5, +\infty)$.

§13.2 **2.** (a) an even function, (b) omitted, (c) $(-\infty, 0]$: decreasing and $[0, +\infty)$: increasing.

3. (a) an odd function, (b) omitted, (c) $(-\infty, +\infty)$: increasing.

4. (a) neither, (b) omitted, (c) $(-\infty, +\infty)$: increasing.

5. (a) neither, (b) omitted, (c) $(-\infty, +\infty)$: decreasing.

6. (a) an even function, (b) omitted, (c) $(-\infty, 0]$: decreasing and $[0, +\infty)$: increasing.

7. (a) an even function, (b) omitted, (c) $(-\infty, 0]$: increasing and $[0, +\infty)$: decreasing.

8. (a) neither, (b) omitted, (c) $(-\infty, 1]$: decreasing.

9. (a) an even function, (b) omitted, (c) $[-1,0]$: increasing and $[0,1]$: decreasing.

10. (a) neither, (b) omitted, (c) $(-\infty,-1]$: increasing and $[-1,+\infty)$: decreasing.

§13.3 **1.** (a) $(3f-4g)(x)=4x^2+3x-5$, where Domain $(3f-4g)=(-\infty,+\infty)$.

(b) $(f\cdot g)(x)=x^3-3x^2+x-3$, where Domain $(f\cdot g)=(-\infty,+\infty)$.

(c) $(\frac{f}{g})(x)=\frac{x-3}{x^2+1}$, where Domain $(\frac{f}{g})=(-\infty,+\infty)$.

(d) $(f\circ g)(x)=x^2-2$, where Domain $(f\circ g)=(-\infty,+\infty)$.

(e) $(g\circ f)(x)=x^2-6x+10$, where Domain $(g\circ f)=(-\infty,+\infty)$.

2. (a) $(3f-4g)(x)=3\sqrt{x-1}-4x^2-12$, where Domain $(3f-4g)=[1,+\infty)$.

(b) $(f\cdot g)(x)=(x^2+3)\sqrt{x-1}$, where Domain $(f\cdot g)=[1,+\infty)$.

(c) $(\frac{f}{g})(x)=\frac{\sqrt{x-1}}{x^2+3}$, where Domain $(\frac{f}{g})=[1,+\infty)$.

(d) $(f\circ g)(x)=\sqrt{x^2+2}$, where Domain $(f\circ g)=(-\infty,+\infty)$.

(e) $(g\circ f)(x)=|x-1|+3$, where Domain $(g\circ f)=(-\infty,+\infty)$.

3. (a) $(3f-4g)(x)=\frac{-4x^3+24x^2-5x+48}{x-6}$, where Domain $(3f-4g)=\{x\in R\mid x\neq 6\}$.

(b) $(f\cdot g)(x)=\frac{x^3+2x}{x-6}$, where Domain $(f\cdot g)=\{x\in R\mid x\neq 6\}$.

(c) $(\frac{f}{g})(x)=\frac{x}{(x-6)(x^2+2)}$, where Domain $(\frac{f}{g})=\{x\in R\mid x\neq 6\}$.

(d) $(f\circ g)(x)=\frac{x^2+2}{x^2-4}$, where Domain $(f\circ g)=\{x\in R\mid x\neq \pm 2\}$.

(e) $(g\circ f)(x)=\frac{3x^2-24x+72}{(x-6)^2}$, where Domain $(g\circ f)=\{x\in R\mid x\neq 6\}$.

4. $\{-25\}$. **5.** $\{309\}$. **6.** $\{\frac{1}{14}\}$. **7.** $\{3\sqrt{3}\}$.

8. $\{7\}$. **9.** $g(x)=x^3$, $h(x)=x+5$.

10. $g(x)=x^{13}$, $h(x)=x^2+1$.

11. $g(x)=|x|$, $h(x)=x^2+5$.

12. $g(x)=\sqrt{x}$, $h(x)=x^2+2$. **13.** $g(x)=\sqrt[3]{x}$, $h(x)=x-2$.

14. $g(x)=x^{\frac{3}{2}}$, $h(x)=x^2+3$. **15.** $g(x)=\dfrac{1}{x}$, $h(x)=x-5$.

§13.4 **2.** (a) Yes, (b) $f^{-1}(x)=\frac{1}{2}(x-5)$, where Domain $(f^{-1})=(-\infty,+\infty)$.

3. (a) No, (b) $f^{-1}(x)$ does not exist.

4. (a) Yes, (b) $f^{-1}(x)=\sqrt{x+3}$, where Domain $(f^{-1})=[-3,+\infty)$.

5. (a) Yes, (b) $f^{-1}(x)=-\sqrt{x+3}$, where Domain $(f^{-1})=[-3,+\infty)$.

6. (a) Yes, (b) $f^{-1}(x)=\sqrt[3]{x-1}$, where Domain $(f^{-1})=(-\infty,+\infty)$.

7. (a) Yes, (b) $f^{-1}(x)=x^2+2$, where Domain $(f^{-1})=(-\infty,+\infty)$.

8. (a) Yes, (b) $f^{-1}(x)=\dfrac{1-3x}{x}$,

where Domain $(f^{-1})=(-\infty,0)\cup(0,+\infty)$.

9. $\{3\}$. **10.** $\{0.5\}$.

Chapter 14

§14.1 **1.** (a) x-intercept: none, y-intercept: $Q(0, 3)$, (b) omitted.

2. (a) x-intercept = y-intercept: $O(0, 0)$, (b) Since the x-intercept and the y-intercept coincide, we need to pick up another point, say to evaluate the y-value at $x = 1$; the plot is omitted.

3. (a) x-intercept: $P(-\frac{4}{3},0)$, y-intercept: $Q(0, 4)$, (b) omitted.

4. (a) x-intercept: $P(\frac{4}{3},0)$, y-intercept: $Q(0, -4)$, (b) omitted.

5. (a) x-intercept: $P(\frac{3}{4},0)$, y-intercept: $Q(0, 3)$, (b) omitted.

6. (a) x-intercept: $P(-\frac{3}{4},0)$, y-intercept: $Q(0, -3)$, (b) omitted.

7. (i) $q = 1.5p + 4.5$, (ii) $q = 13.5$ million bushels per month.

8. $p = \$4$ per bushel.

§14.2 **1.** (a) $f(x)$ has an absolute maximum at $x = 0$, (b) vertex: $V(0, 3)$, line of symmetry: $x = 0$, (c) omitted.

2. (a) $f(x)$ has an absolute minimum at $x = -2$, (b) vertex: $V(-2, 1)$, line of symmetry: $x = -2$, (c) omitted.

3. (a) $f(x)$ has an absolute maximum at $x = 1$, (b) vertex: $V(1, 2)$, line of symmetry: $x = 1$, (c) omitted.
4. (a) $f(x)$ has an absolute maximum at $x = -3$, (b) vertex: $V(-3, -2)$, line of symmetry: $x = -3$, (c) omitted.

5. $f(x) = -\frac{1}{2}(x^2 + 2x - 5)$. **6.** $f(x) = 4x^2 - 16x + 15$.

7. $f(x) = x^2 + 2x$. **8.** $f(x) = \frac{1}{4}x^2 - 2x + 3$. **9.** {2 and 6}.

10. 50 chairs have to be produced with the maximum profit of $5,000.

§14.3 **1.** (a) All three zeroes $\{-2, 0, 2\}$ are with a multiplicity of one, (b) x-intercepts: $P_1(-2, 0)$, $P_2(0, 0)$, $P_3(2, 0)$; y-intercepts: $P_2(0, 0)$, (c) omitted.
2. (a) Zeroes: $\{2\}$ with a multiplicity of one, (b) x-intercepts: $P(2, 0)$; y-intercepts: $Q(0, -8)$, (c) omitted.
3. (a) Zeroes: $\{-2, 0\}$; both have a multiplicity of two, (b) x-intercepts: $P_1(-2, 0)$, $P_2(0, 0)$; y-intercepts: $P_2(0, 0)$, (c) omitted.
4. (a) Zeroes: $\{-2, 1\}$; -2 and 1 have respectively multiplicities of two and three, (b) x-intercepts: $P_1(-2, 0)$, $P_2(1, 0)$; y-intercepts: $Q(0, -4)$, (c) omitted.
5. (a) Zeroes: $\{-1, 1, 2\}$; -1, 1, and 2 have respectively multiplicities of one, two, and three. (b) x-intercepts: $P_1(-1, 0)$, $P_2(1, 0)$, $P_3(2, 0)$; y-intercepts: $P_2(0, -8)$, (c) omitted.

§14.4 **1.** {109}. **2.** {88}. **3.** {– 3}. **4.** {1}.
5. {– 233}. **6.** $q_2(x) = -2x^2 + 7x - 27$; $R = 109$.
7. $q_3(x) = x^3 + 3x^2 + 10x + 30$; $R = 88$.
8. $q_4(x) = 3x^4 - 3x^3 + 2x^2 - x + 3$; $R = -3$.
9. $q_5(x) = -x^5 + x^4 + x^2$; $R = 1$.
10. $q_6(x) = x^6 - 3x^5 + 7x^4 - 16x^3 + 32x^2 - 61x + 117$; $R = -233$. **11.** Yes. **12.** No. **13.** Yes.
14. No. **15.** Yes.

§14.5 **1.** (a) positive zeroes: 2 or none; negative zeroes: 1, (b) upper bound: 5; lower bound: – 3.
2. (a) positive zeroes: 2 or none; negative zeroes: 2 or none, (b) upper bound: 9; lower bound: – 8.

3. (a) positive zeroes: 5, 3 or 1; negative zeroes: none, (b) upper bound: 10; lower bound: – 10.
4. (a) positive zeroes: 1; negative zeroes: 3 or 1, (b) upper bound: 7; lower bound: – 7.
5. (a) positive zeroes: 3 or 1; negative zeroes: 1, (b) upper bound: 10; lower bound: – 7.

6. $\{\frac{1}{2}, \pm i\}$. **7.** $\{-\frac{1}{2}, \frac{1}{3}, \pm 1\}$. **8.** $\{1, 2, 3, \pm i\}$.

9. $\{-\frac{1}{2}, \frac{1}{3}, \pm 1, \pm i\sqrt{2}\}$. **10.** $\{-6, -\frac{1}{2}, 1, \pm\sqrt{2}, \pm i\sqrt{2}\}$.

11. $c_0^* = -2, c_1^* = -1, \Delta^* = 104, A = 1 + \frac{\sqrt{78}}{9} = 1.981307$,
$B = 0.0186932, \sqrt[3]{A} = 1.255983, \sqrt[3]{B} = 0.2653965$,
$x_1 = 1.5213795$, $x_2 = -0.7606898 + i \cdot 0.8578479$, $x_3 = \bar{x}_2$.

12. $c_0^* = 2, c_1^* = 3, \Delta^* = 216, A = -1 + \sqrt{2}, B = -1 - \sqrt{2}$,
$\sqrt[3]{A} = 0.7454324$, $\sqrt[3]{B} = -1.3415$, $x_1 = -0.5960$,
$x_2 = 0.29804 + i \cdot 1.80729$, $x_3 = \bar{x}_2$.

13. $c_0^* = 14, c_1^* = -12, \Delta^* = -1620, \theta = \tan^{-1}(-\frac{2}{3}\sqrt{\frac{5}{7}})$
$= 180^\circ - 29.39848^\circ = 150.6015^\circ, x_1 = 0.56041$,
$x_2 = -5.94164, x_3 = -0.61877$.

14. $c_0^* = -\frac{38}{27}, c_1^* = -\frac{4}{3}, \Delta^* = 44, A = \frac{19}{27} + \frac{1}{3}\sqrt{\frac{11}{3}}$,
$B = \frac{19}{27} - \frac{1}{3}\sqrt{\frac{11}{3}}, \sqrt[3]{A} = 1.1030187$, $\sqrt[3]{B} = 0.402935$,
$x_1 = 1.839287$, $x_2 = -0.41964 + i \cdot 0.60629$, $x_3 = \bar{x}_2$.

15. $c_0^* = 4, c_1^* = -6, c_2^* = -432, \theta = \tan^{-1}(-1) = \dfrac{3\pi}{4}$.
$x_1 = -1 + 2\sqrt{2}\cos(\dfrac{\pi}{4}) = 1, x_2 = -1 + 2\sqrt{2}\cos(\dfrac{11\pi}{12})$
$= -3.73205, x_3 = -1 + 2\sqrt{2}\cos(\dfrac{19\pi}{12}) = -0.26795$.

16. $c_2 = 0, c_1 = -4, c_0 = -1$, resolvent cubic equation:
$y^3 + y - 2 = 0, y_0 = 1,\ d_1 = 0.956145$,
$d_2 = i \cdot 1.383551$, $x_1 = 1.66325$, $x_2 = -0.24904$,
$x_3 = -0.70711 + i \cdot 1.38355$, $x_4 = \bar{x}_3$.

17. $c_2 = -1, c_1 = 1, c_0 = 3$, resolvent cubic equation:
$y^3 - y^2 - \dfrac{11}{4}y - \dfrac{1}{8} = 0, y_0 = 2.248$,

$d_1 = i\cdot 0.92728, d_2 = i\cdot 0.6231, x_1 = 1.06061 + i\cdot 0.92728,$

$x_2 = \bar{x}_1, x_3 = -1.06021 + i\cdot 0.6231, x_4 = \bar{x}_3.$

18. $c_2 = -3.375, c_1 = c_2, c_0 = 0.05078$, resolvent cubic equation:

$y^3 - 3.375y^2 + 2.79688y - 1.42383 = 0, y_0 = 2.47826,$

$d_1 = 1.09834, d_2 = i\cdot 0.55642, x_1 = 2.9615, x_2 = 0.76483,$

$x_3 = -0.36316 + i\cdot 0.55642, x_4 = \bar{x}_3.$

19. $c_2 = 2, c_1 = -2, c_0 = -3,$ resolvent cubic equation:

$y^3 + 2y^2 + 4y - \dfrac{1}{2} = 0, y_0 = 0.11767,$

$d_1 = 1.00126, d_2 = i\cdot 1.76641, x_1 = 1.24382, x_2 = -0.7587,$

$x_3 = -0.24256 + i\cdot 1.76641, x_4 = \bar{x}_3.$

20. $c_2 = -8.375, c_1 = 11.875, c_0 = -5.51172,$ resolvent cubic equation:

$y^3 - 8.375y^2 + 23.04688y - 17.62695 = 0, y_0 = 1.24285,$

$d_1 = i\cdot 0.44712, d_2 = 2.70778, x_1 = 0.03831 + i\cdot 0.44712,$

$x_2 = \bar{x}_1, x_3 = 1.16947, x_4 = -4.24608$

21. $\{1.52124\}$. **22.** $\{1.651\}$. **23.** $\{1.45215, -1.16309\}$.

24. $\{1.21765, -1.74866\}$. **25.** $\{1.29858\}$.

§14.6 **1.** (a) Omitted. (b) $\{\frac{3}{2}, \pm 2i\}$.

2. (a) Omitted. (b) $\{\frac{1}{3}, 2 \pm \frac{\sqrt{11}}{2} i\}$.

3. (a) Omitted. (b) $\{-\frac{3}{2}, \frac{1}{2}, -2 \pm 3i\}$.

4. (a) Omitted. (b) $\{3 \pm \sqrt{3}i, -1 \pm \sqrt{2}i\}$.

5. (a) Omitted. (b) $\{-1, \pm\sqrt{5}i, 1 \pm \sqrt{2}i\}$.

6. $p_3(x) = 2x^3 - 11x^2 + 16x + 11.$

7. $p_3(x) = 3x^3 - 21x^2 + 51x - 45.$

8. $p_4(x) = -4x^4 + x^3 - 4x^2 - 39x + 10.$

9. $p_4(x) = \dfrac{1}{2}x^4 - 2x^3 + \dfrac{31}{2}x^2 - 50x + 75.$

10. $p_5(x) = 3x^5 + 2x^4 - 4x^3 + 124x^2 + 193x - 78.$

Chapter 15

§15.1 **1.** Domain $(r_1) = \{x \in R \mid x \neq -\sqrt{2}\}$.

2. Domain $(r_2) = \{x \in R \mid x \neq \pm\sqrt{\frac{3}{2}}\}$.

3. Domain $(r_3) = (-\infty, +\infty)$.

4. Domain$(r_4) = \{x \in R \mid x \neq \sqrt[3]{\dfrac{9+\sqrt{78}}{9}} + \sqrt[3]{\dfrac{9-\sqrt{78}}{9}}\}$.

5. Domain$(r_5) = \{x \in R \mid x \neq \pm\sqrt{2}, \pm\sqrt{3}\}$.

6. No, $r_6(x) = 3x - 9 + \dfrac{32}{x+3}$.

7. No, $r_7(x) = \dfrac{5}{2} + \dfrac{-19x+7}{4x^2+6x-2}$.

8. No, $r_8(x) = x + 1 + \dfrac{6x+2}{x^2-x-6}$.

9. No, $r_9(x) = 2x^2 + x + 4 + \dfrac{8}{x-3}$.

10. No, $r_{10}(x) = 2x^2 - 5x + 12 + \dfrac{-36x+71}{x^2+x-6}$.

11. $t(x) = \dfrac{-2x^2+49x-3}{4x^2-1}$. **12.** $u(x) = \dfrac{x^2-9}{4x^2-1}$.

13. $v(x) = \dfrac{2x^2-7x+3}{2x^2+7x+3}$. **14.** $f(x) = -x$. **15.** $g(x) = \dfrac{-5x+6}{4x+5}$.

16. {0.014 kg} **17.** {27.73 lb/in^2}.

18. {0.55737705 m^3/kg-mol}.

§15.2 **1.** (a) vertical asymptotes: $x = 0$, (b) horizontal asymptotes: $y = \dfrac{3}{2}$, (c) omitted.

2. (a) vertical asymptotes: $x = -2$, $x = 1$ (b) horizontal asymptotes: $y = 0$, (c) omitted.

3. (a) vertical asymptotes: none, (b) horizontal asymptotes: $y = 3$, (c) omitted.

4. (a) vertical asymptotes: $x = -1$, $x = 1$, (b) horizontal asymptotes: $y = \dfrac{5}{2}$, (c) omitted.

5. (a) vertical asymptotes: $x = 0$, $x = 2$, $x = -2$, (b) horizontal asymptotes: $y = 2$, (c) omitted.

6. (a) vertical asymptotes: $x = 0$, (b) oblique asymptotes: $y = -x + 2$, (c) omitted.

7. (a) vertical asymptotes: $x = -2$, $x = 2$, (b) oblique asymptotes: $y = x + 1$, (c) omitted.

8. (a) vertical asymptotes: $x = 2$, (b) curvilinear asymptotes: $y = x^2 - 1$, (c) omitted.

9. (a) N(5) ≈ 158 eggs; N(15) ≈ 218 eggs; N(30) ≈ 308 eggs, (b) As $t \to +\infty, N(t) \to 1000$ eggs.

10. The average cost per unit is at a minimum when production level = 500 units per month.

§15.3 **1.** $r_1(x) = \frac{5}{2(x-1)} - \frac{5}{2(x+1)}$. **2.** $r_2(x) = \frac{2}{x-3} + \frac{1}{x+2}$.

3. $r_3(x) = \frac{1}{32(x-2)} - \frac{1}{32(x+2)} - \frac{1}{8(x^2+4)}$.

4. $r_4(x) = \frac{7}{8(x-2)} - \frac{1}{8(x+2)} - \frac{3}{4x}$.

5. $r_5(x) = \frac{7x+3}{2(x^2+1)} - \frac{7}{2(x+1)}$.

Chapter 16

§16.1 **1.** (a) $x = \frac{1}{8}$, (b) $y = -1$, (c) $z = \frac{12}{11}$,

(d) $w = e^{\frac{32}{7}} = 98.68213$.

2. Omitted. **3.** Omitted. **4.** Omitted.

5. (a) \$318,551, (b) \$404,958, (c) \$546,636.

6. {7.7 years}. **7.** {9.2 years}. **8.** {17.5%}.

9. {\$10,023.24}.

§16.2 **2.** {10}. **3.** {– 5}. **4.** {– 7}. **5.** {4}.
6. {5}. **7.** {– 4}. **8.** {– 3}. **9.** {4}.
10. $\{\frac{1}{2}\}$. **11.** $\{\frac{3}{2}\}$. **12.** $\{-\frac{1}{2}\}$. **13.** $\{-\frac{5}{2}\}$.
14. {–2}. **15.** $\{\frac{5}{2}\}$. **16.** $\{\frac{1}{3}\}$. **17.** $\{\frac{2}{3}\}$.
18. $\{\frac{4}{3}\}$. **19.** $\{\frac{5}{3}\}$. **20.** $\{-\frac{1}{3}\}$. **21.** $\{-\frac{4}{3}\}$.
22. {– 2}. **23.** $\{\frac{2}{3}\}$. **24.** $\{\frac{5}{3}\}$. **25.** {4}.
26. {– 5}. **27.** {– 3}. **28.** {4}. **29.** $\{\frac{1}{2}\}$.
30. $\{\frac{3}{2}\}$. **31.** $\{-\frac{1}{2}\}$. **32.** $\{-\frac{5}{2}\}$. **33.** {0}.
34. {– 1}. **35.** {0}. **36.** {– 3}. **37.** {0}.
38. {– 5}. **39.** {1.8572}. **40.** {1.1047}. **41.** {– 0.0512}.
42. {– 0.4884}. **43.** {1.2162}. **44.** {0.1790}.

45. $\{4.7554\}$. **46.** $\{1.6826\}$.

47. (a) Domain $(g_1) = (-\infty, 0)$, (b) vertical asymptotes: $x = 0$, (c) x-intercept: $Q(-1, 0)$. (d) omitted.

48. (a) Domain $(g_2) = (1, +\infty)$, (b) vertical asymptotes: $x = 1$, (c) x-intercept: $Q(2, 0)$. (d) omitted.

49. (a) Domain $(g_3) = (-\infty, 2)$, (b) vertical asymptotes: $x = 2$, (c) x-intercept: $Q(1, 0)$. (d) omitted.

50. (a) Domain $(g_4) = (1, +\infty)$, (b) vertical asymptotes: $x = 1$, (c) x-intercept: $Q(3, 0)$. (d) omitted.

51. (a) Domain $(g_5) = (0, +\infty)$, (b) vertical asymptotes: $x = 0$, (c) x-intercept: $Q(e, 0)$. (d) omitted.

52. {10 dB}. **53.** {60 dB}. **54.** {70 dB}.

55. {120 dB}.

§16.3 **1.** $\{\frac{2}{3}\}$. **2.** $\{1.3635\}$. **3.** $\{1.2731\}$. **4.** $\{0.4055\}$.

5. $\{\ln(2+\sqrt{5})\}$. **6.** $\{3\}$. **7.** $\{1\}$. **8.** $\{3\}$.

9. $\{3\}$. **10.** $\{\sqrt{e^5}\}$.

§16.4 **1.** {approximate 8 hours}.

2. {year of 2,061}. **3.** {\$8,046.11}.

4. {approximate 271 days}. **5.** {8.67 grams}.

6. {approximate 13,301 years old}.

7. {34,708 years old}. **8.** {1.77 minutes}.

9. {6:25 p.m.}. **10.** $\{25.3^\circ C\}$.

11. {7,225 units}. **12.** $\{-1.628 \cdot 10^4 J\}$.

Chapter 17

§17.1 **1.** $\{(x, y) = (-2, 2)\}$. **2.** $\{(x, y, z) = (-1, \frac{3}{2}, \frac{5}{2})\}$.

3. No Solution

4. $\{(x, y, z) = (\frac{5}{7}z + \frac{11}{7}, \frac{6}{7}z - \frac{5}{7}, z) \mid z \in R\}$.

5. $\{(x, y) = (-\frac{3}{5}, -\frac{4}{5})\}$. **6.** No Solution

7. $\{(x, y, z) = (z - 11, -2z + 9, z) \mid z \in R\}$.

8. $\{(x, y, z, w) = (-5z - w, -2z + w + 1, z, w) \mid z, w \in R\}$.

9. $\{(x, y, z, w) = (9, \frac{31}{2}, \frac{21}{2}, 2)\}$.

10. {Investment at 5% = \$2,000, at 6% = \$3,000, at 7% = \$4,000}.

§17.2 **1.** Size of $A_1 = 1 \times 1$. **2.** Size of $A_2 = 2 \times 2$.
3. Size of $A_3 = 3 \times 2$. **4.** Size of $A_4 = 1 \times 4$.
5. Size of $A_5 = 3 \times 1$.

6. (a) $A+B=\begin{bmatrix}1\\7\end{bmatrix}$, (b) $A-B=\begin{bmatrix}3\\-1\end{bmatrix}$, (c) $\frac{1}{2}\cdot A=\begin{bmatrix}1\\\frac{3}{2}\end{bmatrix}$,

(d) $\frac{1}{4}\cdot B=\begin{bmatrix}-\frac{1}{4}\\1\end{bmatrix}$, (e) $\frac{1}{2}A-\frac{1}{4}B=\begin{bmatrix}\frac{5}{4}\\\frac{1}{2}\end{bmatrix}$.

7. (a) $A+B=\begin{bmatrix}1 & 1 & 7\\4 & 6 & -9\end{bmatrix}$, (b) $A-B=\begin{bmatrix}-3 & 3 & 1\\2 & -4 & -1\end{bmatrix}$,

(c) $2\cdot A=\begin{bmatrix}-2 & 4 & 8\\6 & 2 & -10\end{bmatrix}$, (d) $3\cdot B=\begin{bmatrix}6 & -3 & 9\\3 & 15 & -12\end{bmatrix}$,

(e) $2A-3B=\begin{bmatrix}-8 & 7 & -1\\3 & -13 & 2\end{bmatrix}$.

8. (a) $A\cdot B=[-17]=-17$, (b) $B\cdot A=\begin{bmatrix}-2 & 6\\5 & -15\end{bmatrix}$.

9. $B\cdot A=\begin{bmatrix}-7 & 1 & 2\end{bmatrix}$. **10.** $A\cdot B=\begin{bmatrix}25\\-23\end{bmatrix}$.

11. $B\cdot A=\begin{bmatrix}-7 & 1 & 2\\-20 & 7 & -15\end{bmatrix}$. **12.** $A\cdot B=\begin{bmatrix}25 & 14\\-23 & -13\end{bmatrix}$.

13. (a) $A\cdot B=\begin{bmatrix}1 & -13\\1 & -22\end{bmatrix}$, (b) $B\cdot A=\begin{bmatrix}-4 & -11\\-7 & -17\end{bmatrix}$,

(c) $A^2=\begin{bmatrix}7 & 18\\12 & 31\end{bmatrix}$, (d) $B^3=\begin{bmatrix}0 & -27\\27 & -54\end{bmatrix}$.

14. (a) $A\cdot B=\begin{bmatrix}-13 & 1 & -1\\-3 & -1 & -3\\-16 & 17 & 5\end{bmatrix}$,

(b) $B\cdot A=\begin{bmatrix}-14 & -1 & -21\\2 & 4 & -6\\1 & 4 & 1\end{bmatrix}$,

(c) $A^3 = \begin{bmatrix} 90 & 22 & 80 \\ 14 & 10 & 12 \\ 49 & 18 & 53 \end{bmatrix}$, (d) $B^2 = \begin{bmatrix} 19 & -2 & 10 \\ 2 & 8 & 12 \\ 2 & -5 & -1 \end{bmatrix}$.

15. (a) $f(\vec{v}_1) = [8 \quad 12]^T = 4 \cdot \vec{v}_1$, (b) $f(\vec{v}_2) = [3 \quad 1]^T$,
(c) $f(\vec{v}_3) = [-1 \quad 1]^T = (-1) \cdot \vec{v}_3$.

Remark

$\vec{v}_1$ and $\vec{v}_3$ are the eigenvectors, 4 and -1 are eigenvalues of the matrix A.

16. $p_2(x) = 2x^2 - 3x + 1$. **17.** $p_3(x) = x^3 - 2x^2 + x + 3$.
18. Omitted.

§17.3 **1.** $\det(A_1) = -2$. **2.** $\det(A_2) = -6$. **3.** $\det(A_3) = 0$.
4. $\det(A_4) = -14$. **5.** $\det(A_5) = 0$. **6.** $\det(A_6) = 0$.
7. $\det(A_{71}) = 3$. **8.** $\det(A_8) = -6$.
9. $\det(A_9) = -315$. **10.** $\det(A_{10}) = 61$.
11. $\det(A_{11}) = 10$. **12.** $\det(A_{12}) = 0$.
13. $\det(A_{13}) = 320$. **14.** $\det(A_{14}) = 1.8$.
15. $\det(A_{15}) = -105$. **16.** $\det(A_{16}) = 110$. **17.** $\{2 \pm \sqrt{5}\}$.
18. $\{1, 3, 5\}$. **19.** $\{(-2, 2)\}$.
20. $\{(-1, \frac{3}{2}, \frac{5}{2})\}$.
21. Cramer's rule does not apply and the system has no solution
22. Cramer's rule does not apply, yet the system has infinitely many solutions
23. $\{(9, \frac{31}{2}, \frac{21}{2}, 2)\}$.

§17.4 **2.** The 2nd matrix is obtained by interchanging row 1 and row 3 of the 1st matrix.
3. The second matrix is obtained by interchanging row 1 and row 2, and successively row 2 and row 3 of the 1st matrix.
4. Row 1 of the 1st matrix is twice as large as that of row 1 in the 2nd matrix.
5. Rows 2 and 3 of the 1st matrix are three and four times as large as the corresponding rows of the 2nd matrix.
6. By using Eq. (17.4.2). **7.** By using Eq. (17.4.5).
8. By using Eq. (17.4.5) twice.
9. $\{-188\}$. **10.** $\{-524\}$. **11.** $\{74\}$. **12.** $\{-264\}$.

13. {1,139}. **14.** {1,550}. **15.** {– 1,826}. **16.** {3}.
17. {2}. **18.** Yes. **19.** No.

§17.5 **4.** $A_4^{-1} = \begin{bmatrix} -0.3 & 0.4 \\ 0.1 & 0.2 \end{bmatrix}$.

5. A_5^{-1} does not exist.

6. $A_6^{-1} = \begin{bmatrix} \frac{1}{2} & 0 & \frac{1}{2} \\ -\frac{3}{4} & \frac{1}{2} & -\frac{1}{4} \\ \frac{5}{4} & -\frac{1}{2} & -\frac{1}{4} \end{bmatrix}$.

7. A_7^{-1} does not exist.

8. $A_8^{-1} = \begin{bmatrix} 2 & -\frac{1}{2} & \frac{3}{2} & \frac{1}{2} \\ \frac{7}{2} & -\frac{5}{4} & \frac{13}{4} & \frac{3}{4} \\ \frac{5}{2} & -\frac{5}{4} & \frac{9}{4} & \frac{3}{4} \\ 0 & -\frac{1}{2} & \frac{1}{2} & \frac{1}{2} \end{bmatrix}$.

9. A_9^{-1} does not exist. **10.** $\{(-3, -1)\}$.

11. $\{(\frac{1}{3}(1-2x_2), x_2) \mid x_2 \in R\}$. **12.** No Solution.

13. $\{(-1, \frac{3}{2}, \frac{5}{2})\}$. **14.** $\{(0.5, 0.25, -1.75)\}$.

15. $\{(2, -2, 1)\}$. **16.** $\{(2(1-x_3), 1, x_3) \mid x_3 \in R\}$.

17. $\{(-2x_3, -2, x_3) \mid x_3 \in R\}$. **18.** No solution.

19. $\{(9, 15.5, 10.5, 2)\}$. **20.** $\{(1, 3, 2, 1)\}$.

21. $\{(5.5, 12.25, 9.25, 2.5)\}$. **22.** No solution.

23. No solution.

24. $\{(4-5x_3+x_4, -2x_3+x_4, x_3, x_4) \mid x_3, x_4 \in R\}$.

25. No solution.

Chapter 18

§18.1 **4.** {– 5, – 3, 7, 11, 15, …}. **5.** {1, 1, 3, 7, 13, 21, …}.

6. {1, – 1, 3, 19, 53, 111, …}.

7. {– 4, – 2, 38, 224, 736, 1826, …}.

8. $\{-\frac{1}{3}, -\frac{3}{2}, 5, \frac{7}{6}, \frac{9}{13}, \frac{1}{2}, \ldots\}$.

9. {1, – 3, 9, – 27, 81, – 243, …}.

10. $\{-3, 1, -\frac{1}{3}, \frac{1}{9}, -\frac{1}{27}, \frac{1}{81}, \ldots\}$.

11. $\{1, \frac{1}{6}, \frac{1}{5{,}040}, \frac{1}{362{,}880}, \frac{1}{39{,}916{,}800}, \ldots\}$.

12. $a_n = (-1)^n, n = 0,1,2,\ldots$ **13.** $a_n = -2n+3, n = 0,1,2,\ldots$.

14. $a_n = 2^{-n}, n = 0,1,2,\ldots$ **15.** $a_n = (-\frac{1}{3})^n, n = 0,1,2,\ldots$.

16. $a_n = [(2n)!]^{-1}, n = 0,1,2,\ldots$ **17.** $\{56\}$.

18. $\{\frac{1}{6,840}\}$. **19.** $\{1,140\}$. **20.** $\{(n^2+5n+6)^{-1}\}$.

21. $\{630\}$. **22.** $\{16,410\}$. **23.** $\{9,860\}$.

24. $\{152,595\}$. **25.** $\{\frac{147}{60}\}$. **26.** $\{-182\}$.

27. $\{\frac{426,457}{362,880}\}$.

28. (a) $10! \approx 3,598,695.62$ ($10! = 3,628,800$), (b) $11! \approx 39,615,541.91$ ($11! = 39,916,800$).
(c) $12! \approx 475,687,486.5$ ($12! = 479,001,600$).

30. $\sqrt{3} \approx a_5 = \frac{18,817}{10,864} \approx 1.73205081$, ($\sqrt{3} = 1.732050808\ldots$).

§18.2 **1.** $\{\frac{55}{6}\}$. **2.** $a_n = \frac{7}{16} + \frac{n}{8}, n = 0,1,2,\ldots,9$.

4. Yes, $a_n = 1.1 - 0.3n, n = 0,1,2,\ldots$.

5. Yes, $b_n = \frac{1}{5} + \frac{2}{3}n, n = -1,0,1,\ldots$.

6. Yes, $c_n = -\frac{1}{3} - \frac{3}{4}n, n = 1,2,\ldots$.

7. No. **8.** Yes, $d_n = n \cdot \ln 2, n = -1,0,1,\ldots$

9. No. **10.** No. **11.** Yes, $e_n = n, n = 5,6,7,\ldots$.

12. Yes, $f_n = n, n = -2,-1,0,\ldots$. **13.** $a_n = -25 + 5n, n = 0,1,2,\ldots$

14. $a_n = 3 + \frac{1}{3}n, n = 0,1,2,\ldots$. **15.** $a_n = 2 - \frac{2}{5}n, n = 0,1,2,\ldots$.

16. $a_n = 2n+1, n = 0,1,2,\ldots$. **17.** $a_n = 3n-1, n = 0,1,2,\ldots$.

18. $a_n = 4n-5, n = 0,1,2,\ldots$ **19.** $\{n^2+n\}$. **20.** $\{n^2\}$.

21. $\{133\}$. **22.** $\{-23\}$. **23.** $\{2,500\}$. **24.** $\{2,550\}$.
25. $\{-1\}$. **26.** {12.5 years}. **27.** {\$202,775}.
28. $\{700\}$.

§18.3 **3.** Yes, $a_n = 3 \cdot 2^n, n = 0,1,2,\ldots$

4. Yes, $b_n = 5 \cdot (-2)^n, n = 0,1,2,\ldots$

5. Yes, $c_n = 3 \cdot (-\frac{1}{2})^n, n = -2,-1,0,1,\ldots$.

6. Yes, $d_n = 3 \cdot (\frac{2}{3})^n, n = -1,0,1,\dots$.

7. No. **8.** No. **9.** Yes, $e_n = (\ln 2)^n, n = 1,2,3,\dots$.

10. No. **11.** Yes, $f_n = e^{2n}, n = 0,1,2,\dots$.

12. No. **13.** Yes, $g_n = (3!)^{-n}, n = 0,1,2,\dots$.

14. $a_n = 5^{n+1}, n = 0,1,2,\dots$. **15.** $a_n = 3^{n-3}, n = 0,1,2,\dots$.

16. $a_n = \frac{3}{5}(\frac{2}{3})^n, n = 0,1,2,\dots$. **17.** $a_n = \frac{2}{9}(\frac{3}{2})^n, n = 0,1,2,\dots$.

18. $a_n = 5(\frac{2}{3})^n, n = 0,1,2,\dots$. **19.** $a_n = (\frac{2}{3})^n, n = 0,1,2,\dots$.

20. $\{\frac{2047}{8}\}$ **21.** $\{8,188\}$. **22.** $\{\frac{3,415}{4}\}$.

23. $\{\frac{17,075}{64}\}$. **24.** $\{\frac{42,981,185}{4,782,969}\}$. **25.** $\{\frac{e^{14}-1}{e^{10}(e-1)}\}$.

26. $\{\frac{\ln 2[1-(\ln 2)^{21}]}{1-\ln 2}\}$. **27.** $\{9\}$. **28.** $\{\frac{16}{9}\}$. **29.** $\{2\}$.

30. $\{-16\}$. **31.** $\{\frac{3}{2}\}$.

32. {approximate \$637,498}. **33.** {approximate \$32,576}.
34. {approximate \$805}. **35.** {approximate \$67,343}.

§18.5 **7.** (a) {78}, (b) {286}, (c) {715}, (d) {1,287}.
8. (a) {91}, (b) {364}, (c) {1,001}, (d) {2,002}.
9. (a) {3,003}, (b) {5,005}, (c) {455}, (d) {1,365}.
10. (a) {1,140}, (b) {4,845}, (c) {15,504}, (d) {38,760}.

11. $(x+y)^8 = x^8 + 8x^7y + 28x^6y^2 + 56x^5y^3 + 70x^4y^4 + 56x^3y^5 + 28x^2y^6 + 8xy^7 + y^8$.

12. $(x+y)^9 = x^9 + 9x^8y + 36x^7y^2 + 84x^6y^3 + 126x^5y^4 + 126x^4y^5 + 84x^3y^6 + 36x^2y^7 + 9xy^8 + y^9$.

13. $(x+y)^{10} = x^{10} + 10x^9y + 45x^8y^2 + 120x^7y^3 + 210x^6y^4 + 252x^5y^5 + 210x^4y^6 + 120x^3y^7 + 45x^2y^8 + 10xy^9 + y^{10}$.

14. $(2x+3y)^4 = 16x^4 + 96x^3y + 216x^2y^2 + 216xy^3 + 81y^4$.

15. $(3x-2y)^4 = 81x^4 - 216x^3y + 324x^2y^2 - 96xy^3 + 16y^4$.

16. $(2x+3)^5 = 32x^5 + 240x^4 + 720x^3 + 1,080x^2 + 810x + 243$.

17. $(3x-2)^5 = 243x^5 - 810^4 + 1080x^3 - 720x^2 + 240x - 32$.

18. $(\sqrt{x}+\sqrt{3})^6 = x^3+6\sqrt{3x^5}+45x^2+60\sqrt{3x^3}+135x$
$+54\sqrt{3x}+27$.

19. $(\sqrt{x}-\sqrt{2})^6 = x^3-6\sqrt{2x^5}+30x^2-40\sqrt{2x^3}+60x-24\sqrt{2x}+8$.

20. $\{70x^4y^4\}$. **21.** $\{-216x^3y\}$. **22.** $\{1080x^2\}$.
23. {70}. **24.** {126}. **25.** {252}. **26.** {216}.
27. {− 96}. **28.** $\{60\sqrt{3}\}$. **29.** $\{6\sqrt{3}\}$. **30.** {30}.

Chapter 19

§19.1 **1.** {Yes}. **2.** {Yes}. **3.** {Yes}. **4.** {No}.
5. {No}. **6.** {Yes}. **7.** {Yes}. **8.** {Yes}.
9. {Yes}. **10.** {Yes}. **11.** {Neither}.
12. {Bernoulli}. **13.** {Neither}.
14. {Binomial}. **15.** {Neither}.
16. {Bernoulli}. **17.** {Binomial}.
18. {Binomial}. **19.** {Bernoulli}.
20. {Binomial}.
21. *S* = {HHH, HHT, HTH, THH, HTT, THT, TTH, TTT}; *E* = {HHT, HTH, THH}.
22. *S*: the same as Example 19.1.2(b); *E* = {(3, 6), (6, 3), (4, 5), (5, 4), (4, 6), (6, 4), (5, 5), (5, 6), (6, 5), (6, 6)}.
23. *S* = {HHHH, HHHT, HHTH, HTHH, THHH, HHTT, HTHT, HTTH, THTH, THHT, TTHH, HTTT, THTT, TTHT, TTTH, TTTT};
E = {HHTT, HTHT, HTTH, THTH, THHT, TTHH}.
24. *S* = {{*B*, *G*}, {*B*, *R*}, {*B*, *W*}, {*B*, *Y*}, {*G*, *R*}, {*G*, *W*}, {*G*, *Y*}, {*R*, *Y*}, {*R*, *W*}, {Y, W}};
E = {{B, W}, {G, W}, {R, W}, {Y, W}}.
25. *S* = {*ab, ac, ad, bc, bd, cd, ba, ca, da, cb, db, dc*}; *E* = {*ab, bc, bd, ba, cb, db*}.

§19.2 **1.** {28}. **2.** {20}. **3.** {240}. **4.** {676}.
5. {6,840}. **6.** {17,576,000}. **7.** {40,320}.
8. {34,650}. **9.** {81}. **10.** {96}. **11.** {2,880}.
12. {90}. **13.** {180}. **14.** {2,520}.
15. {3,491,888,400}.
16. {11,760}. **17.** {1,898,208}.
18. (a) {220}, (b) {56}, (c) {112}, (d) {164}, (e) {52}.
19. (a) {256}, (b) {70}, (c) {93}, (d) {37}.

20. {HHHTTT, HHTHTT, HHTTHT, HHTTTH, HTHHTT, HTHTHT, HTHTTH, HTTHHT, HTTHTH, HTTTHH, THHHTT, THHTHT, THHTTH, THTHHT, THTHTH, THTTHH, TTHHHT, TTHHTH,TTHTHH, TTTHHH}.

§19.3 **6.** (a) $\frac{1}{12}$, (b) {7}. **7.** More likely. **8.** {0.4914}.
9. $P(\{X=10\}) > P(\{X=9\})$. **10.** {0.25}. **11.** {0.5}.
12. $\{\frac{2}{13}\}$. **13.** $\{\frac{36}{2{,}598{,}960}\}$. **14.** $\{\frac{5{,}108}{2{,}598{,}960}\}$.
15. $\{\frac{10{,}200}{2{,}598{,}960}\}$. **16.** $\{\frac{3{,}744}{2{,}598{,}960}\}$.
17. $\{\frac{624}{2{,}598{,}960}\}$. **18.** $\{\frac{54{,}912}{2{,}598{,}960}\}$.
19. $\{\frac{123{,}552}{2{,}598{,}960}\}$. **20.** $\{\frac{1{,}098{,}240}{2{,}598{,}960}\}$.
21. $\{\frac{35}{128}\}$. **22.** $\{\frac{93}{256}\}$. **23.** $\{\frac{37}{256}\}$. **24.** $\{\frac{7}{16}\}$.
25. $\{\frac{3}{10}\}$. **26.** $\{\frac{5}{6}\}$. **27.** $\{\frac{1}{3}\}$.
28. {0.754386}. **29.** {0.4}. **30.** {0.85}.
31. $\{\frac{96}{5{,}040}\}$. **33.** $\{\frac{386}{1{,}024}\}$.
34. $\{\frac{7}{9}\}$. **35.** {0.526771}.

§19.4 **1.** $P(F\mid G)=\frac{1}{14}$, where F = {Freshmen}, G = {Girls}.
2. $P(C\mid M)=0.7$, where C = {like chemistry}, M = {like mathematics}.
3. {0.81818}. **4.** (a) $\{\frac{12}{17}\}$, (b) $\{\frac{12}{19}\}$. **5.** $\{\frac{2}{3}\}$.
6. {0.9}. **7.** Yes, since $P(N\mid B)=\frac{1}{5} > P(N\mid \overline{B})=\frac{1}{22}$.
8. (a) $P(C\mid S)=0.2$, (b) $P(\overline{C}\mid S)=0.8$ (c) $P(C\mid \overline{S})=0.04$, (d) $P(\overline{C}\mid \overline{S})=0.96$.
9. (a) $P(A\mid M)=\frac{3{,}738}{8{,}442}$, (b) $P(A\mid \overline{M})=\frac{1{,}494}{4{,}323}$.
10. $\{\frac{11}{850}\}$. **11.** {0.105884}. **12.** $\{\frac{25}{56}\}$.

13. (a) Not independent, (b) Independent, (c) Independent, (d) Independent, (e) Independent, (f) Independent, (g) Not independent, (h) Not independent, (i) Not independent.
14. Not independent. **15.** Not Independent.
16. (a) Not independent, (b) Not independent.

§19.5 **6.** No. **7.** $\{0.54\}$. **8.** $\{0.97\}$.
9. (a) $\{0.18\}$, (b) $\{0.12\}$. **10.** $\{0.0404235\}$.
11. $\{0.5892857\}$. **12.** $\{0.9375\}$. **13.** $\{0.047619\}$.
14. (a) $\{0.390625\}$, (b) $\{0.5625\}$, (c) $\{0.046875\}$.
15. $\{0.5238095\}$. **16.** $\{0.4929578\}$.
17. $\{0.2559415\}$. **18.** $\{\frac{1}{6}\}$. **19.** Yes. **20.** Yes.
21. (a) $\{0.0000015\}$, (b) $\{0.0034985\}$.
22. (a) $\{0.024\}$, (b) $\{0.806\}$. **23.** $\{0.016\}$.
24. $\{0.99994\}$. **25.** $\{0.398\}$.

Index

Index

www.ingramcontent.com/pod-product-compliance
Lightning Source LLC
Chambersburg PA
CBHW031346210326
41599CB00019B/2666

* 9 7 8 0 6 9 2 3 9 3 1 1 6 *